LANDMARK
Yellow Pages

LANDMARK
Yellow Pages

Where to Find All the Names, Addresses, Facts, and Figures You Need

National Trust for Historic Preservation

PAMELA DWIGHT
GENERAL EDITOR

Preservation Press

John Wiley & Sons, Inc.
New York Chichester Brisbane Toronto Singapore

Printed in the United States of America.
5   4   3   2

Library of Congress Cataloging in Publication Data

Landmark yellow pages : where to find all the names, addresses, facts, and
    figures you need / National Trust for Historic Preservation.—[2nd ed.]
        p.   cm.
    Includes bibliographical references and index.
    ISBN 0-471-14398-7
    1. Cultural property, Protection of—United States—Handbooks,
manuals, etc. 2. Cultural property, Protection of—United States—
Societies, etc.—Directories. 3. Historic buildings—United States—
Conservation and restoration—Handbooks, manuals, etc. 4. Historic
buildings—United States—Conservation and restoration—Societies,
etc.—Directories. I. National Trust for Historic Preservation in the
United States.
E159.L28   1992
363.6'9'973—dc20   91-44501

Cover design by Meadows & Wiser, Washington, D.C.
Typeset by BG Composition, Baltimore, Maryland
Printed on recycled paper by Victor Graphics, Inc.

All text illustrations are from the Historic American Buildings Survey and Historic American Engineering Record, National Park Service, U.S. Department of the Interior.

Cover illustration: North and south sides of Chestnut Street, Philadelphia, about 1879. (*Baxter's Panoramic Views*, 1857–82)

CONTENTS

Acknowledgments   ix

I. ALL ABOUT PRESERVATION

II. PRODUCTS AND SERVICES

III. THE PRESERVATION NETWORK

ACKNOWLEDGMENTS

The National Trust for Historic Preservation was established 40 years ago in a collaborative effort spearheaded by dedicated preservationists from organizations as diverse as the National Park Service, National Gallery of Art, American Historical Association, American Institute of Architects, and American Scenic and Historic Preservation Society—all with the support of congressional leaders, who passed the National Trust's charter, and President Harry S Truman, who signed the legislation on October 26, 1949. That spirit of partnership remains an important part of the preservation movement and continues to be exemplified by this second edition of *Landmark Yellow Pages*.

The concept for the directory began with *The Brown Book: A Directory of Preservation Information* published in 1983 and compiled by Diane Maddex with the assistance of Ellen R. Marsh. That work was updated and somewhat enlarged by the U.S. Committee, International Council on Monuments and Sites (US/ICOMOS), under the direction of Russell V. Keune, AIA, and Terry B. Morton, Hon. AIA, and appeared as Part I of the first edition of *Landmark Yellow Pages* in 1990. This first edition also included the list of key preservation contacts in each state, compiled by US/ICOMOS, plus a listing of the National Trust's member organizations in each state.

The present edition of *Landmark Yellow Pages* greatly expands the previous edition with more detailed information about several preservation activities. In addition, each chapter in Part I and III credits the person or persons who reviewed that particular section. In some cases, substantially new material was added, while in others the earlier text needed only factual updating. Some chapters did not appear at all in the last edition and are completely new to the directory.

Part II adds for the first time an advertising supplement offering a wide range of services to anyone involved in the field as a professional or even as a private homeowner.

Part III discusses the preservation network, highlighting key federal agencies involved in preservation activities, and lists key contacts in each state as well as members of the National Trust's Preservation Forum, a program geared to the preservation professional.

The Preservation Press wishes to thank especially Richard Wagner and Dwight Young for reviewing the directory and Suzanne Dane for editing it. Patricia Flowe assisted with the advertising supplement and word processing.

I. ALL ABOUT PRESERVATION

ARCHITECTURAL STYLES

Antoinette J. Lee and Richard Wagner

"Stylistic designations aid in describing architecture and in relating buildings—perhaps of different chronological periods—to one another. But more than that, stylistic classification acknowledges that building is not just a craft, but an art form that reflects the philosophy, intellectual currents, hopes and aspirations of its time."

John Poppeliers and others, *What Style Is It?*, 1977

Shiplap House, Annapolis, Maryland, a southern colonial. (M. Weil, HABS)

cal extent of the country's various architectural styles has depended on the spread of ideas through transportation routes, the media, and migration.

The architectural styles outlined below represent well-documented designs that have influenced American buildings. Each style is defined by its basic building form, framing system, roof line, materials, floor plan, other construction methods, and ornamentation.

Many significant American buildings represent pure examples of particular architectural styles. But most American buildings either represent local or regional interpretations of national styles, illustrate hybrids of style, or cannot be categorized by any particular style. In many areas, architectural styles have grown out of local conditions, such as readily available building materials or climatic factors.

From the austerity of New England colonials, through the ornate extravagance of Second Empire, to the streamlined shapes of Art Deco, every architectural style possesses its own special characteristics of structure and ornament. This survey of 20 influential styles of American architecture celebrates a rich cultural heritage, representing four centuries of creativity. The text is adapted from the book *What Style Is It?*

"All styles are good except the boring kind."

Voltaire, "L'Enfant Prodigue," 1736

Architectural styles help define a building's time and place. In the United States, building styles have been influenced by dominant cultural forces, influential architectural designers, and evolving aesthetic tastes. American building styles embody both grand ideas that have swept across the nation and small ideas that have grown within a region or locale; all styles reflect the skills and outlook of their architect or builder. The geographi-

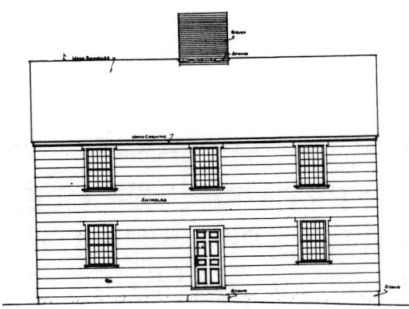

Old Ogden House, Fairfield, Connecticut. (W. Schomburg, HABS)

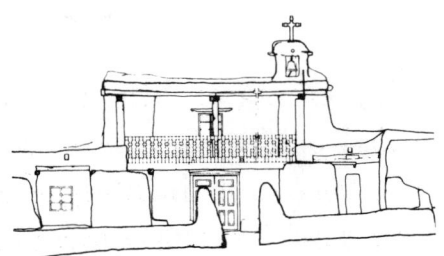

Mission Church, Santa Ana Pueblo, N.M. (J. Shafer, K. Ahn, HABS)

Dr. Upton Scott House, Annapolis, Md. (J. Waite, HABS)

EARLY ENGLISH COLONIAL
1600s–1700

The term "late medieval" perhaps best describes 17th-century English colonial architecture in America. Residences were modeled after the ample but plain houses built in England in the late 1500s, in which medieval forms predominated: steeply pitched roofs, massive chimneys, and small windows with leaded casements. In New England, where hardwoods were plentiful, the substantial half-timbered house was almost universal, with clapboard covering, low ceilings and small rooms to conserve heat in winter. In the southern colonies, airier one-story brick houses were built throughout the 1600s. Settlers from other parts of the world—France, Holland, Germany, Scandinavia and Africa—also brought their own building traditions to America.

New England Colonial
- medieval half-timbered construction with clapboard covering
- steeply pitched gable roof
- tall, massive, central chimney
- small leaded casement windows
- often a "saltbox" silhouette
- second-story overhangs

Southern Colonial
- brick or timber-frame construction
- steeply pitched gable roof
- massive chimneys at ends of houses
- narrow plan, often only one room deep
- patterned, bonded brickwork

SPANISH COLONIAL
1565–1850

From Florida to California and the Southwest, the Spanish left a lasting architectural tradition that ranks only with that of the English. In addition to grid street plans, the most important remaining examples of Spanish colonial architecture in the United States are the mission churches of the Southwest; these were frontier versions of the exuberant baroque style of 16th- and 17th-century Spain, especially as it had developed in Mexico. Missionary priests reproduced the baroque style with whatever materials and labor were at hand. The New Mexico missions, strongly influenced by the building techniques of the Pueblo Indians, are the most austere. Elsewhere, artisans trained in Europe and Mexico produced more elaborate structures.

The Spanish colonial style was revived beginning in the 1890s and into the early 20th century, particularly in California, Florida, and the Southwest.

- adobe or stone construction, often coated with lime wash or plaster
- massive, unadorned, windowless walls
- flat or red tile roof
- projecting roof timbers, sometimes supported by decorative brackets
- twin bell towers
- Spanish baroque ornament applied to bare walls
- curved gable

GEORGIAN
1700–76

Named for the kings who ruled England for most of the 1700s, the Georgian style reflected Renaissance architectural forms made popular in England by the architect Sir Christopher Wren (1632–1723). Wren's work was based on Italian architecture of the 1500s, especially that of Andrea Palladio (1508–80), who freely adapted Roman classical forms. In America, Georgian buildings had a symmetrical, axial composition enriched with classical detail.

The Revolutionary War brought a halt to construction projects and effectively ended the Georgian style in America, although conservative builders continued to use it into the 1800s. The style was revived at the time of the 1876 Centennial, when architects were moved by patriotic zeal to look to the American past for models.

- symmetry in plan and exterior design
- symmetrical arrangement of building parts on an axis
- geometrical proportions
- hipped roof
- main entrances emphasized with columns, pilasters, and broken pediment
- sash windows
- Palladian windows
- classical decorative details

Amory Tickner House, Boston, Mass. (J. Dudley, HABS)

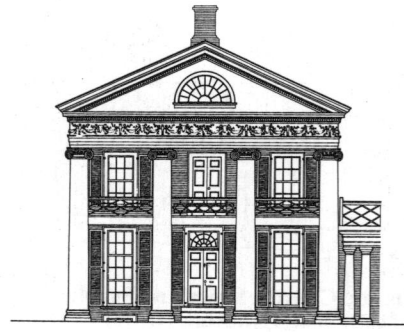

Pavilion II, University of Virginia, Charlottesville, Virginia. (M.D. Sullivan, HABS)

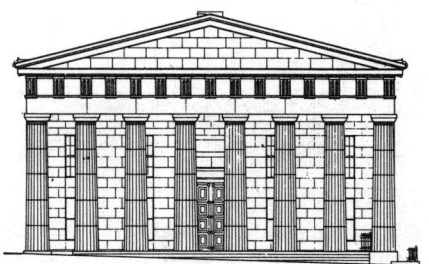

United States Sub-Treasury, New York, New York. (T. Rachelle, HABS)

FEDERAL 1780–1820

By 1776 a new style created in Britain by the Adam brothers had surpassed Georgian Palladianism in popularity. The Adamesque style combined Renaissance and Palladian forms, the delicacy of French rococo and features from recently excavated houses and villas of ancient Rome. In America it was called the Federal style because it flowered in the early decades of the new nation.

The Federal style differs most strikingly from the Georgian in its interior design and use of pastel colors. Many rooms were oval, circular, or octagonal in shape. Mantels, cornices, doorways, and ceilings were decorated with delicate rosettes, urns, swags, and garlands. Federal-style buildings are found especially throughout the cities and towns of the eastern seaboard.

- low-pitched roof
- smooth facade
- large window panes
- exterior decoration confined to porch or entrance, such as fanlight over doorway
- delicate columns and molding
- louvered shutters
- circular, oval, or octagonal rooms
- interior wall decoration of garlands, swags, urns, and rosettes
- pastel colors

JEFFERSONIAN 1790–1830

In the 1780s the state of Virginia asked Thomas Jefferson to find an architect to design a new state capitol. Instead, Jefferson designed the building himself. Inspired by the Maison Carrée, a Roman temple at Nîmes in southern France, he created the first pure temple form in American architecture. For Jefferson, the Roman orders symbolized Rome's republican form of government, which he saw being revived in the New World. Jefferson's architectural theories can be found in many red brick houses and courthouses in Virginia and places where Virginians settled.

- red brick construction
- raised first floor
- slender columns with smooth shafts
- pedimented portico
- classical moldings left plain and painted white
- lunette enclosed in pediment

GREEK REVIVAL 1820–60

The Greek Revival style symbolized for many the idea that America, with its democratic ideals, was the spiritual successor of ancient Greece. By the mid-1840s the Grecian motifs were used throughout the country for churches, banks, courthouses, and other public buildings as well as for houses.

The most easily identified features are the columns and pedimented porch resembling a Greek temple. (Not every Greek Revival structure had these features, however.) Because ancient Greek buildings did not use arches, Greek Revival architects abandoned the arched entrances and fan windows so common in the Federal and Jeffersonian styles.

- pared-down simplicity
- columns (often fluted) and capitals
- pedimented roof
- tall first-floor windows
- heavy cornice
- rectangular transom over entrance
- plain frieze

Christ Episcopal Church, Raleigh, North Carolina. (G. Small, P. Wilday, E. Jenkins, HABS)

Public Square, Nashville, Tennessee. (R. Dunay, HABS)

Oriental Greek Orthodox Church, Jacksonville, Florida. (R. Moje, HABS)

GOTHIC REVIVAL 1830–90

The Gothic Revival was fostered by literature's Romantic movement of the late 1700s and early 1800s, which glorified the medieval past and was brought to America from England. By the 1830s a growing taste for the romantic—coupled with dissatisfaction with the restraints of classical architecture—turned the Gothic Revival into a popular movement.

It was an enduring style. The invention of the jigsaw made possible the fanciful wooden scrollwork known as Carpenter Gothic. After the Civil War, architects produced the eclectic High Victorian Gothic style, which drew on Italian and German as well as English Gothic models. In the late 19th century more authentic Gothic designs emerged in Collegiate Gothic, which left its stamp on many college campuses. Gothic Revival remained the most influential style for churches well into the 1900s.

- steep gabled roof
- pointed arches
- picturesque silhouette
- towers and battlements
- bay and oriel windows
- leaded stained glass
- crenellation
- "gingerbread" trim on eaves and gable ends

ITALIANATE 1830–80

The architecture of Italy inspired this building style, which enjoyed immense popularity in the 1850s. Also known as the Tuscan, Lombard, round, bracketed, and even the American style, the Italianate could be as picturesque as Gothic or as restrained as the classical. In the 1850s it was very nearly America's national style.

The development of cast-iron and pressed metal technology in the mid-1800s permitted the mass production of such decorative features as bracketed cornices and window moldings. These features were applied to a variety of commercial buildings and urban row houses. New York, St. Louis, and Portland, Oregon, had districts of cast-iron buildings in the Italianate style. Towns across America still boast stores with cast-iron fronts masquerading as Italian palaces.

- low-pitched or flat roof
- round arches
- heavily decorated, bracketed cornices and eaves
- scroll-shaped brackets
- tall first-floor windows
- hood moldings over windows
- cupola
- ample porches or verandas
- cast-iron facades on some commercial buildings

EXOTIC REVIVALS 1830–1930

Reflecting the period's romantic turn of mind, 19th-century architects explored exotic historic styles in search of appropriate forms.

The Egyptian Revival was inspired by French archeological work in Egypt. This massive style, with its heavy sense of permanence, was considered appropriate for prisons, mausoleums, cemetery gates, churches, and monuments. In the 1920s the style was revived once more for movie theaters.

Near Eastern architectural forms were adopted in the Moorish Revival style, chiefly for garden kiosks, clubs, hotels, theaters, and a few ostentatious mansions. In the mid-1800s the style was also associated with the Jewish Reform movement in America and was used in the design of synagogues.

Egyptian Revival
- battered (sloping) walls
- battered window and door frames
- columns topped with palm or lotus capitals
- concave cornice
- winged-disk motif

Moorish Revival
- Moorish arches
- domes of various sizes and shapes
- minaret-like spire
- intricate surface decoration, including mosaics and tiles

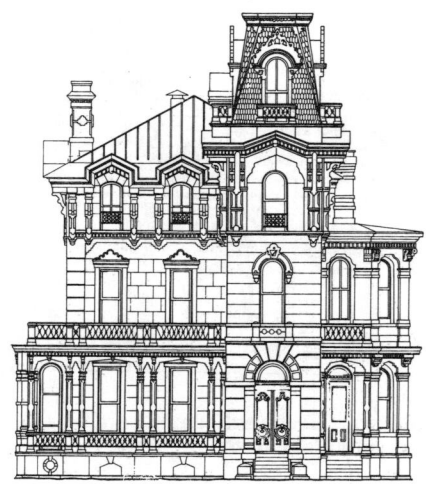

Goyer-Lee House, Memphis, Tennessee. (M.P. Frederickson, D.K. Pattison, HABS)

John Griswold House, Newport, Rhode Island. (J. Chimura, HABS)

John Houghton House, Austin, Texas. (D. Yturralde, HABS)

SECOND EMPIRE 1860–90

Picturesqueness, asymmetry, and eclecticism marked the architecture of the mid- to late 19th century. Architects borrowed freely from a variety of styles, placing great emphasis on character and a sense of permanence.

The Second Empire style takes its name from French designs built during the reign of the emperor Napoleon III (1852–70). The hallmark of the Second Empire style is the mansard roof, adopted from the 17th-century French architect François Mansart (1598–1666). The style, which aspired to a monumental and ornate look, was used widely for public buildings and many houses.

- mansard roof
- prominent projecting and receding surfaces
- paired columns
- projecting central bay
- classical pediments and balustrades
- windows flanked by columns or pilasters
- arched windows with pediments and molded surrounds
- tall first-floor windows

STICK STYLE 1860–90

The Stick Style evolved from the Carpenter Gothic to flourish in the mid- to late 1800s. Embodying the idea that architecture should be truthful, Stick Style houses expressed the building's inner structure through its exterior ornament. A series of boards was applied over the clapboard surface—most often on gable ends and upper stories—to symbolize the structural skeleton. Sometimes diagonal boards were incorporated to resemble Tudor-style half-timbering.

- wood construction
- vertical, horizontal, or diagonal boards applied over clapboard siding
- angularity, asymmetry, verticality
- roof composed of steep intersecting gables
- large veranda or porch
- simple corner posts, roof rafters, brackets, porch posts, and railings

QUEEN ANNE 1880–1900

Eclecticism is the keynote of the Queen Anne style. The name was coined in England to describe buildings that grafted classical ornament onto medieval forms. The style is varied and decoratively rich, with picturesque and asymmetrical silhouettes shaped by turrets, towers, gables, and bays. First floors were often brick or stone, while upper floors were of stucco, clapboard, or decorative shingles. The picturesque effects were best employed in sprawling, free-standing residences, but the Queen Anne style also had a major impact on the urban row house. The typical projecting bay front topped by a gable or pinnacle roof is found in cities from Boston to San Francisco.

- rambling, asymmetrical silhouette
- corner towers or turrets
- steep gable or hipped roof with dormers
- huge "medieval" chimneys
- verandas and balconies
- contrasting materials and colors
- second-story overhangs
- gable ends decorated with half-timbering or stylized relief decoration
- molded bricks as decorative accents
- stained-glass window accents

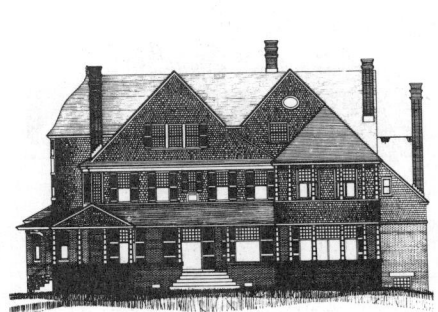

Isaac Bell House, Newport, Rhode Island. (T. Schubert, HABS)

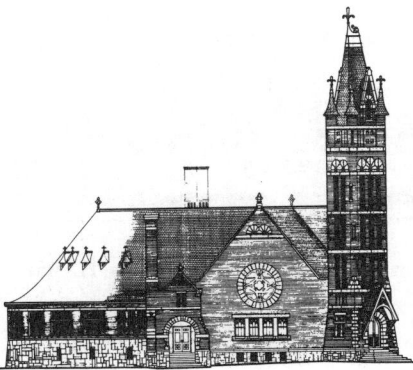

First Presbyterian Church, Salisbury, North Carolina. (G. Anastes, E. Mills, HABS)

Grand Riviera Theatre, Detroit, Michigan. (C. Morrison, HABS)

SHINGLE STYLE 1880–1900

A completely American style that grew out of the Queen Anne, the Shingle Style was born in New England. It reflected the post-Centennial interest in American colonial architecture, especially the shingle architecture of coastal towns that were being rediscovered as fashionable resorts.

Less ornate and more horizontal than the Queen Anne house, the Shingle Style house is a rambling two- or three-story structure entirely covered with unpainted wooden shingles. The first examples of the fully developed style appeared in the 1880s. Among the most important practitioners were H. H. Richardson, Bruce Price, and McKim, Mead and White. Some of Frank Lloyd Wright's earliest work was in the Shingle Style.

- unpainted wood shingles entirely covering the exterior
- prominent roofs, either steeply pitched or with long slopes
- rough-surfaced stone or field rubble used as contrasting materials
- turrets and verandas integrated into the overall design
- eaves close to walls
- reduced ornament

RICHARDSONIAN ROMANESQUE 1870–1900

Architects had experimented with the Romanesque Revival for public buildings in the 1840s and 1850s, borrowing round arches and other features from the pre-Gothic architecture of Europe. As interpreted by H. H. Richardson (1836–86), however, Romanesque became a different and uniquely American style.

Richardsonian Romanesque was favored for churches, university buildings, train stations and courthouses. Although Richardson produced few houses in this style, elements of his work found their way into many residences of the period.

- massiveness
- stone construction with rock-faced finish
- broad round arches
- towers (often a single massive tower)
- broad roof planes
- eyebrow dormers
- deep-set windows; bands of windows
- cavernous door openings
- doors and windows defined by contrasting color or short, robust columns
- eaves close to walls
- little carved or applied ornament

BEAUX ARTS 1890–1920

Les beaux-arts ("the fine arts") refers to the aesthetic principles of the Ecole des Beaux-Arts in Paris. American architects who studied at the Ecole or who were trained by its graduates were influenced by the school's academic design principles, which emphasized the study of Greek and Roman structures, composition, and symmetry.

Beaux Arts architecture is characterized by large and grandiose symmetrical classical compositions with a wealth of exuberant detail and a variety of stone finishes. American Beaux Arts designs generally were for colossal public buildings. About 1900, however, such designs gave way to more sedate forms, which were used for the town houses and country and resort villas of the wealthy.

- grandiose composition
- imposing stairway
- large arched openings
- variety of stone finishes
- projecting facades or pavilions
- monumental columns
- classical ornament
- enriched entablature topped with a tall parapet, balustrade, or attic story
- pronounced cornice
- decorative swags, medallions, cartouches, and sculpture

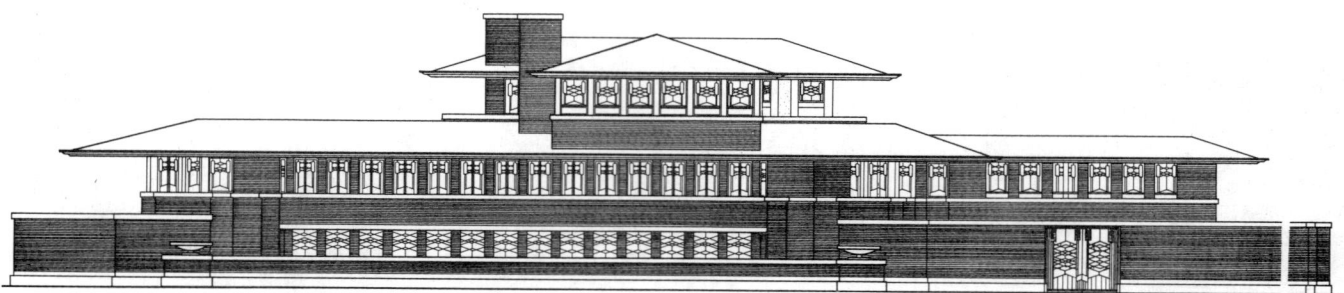

Frederick C. Robie House, Chicago, Illinois, in the Prairie Style. (J.J. Erins, HABS)

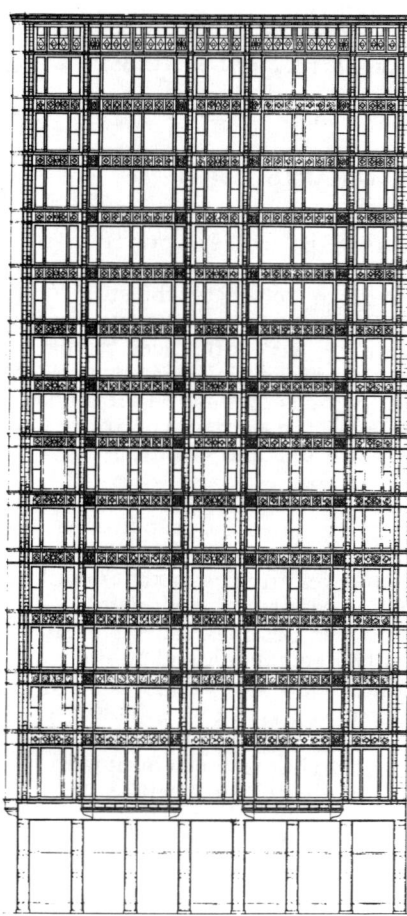

Reliance Building, Chicago, Illinois, in the Chicago Style. (P. Borchers, N. Clouten, HABS)

CHICAGO STYLE 1875–1910

In the late 1800s Chicago-based architects and engineers exploited new construction technologies to produce the tall commercial buildings that would transform cities around the world. Until then, building height had been limited by the ability of masonry walls to support upper stories. By using a cast-iron or wrought-iron skeleton frame—coupled with improvements in fireproofing, wind bracing and foundation technology and the invention of the elevator—Chicago architects began to create commercial buildings of 6 to 20 stories.

The best-known architect of this school was Louis Sullivan (1856–1924). His buildings are easily identified by their distinctive low-relief decoration of intricately interwoven leaf designs around the entranceway, cornice, and windows.

- tall rectangular buildings of 6 to 20 stories
- three-part construction: one- or two-story base with large display windows; shaft housing identical floors of offices; elaborate cornice
- gridlike exterior mimicking the steel skeleton
- large areas of glass, terra cotta, or other nonsupporting material
- vertical piers between windows, emphasizing height
- stripped, no-nonsense exterior

PRAIRIE STYLE 1900–20

About 1900 another group of Chicago architects developed a distinctive midwestern residential style, known as the Prairie Style. Their acknowledged leader was Frank Lloyd Wright (1867–1959). Rejecting the currently popular revivals of historic styles, they sought to create buildings that harmonized with the midwestern prairie. The Prairie Style house had a strongly horizontal appearance, emphasized by porches, walls, and terraces extending from the main structure. Windows were arranged in horizontal ribbons and often featured stained glass in stylized floral or geometric patterns.

Interiors were as innovative as exteriors; Prairie School architects often designed furnishings for their houses. The style had its greatest influence in the Chicago area, but examples can be found as far afield as Rochester, New York, the West Coast, and Puerto Rico.

- low, horizontal silhouette
- wide overhanging eaves
- porches, walls, and terraces extending from the main house to emphasize horizontal lines
- broad, low-pitched roof
- large, low, plain rectangular chimney
- walls of light-colored brick or stucco and wood
- horizontal ribbons of casement windows; stained-glass accents in stylistic floral or geometric designs
- walls at right angles; no curves

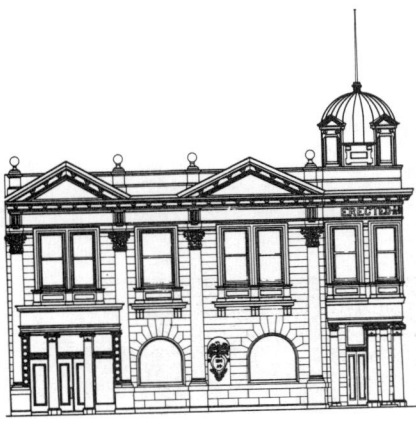

Bank of San Mateo, Redwood City, California. (A. Weinstein, HABS)

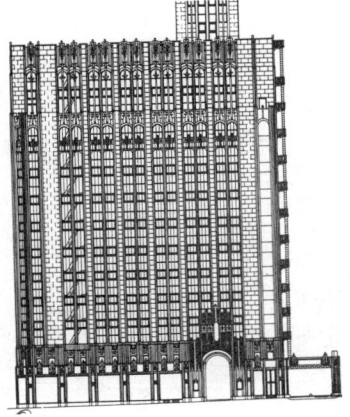

Richfield Oil Building, Los Angeles, California. (R. Giebner, HABS)

Design for a modern house. (*The Modern House*, New York, 1933. Adapted and delineated by S. Bauer)

CLASSICAL REVIVAL 1900–20

The later, more refined stage of the Beaux Arts style influenced the last, or 20th-century, phase of the Classical Revival in the United States. In the late 1800s and early 1900s commissions for public buildings and grand houses generally went to architects trained in the Beaux Arts tradition, who produced designs based on classical and Renaissance models.

Less theatrical than Beaux Arts, the Classical Revival style is based more on Greek than Roman architectural orders. Consequently, the arch is not often used, and highly decorated moldings are rare. The style was used primarily to create massive public buildings on a grand scale.

- monumental size
- symmetry
- giant columns
- Greek and some Roman classical forms
- smooth or polished stone surface
- unadorned entablature and roof line

ART DECO 1925–40

Art Deco took its name from the Exposition Internationale des Arts Decoratifs et Industriels Modernes, held in Paris in 1925 as a showcase for works of "new inspiration and real originality." Art Deco and its derivation, Art Moderne (also called Streamline Moderne or Modernistic) were the first popular styles in the United States to break with the tradition of reviving historical styles.

Art Deco consciously strove for modernity, an artistic expression of the machine age, and the suggestion of motion. Its forms were simplified and streamlined. Essentially, Art Deco was a style of decoration and was applied to jewelry, clothing, furniture, and handicrafts as well as buildings. At its best, Art Deco architecture was a harmonious collaboration of architects, painters, sculptors, and designers.

Art Deco
- surfaces of concrete, stucco, or smooth-faced stone
- vertical emphasis
- facades often arranged in a stepped series of setbacks
- hard-edged, low-relief geometrical designs and stylized figures or floral motifs; designs multicolored, often vivid
- accents in terra cotta, glass, and colored mirror

Art Moderne
- surfaces of concrete, stucco, or metal
- horizontal emphasis
- facades asymmetrically composed
- accents in terra cotta, glass block
- curved corners and other details suggesting motion

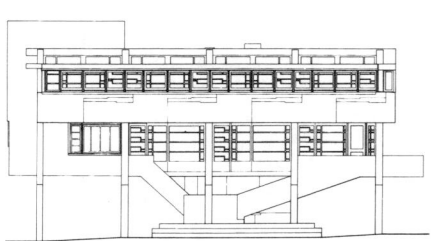

Lovell Beach House, Newport Beach, California. (R.H. Nagata, S.A. Westfall, HABS)

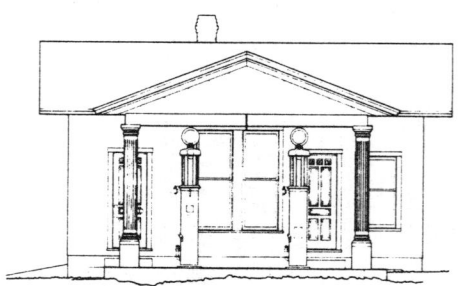

Rockfish Service Station, Augusta County, Virginia. (D. Donovan, HABS)

INTERNATIONAL STYLE 1920–45

The hallmarks of the International Style are stark simplicity, vigorous functionalism, and flexible planning, all based on modern structural principles and materials. Whereas Chicago School architects merely revealed skeleton-frame construction, International Style architects reveled in it.

Ribbons of windows became an important design feature, creating a horizontal feeling even in high-rise buildings. Artificial symmetry was studiously avoided, but balance and regularity were fostered. Mundane building components such as elevator shafts and air-conditioning machinery became highly visible aspects of design. Many of the most famous architects working in 20th-century America—such as Walter Gropius, Ludwig Mies van der Rohe, Richard Neutra, and Marcel Breuer—designed in the International Style.

- concrete, glass, and steel construction
- complete absence of ornamentation
- asymmetrical but balanced composition
- horizontal emphasis
- flat roof
- smooth and uniform wall surface
- mundane building components incorporated into the visual design
- horizontal bands of windows
- corner windows
- windows set flush to the wall

VERNACULAR ARCHITECTURE

As preservationists across the country expand architectural and historical surveys of buildings to small communities and rural areas, new architectural terminology has evolved to depict local building practices that do not conform to well-known styles. In such instances, buildings may be classified as "vernacular." Vernacular architecture has been described as common, ordinary buildings or landscapes fashioned by anonymous people for functional purposes. While vernacular architecture may incorporate one or more styles, it is not a style in itself. Rather, it is a classification or a way of studying a building, a group of buildings, or a landscape. Vernacular buildings result from (1) commonly agreed upon forms and elements that are passed among generations; (2) practices learned from the Old World; (3) community or group tastes rather than those of an individual designer or architect; or (4) use of locally available building materials. To analyze vernacular architecture one must study social history, folklore, cultural geography, and anthropology to provide insight into the ethnic roots and lifeways of the designers and occupants of these buildings. Early vernacular studies focused on the analysis of farmhouses and related buildings. More recently, vernacular studies have expanded to include rural schoolhouses, tract houses, urban dwellings, field patterns, town plans, factories, and commercial strips.

FURTHER READING

American Architecture Since 1780: A Guide to the Styles. Marcus Whiffen. 1969. Rev.ed. Cambridge, Mass.: MIT Press, 1992.

American Shelter: An Illustrated Encyclopedia of the American Home. Lester Walker. New York: Overlook Press, 1981.

America's Architectural Roots: Ethnic Groups That Built America. Dell Upton, ed. Washington, D.C.: Preservation Press, 1986.

Architecture in the United States: A Survey of Architectural Styles Since 1776. Ralph W. Hammett. New York: John Wiley, 1976.

A Field Guide to American Architecture. Carole Rifkind. New York: New American Library, 1980.

A Field Guide to American Houses. Virginia and Lee McAlester. New York: Alfred A. Knopf, 1984.

Identifying American Architecture: A Pictorial Guide to Styles and Terms, 1600–1945. John J. G. Blumenson. Nashville: American Association for State and Local History, 1977. Rev. ed. New York: W. W. Norton, 1981.

Old House Dictionary. An Illustrated Guide to American Domestic Architecture 1600–1940. Steven J. Phillips. Lakewood, Colo.: American Source Books, 1989.

What Style Is It? A Guide to American Architecture. John Poppeliers, S. Allen Chambers and Nancy B. Schwartz, Historic American Buildings Survey. 1977. Rev. ed. Washington, D.C.: Preservation Press, 1983.

RESEARCHING A BUILDING'S HISTORY

Elizabeth Fitzpatrick (Penny) Jones

"Research on structures falls into four basic phases or types:
1. The study of the physical evidence to be found in the structure itself
2. Complete investigation of legal records to provide lists of names, dates, and transactions which are vital pieces of the building's past
3. Research of the original documents that are found in libraries and archives to supply facts that might pertain to the building or its owners
4. Comparative research, which involves structures similar in type or style, to broaden the perspective of the researcher and put the structure into an historical framework."

Cynthia Durko, "Researching a Building." In *Preservation Illinois: A Guide to State and Local Resources*, 1977.

WHY RESEARCH?

Since research can be time consuming and expensive, one may question the need to delve into the history of a building. But research is important and can prevent many unfortunate mistakes in a building restoration or rehabilitation project. Researching an old building also aids the historic preservation movement. The following list of benefits helps justify the costs, in time and money, required by a thorough investigation of a historic structure.

1. Research provides information to ensure that a building restoration accurately reflects the building's history. Information gathered can help determine which parts of a building should be preserved and which parts can be changed.

2. Research documents a building's history for local or state surveys or for nomination to the National Register of Historic Places.

3. Research interprets the building for visitors. Written and oral interpretations should be based on sound research that will contribute to the cumulative knowledge of our country's heritage.

4. Research connects residents to their building's past, providing a link to the people who have lived there before them and to the events that have occurred there.

5. Research gives credibility to the historic preservation movement. History brings a building to life. When people know the history of a building, it becomes more than the bricks and mortar that have been used to build it. Building preservation, then, preserves more than the structure—it maintains a tangible link to the past.

6. Research can be fun. Researching the history of a building is often compared to putting together a jigsaw puzzle. The researcher experiences the elation of finding new bits of information that fit into the research puzzle to provide a meaningful picture of the building's history.

Keep in mind your purpose when researching the general history, dates, and architect of an individual house or commercial building. You will need different information than that needed to prepare historic district surveys or to submit a nomination to a landmarks register. Consult your state historic preservation office and published guides for conducting various levels of preservation-related research.

Don't duplicate work that has already been done. Go to the library and local preservation and historical organizations as well as to the state historic preservation office to find out what research has been done on your building. If the building has been included in a local, state, or national register or survey, much research may have been completed already, such as the date of construction, the style of the building, and facts on its builder or designer.

WRITTEN DOCUMENTATION

Much information about a building is in the public record and readily available. Other facts can be found in private hands through investigative research.

Title Search

A title search produces information on the transfer of property ownership and associated details. Although the search provides information about the land a building stands on, not about the building itself, the data gathered at this stage is key to locating other vital information in tax records, wills, and related documents.

Begin your title search by consulting the property abstract, which outlines the chain of title including information on all pertinent deeds, mortgages, wills, probate records, litigation, and tax sales. If an abstract is not available you should go to the county courthouse to inquire at the register of deeds, county clerk, probate office, or tax assessor.

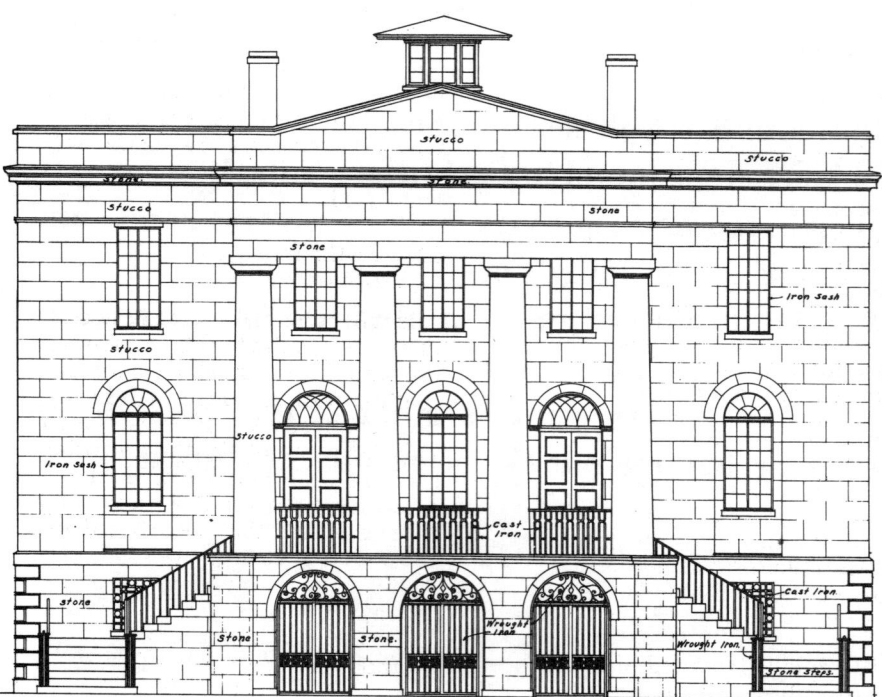

County Records Building, Charleston, South Carolina. (M. Halsey, HABS)

Work backward through the deed books from the present owner to the earliest available citation. Make note of useful information, such as the deed book volume and page numbers, dates, full names of the grantor and grantee, property descriptions (compare newer descriptions to previous ones), kind of deed, mortgages or covenants, and price. If title passed through inheritance rather than through sale, probate records also should be checked. If a parcel was subdivided from a larger property, records for the abutting properties may yield information. Boundary and other topographic changes may require that you research records in a different jurisdiction.

Other Government Documents
Perseverance is needed as you move beyond the deed search to other recorded documents. There are many types of documents to pursue, and you may have to go to several different government offices before you find the information you are seeking. Any of the following documents could provide information valuable to your search: will books, inventory and settlement

books, probate records, death certificates, marriage records, birth records, court cases and indexes, building permits, subdivision and plot books, tax records, and the minutes from meetings of public bodies, such as the city council. These types of records can be found in a variety of places—the county courthouse, city hall, planning departments, and state archives, for instance.

You could learn many varied and useful pieces of information from this research. From inventory and settlement books you may find a room-by-room listing (with prices) of the furniture in the house when the owner died. Census records have information about the people and machinery on site at the time of the survey. Tax records include information on building age, improvements, ownership, assessed value, and occupancy. Building permits provide a record of building alterations over time.

Directories
City directories and crisscross directories often contain owner or occupant names for both residen-

tial and commercial buildings. These directories also can help date buildings in urban settings, although you must watch for changes in numbering systems and street names. Some communities have indexes to these changes.

R. L. Polk and Company, the most prolific publisher of city directories, has produced these listings for more than a century. City planning offices and libraries usually have current and past editions that may be useful in your research, or you can contact the Polk Company in Detroit (See Contacts, below.)

Newspapers
Using old newspapers to research a building is very time consuming but may provide information not available elsewhere: notices about previous owners, advertisements for selling the building, information on the development of a subdivision or building, articles about the restoration of a building, and stories about fires or other disasters that may have affected the neighborhood or even the particular building.

In addition to dailies, consult weekly and specialized newspapers that deal with the neighborhood or subject. But before beginning to look through any of these newspapers, ask if an index is available; an index could save many research hours. Newspapers often are found on microfilm at local libraries and at historical societies. You also may request information directly from your local newspaper.

Maps
County atlases, city maps, railroad maps, and bird's-eye-view maps often show the location of buildings as well as property lines and roads. Insurance maps prepared by the Sanborn Map Company, which date back to 1867, are invaluable for researching buildings in urban areas. These maps are available for more than 12,000 cities and towns in the United States. The location and actual "foot print" of buildings are shown as well as information such as building materials used and the number of stories in the building. Sanborn maps sometimes were updated by pasting a change over

original pages. You may still find the maps useful—it is sometimes possible to see through the additions, or you may find volumes that were never updated.

The Library of Congress, state archives, local historical societies, and local and university libraries are places to find these maps. The Library of Congress maintains the largest and most complete set of Sanborn maps in the world. You can obtain slides and black-and-white prints from its collection, which is listed by city and town in the publication *Fire Insurance Maps in the Library of Congress*.

Architectural Drawings and Specifications

Architectural drawings, building specifications, blueprints, and financial records occasionally are transferred from owner to owner as a building changes hands. A more likely source, however, would be the architectural or construction firm that worked on the building. Other sources include state or city archives, university collections, and local historical societies. In addition, descendants and other relatives of the architect may have saved the plans.

Pattern books are another source of architectural plans. First developed in the 1830s, pattern books contain architectural drawings and specifications for homes of many styles and price ranges. Prospective homeowners used the books to select homes that fit their budgets and lifestyles. One of the first pattern books was *The Beauties of Modern Architecture* (1835) by Minard Lafever.

Since the 19th century, many manufacturers also have published pattern books to encourage the use of their products. Some companies, notably Sears, Roebuck and Company, sold all the materials as well as the plans for building a house. These catalog houses are the subject of much historical interest today. A good source of information on catalog houses is *Houses by Mail*, published by the Preservation Press. (See Further Reading, below.)

Photographs, Postcards, and Illustrations

Whether they are professional photographs in archival collections or snapshots in family albums, photographs can give an array of personal, site, and architectural information. Prints, postcards, drawings, and paintings also may exist, especially if the building is prominent in the community. These visual records can provide invaluable information for restoring a building to its original or some subsequent design.

If sent through the mail, postcards often can be approximately dated from the postmark. Otherwise, you can date a postcard or photograph by determining the model years of automobiles or the approximate date of other technological innovations in the photograph.

There are many local sources of photographs and illustrations, such as historical societies, libraries, photographic studios, and previous owners of the building. Utility companies also may have collections of photographs of buildings and streetscapes. There are several large national collections of photographs and illustrations. The Library of Congress has a general collection of photographs begun in 1846, many of which document the evolution of American architecture. The National Archives, which stores inactive federal records, has collected about six million pictures since its inception in 1934.

Personal and Business Records

Personal papers may reveal additional information about the building you are researching as well as about the people who have lived there. Look for diaries, letters, bills, account books, insurance policies, scrapbooks, and journals. Building owners sometimes described their homes or commercial buildings in letters or wrote about the purchase of specific fixtures. Bills often accurately describe the work done to the building, giving the date and naming the architect, carpenter, or builder. Personal papers may be found by contacting relatives of

previous owners or by going to depositories such as historic societies, libraries, or archives. Records from local businesses can be useful too. A building materials supplier may have records showing the date of construction and what materials were used.

THE BUILDING AS A LABORATORY

The most obvious source of documentation is the building itself (as long as it still stands). Remember that many older houses will have been changed over time: parts may have been demolished, while other sections were added. It will take some sharp detective work to determine the parts that are original and those that have been altered. Analyze the floor plan to see if alterations have occurred; also look for changes in construction methods and materials.

Analyzing the building is important before a restoration or rehabilitation begins to determine what time period is most important to the building's history, what architectural elements should be saved, what should be reproduced, and what, if anything, should be eliminated. Remember that it may not be necessary to get rid of all changes that have been made to the building as they are also part of the building's historical development. In any event, the changes the building has undergone, though sometimes yielding unpleasant results, are themselves interesting objects of study.

Architectural Style

The architectural style can lead you to the period in which the building was constructed and how it fits into the history of architecture. Be alert to building techniques and materials, however, because styles have been revived at various periods in American history.

Compare the building to illustrations in style books, remembering that most buildings are not pure examples of a style. Also, compare the building to similar structures in the neighborhood or region if they exist. Determine what aspects of the building characterize its style and what makes it unique. An appropriate restoration would retain the elements of style and not obscure them by using inappropriate building materials or methods.

Construction Materials and Methods

Construction materials and methods have changed over time and can provide many clues to the building's age and history. For example, with advances in technology, window panes have grown larger. The small, wavy-glassed panes used in colonial houses were not only a style but a technological requirement. The wood muntins used to hold the glass in place created the traditional four-over-four or six-over-six window-pane pattern. As glass manufacturing technology grew more sophisticated, larger panes could be produced, resulting in different styles of windows.

Examine the beams in the cellar and attic for the marks of a broadax, handsaw, or a circular saw. Measure the dimensions of the two-by-four timbers. Before 1930 these pieces of wood were full-size, that is 2 inches wide by 4 inches thick. After that date, the dimensions of most two-by-fours were reduced to 1¾ inches by 3½ inches. Next, determine whether pegs, handmade nails, or machine-made nails were used.

Indoor plaster has gone through easily recognizable changes. Until the mid-1940s plaster generally was composed of cement and water, with animal hair—usually horse or cattle hair—as a binder. The usual method of construction was to apply three coats of plaster to a wood lath. In the 1950s metal lath began to replace wood, and the number of coats of plaster was reduced to two. Since the 1960s, the use of drywall plaster, which comes in large sheets that are screwed or nailed directly to the structure, has become common.

Observing exterior construction also can add evidence to your search. Clapboards on colonial and Federal houses, for example, tend to be fairly wide and rough. This was caused by the primitive technology used to rip them from the timber. After mechanical rippers and sanders were developed in the mid-19th century, clapboards became more standardized, usually 4 inches wide with a smooth finish. Since the 1960s the use of aluminum and vinyl siding has increased dramatically, with wood clapboards being installed only on more expensive houses.

Sophisticated scientific methods can provide valuable information about a house that is otherwise unobtainable. The science of archaeometry, which often is used by art historians to distinguish forgeries, is also used by laboratories to analyze paint chips for building restoration. Dendrochrenology, or wood dating, can help date buildings. Archeological investigations of the site uncover many interesting facts about a building and its residents, even if the building no longer stands. Some researchers use metal detectors to find interesting artifacts in their backyards. For example, Virginia residents often still find bullets from the Civil War.

Social and Historical Contexts

Examine the neighborhood where your building is located. Determine how the buildings in the neighborhood differ from other structures in the city or region and how they are similar to them. Notice how the buildings relate to each other and to the rest of the neighborhood. What is the scale of the buildings? Are they small single-family dwellings, imposing mansions, or apartment buildings? Does the neighborhood have a variety of architectural styles or only a few? Look for physical and social focal points: a church, a school, or a government building may be of architectural as well as cultural importance to the neighborhood.

Study the history of the neighborhood. The evolution of transportation has been a major factor shaping the configuration of our towns and cities and the location of neighborhoods. Before the advent of the railroad, almost all city dwellers lived within walking distance of their jobs and shops. Train travel gave people the opportunity to commute to work from areas outside the city, and the first commuter suburbs were established between 1850 and 1870. Many of these first suburbs are now our historic inner-city neighborhoods.

Inner-city neighborhoods declined through the 1950s and especially in the 1960s when urban renewal caused many inner-city buildings to be demolished. In the 1970s, however, increasing energy costs, highway congestion, delays in having families, and a rise in suburban housing prices encouraged some people to move back into town. As the neighborhood has changed so have the people who have lived there. Different ethnic and socioeconomic groups have passed through as the neighborhood has declined and as it has been renewed. These are all interesting pieces of history for your research puzzle.

Oral Research

Talking to people who know about your building can provide valuable information. A title search provides the names of people who have been associated with your building, or you can ask current neighbors what they know about the building and its previous owners. Even if there has been a large shift in the neighborhood population, you probably will be able to find at least one old-timer. If not, ask neighbors for names of former residents you might contact.

Key questions you should ask include:
Whom do you remember living here? What were their names? What was their family life like? What was their financial situation?

What changes to the house do you remember? Was any part of the house demolished? Was something added? Who made these changes? What was the house like before the changes?

What was the outside of the house like? When and why was new siding added? When was a new roof put on? What different colors has the house been painted? Was there a porch?

What about the interior of the house? What kind of furniture did the owners have? What did the walls and floors look like? Was the inside in good repair?

How has the neighborhood changed while you've lived here? Have many buildings been demolished? How have the neighborhood residents changed over time?

Do you know of other people who remember this house or its residents?

For best results, take the following steps when conducting oral research about your building:

Prepare your list of questions before the interview so you can direct the interviewee to what you consider most important. A standard list of questions also will allow you to compare the answers of the interviewees.

Conduct the interview in the building you are researching in order to stir the memory of the interviewee.

Use a tape recorder so that you won't miss anything.

Be patient and establish a trusting rapport with the interviewee. If the interviewee doesn't feel comfortable with you, you probably won't learn much useful information. And, it may take a person a long time to remember events in the distant past.

Compare the interviewees' answers with written documentation.

Analyze the results and write a brief report for your records and for the local historical society, the library, and the state historic preservation office.

WHERE TO GO, WHAT TO LOOK FOR

County records office
Building permits
Deeds
Liens
Mortgages
Tax records
Will extracts

Probate office
Inventories
Wills

City hall
Annual reports
Building inspector's files
Cemetery records
Maps, plans, and atlases
Property tax records

Library and historical society
Annual reports
Architectural surveys
Biographical publications
Cemetery records
Census data
Commemoratives
Directories
Genealogies
Insurance records and maps
Local histories
Manuscripts (private papers)
Maps, plans, and atlases
Newspapers
Photographs, drawings, prints, and postcards
Publications
Scrapbooks
Tax records
Vital statistics

Local newspaper
News articles
Obituaries
Photographs

Neighbors
Memories
Photographs
From *Researching the Old House*

LIBRARIES AND RESEARCH CENTERS

American Antiquarian Society
185 Salisbury Street
Worcester, MA 01609
(508) 755-5221

American Institute of Architects
Information Center
AIA Archives
1735 New York Avenue, N.W.
Washington, DC 20006
(202) 626-7493

Archives of American Art
National Museum of American Art and the National Portrait Gallery
Smithsonian Institution
8th and F Streets, N.W.
Washington, DC 20560
(202) 357-2781

Art Institute of Chicago
Ryerson and Burnham Libraries of Art and Architecture
Michigan Avenue at Adams Street
Chicago, IL 60603
(312) 443-3666

The Athenaeum of Philadelphia
219 South 6th Street
Philadelphia, PA 19106
(215) 925-2688

Avery Architectural and Fine Arts Library
Columbia University
New York, NY 10027
(212) 854-3501

Cooperative Preservation of Architectural Records
Prints and Photographs Division
Library of Congress
Washington, DC 20540
(202) 707-6399

Massachusetts COPAR
P.O. Box 129
Cambridge, MA 02142

New York COPAR
New York Chapter, American Institute of Architects
457 Madison Avenue
New York, NY 10022

Cooper-Hewitt Museum
National Museum of Design
Smithsonian Museum
2 East 91st Street
New York, NY 10128
(212) 860-6868

The Information Exchange
Municipal Art Society of New York
The Urban Center
457 Madison Avenue
New York, NY 10022
(212) 935-3960

Library of Congress
Manuscript Division
Washington, DC 20540
(202) 707-5287

National Trust for Historic Preservation
Library
University of Maryland
Architecture Building
College Park, MD 20742
(301) 405-6320

Northwest Architectural Archives
University of Minnesota Libraries
826 Berry Street
St. Paul, MN 55114
(612) 627-4199

Partners for Livable Places
Livability Clearinghouse
1429 21st Street, N.W.
Washington, DC 20036
(202) 887-5990

Society for the Preservation of New England Antiquities
Library
141 Cambridge Street
Boston, MA 02114
(617) 227-3956

CONTACTS

Art Libraries Society of North America
3900 East Timrod Street
Tucson, AZ 85711
(602) 881-8479

Association of Architectural Librarians
American Institute of Architects
1735 New York Avenue, N.W.
Washington, DC 20006
(202) 626-7499

Council of Planning Librarians
1313 East 60th Street
Chicago, IL 60637
(312) 947-2163

R. L. Polk & Co.
431 Howard Street
Detroit, MI 48226
(313) 393-0880

Society of American Archives
600 South Federal Street
Suite 504
Chicago, IL 60605
(312) 922-0140
Historic Building Illustration Sources

SOURCES FOR ILLUSTRATIONS

American Institute of Architects Foundation
Prints and Drawings Collection
1735 New York Avenue, N.W.
Washington, DC 20006
(202) 626-7571
Achitects' drawings
Richard Morris Hunt Collection

19th-century photographs of American and European architecture

Bettman Archive
209 Broadway
5th Floor
New York, NY 10010
(212) 777-6200
American history (5 million pieces)

Historic New Orleans Collection
533 Royal Street
New Orleans, LA 70130
(504) 523-4662
Louisiana history and culture (150,000 pieces)

International Museum of Photography
George Eastman House
900 East Avenue
Rochester, NY 14607
(716) 271-3361

19th- and 20th-century photographs (250,000 photos)

Motion pictures and publicity stills (2 million pieces)

League of Historic American Theatres
Archives
Princeton University Library
Princeton, NJ 08540
(609) 258-3223
Chesley Collection

Library of Congress
Prints and Photographs Division
Washington, DC 20540
(202) 707-6394
Historic American Buildings Survey
(20,000 structures)

Historic American Engineering Record
(1,750 structures)

Farm Security Administration and Office of War Information (100,000 pieces)

Carnegie Survey of the Architecture of the South (7,000 pieces)

Pictorial Archives of Early American Architecture (6,500 pictures)

Detroit Publishing Company Collection
(22,000 pieces)

Seagram County Court House Archives
(11,000 pieces)

James Grode Collection: altered or demolished structures (2,000 pieces)

National Gallery of Art
Photographic Archives Department
Constitution Avenue and 4th Street, N.W.
Washington, DC 20565
(202) 842-6030
Dunlap and other collections (4.5 million pieces)

Society for the Preservation of New England Antiquities
Library
141 Cambridge Street
Boston, MA 02114
(617) 227-3956
New England architecture, street views, and transportation (500,000 pieces)

Theatre Historical Society of America
Archives
152 North York Road
2nd Floor
Elmhurst, IL 60148
(708) 782-1800
Photographs of historic movie theaters

Underwood Photo Archives
3109 Fillmore Street
San Francisco, CA 94123
(415) 346-2292
Collections, 1870–1970 (110,000 photographs)

FURTHER READING

Directory of Art Libraries and Visual Resource Collections in North America. Judith A. Hoffberg and Stanley W. Hess, Art Libraries Society of North America. New York: Neal-Schuman Publishers, 1978.

Fire Insurance Maps in the Library of Congress. Library of Congress, Geography and Map Division, Reference and Bibliography Section. Washington, D.C.: U.S. Government Printing Office, 1981.

The History of a House: How to Trace It. Linda Ellsworth. Nashville: American Association for State and Local History, 1976.

House Histories: A Guide to Tracing the Genealogy of Your Home. Sally Light. Spencertown, N.Y.: Golden Hills Press, 1989.

Houses and Homes: Exploring Their History. Barbara J. Howe, Dolores A. Fleming, Emory L. Kemp and Ruth Ann Overbeck. Nashville: American Association for State and Local History, 1987.

Houses by Mail: A Guide to Houses from Sears, Roebuck and Company. Katherine Cole Stevenson and H. Ward Jandl. Washington, D.C.: Preservation Press, 1986.

How Old Is This House? Skeleton Key to Dating and Identifying Three Centuries of American Houses. Hugh Howard, Home Renovation Associates. New York: Farrar, Straus and Giroux, 1989.

"How to Date a House," David M. Hart. *Yankee Magazine,* July 1976 and November 1976.

Index to American Photographic Collections, James McQuaid, ed. Boston: G.K. Hall Library Catalogs, 1982.

Researching the Old House. Lydia B. Summers and Mary Eliza Wengren. Portland, Maine: Greater Portland Landmarks, 1981.

REHABILITATION AND MAINTENANCE

Richard Wagner

"The National Park Service recommends a three-pronged approach to rehabilitation together with consideration of the Secretary of the Interior's Standards for Rehabilitation: (1) If the work required goes beyond simple maintenance, the best approach is to repair, not replace, distinctive architectural features and materials. (2) If repair is not possible, because of very severe deterioration, then the form and details of the deteriorated feature or material should dictate the appearance of its replacement. (3) If the original appearance is not known, the replacement feature should be a new design that is visually compatible with the remaining historic features of the house."

Judith L. Kitchen, *Caring for Your Old House: A Guide for Owners and Residents*

In the past 25 years historic preservation has become a popular and rewarding activity throughout America. Merchants have brought traditional downtowns and neighborhood commercial centers back to life. Homeowners have reclaimed their neighborhoods by preserving their houses and open spaces. Communities have discovered the appeal their historic buildings hold for tourists and citizens. And local governments have supported preservation by creating historic districts and design review ordinances.

These examples illustrate the philosophy of historic preservation, that of maintaining and enhancing the existing character of a building

or landscape. To understand how preservation works, one must first know some of its basic terms. Stabilization, restoration, and rehabilitation refer to distinct types of preservation activities.

Stabilization, often the first step in preservation and reuse, protects a building from deterioration. Stabilization might reestablish a building's weather resistance or strengthen the building to prevent its collapse. Or, stabilization might involve planting a landscape to control soil erosion.

Restoration returns a building or landscape to a particular period in time, which may be chosen for historical as well as architectural

reasons. Building restoration may include the removal of later additions or changes, the repair of deteriorated elements, or the replacement of missing features. A landscape may be restored by replanting with historically accurate plants or by replicating garden walks, walls, and pavilions. Because of its expense, restoration is normally confined to the most important buildings and landscapes in a community.

Rehabilitation, a far more common activity than restoration, returns a building or landscape to a useful state. Buildings are repaired, parts are replaced, mechanical systems are upgraded, alterations are made, and additions are constructed in

Schermerhorn Row, New York, New York. (A. Morrison, HABS)

rehabilitation projects, often to adapt a building or landscape to new uses. The changes, however, must respect the historic character of the property if rehabilitation is to be considered preservation.

Most owners of older and historic properties will be faced with the need to stabilize, restore, or rehabilitate their buildings or landscapes at some time. The terminology, procedures, legal requirements, and economic benefits of preservation can be daunting. While each stabilization, restoration, or rehabilitation project will be different, this chapter is intended to lead a property owner through the general steps of a preservation project from its inception to its continued maintenance, explaining procedures, legal requirements, and other issues likely to be encountered.

RESEARCH IS IMPORTANT

Before beginning a restoration or rehabilitation project, a property owner will need to research two broad types of information to avoid potentially costly mistakes: (1) historical and architectural information about the construction and development of the building and its site; and (2) legal controls, such as building and zoning codes, and requirements, such as design guidelines, that will govern the manner in which the project is executed.

A property owner should gather as much information as possible about the history, architecture, and development of the property. This information will help determine what time period is most important to the property's history and which features are significant and should be retained. This information, along with the intended use of the property and the amount of financing available will determine which preservation procedure is most appropriate.

A number of facts constitute the history and architectural development of a building or landscape: Who owned the property over the years? What significant events occurred there? How has the property been altered over time? Who

were the designers, architects, and builders? What construction methods and materials were used? Thorough research also should interpret the information about a particular building or landscape, placing it within its historical and architectural period, relating it to local, regional, and national patterns.

In some cases, particularly if the building is listed in the National Register of Historic Places or recognized in a local or state register, its history and architectural development have been researched already. In other cases, the property owner should undertake the research using deeds, historical documents, newspapers, old photographs and drawings, and other materials that show the history and changes to the property. (See the chapter on Researching a Building's History.)

WHAT LEGAL CONTROLS ARE RELEVANT?

Building codes are designed to ensure public safety while zoning codes and ordinances are meant to protect public welfare and property values. Together they constitute the major forms of legal controls affecting the restoration and rehabilitation of property. In addition, many communities have enacted local design guidelines or adopted national guidelines for the rehabilitation of historic buildings.

Building Codes

Building codes are enacted and enforced at the local level to regulate the methods and materials of construction, egress requirements, and other issues related to public safety. These codes reflect building occupancy or use, having different requirements for a single-family house, for example, than for an office building. Most local governments have adopted one of the model building codes written by private building code organizations, usually making changes to accommodate local conditions. The principal model building code organizations, and the codes they publish are:

Building Officials and Code Administrators, *Basic Building Code*

International Conference of Building Officials, *Uniform Building Code*

Southern Building Code Congress, *Standard Building Code*

Each of the model building codes, as well as model specialty codes for electrical, mechanical, and plumbing systems, are written primarily for new construction. Thus, compliance with such a code in a restoration or rehabilitation project can be problematic and can lead to additional expense. Common code-compliance issues involve methods and materials of construction to meet fire safety and egress requirements and may result in the alteration of historic fabric, thereby bringing preservation into potential conflict with safety.

Usually it is not a problem if a historic building does not comply with current codes, except for certain seismic safety and handicap accessibility codes, as long as the building's use does not change. Code officials, charged with ensuring public safety, recognize that these buildings have provided safe environments for many years. As long as the use does not change, and as long as substantial rehabilitation work is not undertaken, the building can continue to be occupied without being brought into compliance with current code requirements.

If substantial rehabilitation work is undertaken, however, or if the use of a historic or old building is changed, most building codes require that the building be upgraded to current code standards. Substantial rehabilitation is defined, in most building codes, by the "25/50 percent rule": If the cost of rehabilitation work is more than 50 percent of the building's assessed value before rehabilitation (even if it is to continue in the same use), the entire building must be brought into compliance with the building code. If the rehabilitation work is between 25 and 50 percent of the assessed value, the code official uses his or her judgment about what portions of the building must be brought up to code. If rehabilitation work is less than 25 percent of the assessed value, and the building's use does not change, only the work done must meet code.

In practice, since many old buildings have low assessed values before rehabilitation, the property owners are faced with bringing their entire buildings into compliance with the current code. Unfortunately, this sometimes requires the alteration of historic fabric, spaces, or features, placing compliance with the code in potential conflict with appropriate restoration and rehabilitation practices. For example, many stairways in historic buildings do not meet current code requirements—they are too steep or too narrow or have too many steps without a landing. To meet the current code, they would have to be altered. Conflicts between preservation and code compliance also regularly occur over the location of fire breaks and doors and windows, fire resistance of materials, and other historic features and methods of construction.

To address this problem, the International Conference of Building Officials created the Uniform Code for Building Conservation (UCBC). Unlike current codes, which are written for new buildings, the UCBC interprets the safety of historic methods and materials of construction. The stated purpose of the code is to "encourage the continued use or reuse of legally existing buildings and structures . . . [by] offering alternative methods of achieving safety so that the inventory of existing buildings can be preserved." Many local governments are beginning to adopt the UCBC for use in the restoration and rehabilitation of historic buildings that are listed in local or state registers or in the National Register, or located within historic districts. A few states also issue state building codes specifically designed for use in the restoration or rehabilitation of historic buildings.

Zoning Ordinances
When first created at the beginning of the 20th century, zoning codes and ordinances were designed to protect the public welfare by separating obnoxious and polluting uses of land and buildings from residential, commercial, educational, and religious uses. Thus, factories, slaughterhouses, and the like were forced to locate in certain districts rather than be allowed to spread smoke and effluence throughout the community. Over the years, as manufacturing and other obnoxious uses have become cleaner, zoning has become a tool of local governments to protect and enhance the value of private property. Preservationists have taken advantage of this change in the use of zoning codes and ordinances to increase legislative protection for historic buildings. Historic overlay zones or districts have been created in many communities to regulate changes to the existing buildings and landscapes. Often these zones allow uses in the historic buildings that normally would be excluded under ordinary zoning. For example, many historic overlay zones in residential areas allow certain professional offices, funeral homes, and other nonresidential users to occupy buildings. They are allowed because the negative impacts of the nonresidential uses in the area are outweighed by the benefits of finding uses for large historic houses that may no longer be wanted as residences, thereby enhancing their chances of preservation.

Design Guidelines
If a building or site is located within a historic zone or district, alterations or new construction may be subject to some form of design review. In some districts only those properties designated as contributing to the historic or architectural character of the area will be subject to review, while in other districts all proposed changes must be approved. Often the public agency responsible for the review will adopt a set of design guidelines to help property owners understand what types of changes are considered acceptable.

Design guidelines are employed in one of three ways:

Advisory guidelines. These are used in communities that are not yet politically ready to legislate design. They offer advice and guidance to the property owner but are not legal requirements that must be followed.

Incentive guidelines. Many communities have developed design guidelines for historic districts that are specifically linked to certain financial or other types of incentives. If a property owner wishes to take advantage of the incentives, the building's design must comply with the guidelines.

Mandatory guidelines. More and more communities are legislating mandatory design guidelines for restoration and rehabilitation of historic buildings and districts. Like building and zoning codes, the requirements of the guidelines must be complied with before changes will be allowed.

While local guidelines will differ, almost all contain a section that describes the important visual characteristics of the existing buildings and environment that should be preserved. These characteristics include the following:

- orientation, setback, spacing, and site coverage of buildings
- height, width, and massing of buildings
- size, shape, and proportion of building fenestration
- materials, textures, colors, and details of facades
- roof forms and cornice lines
- open space—vistas, vegetation, edges of open spaces

In addition to the above characteristics, design guidelines should address unique features found in particular areas. Guidelines written for a residential neighborhood may, for example, include characteristics of porches, driveways, and garages. Those written for a downtown may contain a section covering business signs, awnings, and canopies, as well as street elements such as bus shelters, kiosks, and informational signs. Often guidelines will contain separate sections on existing buildings and new construction, allowing the special characteristics of each to be presented in detail.

National Standards and Guidelines for Rehabilitation

The United States is the only country with national standards and guidelines for the restoration and rehabilitation of privately owned historic property—*The Secretary of the Interior's Standards for Rehabilitation and Guidelines for Rehabilitating Historic Buildings.* The standards and guidelines must be followed by the owner of a National Register building who wishes to take advantage of the federal investment tax credits for rehabilitating historic income-producing buildings, or if federal money is used in the project. Many states also require compliance with the standards and guidelines if an owner wishes to use state financial incentives. The Secretary of the Interior's Standards and Guidelines often are used as the basis for local design guidelines. For a detailed explanation of the standards and guidelines, see the chapter on Rehabilitation Standards.

INSPECT THE BUILDING

After researching a building's history and architecture and the codes, ordinances, and guidelines that will regulate its restoration or rehabilitation, the next step in the project is to thoroughly inspect the building and to analyze its potential. Depending on the complexity of the preservation project, the property owner may perform the analysis, or experienced architects, engineers, materials conservators, and other specialists may be called in.

For most preservation projects, four types of analyses should be performed.

1. Spatial. This analysis describes the physical characteristics of all spaces in the existing building—their shape, dimensions, location, and access—to determine if the building will accommodate the proposed use as it exists or if alterations and additions will be required.

2. Structural. The ability of the existing building to accommodate new or continued uses is dependent on its structural soundness as well as on the amount of weight that can be supported by floors, columns, and walls. If the building's structure is unstable or must be strengthened to support additional weight, the cost of these modifications may significantly increase the project budget.

3. Code. A code analysis will inform the property owner of the changes that must be made to bring the building into compliance with the local building code. Depending on the building's existing conditions, its proposed use, the local code, and its interpretation by the code official, upgrading the building to comply with the code may be expensive and may require that the building's architectural integrity be compromised.

4. Materials. Documenting the physical state of a building's materials and understanding the extent and causes of deterioration found are the first steps in addressing the repair or replacement of materials. Knowledgeable professionals may inspect the building visually or it may be necessary to test the materials in a laboratory.

In addition to these four major analyses, a property owner also may undertake an energy audit of the building to help determine appropriate energy conservation techniques. Typically, the audit analyzes the thermal performance properties of the building's exterior; infiltration rates around doors, windows, and other openings; and the efficiency of the building's mechanical and electrical systems.

FINANCIAL ANALYSIS

Before proceeding with the project, an owner should understand how much the restoration or rehabilitation is likely to cost and should research the availability of financial incentives. If the building is to be rented or sold on completion, the owner also should understand the amount of income that he or she can expect to receive.

Restoration or Rehabilitation Costs

Various types of costs are associated with restoration and rehabilitation projects. Typically they include the following:

- Professional fees for architects, engineers, landscape architects, and other members of the design team. If the building is not already listed in the National Register, professional fees may be paid to an architectural historian to research the building and complete the nomination.

- Property costs. These are the costs of acquiring the building or site, if not already owned.

- Permit fees. Building permits, zoning changes, waivers, water and sewer connections, electrical hookups, and other fees usually are paid to the local government and utility companies as part of the rehabilitation or restoration process.

- Construction costs. The largest portion of a restoration or rehabilitation project budget will be consumed by the fees paid to contractors and subcontractors and by the costs of materials and supplies.

- Cost of capital. Owners usually must borrow money for construction and other costs from a bank, a savings and loan, an insurance company, a pension fund, or some other source. The cost of that money—the rate of interest charged and the term of the loan—will dramatically affect the total cost of a project.

If the project is to be rented or sold when completed, the expected income should be estimated early in the process. Income will depend on a number of factors:

- location of the property
- supply and demand for the type of space
- attractiveness
- features of the project, such as on-site parking and security, that make the rehabilitated building more desirable than other similar buildings

Comparing expected income to expenses allows the owner to estimate whether a project will be financially feasible or not. For projects that are to be sold on completion, the comparison is fairly simple:

Rehabilitation or restoration costs + profit = sales price

For projects that are to be rented on completion, the comparison is more complex:

Rehabilitation or restoration costs + annual operating costs + profit = rent

A standardized method of making these comparisons, known as feasibility analysis or pro forma analysis, is used to calculate the financial success or failure of the building. The analysis allows the owner to investigate various financial scenarios and helps determine how much money should be spent on the restoration or rehabilitation.

Financial Incentives
Financial incentives often make the difference between the financial success or failure of a rehabilitation or restoration project. Administered at the national, state, and local levels, the major incentives available are grants, low-interest loans, and various types of tax incentives.

Depending on restrictions placed by the agency or organization offering them, incentives can be used to assist with research, design, construction, or other aspects of the property's preservation. Many of these programs, such as those offered by the National Trust, are available only to owners of registered historic properties. Others, such as the community development funds from the U.S. Department of Housing and Urban Development (HUD) or housing rehabilitation funds from Farmers Home Administration, can be used for new construction and the rehabilitation of nonhistoric older buildings as well as for registered historic properties.

Most tax incentives reduce the owner's tax liabilities after a rehabilitation or restoration has occurred. Since 1976 the federal government has offered various forms of income tax deductions and credits for rehabilitating National Register income-producing buildings. Several states offer similar income tax incentives for rehabilitation of national or state register income-producing buildings. And many local governments offer property tax abatement programs for rehabilitation and restoration. Some local governments and nonprofit organizations accept easements on historic properties that allow the owner to further reduce his or her tax burden. Many successful rehabilitation and restoration projects are possible only with the use of financial incentives—without the incentives, costs often exceed income.

For additional information on the costs and benefits of preservation, see the following chapters: Easements, Federal Programs, Fund Raising, National and Regional Contacts, Revolving Funds, and Tax Incentives.

CHOOSING DESIGN PROFESSIONALS

Except for the simplest of projects, it is best to engage an architect or other design professional during the investigation of the work. Engaging a qualified professional who understands the design, construction, and materials of historic buildings can be cost effective. The professional will know what methods and treatments will best stabilize or repair deteriorated features, how the design should be developed to comply with local design guidelines or the Secretary of the Interior's Standards, and what will be required to bring the building into compliance with the local building code. A qualified architect should also be able to create a cost-effective design that effectively utilizes existing spaces for the proposed new use while retaining important historical or architectural features. With the assistance of an engineer versed in old and historic buildings, the architect will be able to correct existing structural problems as well as to design new mechanical, electrical, and plumbing systems for the project.

This is not to suggest that any licensed architect should be engaged to restore or rehabilitate a historic building. Just as attorneys or physicians specialize in the various branches of law and medicine, certain architects have far greater skills in restoration and rehabilitation than do others.

Finding a qualified architect need not be difficult. Although state historic preservation offices will not recommend one architect or firm, most will provide a list of architects who are familiar with historic buildings and their construction. Lists of architects who consider their firms to have restoration and rehabilitation experience also can be obtained from the local or state chapter of the American Institute of Architects. These lists typically are generated from questionnaires about the type of work an architect has recently completed. Similar lists can be obtained from the American Society of Landscape Architects, the American Society of Interior Designers, the Association for Preservation Technology, and other organizations to which architects and designers belong.

The architect hired to supervise the restoration or rehabilitation project normally will secure all other professionals required to design the project. This may include engineers as well as landscape architects and interior designers. A contractor also may be part of a design team.

DESIGN STAGES

For most projects, the property owner should expect the design to evolve in three major stages: (1) design schematics, (2) design development, and (3) production of contract documents.

Design Schematics
Design schematics that incorporate information obtained from the spatial, structural, code, and materials analyses show how the space is to be utilized and how codes are to be met. The schematics will show how existing exteriors are to be treated and, if additions are required, their massing, proportions, and fenestration. If design review at the local level is required,

it is appropriate to obtain initial comment from the review board at this stage in the design process. If the building is subject to state or national review, if it is a complex design, or if the schematics require a possible compromise of historic or architectural fabrics, an initial review by the state historic preservation office or the Preservation Assistance Division of the appropriate National Park Service regional office is advisable as well. During this initial phase, the architect should supply the owner with a preliminary construction cost estimate, which can be used to estimate the cost of the project and determine its financial feasibility. Some local, state, and national financial assistance programs require submission of design schematics as part of their application process.

Design Development
After the schematics have been reviewed and approved by all parties, the architect will proceed with the design development phase of the work. The owner will receive details on methods of treating deteriorated fabric, materials to be used to construct the building, and appropriate finishes to be used. The architect also will develop a detailed cost estimate for construction and may submit the design for preliminary approval to the building code official. If the design is subject to local, state, or national design review and was not previously submitted for comment, it should be at this time. During this stage, the owner should secure financing for the project and, if the property is to be rented, should consider utilizing the drawings and plans to secure tenants.

Contract Documents
The final stage of the design is the production of contract documents—working drawings and specifications. These documents are used by contractors to prepare bids for the project and, later, by the successful bidder to direct the building's restoration or rehabilitation. Contract documents are used to secure a building permit from the building code and permit office, and the local design review board may require that the documents

meet their approval. Contract documents are part of the legal contract between the property owner and the contractor, specifying exactly how the work is to be performed. Properly executed working drawings and specifications are important in reducing disputes or misunderstandings that may arise during construction.

Nonprofit preservation organizations and universities are two more entities that may play roles in the design of a restoration or rehabilitation project. Some preservation organizations provide grants for analyses or for schematic designs, or in the case of particularly significant property, a nonprofit organization may undertake the project itself in the hope of preserving the building or site. Other preservation organizations, such as local Main Street revitalization groups, may have access to architects who will provide free schematic designs for buildings located within their districts. Universities, particularly those with preservation programs based in departments of architecture or design, are often willing to allow students to develop schematic designs for buildings as part of their education. And a few local chapters of the American Institute of Architects sponsor community design centers that provide low-cost schematic designs for projects meeting their criteria.

CONSTRUCTION BEGINS

For most property owners, the construction phase of a rehabilitation or restoration project is the most exciting as well as the most nerve-racking part. The walls, floors, and roof may be stripped bare; old wiring and mechanical systems may be removed; and openings in the walls may be created to allow equipment and material into the building. Work may proceed at a rapid rate or may come to a complete halt awaiting the arrival of materials or subcontractors. Changes to the contract documents may be required as construction work uncovers previously unknown historic fabric that should be saved. The list of apparent and real problems during construction is endless. Yet, many

projects, if managed by a reputable contractor and overseen by a knowledgeable architect, will be completed on time and within the stated budget.

Selecting the right contractor is critical to completing the project successfully. As with architects and other design professionals, some contractors are better qualified than others to undertake restoration and rehabilitation projects because of their training and experience. The contractor must be sensitive to historic fabric and understand the importance of protecting significant features during construction. If unknown historic fabric is uncovered during work, a contractor should recognize its importance and ensure that it is not disturbed until decisions can be made about whether it will be saved. Contractors usually prepare competitive bids of their cost proposals to perform the construction based on an examination of the contract documents. While the cost proposal is important, selecting a contractor also should be based on the firm's reputation and experience with historic preservation projects and its proposed schedule to complete the work. Often, when these and other factors are considered, the contractor submitting the lowest cost proposal does not get the job.

If the restoration or rehabilitation project is complex or involves highly specialized skills, it is often beneficial to involve a contractor or subcontractor in the schematic and design development phases of the project. When this is done, the contractor is usually selected based on reputation and skills, with the cost of the work negotiated after completion of the contract documents. Architects often will help owners select a contractor, either early in the process or by reviewing the bids submitted. In addition, most state historic preservation offices will supply lists of contractors that have worked on historic properties, but the office will not recommend any one firm nor consider a firm's inclusion on the list as an endorsement of its work.

Potential Problems

No matter how experienced the contractor or the architect, problems can arise during the construction phase of a restoration or rehabilitation project. Some of the most common problems are discussed below.

Underestimating the time or budget required to complete the project. The time and budget for a new building can be estimated with some accuracy because all components, systems, and conditions are known. This is not so true of rehabilitation and restoration work. Structures that were sound may weaken when construction is started. Accidents may cause historic fabric to be destroyed, requiring its duplication. Historic fabric hidden for years may be uncovered and work halted until the discovery can be incorporated into the design. Materials promised for delivery on a certain date may be delayed, or scheduling difficulties may be encountered with subcontractors and skilled tradespeople. Thus, most restoration and rehabilitation projects should allow for a flexible construction schedule and a healthy contingency budget.

Testing conservation techniques. Although knowledge of historic materials treatment has increased significantly in the past 25 years, it is still advisable to test conservation methods during construction. For example, if the exterior of a building is to be cleaned, it is important to test various cleaning methods on small unobtrusive areas before deciding how to treat the entire structure. Even after the best method has been selected, certain areas may require an alternative treatment because of differences in pollutants or materials. Although these tests may prolong the construction period, it is always best to evaluate the proposed technique on a limited area of the building before completing the work.

In the past two decades, dozens of new materials have been developed to take the place of deteriorated historic fabric. Based on complex chemical compounds, the substitute materials typically are lighter, stronger, more fire and weather resistant, and less expensive than the original. Fiberglass and polyester compounds often are used as substitutes for cast iron, and chemically hardened gypsum compounds are used to replace or repair deteriorated stone. The list of substitute materials grows each year. Often it is necessary to test the material during construction to determine if it matches the appearance of the original material and to ensure that it will not harm adjoining original material and that it will weather appropriately.

Miscommunication. Even the most straightforward restoration or rehabilitation project can run afoul because of miscommunication, primarily due to the large number of people involved and the technical nature of the process. For example, the owner of the building, the design review board members, the general contractor, and the paint contractor may all agree on the color of a wall, choosing the same paint chip. But once the wall has been painted and the color covers a large surface rather than a small chip, some participants may be dissatisfied with the result.

In addition to these common problems, hundreds of other conflicts can arise during project design and construction. These problems can be minimized, however, if the following guidelines are heeded:

- Be realistic about the time and expense required to complete the project.
- Conduct appropriate background research that will assist in the design of the project.
- Work closely with review boards, building officials, and others who must approve the design materials and construction methods.
- Engage qualified professionals.
- Be flexible to necessary changes that arise during construction.

When the project is completed, the architect, contractor, and owner should go over every aspect of the job using a "punch list" to ensure that it was constructed as specified. Since problems can arise after the punch list is completed and after the building is occupied, property owners should request workmanship and materials guarantees from the contractor to cover any remedial work required for up to one year after occupancy.

MAINTAINING THE COMPLETED PROJECT

Once the project has been completed, it is important that the owner's investment be protected by proper maintenance. Regular cleaning of the building and an annual inspection to determine if materials or systems are deteriorating are recommended. In addition, property owners should consider establishing a replacement and repair fund, sometimes known as a "sinking fund," so that money is available when repair work is needed.

Regular cleaning of the building can be on a short- or long-term cycle. For example, interior surfaces can be dusted once a week and washed every three months, or gutters can be cleaned every four months and windows cleaned semiannually. At the start of both the heating and cooling seasons, filters in furnaces or air conditioners should be cleaned and chimneys swept. Cleaning not only provides a sanitary environment, it also helps remove pollutants and organic growth that contribute to deterioration.

Every year a property owner should inspect the building thoroughly for signs of deterioration. Often the type of deterioration and its time frame can be anticipated. Most exterior paint will last 7 to 10 years depending on weather conditions and exposure. Window putty can be expected to deteriorate in 10 to 15 years and caulking in 15 to 20 years. Roofing materials probably will need replacement every 20 to 40 years, while mortar joints may last 50 or more years before requiring tuckpointing.

MAINTENANCE INSPECTION CHECKLIST

Many property owners use a checklist to record the condition of a building and to ensure that a regular maintenance cycle is established. If the checklist is used on a regular basis, the property owner will have comparative data on the building's condition over time, which will help estimate the rate of deterioration and the time remaining before repair or replacement is needed.

A maintenance checklist should be tailored to each property. It should itemize the materials and systems contained in the building and record their condition. In addition to a written checklist, plans of the building that note the exact location of problems as well as photographic records can be valuable when taking corrective measures. The following is a list of the typical sections included in a maintenance inspection checklist and some of the maintenance problems to watch for:

Roof
1. Does the roof membrane (shingles, slates, metal, etc.) appear to be thin, worn or cracked? Are parts of the membrane missing?

2. Does part of the roof sag?

3. Are the flashing, gutters, or downspouts rusting, sagging, or missing? Are leads to storm sewers blocked?

4. Are cresting, finials, and other decorative roof elements loose, rusting, or missing?

5. Is the chimney sagging? Are the mortar joints tight? Is the spark cap rusting or missing?

Exterior Walls
1. Are the walls plumb and without bulges? Are cracks evident?

2. Are the doors, windows, and other opening frames square and true?

3. Is masonry spalling (flaking), crumbling, or missing? Are the mortar joints tight?

4. Is the wood siding cracked, loosened, rotted, or split?

5. Is the cast iron or pressed metal rusting, pitted, or missing?

Windows
1. Do the windows open and close smoothly?

2. Is the glass broken? Is the putty sound?

3. Do the frame, muntins, and sash show signs of rust, rot, or insect damage?

4. Is the window hardware (latches, counterweights, pulleys, etc.) in good repair?

5. Is the weatherstripping in good repair? Do storm windows fit tightly? Are the screens damaged?

Doors
1. Do the doors open and close smoothly?

2. Do frames or doors show signs of rust, rot, or insect damage?

3. Is the door hardware (hinges, handles, locks, etc.) in good repair?

4. Is the threshold rotted?

5. Do glass panels in the doors fit tightly? Are the stops or putty in good repair?

Exterior Features
1. Are porches, stairs, railings, cornices, brackets, and other exterior features in good repair? Are elements missing?

2. Is the paint cracked, faded, or peeling?

Foundation
1. Does water drain away from the foundation?

2. Does the stone, brick, or concrete show signs of spalling, crumbling, or cracking?

3. Is the mortar tight?

4. Is algae or moss growing on the foundation?

Interior
1. Is the basement damp or subject to periodic standing water?

2. Do floors sag? Do floorboards squeak? Is the floor covering (finished wood flooring, tiles, stone, terrazzo, carpet, etc.) in good repair?

3. Is the plaster damp, cracked, or loose?

4. Are baseboards and chair rails in good repair?

5. Are the crown molding, plaster medallions, and other wall and ceiling decorative elements in good repair?

6. Do stairs sag or squeak? Is the railing in good repair?

7. Are the paint, wallpaper, and other wall and ceiling finishes in good repair?

8. Can the sky be seen from the attic? Is the ceiling insulation damp or compacted? Do vents have screens or louvers? Is there evidence of water stains on rafters or joints?

Heating and Cooling Systems
1. Is the heating or cooling plant more than 30 years old? (If so, it may need replacement.) Has energy consumption increased over the past five years?

2. Is the distribution system in good repair? Are registers and vents in good repair? Do the radiators need repair or repainting?

3. Have the filters been cleaned or changed? Have the humidity control and air conditioning coolant been checked?

Plumbing Systems
1. Is the hot water heater more than 30 years old? Does it show signs of rust?

2. Is the tap pressure less than last year? (If so, the pipes may have built up deposits.)

3. Do the pipes sweat? Do the joints between pipes show signs of rust? Are the joints tight?

4. Are the fixtures (sinks, tubs, toilets, etc.) in good repair? Do the faucets leak? Does the toilet run without being flushed?

5. Do outside hose connections drip?

6. Has water consumption increased over the past five years?

Electrical System
1. Is the wiring more than 50 years old? Is it split or frayed? Is it lacking insulation?

2. Do all outlets, switches, and fixtures work?

3. Is the fuse or breaker box in good repair?

4. Has electricity consumption increased over the past five years? (If so, is the wiring still adequate to carry the load?)

CONTACTS

American Institute of Architects
1735 New York Avenue, N.W.
Washington, DC 20006
(202) 626-7300

American Society of Interior Designers
608 Massachusetts Avenue, N.E.
Washington, DC 20002
(202) 546-3480

American Society of Landscape Architects
4401 Connecticut Avenue, N.W.
Washington, DC 20008-2302
(202) 686-2752

Association for Preservation Technology
904 Princess Anne Street
P.O. Box 8178
Fredericksburg, VA 22404
(703)373-1621, 373-1622

FURTHER READING

The Architecture of Form. Lionel March. New York: Van Nostrand Reinhold, 1977.

Buying and Renovating a House in the City: A Practical Guide. New York: Alfred A. Knopf, 1972.

Caring for Your Old House: A Guide for Owners and Residents. Judith L. Kitchen. Washington, D.C.: The Preservation Press, 1991.

Design and Development: Infill Housing Compatible with Historic Neighborhoods. Ellen Beasley. Information Series. Washington, D.C.: National Trust for Historic Preservation, 1988.

The Design Process. Ellen Shoshkes. New York: Whitney Library of Design, 1989.

Developing Downtown Design Guidelines. California Main Street Program. Sacramento: California Department of Commerce, n.d.

Fabrics for Historic Buildings. Jane C. Nylander. Rev. ed. Washington, D.C.: The Preservation Press, 1990.

Floor Coverings. Helene Von Rosensteil and Gail Caskey Winkler. Washington, D.C.: The Preservation Press, 1988.

Guidelines on Fire Rating of Archaic Materials and Assemblies. Washington, D.C.: U.S. Department of Housing and Urban Development, n.d.

Guiding Design on Main Street. National Main Street Center. Washington, D.C.: National Trust for Historic Preservation, 1988.

How Architects Visualize. Tom Porter. New York: Van Nostrand Reinhold, 1979.

How Designers Think. Bryan Lawson. London: Architectural Press, 1980.

Interpreting the Secretary of the Interior's Standards for Rehabilitation. Washington, D.C.: Technical Preservation Services, U.S. Department of the Interior, 1980-. Sets and occasional bulletins.

Lighting for Historic Buildings. Roger W. Moss. Washington, D.C.: The Preservation Press, 1988.

Masonry: How to Care for Old and Historic Brick and Stone. Mark London. Washington, D.C.: The Preservation Press, 1988.

New Life for Old Houses. George Stephen. Washington, D.C.: The Preservation Press, 1989.

Preservation Briefs Series. Technical Preservation Services, U.S. Department of the Interior. Washington, D.C.: U.S. Government Printing Office, 1975-.

Preservation Tech Notes Series. Technical Preservation Services, U.S. Department of the Interior. Springfield, Va.: National Technical Information Service, 1985-.

Principles of Design in Architecture. K. W. Smithies. New York: Van Nostrand Reinhold, 1981.

Rehabilitation and Pro Forma Analysis. Donovan D. Rypkema for the National Main Street Center. Washington, D.C.: National Trust for Historic Preservation, 1987.

Remodeling Old Houses Without Destroying Their Character. George Stephen. New York: Alfred A. Knopf, 1972.

Repairing Old and Historic Windows: A Manual for Architects and Homeowners. New York Landmarks Conservancy. Washington, D.C.: The Preservation Press, 1991.

Respectful Rehabilitation: Answers to Your Questions About Old Buildings. National Park Service. Washington, D.C.: The Preservation Press, 1982.

Scale in Architecture. Frank Orr. New York: Van Nostrand Reinhold, 1979.

The Secretary of the Interior's Standards for Rehabilitation and Guidelines for Rehabilitating Historic Buildings. W. Brown Morton III and Gary L. Hume. 1979. Rev. ed. Washington, D.C.: U.S. Department of the Interior, National Park Service, 1990.

So You Want To Fix Up An Old House. Peter Hotton. Boston: Little, Brown, 1979.

Uniform Code for Building Conservation. Whittier, Calif.: International Conference of Building Officials, 1987.

Wallpapers for Historic Buildings. Richard C. Nylander. Rev. ed. Washington, D.C.: The Preservation Press, 1992.

Walls and Molding: How to Care for Old and Historic Wood and Plaster. Natalie Shivers. Washington, D.C.: The Preservation Press, 1990.

ADAPTIVE USE

Richard Wagner

"You cannot hang a building on a wall like a painting; you have to find a use for it."

Arthur P. Ziegler, Jr., *New Life for Old Buildings*, 1977

What is adaptive use?
Adaptive use generally refers to altering the physical characteristics of a building to accommodate a new use. The alterations may change room sizes and arrangements, replace mechanical and electrical systems, add new stairs or elevators, create new entries, or make a host of other changes to meet the requirements of the new use.

Why does a building's use change?
A change in a building's use may occur for a number of reasons. One is that the need for a particular use at the building's location may change. For example, owners have abandoned large Victorian town houses in inner cities because the neighborhoods are no longer considered desirable places to live. Many of these houses have been adapted successfully to such commercial uses as offices, retail shops, and funeral homes.

The spatial needs of a particular use may no longer fit with the original building. For example, the space needed for business offices has changed substantially in the past 100 years. Small separate offices for workers have been replaced with large open floor plans that are divided into work stations by low screens. Thus, many of the late 19th-century office buildings in our communities have been adapted to other uses.

Changes in technology also cause changes in the way buildings are used. The technology used in hospitals has changed so dramatically in the past 50 years that few patient wards built more than a half century ago are still being used for that purpose.

A building's use also may change because the reason the building was constructed no longer exists. Hundreds of places that were built to manufacture horse-drawn buggies or hand-rolled cigars, to house telegraph offices or vaudeville theaters, or to serve dozens of other uses that no longer exist have been adapted for other purposes.

What are the benefits of adaptive use?
First, putting an old building to new use extends the building's life since occupied structures are less likely to be demolished than vacant ones. Second, using old buildings for new purposes contributes to the economic health of both the community and the structure because inhabited buildings generate taxes for the public sector and provide places of employment or residence.

Third, adapted buildings may provide space for cultural or recreational organizations, serve educational institutions, or house a host of other uses that enrich a community.

And, finally, the continued existence of an older building by adapting it to new uses helps preserve the character of its neighborhood.

What are the problems with adaptive use?
Attempting to adapt an old building to new uses may be physically as well as economically difficult. For example, upgrading a building to meet current building code requirements for egress, accessibility, and safety can be physically difficult as well as expensive. In certain regions of the country upgrading buildings to meet seismic code requirements makes adaptive use extremely difficult. In addition, the building's new use may require that its structural system be strengthened. While this is almost always technically possible, it is often expensive.

Another problem faced when adapting buildings to new uses is that the large-sized rooms in older buildings often make meeting today's standards of heating and cooling more expensive than in new construction. And if the building contains historically significant interiors, the problem of adding new mechanical systems without altering the appearance of the rooms or disturbing historic fabric is even greater.

What is the adaptive-use potential of my building?
The adaptive-use potential of any building is determined by its location, physical integrity, spatial composition, and the amount of money that will be required to adapt it to a new use. Many property owners commission market studies to determine what types of uses could be supported within their buildings. Owners also may commission analyses to determine the building's physical integrity and, if the building is found to be deficient, the cost to upgrade it. Often it is useful to commission schematic designs for different uses in the building to determine which has the best spatial fit, what types of building code problems exist, and how much the adaptation will cost.

What if no new use for a building can be found?
If possible, mothball the building and wait for circumstances to change. Locational characteristics that make the building difficult to reuse today may change in a few years. Consider the hundreds of old school buildings that were closed in the late 1960s and 1970s as the school-age population declined. Today, many that were not torn down are being adapted to housing for the elderly as that sector of the population grows.

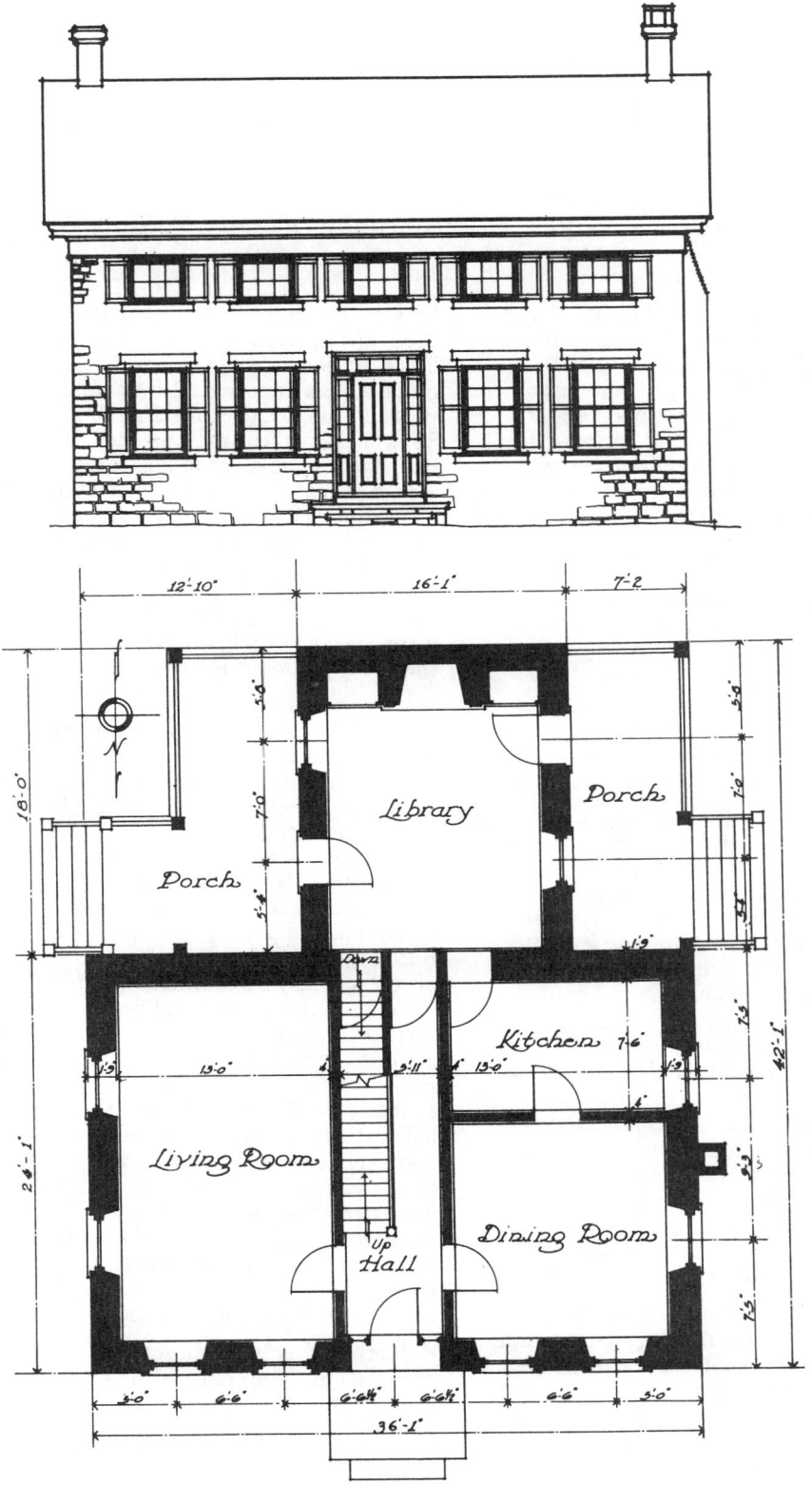

Charles Miller House, Menomonee Falls vicinity, Wisconsin. (R. Adams, HABS)

Are there specific requirements for adaptive use of National Register buildings?
If a National Register building is to be adapted for an income-producing purpose and the owner wishes to take advantage of current federal tax benefits, then the Secretary of the Interior's Standards for Rehabilitation must be followed.

The first standard states:

A property shall be used for its historic purpose or placed in a new use that requires minimal change to the defining characteristics of the building and its site and environment.

Is adaptive use more or less expensive than new construction?
The cost of an adaptive-use project depends on the physical integrity and spatial arrangement of the building and the requirements of the new use. In some cases, where only a few changes are required, it may be very inexpensive to adapt the building to the new use. In other instances, the changes required to fit the new use may make adapting the existing building more expensive than constructing a new one. If that is the case, a different use that better fits the existing conditions should be considered.

FURTHER READING

Adaptive Reuse: Issues and Case Studies in Building Preservation. David G. Woodcock, W. Cecil Steward, and R. Alan Forrester, eds. New York: Van Nostrand Reinhold, 1988.

Buildings Reborn: New Uses, Old Places. Barbaralee Diamonstein. New York: Harper and Row, 1978.

Change of Use. Pamela Cunningham. London: Alpha Books, 1988.

Great Adaptations: New Residential Uses for Older Buildings. Jill Herbers. New York: Whitney Library of Design, 1990.

New Life for Old Buildings. Mildred F. Schmertz and editors of *Architectural Record.* New York: McGraw-Hill, 1982.

Re-Architecture: Old Buildings/New Uses. Sherban Cantacuzino. New York: Abbeville Press, 1989.

Remaking America: New Uses, Old Places. Barbaralee Diamonstein. New York: Crown, 1986.

COMPARING REHABILITATION AND NEW CONSTRUCTION

Donovan D. Rypkema

"It has wonderful bones."

Robert Metzger, describing a Georgian-style Pennsylvania fieldstone house, *Architectural Digest*, March 1985.

Over the last 25 years historic preservationists and their adversaries have waged vigorous arguments over the economics of rehabilitation. In many cases both sides were armed with myth rather than reality. Yet, within the framework of real estate economics, it is possible to make direct comparisons between the economic benefits of rehabilitation and the like benefits of new construction. Set forth below are the fallacies and truths about these economic arguments, applied most often to office and retail properties and sometimes to housing.

Myth #1. The cost of rehabilitation saves 25 to 33 percent over comparable new construction.

Myth #2. The cost of rehabilitation in the end is nearly always more than new construction.

Myth #3. Because each project is site specific, there is no way to know whether or not rehabilitation will be cheaper than new construction.
What is to be concluded from these conflicting myths? Obviously, the cost of any particular project is site specific. Yet, certain generalizations can be made. First, make sure apples and oranges are not being compared. Cost comparisons between rehabilitation and new construction vary greatly, depending on whether or not the new building requires the demolition of an existing structure.

A broad spectrum of studies on this topic has been conducted by the National Trust for Historic Preservation, the Urban Land Institute, the Department of Commerce, the Department of the Interior, and the Institute for Health Planning, among others. Construction cost manuals also provide line-item comparisons for rehabilitation and new construction. If no demolition is required, a major commercial rehabilitation will probably cost from 12 percent less to 9 percent more than the cost of comparable new construction, with the typical rehabilitation cost being about 4 percent below new construction.

On the other hand, if new construction would incur the cost of razing an existing building, cost savings from rehabilitation should range from 3 to 16 percent. Furthermore, whenever major elements of the building can be reused or repaired (mechanical, plumbing or electrical systems, windows, the roof) cost savings will increase significantly.

Myth #4. It will take longer to complete a rehabilitation project than it will to construct a comparable new building.
An extensive series of analyses of real estate projects by the federal government revealed that rehabilitation often can reduce construction time by up to 18 percent. Rehabilitation may garner additional savings if regulatory hurdles must be overcome for new construction. Construction time savings mean interest savings and an earlier receipt of income. Moreover, it is often possible in rehabilitation projects to continue to generate rents while work is in progress, which generally is not an option for new construction.

Myth #5. Old buildings suffer from functional obsolescence resulting from their inefficiency.
Architects, builders, engineers, and developers often cite inappropriate floor size, layout, and ceiling height and inadequate parking space, loading docks, floor loads, and mechanical systems as reasons for the inefficiency of old buildings.

Preservationists often deny that these are legitimate issues—but certainly they are. If the building's proposed use requires 12-foot ceilings and the existing building has 8-foot ceilings, the use and the building probably are incompatible. The most cost-effective new use for a building is the same or similar to the one for which the building was constructed. Additionally, the building's original (or most recent) use has a greater influence on the ultimate cost of rehabilitation than the use to which it is being converted.

Most often, however, the argument of functional obsolescence stems from an erroneous perception held by the architect, engineer, or developer. Functional obsolescence is not defined by this year's volume of the *Architect's Standards for Current Construction* or the most recent edition of the *Uniform Building Code*. Functional obsolescence—the diminution of value resulting from building inefficiency—is measured by the actions of the marketplace. Do users of office space pay less, for example, for buildings where the hallways are seven-feet wide instead of ten? If not, functional obsolescence, by definition, does not exist. It is presumptuous of architects, engineers, or builders to claim that a building no longer meets the demands of the marketplace without having proof.

If, on the other hand, functional obsolescence does exist—and that is a determinable condition—demolition is only one of many options. The solution may lie not in hiring wrecking contractors but in finding a more creative design professional or developer.

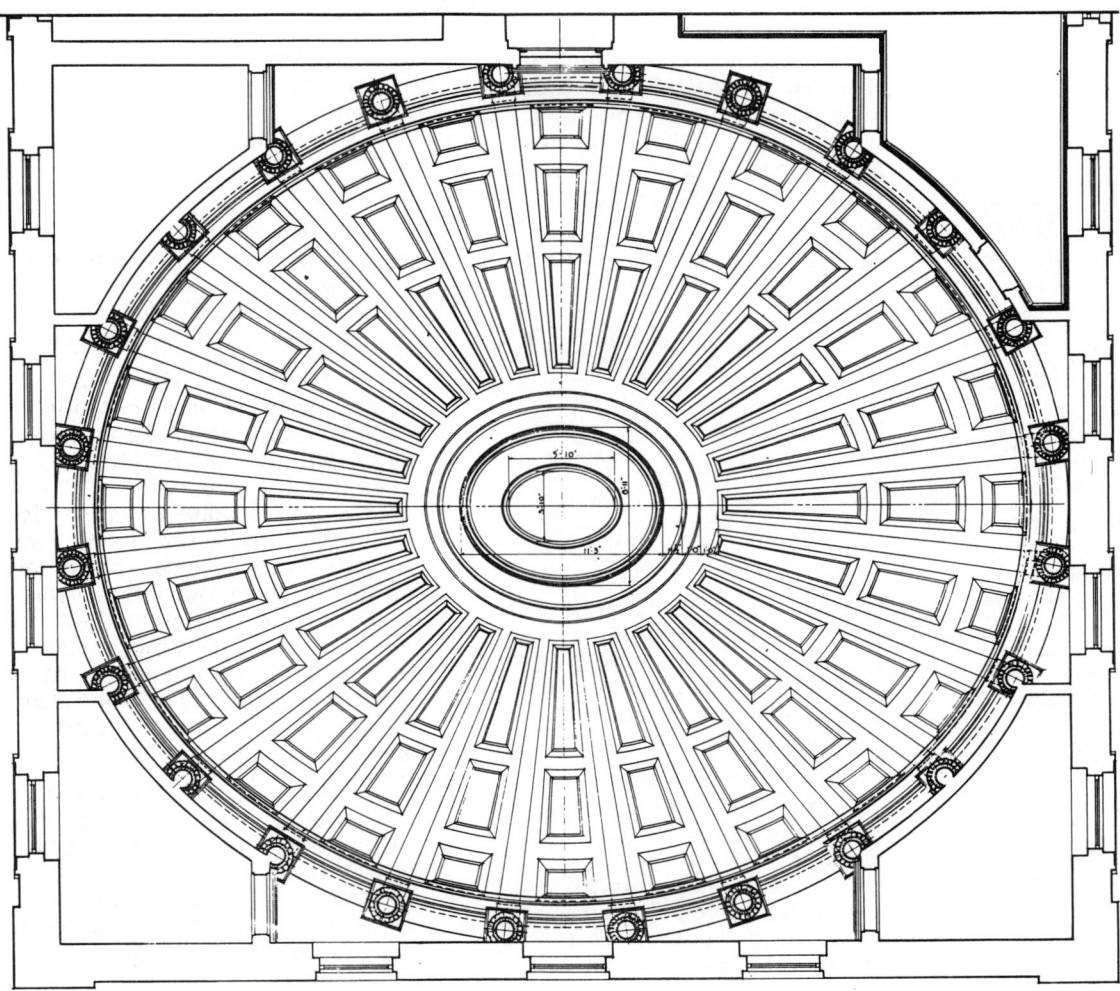

Ceiling, West Court Room, Old St. Louis Courthouse, St. Louis, Missouri. (F.R. Leslie, HABS)

Assume, however, that the market does penalize a particular historic structure—that because of ceiling height, hall width, and floor layout, the space can generate rents of only $17 per square foot instead of $20 for a comparable new building. Is the only alternative to tear the building down? No. The rehabilitated building is worth less to the investor—perhaps $100 per square foot instead of $120—but compensation can be made for this difference. Construction or acquisition costs need to be reduced; more favorable financing must be obtained; alternative uses must be considered; public incentives must be garnered; time savings need to be factored in; or any one of literally dozens of other variables could be altered.

In a recent *New York Times* advertisement, one of the amenities touted for a new Manhattan office building was operable windows, an attribute often cited in the recent past by architects of modern buildings and mechanical engineers as an example of functional obsolescence.

Myth #6. The cost of trying to retrofit an existing building to meet the standards of today's tenants makes rehabilitation too costly to pursue.
With the geometric growth in rehabilitation activities that occurred from 1982 through 1986, there was a dramatic evolution in the cost-effective technology of building rehabilitation. Windows,

mechanical systems, and energy conservation methods have been developed that make rehabilitation more financially competitive.

Myth #7. The ratio of rentable square feet to gross square feet (net-to-gross ratio) is significantly lower in old buildings, making new construction a more cost-effective alternative.
It is undeniably true that some old buildings have a lower percentage of rentable space than some new ones. It should be kept in mind, however, that spacious public areas such as lobbies, hallways, and foyers often are considered amenities for which tenants may be willing to pay extra. The highest

quality buildings seldom skimp on public spaces.

The most comprehensive analysis of the net-to-gross comparison between old and new buildings is undertaken annually by the Building Owners and Managers Associations (BOMA). Their *Experience Exchange Report* gathers detailed information on 4,000 buildings in all parts of the United States and Canada—buildings of all ages, locations, and sizes. Year after year their studies show that there is little, if any, difference in the ratio of rentable square feet to gross square feet for buildings up to 9 years old and for those 50 years old and older. In some reports new buildings have a 92 percent net-to-gross ratio and old buildings an 89 percent ratio; in other years the percentage for both is identical.

Furthermore, a quality design professional who is sensitive to historic structures usually can find ways to mitigate a less desirable net-to-gross ratio without diminishing the architectural integrity of the building.

Myth #8. The rents that can be obtained from rehabilitated space are much lower than those from a new building.

In some markets quality rehabilitated space will command the highest rents in the community. In most areas, however, rehabilitated office space seems to rent for 85 to 90 percent of the rent levels of comparable quality new construction. Rehabilitated retail space will range from 95 to 100 percent of comparative new space.

BOMA data indicate the average total income per square foot for older buildings (50 or more years old) to be around 90 percent of that of properties built within the last nine years. But, their studies do not distinguish between buildings that have been rehabilitated and those that have not. One can reasonably assume that income per square foot for rehabilitated buildings is higher than the BOMA average.

Myth #9. Floor layouts of older buildings will not work for large corporate tenants.

Certainly not every building, old or new, will be suitable for every

tenant. But, even new buildings are not dominated by corporate tenants that rent entire floors. The average office tenant in a new building leases around 10,000 square feet. Many rehabilitated historic structures can accommodate a tenant of this size without difficulty.

Myth #10. The vacancy rate for old buildings is much higher than for new ones.

In very weak markets old buildings seem to have somewhat higher vacancy rates, primarily as a result of the extraordinary incentives offered by the owners of unoccupied new buildings who are often desperate developers or banks that have foreclosed on a property. There is no evidence, however, that long-term vacancy is greater in quality rehabilitated buildings than in comparable new buildings. In fact, vacancy in older buildings has been less than that of new buildings consistently for the past several years of BOMA reports.

Myth #11. The remaining economic life of rehabilitated projects is shorter than for a new building.

There is an old adage in real estate that says, "more buildings are torn down than fall down." This is a result of the difference between a property's physical life and its economic life. Remaining economic life is the period over which a building can be expected to contribute to the value of the entire property. But, the rehabilitation of a structure gives birth to a new economic life and extends the building's physical life. There is no evidence that a building that has undergone a quality rehabilitation will have a shorter remaining economic life than an equivalent new building.

Myth #12. The cost of operating old buildings is much greater than it is for energy efficient new buildings.

This is a myth for which the answers true, false, and both of the above would all be correct. In a direct comparison between buildings 50 or more years old (including both rehabilitated and nonrehabilitated buildings) and those built

within the last 9 years, old buildings have lower total expenses per square foot, but the expenses represent a higher percentage of income than for new buildings.

This divergence comes primarily from utility costs and real property taxes. There is little difference in utility operating costs between old and new buildings when rehabilitation has upgraded mechanical systems and included other energy efficient measures.

Property taxes are another matter. On the average, real property taxes for old buildings are much higher as a percentage of total income and of net operating income than for new buildings. Most likely this is a reflection of land being taxed at its highest and best use, which might be as developed to a 50-story high rise, instead of its current use. In the long run, this discrepancy cannot help but encourage demolition. Preservationists need to pay more attention to ensuring that property taxes are fairly assessed.

On the other hand, many states freeze tax assessments of rehabilitated historic properties at the prerehabilitation level for several years. Reducing a building's expenses has the same effect as increasing rents: greater net income. Knowledgeable developers will include these local incentives in an overall analysis of a property's feasibility.

Myth #13. Financing available for rehabilitation projects is not as attractive as that for new construction.

Unfortunately this myth often has some validity, because some rehabilitation projects do present more risk. Rarely, however, would a rehabilitation project receive more favorable conventional financing than new construction, because most lenders automatically perceive all rehabilitation projects as greater risks. While the downturn in the real estate and financial industry in the 1990s is a result of too much money chasing after new construction in the 1980s, troubled rehabilitation projects form an insignificant percentage of total dollars for bad loans.

Myth #14. When the land and the building can be purchased for the value of the land, the building has no value, therefore there is no economic reason not to raze the structure.
This is the "end-of-economic-life" argument—that the building no longer contributes value to the property and that one way to free up the value of the land is to demolish the building. In fact, this may push the cost of the land far above its value. The cost of demolishing a building is commonly underestimated by developers and particularly by city governments. A demolition cost of $5 to $10 per square foot for a masonry building would not be uncommon and could be even higher in a densely developed urban site. Demolition can easily add from five to eight percent to the total cost of a project. Reinvestment in the building instead of demolition also frees up the land value.

Myth #15. It is necessary to attract large firms as tenants for a real estate deal to work.
In fact, nearly all net job growth in the 1990s will be from small firms. Of the hundreds of business classifications, only 20 of them will provide 40 percent of all the net job growth this decade. What is the average size of these rapidly growing firms? From 11 to 14 employees. What is the typical space need in those industries? About 250 square feet per person. What is the average amount of space used by an office tenant in a building 50 or more years old? About 4,500 square feet. Space in rehabilitated historic buildings is often the ideal size to meet the needs of America's fastest growing businesses.

Myth #16. It is in the suburbs where money is being made in commercial real estate investment, not in downtowns where most old buildings are located.
Talk of the decline of the central city notwithstanding, BOMA data consistently show that downtown property commands higher rents, has lower vacancy rates, and has higher net operating income than its suburban counterpart.

Myth #17. Old buildings will be rehabilitated with or without historic rehabilitation tax credits.
Annually, the National Park Service surveys property owners who have completed historic rehabilitation tax credit projects in the preceding year. Every year, responses show that the projects would not have gone forward without these tax credits. Tax savings are an important component of the total return in most real estate transactions and are crucial in attracting private capital to historic preservation.

Myth #18. There is no way of knowing at the outset of a rehabilitation project which building components are going to save money and which are going to cost more than new construction.
Any given project will be subject to situational variables. Based on overall patterns of experience, however, there is a degree of predictability; certain building components cost relatively more in new construction while others cost less.

- Building components most likely to be *less* expensive in rehabilitation than new construction include the foundation, superstructure, exterior cladding, and roof.
- Building components most likely to be *more* expensive in rehabilitation than in new construction include the interior construction, mechanical systems, elevators, and the general conditions of the construction contract.
- Building components most likely to be equivalent in rehabilitation and new construction include the substructure and the electrical system. Design costs also are similar for both new construction and rehabilitation projects.

Myth #19. The only way to rehabilitate an old building is to do the whole thing over. You might as well build a new building as end up with a new building stuck in an old shell.
It is true that in many preservation projects almost every building element is replaced, updated, rebuilt, or restored. The cost comparisons following myths 1, 2, and 3 reflect that assumption. But, there are thousands of rehabilitation projects that require a much smaller investment. In the heyday of rehabilitation activity, from 1982 to 1986, 80 percent of all historic rehabilitation tax credit projects cost less than $500,000 and nearly half cost less than $100,000.

While it is somewhat more difficult to generalize about smaller rehabilitation projects, certain basic cost ranges can be made. Assume that minor rehabilitation would include the following tasks: interior demolition and partitioning; repointing exterior walls (without window replacement); using existing elevator systems; updating but not replacing mechanical and plumbing systems; replacing the roof cover but not the roof structure; and installing a new electrical system. This type of rehabilitation can be undertaken for 40 to 50 percent of the cost of building a comparable new structure. Much of this cost savings might be offset, however, by the higher cost of acquiring a building with so many elements in useable condition.

The arguments for historic preservation are many. They include architectural distinction, social continuity, cultural context, urban planning, historical consciousness, environmental protection, and aesthetic excellence. All of these arguments are valid ones and should continue to be made. But, increasingly preservationists and their allies are called upon to make an economic case for preservation. And, as shown above, historic preservation makes sense in large measure because it saves dollars and cents.

This chapter is excerpted from *The Economics of Rehabilitation* prepared by Donovan D. Rypkema for the National Trust's Information Series.

FURTHER READING

See Further Reading at the end of "Public Benefits" on page 37.

PUBLIC BENEFITS

Bridget Hartman

"Here is your country. Do not let anyone take it or its glory away from you. Do not let selfish men or greedy interests skin your country of its beauty, its riches, or its romance. The world, the future, and your children shall judge you accordingly as you deal with this sacred trust."

President Theodore Roosevelt, Antiquities Act of 1906

THE CASE FOR PRESERVATION

"Historic preservation is bad for my business."

"Preservation regulations threaten our community's economic growth."

"Historic district designation will cause my property to decline in value."

"Preservation incentives are just tax breaks for a special interest group."

How often have preservation advocates heard these statements—from city council members, chamber of commerce officials, developers, realtors and property owners—when arguing for passage of a local landmarks ordinance or implementation of a financial incentive program? Yet, the data collected in many communities across the country document that preservation is good for business. The economic consequences of preservation ordinances are often positive. Well-designed tax incentive programs for historic properties can enhance not only the physical character of an area but also its fiscal health. And strong heritage tourism programs demonstrate the importance of historic preservation as an effective tool to stimulate economic development.

Aesthetic, environmental, educational, social, and psychological reasons—all excellent answers to the question "Why historic preservation?" Equally important, but often more difficult to articulate is the economic rationale for saving and recycling community landmarks. Crafting strong economic arguments in favor of preservation, however, is crucial as state and local officials look for ways to strengthen their economies and increase revenues and as private property rights advocates intensify their assault on historic preservation laws at the local, state, and national levels.

A strong economic justification should be an integral part of making the case for historic preservation. Preservation's main commodity is historic buildings. These older buildings are, like all other buildings, parcels of real estate. As Donovan Rypkema, principal in the Real Estate Services Group of Washington, D.C., observes:

"Regardless of whatever other attributes it might have, a historic building is, ultimately, real estate. Its economic capacity, measures of return, value as an asset, and attractiveness as an investment are all going to be calculated by investors, bankers, and developers in the same fashion as any other parcel of real estate."

And, as with most economic activity, historic buildings can provide many benefits both to the private property owner and to the public sector.

INCREASED PUBLIC REVENUES

The taxing powers of state and local governments have proven to be effective mechanisms for stimulating private investment in historic properties. Today 37 states authorize legislation that provides some form of tax relief, such as property tax abatements, credits, or freezes for historic preservation purposes. These incentive programs not only encourage private property owners to maintain, renovate, and improve their buildings, but they also provide state and local governments with an increased source of revenues.

Appomattox Iron Works, Petersburg, Virginia. (A. Jones, M. Chrisney, HAER)

Potential tax receipts are generated by the increased construction activity that occurs during a building's renovation as well as the increased economic activity that follows the rehabilitation. These revenues can come from sales and property taxes, building permit charges, new business license fees, and lodging taxes.

Washington State's historic properties tax exemption program has contributed to community revitalization and historic preservation efforts, according to a 1991 report conducted jointly by the state's Department of Community Development, Department of Revenue, and the State Attorney General's Office. Moveover, the report estimates that the tax relief measure, created in 1985, will not have a negative long-term impact on the tax base in the state.

Since 1985, when the Washington program began, 122 historic buildings have received the special property tax benefit. The property owners have invested more than $172 million in rehabilitation activity. The Department of Revenue calculates that state and local governments will enjoy a combined $10 million in additional tax revenues. Although $3 million in property tax revenues will be lost over the life of the program, more than $13 million will be gained from additional business and occupational and sales tax revenues resulting from the increased rehabilitation construction activity.

JOB CREATION

"Jobs and Preservation: It Works!" was the recent rallying cry of preservation advocates across the country as they lobbied Congress to pass the Dire Emergency Jobs Bill. This bill, which sought to stimulate the nation's economy through job creation, included $40 million for historic preservation. Although Congress did not enact this and other similar economic growth packages in the 1992 session, the relationship between historic preservation and job creation is indeed valid.

Jobs created as a result of historic preservation fall into two broad areas. First, new construction jobs are created as a result of increased rehabilitation activity. Second, as downtown and neighborhood business districts are revitalized, the new—and particularly small—businesses housed in these older buildings produce new jobs.

The link between urban economic health, older commercial buildings, and small businesses is documented in a 1987 study prepared for the National Trust on small business retention, expansion, and recruitment. Small businesses generate between 51 and 66 percent of all net new private-sector jobs according to the study. These small businesses are more likely to occupy older buildings, which can offer more affordable rents than new buildings and can serve as incubators for start-up businesses, allowing tenants to share equipment and services.

The work of the National Trust's National Main Street Center in hundreds of older downtowns across the country demonstrates that historic preservation makes economic sense. Since 1980, when the program began, more than 700 communities have invested approximately $2 billion of private reinvestment in commercial revitalization activities. The economic payback has been significant. These communities have seen a combined net increase of 51,710 new jobs, a net increase of 14,880 new businesses, and 20,959 building rehabilitations.

For example, in West Virginia, the Main Street program is now in its fifth year with 15 participating towns. To date, 223 new business have opened, relocated, or expanded into the central business areas of these communities as the result of historic preservation and downtown revitalization strategies. The net gain in jobs totals 893.

TOURISM

Tourism is the number one industry in the world today. In the United States, tourism contributes $330 billion annually to our nation's economy, according to a 1991 study by the Travel Industry of America and the U.S. Travel Data Center. Tourism is responsible for six million jobs and supplies $43.6 billion in total tax revenues.

Arthur Frommer, well-known travel guide author, speaks of the strong link between historic preservation and tourism when he writes,

"Every study of travel motivations has shown that an interest in the achievements of the past is among the three major reasons why people travel. The other two are rest and recreation and the desire to view great natural sights. Tourism does not go to a city that has lost its soul. Among cities with no particular recreational appeal, those that have substantially preserved their past continue to enjoy tourism. Those that haven't receive no tourism at all."

In his article "Selling America's Heritage . . . Without Selling Out," Richard J. Roddewig, president of Clarion Associates, a Chicago real estate consulting firm, offers eight reasons why historic sites are so attractive to tourists.

■ In an era of nationally franchised gas stations, restaurants, department stores and look-alike strip center developments, one place is beginning to look just like every other place in America. Thus, travelers enjoy the contrast that historic sites create and the unique sense of place that comes with them.

■ Americans are attracted to historic sites because they are genuine. Their interest in such unique sites stems from a reaction to the plasticized nature of so much American culture.

■ Americans want to visit historic sites because they give tourists a sense of continuity with the past that allows them to better understand themselves and who they are as a people.

■ The baby-boomers, now reaching middle age, are intensely interested in the education of their offspring, and they want to teach their children about American history and architecture.

■ Americans today, especially the baby boomers, are better educated than those of earlier generations and seek education as well as recreation when traveling.

- Americans increasingly want to experience regional differences by staying in places other than chain hotels, thus the welcome growth of historic inns and bed-and-breakfast homes across the country that offer personalized service and unique atmospheres.

- The current trend of dining on regional American cuisines calls for authentic settings, which are best found in historic buildings and districts.

- More and more people are taking short three-to-five day vacations, which are best suited to visiting historic sites near metropolitan areas rather than grand scenic areas or national parks.

REDUCTION OF INFRASTRUCTURE AND SERVICE DEMANDS

Historic preservation activities can maximize the use of existing public infrastructure. Sewer, water, and power systems, curbs, sidewalks, and roads often are in place and require no or modest expansion to accommodate preservation activity, whereas new construction often demands new roads and expanded or new delivery systems.

A recent fiscal impact study prepared by the Government Finance Research Center (GFRC), the research arm of the Government Finance Officers Association, contends that a 1,475-acre corporate business park proposed for the Civil War battlefield site in Brandy Station, Virginia, would be an almost immediate financial drain on the local jurisdiction. The study reports that, while the development may in itself create a positive fiscal cash flow, the project could cost the county money in terms of required government services and capital facilities, starting from the very first year. Eventually, the business park could also cause the county's population to more than triple its current size, from 27,560 today to 97,019 by the year 2011. This population explosion would

severely strain county services and leave the county with deficits which, over a 20-year life of the project, would average approximately $3.5 million per year.

The GFRC report examines an alternative use for Brandy Station as a battlefield park, estimating that a park there could attract between 70,000 and 100,000 nonlocal visitors each year. A park would place nominal demands on existing county infrastructure. The cost for constructing tourist facilities and roads within the park would be modest in comparison to increased revenues for tourism activity the county would receive. In addition, a park facility would create $3.7 million in added annual sales, $1.2 million in added local income and 59 new jobs according to GFRC's study.

State growth-management programs also can serve historic preservation and economic development goals alike. When these programs steer new growth and development into existing cities and away from farmland, they maximize existing infrastructure and minimize the cost of new public service systems in outlying areas. When coupled with appropriate preservation controls, this approach can lead to the revitalization of older cities which often boast large collections of historic buildings.

New Jersey is one of more than a dozen states that has enacted or is considering passage of comprehensive growth-management legislation. A recent economic impact report by Rutgers University's Center for Urban Policy Studies asserts that the state's pending plan offers numerous benefits. Although the same number of jobs would be produced with or without the plan, the Rutgers report estimates that between 1992 and 2010, state and local jurisdictions would save a combined $1.3 billion in infrastructure construction, including new roads, sewer, and water construction. In addition, municipalities would save $122 million a year by maintaining fewer roads and serving more densely populated areas.

NEIGHBORHOOD CONSERVATION AND STABILITY

The rehabilitation of historic buildings contributes significantly to the stabilization and rejuvenation of America's urban areas. A *Time* magazine cover story (November 23, 1987), "Spiffing Up the Urban Heritage," celebrates the comeback of many American cities:

"How did Americans manage to forget for so many years that downtowns are invigorating and old cities grand? That the dignity and *Gemutlichkeit* of 18th-century buildings and 19th-century streets are incomparable. . . . [T]he nation has had a change of heart. The change has been so complete that it is difficult today to remember how recently people were blithely ripping out and throwing away the warp and woof of America's cities."

It is this enhanced quality of life and sense of neighborhood that sets historic preservation apart from most economic initiatives. While any development project could attract new businesses, stimulate private investment, raise property values, and generate increased sales, only the preservation of old buildings can enhance neighborhood pride and give a community its special identity.

New businesses would be formed, private investment stimulated, property values, sales, and taxes increased, and so forth. The one exception is the argument that historic preservation can more effectively enhance neighborhood and community pride than will urban renewal-type new construction.

In the Jamaica Plain neighborhood of Boston, for example, the historic Haffenreffer Brewery closed in 1965, putting some 250 Jamaica Plain residents out of work. As a result, the buildings fell into disrepair, unemployment increased, and the quality of life in the neighborhood deteriorated significantly throughout the 1970s and 1980s. In the late eighties, local activists formed a nonprofit development corporation and bought the complex. The neighborhood group renovated the brewery as an incubator for small business that would hire and train people from the community. Today the old brewery houses 26 small businesses, one-third of them minority- or women-owned, and has been a catalyst for neighborhood redevelopment. Nearly two-thirds of the workers come from the neighborhood and the average salary is 150 percent of the minimum wage.

In Cleveland the recent renovation of the 1920 historic Gordon Square Arcade as 64 low-income apartments and 28 commercial spaces by a neighborhood nonprofit organization has contributed significantly to the stabilization of the Detroit Shoreway neighborhood. In recent decades the neighborhood had deteriorated, resulting in a loss of population and escalating vacancies. By 1980, 28 percent of the population was below the poverty level and the median income was $10,824. The neighborhood's commercial decline reflected this residential decline, with residents traveling to remote suburban centers for basic purchases. The now completed Gordon Square Arcade, a mixed-use community development project, directly employed more than 60 construction workers and created more than 40 permanent jobs in new community-oriented businesses. Moreover, the refurbished arcade has caused a resurgence in the neighborhood's residential real estate market and spurred additional housing rehabilitation in older buildings.

Ten years ago the 17-block warehouse and commercial district in St. Paul's historic Lowertown stood empty, the buildings largely abandoned and the area dominated by parking lots. Today, due in large part to the Lowertown Redevelopment Corporation (LRC), which serves as a catalyst for investment, development, and marketing, Lowertown is one of the fastest growing neighborhoods in the Twin Cities area. Its historic buildings, the human scale of the neighborhood, and the combination of parks, historic lighting, and trees have helped to create "a new urban village" according to a recent article published by LRC. The warehouses and commercial buildings have been

transformed into apartments, condominiums, and artists' studios and lofts for nearly 3,000 residents. A wide range of shops, restaurants, and entertainment serves the neighborhood and attracts visitors from the metropolitan area. Galtier Plaza, a mixed-use complex, alone draws 40,000 people a month to Lowertown.

To date, more than $350 million of new investment in Lowertown has created 2,900 construction jobs and 4,600 permanent jobs, according to the LRC article. Property taxes in the area have increased approximately five times. Galtier Plaza boasts an office occupancy of 90 percent while the overall downtown vacancy rate is at 25 percent. More than 75 percent of the Galtier Plaza condominiums have been sold and 85 percent of the retail space is now leased.

CALCULATING THE ECONOMIC BENEFITS

Calculations of the economic benefits of historic preservation, however, cannot be derived from a single source of data or reduced to a simple algebraic formula. Government Finance Research Center, in a recent project sponsored by the National Trust in cooperation with Scenic America, has developed a step-by-step guide for assessing the economic impacts of historic preservation. The guidebook *The Economic Benefits of Preserving Community Character: A Practical Methodology* examines the consequences of preservation regulations and incentives on a community's economy and their effect on a local government's fiscal condition in three broad areas: construction and rehabilitation activity, real estate market activity, and commercial activity.

Construction and Rehabilitation Activity

Rehabilitating historic properties provides short-term as well as long-term benefits to a local economy. When renovating or adding to an old building, the property owner typically hires a contractor. The contractor in turn employs construction workers. Construction

requires the purchase of supplies and materials. More often than not, these purchases are from area suppliers—local hardware stores, paint stores, or lumber yards. By determining the total number of rehabilitation projects and the total dollar value of these projects over a specific period of time, one can begin to quantify the number of construction, manufacturing, and sales jobs created, as well as calculate the local sales and use tax revenue from construction expenditures.

In Denver's historic Lower Downtown, the passage of a preservation ordinance in 1988 coupled with a number of public-sector incentive and improvement programs accounted for $3.1 million in rehabilitation and new construction work and the creation of 74 full-time construction jobs between 1988 and 1990, according to a recent study by Hamer Siler George Associates, a nationally recognized economic and development consulting firm. New businesses in Lower Downtown numbered 114 and generated an estimated 450 new jobs and $7 million of investment. The majority of the new businesses are small concerns, such as architects, planners, and landscape architects, that have found the low-rise older commercial buildings particularly well-suited to their needs. Twenty new retail businesses, primarily galleries and restaurants, now occupy first-level streetfront locations. Of significant impact has been the development of 100 new residential units in an area that had been predominately for retail and office uses. A decline in land values and lack of demand for office space helped set the stage for the conversion of the warehouse buildings into loft and apartment housing.

The city of Denver and its neighboring jurisdictions saw a substantial increase in revenues, specifically through sales and property taxes, new employment head tax, and parking meter revenues. It is estimated that $1.2 million in taxes and user fees have been generated each year as a result of Lower Downtown's increased business activity.

Considering the overall condition of Denver's economy and compared with other commercial districts in the city, Lower Downtown has enjoyed substantial preservation activity with positive economic impact.

Real Estate Market Activity
Perhaps one of the most frequent arguments against a local preservation ordinance is that its regulations will decrease the value of properties located within the historic district. Yet the majority of research published to date contradicts this claim. A local historic district should be viewed as a protective rather than restrictive preservation tool. As Rypkema states:

"In virtually every objective evaluation of the economic impact of land use controls—and particularly of historic districts—the composite value of the affected property was protected at worst and significantly enhanced at best."

A recent study by the Virginia Department of Historic Resources on the financial impact of historic designation reports that 84 of the 87 local assessors and commissioners of revenue believed that state or federal historic designation did not lower a property's existing value. L. L. Barbour, real estate assessor for the city of Danville, comments:

"In general . . . we have never been able to document market data which substantiates that a state or federal historical designation has any bearing on value. Properties with such designation tend to receive a higher degree of maintenance due to pride of ownership and thus maintain or increase in value better than those of comparable actual age without a designation of historical significance."

William A. Diggs, director of the department of real estate assessments for Chesterfield County, sums up the position held by many Virginia officials who were surveyed for the report:

"I am of the opinion that the problem being experienced is one of perception; not knowing the true facts concerning the [state and federal] Registers [of historic places]."

An informal survey of realtors in Haddonfield, New Jersey, by the local preservation group produced similar favorable results. George Roney, a local realtor with Roney, Vermaat and Leonard says:

"Historic District zoning has proved to be good economics. [Property values] in the district have increased substantially, better than 60 percent."

Jack Leonard, of the same real estate company, emphasizes the collective economic benefits enjoyed by all the property owners within the Haddonfield historic district when he comments:

"The preservation of historic buildings in the residential and commercial district has made Haddonfield the most desirable town in South Jersey for both business and residential homeowners. The strong real estate values that Haddonfield residents each share can be directly attributed to the preservation of the original architectural designs. . . ."

In its 1990 economic impact study of the historic district designation of Denver's Lower Downtown, Hamer, Siler, George Associates concludes:

"The Lower Downtown district designation was controversial and passed with some uncertainty on the part of critics regarding its impact on land and property values and development and redevelopment potentials. Our analysis has revealed its positive impact in creating a unique and successful arts/entertainment and design district."

There exist several different ways to analyze the benefits of real estate market activity. The most common method is to document the changes in the value of a set of properties over a specific period of time. The properties could be historic buildings within a historic district, historic buildings scattered throughout the community, or historic properties that benefitted from some type of financial incentive program for historic preservation. Assessment records provide the most complete source of information. The figures are compared

with changes of value of comparable properties not affected by the preservation regulation or incentive program.

Applying its methodology in Fredericksburg, Virginia, and Galveston, Texas, Government Finance Research Center analyzed the impact of preservation activities in both the residential and commercial areas of these communities. The results of the research convincingly refute claims that preservation regulations hurt property values.

GFRC found that in Fredericksburg, between 1971 and 1990, the average increase in the value of property located in the historic district was substantially higher than for properties located outside the district. In 1971, the value of single residential properties within the historic district and throughout the rest of the city was nearly identical—approximately $17,500. By 1990, the average values were $138,697 for properties located in the historic district and $87,011 for properties elsewhere in the city.

The research produced similar findings in Galveston. Between 1975 and 1991, properties increased by an average of 440 percent in the East End Historic District and by 165 percent in the Silk Stocking Historic District. In comparison, values outside these areas increased over the same period of time by an average of 80 percent.

The change in property values is only one way to measure activity in a real estate market. Measuring changes in vacancy rates, the number of substandard properties or the number of property sales also can produce additional compelling data to support the case for historic preservation. For example, Washington State's historic property tax exemption program is the impetus for reinvestment in many older and neglected urban areas. Of the 122 properties receiving the special tax valuation, 60 percent originally were either vacant or completely abandoned buildings located in blighted neighborhoods according to a recent survey by the state's Office of Archeology and

Historic Preservation. When asked if the rehabilitation project encouraged other similar projects, more than 70 percent of the property owners said yes.

Other examples of measuring the effect of historic preservation initiatives are offered by small communities in West Virginia and Wisconsin. The community of New Martinsville, West Virginia, has had an active downtown revitalization program for five years. The results are impressive. In 1988, 25 percent of the buildings downtown were vacant, while today an aggressive Main Street program can claim that the nine-block historic area is nearly full. Ripon, Wisconsin, boasts similar successes. In a two-year period, an aggressive Main Street program has helped to decrease downtown vacancies from 28 percent to below 4 percent.

Commercial Activity
Although more difficult to quantify, the economic benefits of historic preservation brought on by travel and tourism are many. Tourism brings new money into a community. A visitor spends money in local shops, restaurants, and places of lodging. These funds are then respent as shop owners pay their employees and restaurants and hotels restock their food and supplies. By attracting money from outside the community, tourism also creates jobs for individuals within the communities. Tourism spurs new businesses—shops, restaurants, and hotels—that cater specifically to the traveler. State and local governments benefit, too, through increased tax revenues.

State and local economic development or tourism offices, the local chamber of commerce, convention bureaus, the U.S. Travel Data Center, and the American Automobile Association are sources of information for estimating the dollar amount of tourist expenditures. Another approach is to conduct a survey of local businesses in a historic area, focusing on total retail sales, number of employees, and employment payroll.

The Art Deco landmarks of Miami Beach, the 18th-century Spanish

influence in St. Augustine, and the archeological treasures of the DeSoto Trail are but a few of the historic sites that bolster Florida's economy. According to a 1988 study published by the Florida Departments of State, Natural Resources, and Commerce, for every one million trips by out-of-state visitors, state-identified historic sites support 45,618 Florida jobs, with a payroll of $438.9 million. State revenues connected with heritage tourism total $69 million in taxes. Local governments collect $17.2 million.

The historic sites and overall historic character of the downtown as well as the proximity of four major Civil War battlefields all contribute to the economic vitality of Fredericksburg, Virginia, according to a 1989 tourism survey by Government Finance Research Center. GFRC estimates that tourists spent approximately $11.7 million visiting the historic district and another $17.4 million outside the district. Secondary impacts of these activities total an additional $13.8 million. The city enjoyed an estimated $1.1 million in revenues from these sales.

Historic sites and cultural events themselves offer another way to measure tourism activity in a community. The Spoleto Festival in Charleston, South Carolina; the Civil War battlefield in Gettysburg, Pennsylvania; the copper mines in Butte, Montana; and the Sloss Furnaces in Birmingham, Alabama, earn significant revenues for their respective jurisdictions.

In a recent survey on the impact of tourism, GFRC estimated that from July 1989 through June 1990 approximately 180,000 people visited the Strand Visitors' Center in Galveston, Texas, 100,000 visitors attended the Dickens on the Strand festival, 5,000 went on the historic homes tour, and 66,000 attended performances at the historic Grand Opera House. GFRC estimated that these visitors spent approximately $18 million during their stay in Galveston. Secondary impacts total an additional $11 million in sales. The state received approximately $1.1 million in sales tax revenues

while the city collected approximately $0.5 million. The study concluded:

"While Galveston's sun and ocean beaches have long attracted tourists, it is apparent that the city's historic attractions are making a significant contribution to the local economy."

On a smaller yet as impressive scale, the town of Mineral Point, Wisconsin, has linked historic preservation with tourism infusing new life into its economy, according to a report prepared by Economic Research Associates, a Los Angeles-based national real estate consulting firm. The town's restored stone and log buildings dating from 1830 to 1870 now serve as popular arts-and-crafts outlets and bed-and-breakfast establishments. In 1987 Mineral Point attracted 27,900 visitors who spent $1.8 million in the community. Mineral Point gained 30 new full-time jobs and 15 new businesses as a result of heritage tourism.

Historic preservation is and can be a powerful economic engine for many communities. Historic preservation and economic growth not only can coexist but preservation can mean big business and sound fiscal policies for communities across the country.

In recent surveys of 322 towns and cities across the country, the National League of Cities documents strong links between historic preservation and economic development. Mayors contacted in the surveys cited economic development as one of their top three priorities and identified historic preservation as a major tool for fostering development.

FURTHER READING

"Attitudes of Virginians Regarding Growth and Development." Mason-Dixon Opinion Research, Inc. Public opinion survey. Warrenton, Va.: Piedmont Environmental Council, 1990.

The Cost of Community Services in Deerfield, Massachusetts. American Farmland Trust. Washington, D.C.: American Farmland Trust, 1991.

Density-Related Public Costs. American Farmland Trust. Washington, D.C.: American Farmland Trust, 1986.

Economic Benefits from the Rehabilitation of Certified Historic Buildings in Georgia (Case Studies). Center for Business and Economic Studies. Atlanta: Georgia Department of Natural Resources, Historic Preservation Section, 1987.

Economic Benefits from the Rehabilitation of Certified Historic Structures in Texas: Final Report. Shlaes and Co. Austin, Tex.: Texas Historical Commission, 1985.

Economic Benefits of Preserving Community Character: A Practical Methodology. Government Finance Research Center. Washington, D.C.: National Trust for Historic Preservation, 1992.

Economic Benefits of Preserving Community Character: Fredericksburg, Virginia, Case Study. Government Finance Research Center. Washington, D.C.: National Trust for Historic Preservation, 1992.

Economic Benefits of Preserving Community Character: Galveston, Texas, Case Study. Government Finance Research Center. Washington, D.C.: National Trust for Historic Preservation, 1992.

The Economic Impact of Colonial Williamsburg. Roy L. Pearson and Donald J. Messmer. Williamsburg, Va.: Mid-Atlantic Research, Inc., 1989.

Economic Impacts of Protecting Rivers, Trails, and Greenway Corridors, A Resource Book. Rivers and Trails Conservation Assistance Program. Washington, D.C.: U.S. Department of the Interior, National Park Service, 1990.

Economic Incentives for Historic Preservation. Richard J. Roddewig for the City of Atlanta. Washington, D.C.: National Trust for Historic Preservation, 1987.

Economics of Community Character Preservation: An Annotated Bibliography. Government Finance Research Center and Scenic America. Washington, D.C.: National Trust for Historic Preservation, 1991.

The Economics of Rehabilitation. Donovan D. Rypkema. Information Series, National Trust for Historic Preservation. Washington, D.C.: Preservation Press, 1991.

Elkwood Downs: Its Fiscal Implications for Culpeper County, Virginia. Government Finance Research Center. Washington, D.C.: National Trust for Historic Preservation, 1990.

Final Case Study for the National Scenic Byways Study, the Economic Impact of Travel on Scenic Byways. U.S. Travel Data Center. Washington, D.C.: U.S. Department of Transportation, Federal Highway Administration, 1990.

The Financial Impact of Historic Designation. Richmond, Va.: Virginia Department of Historic Resources, 1991.

Fiscal Incentives for Historic Preservation. Susan Robinson and John E. Petersen. Chicago: Government Finance Officers Association, 1989.

"Historic Preservation and Tourism," Arthur Frommer. *Preservation Forum.* National Trust for Historic Preservation, Fall 1988.

"The Impacts of Historic District Designation in Washington, D.C.," Dennis E. Gale. *Journal of the American Planning Association*, Summer 1991.

Lower Downtown Economic Impacts of Historic District Designation. Hamer, Siler, George Associates. Denver: City and County of Denver, Office of Planning and Community Development, 1990.

"Miracle on Main Street," Daniel Kehrer. *IB Magazine*, January-February 1992.

"Property Rights/Property Value: The Economic Misunderstanding of the 'Property Rights' Movement," Donovan D. Rypkema, an unpublished paper. Washington, D.C.: The Real Estate Services Group, 1992.

"Real Estate, Economics and Historic Preservation," Donovan D. Rypkema. In *Past Meets Future: Saving America's Historic Environment.* Washington, D.C.: Preservation Press, 1992.

"Rehab Takes a Fall," Donovan D. Rypkema and Ian Spatz. *Historic Preservation*, August 1990.

"Selling America's Heritage . . . Without Selling Out," Richard J. Roddewig. *Preservation Forum.* National Trust for Historic Preservation, Fall 1988.

"Selling History," Jonathan Walters. *Governing*, June 1988.

Tourism and Historic Preservation in the South. Tracy Hayes. Washington, D.C.: The National Trust for Historic Preservation, 1987.

FUND RAISING

Ray Foote and Allison Fabian

"There are obviously only a limited number of 'charitable' dollars available from individuals, corporations, and foundations for the support of private, nonprofit groups. . . . It is also obvious that there is, and probably always will be, intense competition for these dollars, with cultural organizations competing both with each other and with other types of 'nonprofits' for available funds. . . . This simply means that preservation organizations must be prepared to make a strong case for their share of the charitable dollars and must understand the basic principles of successful fund-raising before undertaking any fund-raising program."

Lawrence J. Biddle, *Private Funds for Historic Preservation*, 1979

ELIGIBILITY FOR FINANCIAL ASSISTANCE

Eligibility varies greatly from one funding source to another, but most funding goes to nonprofit organizations, not to individuals or for-profit groups. A private individual should not be discouraged, however, from preserving an important building or from carrying out some other useful preservation project. If the cause is worthwhile, a non-profit organizational structure can be created specifically for that cause, or the cause can be brought to an appropriate organization.

Because most preservationists seeking funding are doing so on behalf of a nonprofit organization, fund-raising strategies and funding sources for nonprofit preservation groups are the focus of this section. Individuals and for-profit entities, however, may be eligible for certain types of financial assistance. The most common form of assistance for for-profit groups is through tax benefits received for work on historic, income-producing properties. And, while few grants exist for individuals rehabilitating private residences, low-interest loans for preservation work are sometimes available. Contact statewide preservation organizations for information on loan funds and lenders sympathetic to preservation work.

A MIX OF FUNDING SOURCES

Few preservation organizations receive all their financial support from one donor or even from one type of donor. Rather, they piece together money from many sources to meet their budgets. The challenge for preservationists is to become familiar with the different sources of funding and to choose the blend that is right for their particular organization. As with investments, it is best to be diversified in order to withstand the inevitable fluctuations in giving patterns and guidelines, markets, and politics.

Funding is available for general operating expenses (unrestricted funding) as well as for specific projects (restricted funding), both being important to the health of a preservation organization. Donors often prefer to restrict their contributions to a specific cause or project, which gives them a feeling of greater control over what will be accomplished with their money. Often corporations, foundations, and the government will allow a percentage of restricted funds to count toward administrative overhead costs, in effect making them unrestricted.

Unrestricted monies tend to be more difficult to secure than restricted or "program" funds.

While government agencies, foundations, and corporations will make unrestricted donations, it is contributions from individuals and money raised by special events that can be especially important sources of unrestricted funds.

REQUIREMENTS FOR SUCCESS

Because the number of nonprofit organizations seeking funds during the last decade has increased much more sharply than the funds available, fund-raising success will be greatly affected by how well an organization distinguishes its projects from others competing for the same funds.

Establishing the group's credibility with prospective donors is perhaps the single most important factor in successful fund raising, primarily for two reasons. First, the rise in the number of nonprofit organizations and the consequent competition among them has caused donors to look more closely at each group's legitimacy. Second, individuals born during the baby-boom years, now between roughly 30 and 45 years of age, are becoming the dominant philanthropic force in American society. They tend to be selective about what causes they support, and they seek detailed information about how groups use their money. As a result, preservation organizations must prove they are well run and pursue a necessary mission.

Establish credibility. Establishing credibility requires answers to basic questions about an organization. Prospective donors will be interested to know the following information:

- status of the organization—its name, tax-exempt status, location, date founded, and, if it has a board of directors, who is on it
- organization's formal mission
- its accomplishments toward stated goals
- its plans for the future and how it will accomplish them

Define the project. When looking for funds to finance a particular project, first define the ultimate goal of the project and analyze it carefully. Such an analysis may uncover program elements that would qualify the project for funding from an organization not considered previously. A preservation project, for example, may have a strong educational component, making it eligible to receive funds from the national or state department of education. Also, keep an open mind. Adjusting the project slightly may make it possible to apply for funds not typically considered. Become familiar with the different types of funding available and identify those most appropriate for the project in mind. Knowing the guidelines of a particular funding organization may provide ideas for new projects.

PRIVATE SECTOR FUNDING

Most financial support for preservation projects comes from three private-sector sources—individuals, corporations, and foundations—each having their own particular way of giving.

Individuals

Giving by individuals makes up the vast majority of contributions in America—more than $100 billion of the $122 billion given by all sources in 1990–and should be part of the overall development strategy for most preservation groups.

Many small preservation organizations worry that their lack of contacts with wealthy individuals may limit the success of their fund-raising efforts. Such a concern should not discourage an organization from seeking funds. Individuals of modest means tend to be the most steady givers. A solid fund-raising strategy can be built upon many small contributions with a sprinkling of larger gifts.

Individual fund-raising approaches range from personal visits to broad-based appeals. Large mailings solicit many small donations, while personal approaches to a handful of donors typically bring in a few large gifts. Both methods are entirely workable, depending on an organization's goals and the pool of potential donors available. In reality, most preservation organizations use a combination of these

Entrance, Grove Street Cemetery, New Haven, Connecticut. (A.H.M. Gottschalk, HABS)

methods, developing a healthy diversity of donors. Small organizations contact individuals chiefly by mail and through in-person solicitations. Telemarketing, a coordinated campaign of telephone calling to potential donors, is often too expensive.

The personal approach. Do not assume that because individuals make the majority of contributions it will be easy to raise resources from them. Fund raising from this sector is difficult because people—unlike foundations and corporations—do not announce or publicize their philanthropic interests. As a result, research is a central aspect of soliciting donations from individuals. Determine who is supporting other local or regional organizations, and pay attention to who serves on local boards of directors or chairs other fund-raising events. These efforts signal an individual's appreciation for and willingness to support nonprofit causes.

Successful fund raising from individuals begins with the identification of prospects and is followed by strategies to educate and excite them about a group's work and successes. This cultivation process can include any combination of mailings, publicity, special events, meetings, and other means of conveying an organization's message. An often overlooked way of spreading the word about an organization's financial needs is to ask top donors to contact their acquaintances on behalf of the organization. This is an excellent way to begin building a "friends" group of donors and volunteers. Members of such a group feel they are specially connected with the organization. A friends' newsletter, special events, and public recognition are ways of maintaining commitment. Cultivation continues even after an individual becomes a donor. Moving supporters up to higher levels of giving and involvement is an important way to increase an organization's revenue and strengthen its renewable donor base.

Broad-based appeals. Mass appeals to individuals also require careful research and planning. Printing and mailing costs would be prohibitive and the return on the investment unacceptably low if target groups were not identified before sending out large mailings.

Through research the fund raiser should determine what types of individuals are most likely to support the organization's cause. Are they old or young? Rich or poor? Male or female? Do they live in a certain type of neighborhood? Research of this sort is generally done by sending out test marketing pieces and analyzing the characteristics of those who make donations. Different messages and methods of presentation also can be tested; a variety of fund-raising pieces can be mailed to determine which brings in the most donations.

Targeting mailings to those who are most likely to give enables an organization to tailor its message to a particular group or to several subgroups, boosting donations markedly. While research of this type will significantly increase the efficiency of the appeals, the percentage of return still seldom rises above 2 percent, and a return of 1½ percent is considered good. Without research, the return would be much lower.

Finally, organizations with membership programs are at an advantage because their membership comprises a preexisting pool of interested individuals who can be cultivated to become donors. Again, this can be done through newsletters, events, and the like.

Corporations

American corporations give grants totaling approximately $6 billion annually. Most corporations have a "direct-giving" program—a department within the company (usually public affairs, community relations, or corporate contributions) that is charged with giving out however much money the company has budgeted that year for contributions.

The largest corporations often give through a foundation endowed by the company but which is legally distinct from the company itself.

Such foundations publish annual reports containing guidelines that describe how an organization should approach the foundation and what types of organizations are eligible. Because they are somewhat insulated from the daily business dealings of their affiliated company, corporate foundations are less motivated by the business ramifications of making a particular grant. Nevertheless, a review of the boards of major company foundations reveals that many of the same names appear on each company's corporate board.

Some companies with foundations also use direct giving in their philanthropic efforts. Corporate-giving reference books, such as those from the Taft Group and the Foundation Center, indicate what methods corporations use to support nonprofit organizations.

It is a mistake to focus only on the large Fortune 500 companies. Thousands of small, often privately owned, companies contribute millions of dollars to nonprofit groups every year. Their reasons for making donations are varied but patterns do exist. A recent poll of owners of businesses with fewer than 100 employees revealed their four top reasons for contributing (in descending order): (1) to get involved in their community; (2) for public relations value; (3) because they were asked by a friend or colleague; and (4) because they know the organization. Based on these four reasons for making contributions, it follows that local groups have a better chance of securing funds from small local companies than do outside groups. An organization, therefore, should look first to its own community for financial support. Once the organization receives support from local groups, it then will have a track record to present to larger, more distant companies.

The way an organization approaches corporate-giving officers is important. Corporations rarely make awards in response to mass mailings, even if the letters have personalized salutations. There simply is too much competition from other grant seekers who have taken the more difficult but necessary step to establish personal contact with the giving office and to prove to the corporation that there is a logical reason for giving to their cause. The same rules mentioned above apply here: establish an organization's credibility and define its work clearly. When making initial contact by letter (which most companies prefer), be straightforward and concise and limit the letter to one or two pages; such a letter will stand a much better chance of being read than a longer letter or proposal. Follow these guidelines: (1) write to the correct person; (2) enclose materials that clarify the letter (such as an organization's annual report), but do not overload the recipient with materials; and (3) follow up in a timely manner, perhaps requesting a brief meeting.

When soliciting a company, it is worthwhile to consider how the company would benefit from making a contribution to the nonprofit organization. For instance, will a gift provide the company with public relations opportunities? Or does the organization improve the quality of life for employees of the company? Perhaps supporting the organization will give the company's executives an opportunity to become more involved with a cause they care about personally. In a general sense, companies contribute to nonprofit causes because corporate philanthropy is a well-established and respected business practice and because the company may realize tax advantages.

Finally, corporate fund raising provides excellent opportunities for creative tie-ins between companies and nonprofit groups. For instance, The Nature Company, a small chain of stores selling outdoor and conservation-related items, agreed to display literature and membership applications for The Nature Conservancy, a national, nonprofit conservation organization. The Nature Company also donates a small portion of its revenues to the nonprofit. The for-profit Nature Company receives much good will from the community in return, and its operation is legitimized by its association with the nonprofit organization. The Nature Conservancy benefits by having its message spread to those who shop at the Nature Company, many of whom become members of the conservancy, as well as by receiving a percentage of the store's profits. The fit between these two organizations is excellent.

In a similar way, the National Trust's Historic Hotels of America program has linked its mission with the cause of promoting historic hotels. A membership fee entitles eligible hotels to be listed in the National Trust's directory of Historic Hotels of America and to receive other benefits from their association with the National Trust. The National Trust benefits not only by receiving fees, but also by gaining greater public exposure and name recognition from the program. Historic Hotels of America uses a toll-free reservation number operated by a private travel company, which has agreed to share part of the revenue it receives from commissions on travel bookings in return for having its name listed in the program's promotional materials.

Often these relationships open the door to areas of corporate budgets that, strictly speaking, are not contributions. Depending on the benefits, larger companies may choose to use funds from their marketing or advertising budgets, which are always larger than budgets allocated to nonprofit causes. In such cases, bear in mind that the company's relationship with an organization is probably more of a business deal, and the case for support should reflect this.

There are many imaginative ways to structure tie-ins, but remember that the essence of such partnerships is gain for each participant. As long as the mission of the nonprofit is promoted and not damaged by its association with the for-profit company and there is a reasonable financial reward, tie-ins make sense.

Foundations
Foundations are legal entities established for the purpose of awarding funds to organizations that meet their criteria. The two most common types of foundations are independent and family foundations whose endowments come from a single source or from members of the same family, respectively. A third common type, community foundations, builds its endowments from many donors, typically from the same city or region, and usually restrict their giving to organizations operating in that same area. As with corporations, foundations of all types publish annual reports containing conditions of eligibility and information about how a group should approach them for grants.

In selecting foundations to solicit for support, first determine whether the program is eligible according to the foundation's stated guidelines. The easiest way for a foundation (or a corporation for that matter) to decline a grant is to decide that a proposed project or an organization's mission is outside its criteria. Next determine whether there is a logical match between the organization and the funding source. The reason for selecting any foundation should be logical. For instance, do the organization's programs benefit people in the geographic area where the foundation has interests? Or, has the foundation announced a special focus or theme for the year that meets some of the nonprofit group's programmatic goals?

It can be difficult to learn what causes smaller foundations, particularly family foundations, support because they often do not publish this information. In such cases, consult IRS 990 forms. These public documents, which list grant recipients and the dollar amounts donated by organizations in the previous tax year, are available at the Foundation Center. By determining a pattern of giving from grants made, a project's appeal to the foundation in question can be estimated. Many of these smaller, family foundations function in much the same way as individual donors and should be approached accordingly.

Special Events

Long used as major sources of revenue for organizations, events can be lucrative while also building awareness of an organization's work. The basic premises for any fund-raising event are (1) to minimize (or even eliminate) costs by having facilities, food, entertainment, and other supplies and services donated and (2) to maximize revenue through strong ticket sales and donations. Of course, this is an oversimplification, but the basic idea is sound.

Cash donations in support of an event have always been a favorite means of raising funds. Generally the donors receive complimentary tickets or a table at the event. With the promise of publicity and the potential for a public display of goodwill, many local businesses often enthusiastically support a worthwhile cause, especially if there is a logical pairing between the company and the event they will underwrite. While individuals are good prospects for supporting events, foundations usually do not support organizations in this manner.

Events range from formal black-tie affairs to summertime picnics at the site of a building restoration. Historic preservation groups often are in an advantageous position to benefit from events because the site of the event may actually be the beneficiary. Additionally, historic sites are among the favorites of special-event goers because such venues tend to be more interesting visually and more memorable than banquet rooms. At the National Trust's historic house museums, functions range from the annual Woodlawn Plantation needlepoint exhibition to Lyndhurst's summertime lawn concerts. These events raise money but also introduce many newcomers to the properties, as well as to the National Trust, bringing in new members for the organization and for the properties' friends groups.

FEDERAL FUNDING

Federal government programs offer loans, cooperative agreements, and grants, the latter being the most prevalent form of aid. The two most common types of grants are discretionary grants and entitlement grants. Discretionary grants are fairly open ended, allowing funds to be spent on a wide range of activities, while entitlement grant awards are based on absolute criteria and must be spent on specific projects.

Typically the federal government allocates its grant funds to state and local governments, which then serve as intermediary agencies by subgranting these funds to eligible applicants.

While the majority of government funding is obtained through the funneling process, a few grants are allocated directly by the federal government to grantees. These may be either discretionary or entitlement grants and are awarded on a competitive basis. It is common for these funds to be awarded as matching grants, which hold the grantee responsible for covering at least one-half the project's cost.

Research is crucial to understanding the broad assortment of federal funding programs available. The first step is to examine the *Catalog of Federal Domestic Assistance* (CFDA) or the *Guide to Federal Funding for Governments and Nonprofits*. Both manuals, which are published annually, describe the type of aid offered by each federal assistance program along with eligibility guidelines. Most state and local government offices, as well as most large libraries, have complete sets of these reference books. The General Services Administration (GSA) operates the Financial Assistance Retrieval Program (FARP), which provides a computer printout of government programs pertaining to a project.

A number of federal agencies are recognized specifically for their support of preservation and rehabilitation efforts. The U.S. Department of Housing and Urban Development, the National Park Service, the National Endowment for the Arts, and the National Endowment for the Humanities are among the most popular supporters of historic preservation and offer a wide variety of assistance programs for preservation-related work. Contact each agency's state field office for more information.

Congress is another useful source of information. An organization should contact the federal aid coordinator in its representative's local district office. Statewide and local preservation groups, as well as state historic preservation officers (SHPOs), may provide additional assistance.

When looking for appropriate funding programs, keep an open mind. Do not limit the search to familiar agencies or to programs that specialize in projects. Creative fund raisers can find ways of using funds from agencies not often used to fund preservation projects.

Be persistent in searching for federal funding, and do not become discouraged if the organization or project is ineligible for a particular program. If the organization is ineligible on its own, it may be possible to obtain funds by forming an alliance with another organization.

Following is a brief listing of federally funded programs for preservation, restoration, development, and rehabilitation projects. The list serves as a starting point and includes only the most relevant programs with substantial endowments.

EXECUTIVE BRANCH

U.S. Department of Agriculture
14th Street and Independence Avenue, S.W.
Washington, DC 20250
(202) 720-8732

Farmers Home Administration
14th Street and Independence Avenue, S.W.
Room 5037
Washington, DC 20250
(202) 720-4323

Farmers Home Administration (FmHA) programs provide housing assistance exclusively in rural areas. The administrations funds the acquisition, construction, repair, or rehabilitation of homes for low- and moderate-income individuals.

- Community Facilities Loans: Loans for the construction, enlargement, or improvement of community facilities for health care, public safety, and public services. Contact FmHA district office.

- Business and Industrial Loan Guarantees Program: Loans for businesses and industrial development projects in rural communities for the repair and purchase of equipment and facilities. Contact FmHA district office.

- Home Ownership Loans: Loans for the construction, improvement, repair, or rehabilitation of rural homes and related facilities. Contact FmHA district office.

- Migrant Farmworker Program: Grants for the acquisition, construction, rehabilitation, or operation of housing for migrant farmworkers or the homeless. Contact FmHA district office.

- Nonprofit National Corporations Loan and Grant Program: Funds for activities that improve business, industry, and employment opportunities in rural areas. Contact FmHA district office.

- Rural Economic Development Loans and Grants: Funds for projects that will generate economic development and create jobs. Contact FmHA district office.

- Rural Rental Housing Loans: Loans to build, purchase, or repair apartment-style housing and related facilities. Contact FmHA district office.

- Rural Housing Preservation Grants: Grants for rehabilitation, preservation, or modernization efforts for low-income housing. Contact FmHA state or district office.

- Rural Industrial Development Grant Program: Grants for construction and conversion activities that support the development of small and emerging businesses in rural areas. Contact FmHA district office.

- Rural Water and Waste Disposal Loans and Grants: Funds for implementing domestic water and waste disposal systems for rural communities. Contact FmHA county supervisor.

- Self-Help Housing Loans: Loans for site development, construction, and technical assistance for building individual homes for low-income families. Contact FmHA district office.

Rural Electrification Administration
14th Street and Independence Avenue, S.W.
Room 4025
(202) 720-9570

- Rural Economic Development Loans and Grants: Funds for projects that foster economic development or create jobs in rural areas.

U.S. Department of Commerce
Herbert C. Hoover Building
14th Street and Constitution Avenue, N.W.
Washington, DC 20230
(202) 377-2000

Economic Development Administration
Herbert C. Hoover Building
14th Street and Constitution Avenue, N.W.
Washington, DC 20230
Room 7804
(202) 377-5113

Economic Development Administration (EDA) programs concentrate on alleviating substantial and persistent unemployment and underemployment in economically distressed areas.

- Business Development Assistance Program: Loans for activities in economically distressed areas to generate industrial and commercial expansion. Contact EDA regional office.

- Economic Adjustment Assistance Program: Grants for economic development projects in areas experiencing long-term economic deterioration or sudden and severe economic dislocation. Contact EDA regional office or (202) 377-2659.

- Public Works and Development Facilities Assistance Program: Grants for construction of public works that also create or retain jobs. Contact EDA regional office or (202) 377-5265.

- Public Works Impact Projects: Grants for the renovation or construction of public works and development facilities that stimulate immediate employment in the project area. Contact EDA regional office or (202) 377-2873.

National Oceanic/Atmospheric Administration
1825 Connecticut Avenue, N.W.
Room 714
Washington, DC 20235
(202) 606-4126

- Marine Sanctuary Program: Grants for the conservation and management of areas identified as having national significance owing to their resources or human-use value.

U.S. Department of Education
400 Maryland Avenue, S.W.
Washington, DC 20202
(202) 708-5366

- Institutional Conservation Program: Grants for the analysis of ideal conservation methods in a specific public building, school, or hospital, as well as the actual implementation of such conservation. Contact state energy office.

- Public Library Construction and Technology Enhancement Program: Grants for the acquisition, expansion, renovation, and equipping of buildings that will serve as public libraries. Contact (202) 219-2293.

U.S. Department of Energy
1000 Independence Avenue, S.W.
Washington, DC 20585
(202) 586-5000

Office of Conservation and Renewable Energy
1000 Independence Avenue, S.W.
Room 5G-045
Washington, DC 20585
(202) 586-8034

- State Energy Conservation Programs: Grants for energy conservation in schools and hospitals. Contact Department of Energy (DOE) regional office.

Office of Weatherization
1000 Independence Avenue, S.W.
Room 5G-023
Washington, DC 20585
(202) 586-2204

■ Low Income Household Weatherization Program: Grants fund weather resistance projects for homes of low-income individuals. Contact DOE regional office.

U.S. Department of Health and Human Services
330 Independence Avenue, S.W.
Washington, DC 20447
(202) 619-0257

Administration on Aging
3330 Independence Avenue, S.W.
Washington, DC 20447
(202) 619-0011

■ Supportive Services and Senior Centers: Grants for the construction or renovation of multipurpose facilities in homes for older people. Contact state and area agencies on aging.

Office of Community Services
370 L'Enfant Promenade, S.W.
Washington, DC 20447
(202) 401-9333 or (202) 401-2333

Office of Community Services (OCS) programs are designed to encourage urban and rural development. Funding is used to increase self-help, employment, and ownership opportunities for low-income individuals.

■ Community Services Block Grant: Grants for activities to alleviate the causes of poverty by helping people obtain sufficient jobs, education, and housing. Contact state agency.

■ Discretionary Grants Program: Grants for projects that provide business and employment opportunities for low-income people. Contact (202) 401-9345.

■ Job Opportunities for Low-income Individuals: Grants for projects that create new employment and business opportunities for low-income individuals. Contact (202) 401-2333.

■ Social Services Block Grant: Grants for social service projects designed to reduce the economic dependency of the poor. Contact state agency.

Office of Energy Assistance
901 D Street, S.W.
Washington, DC 20447
(202) 401-9351

■ Low Income Home Energy Assistance Block Grant: Grants help the poor, elderly, and handicapped finance home energy costs and the required technical assistance. Contact regional social services office.

Health Resources and Services Administration
Parklawn Building
5600 Fishers Lane
Rockville, MD 20857
(301) 443-3376

■ AIDS Facilities and Construction Grants: Grants for the establishment or renovation of health care facilities for patients with AIDS or other HIV-related conditions. Contact (301) 443-1440.

U.S. Department of Housing and Urban Development
451 7th Street, S.W.
Washington, DC 20410
(202) 708-1112

Housing and Urban Development (HUD) programs focus on improving housing in both urban and rural communities through acquisition, rehabilitation, and new construction.

■ Community Development Block Grant Program: Grants for a wide range of activities, such as restoration and preservation, designed to improve community development. Contact local community development office.

■ Emergency Shelter Grants: Grants for the renovation, rehabilitation, or conversion of buildings for use as emergency shelters for the homeless. Contact HUD regional office.

■ Historic Preservation Loans: Loans for the preservation, rehabilitation, or restoration of historic residential structures. Contact HUD regional office.

■ HOME Investment Trust Fund: Funds are first allocated as matching block grants to states and large cities and then redistributed. HOME offers a number of programs, including rental rehabilitation, home repair for older and disabled homeowners, urban homesteading, and rental housing production. Contact HUD regional office.

■ Public Housing Comprehensive Improvement Assistance Program: Grants for reconstructing and modernizing housing units to improve their physical condition. Contact local HUD office.

■ Supportive Housing for the Elderly: Grants for construction, reconstruction, and rehabilitation of housing designed to meet the special physical needs of elderly persons. Contact HUD regional office.

U.S. Department of the Interior
1849 C Street, N.W.
Washington, DC 20240
(202) 208-3100

National Park Service
P.O. Box 37127
Washington, D.C. 20013-7127
(202) 208-4747

National Park Service (NPS) programs that focus on the rehabilitation, innovation, and recovery of parks, recreation facilities, and historic buildings are listed below.

■ Historic Preservation Grants-In-Aid Program: Grants for equipment, materials, architectural planning, and construction necessary for the restoration, acquisition, or development of historic properties. Contact state historic preservation officer.

■ National Historic Landmark Program: Advisory services and counseling to study and identify historic landmarks. A bronze plaque and certificate are presented to selected landmarks at a formal ceremony. Contact regional Park Service office.

■ National Natural Landmarks Program: Technical information to help designate natural landmarks and to preserve those already selected. Contact regional Park Service office.

■ National Register of Historic Places: Advisory services and counseling to determine a property's eligibility for inclusion in the National Register. Contact regional Park Service office or state historic preservation officer.

■ Outdoor Recreation Grants: Grants for the acquisition and development of public outdoor recreation and facilities such as inner-city parks, swimming pools, and picnic areas. Contact regional Park Service office and ask for the state liaison officer.

■ Urban Park and Recreation Recovery Program: Funds for the rehabilitation and improvement of existing park and recreation systems. Contact regional Park Service office.

U.S. Department of Labor
200 Constitution Avenue, N.W.
Washington, DC 20210
(202) 523-6666

Employment and Training Administration
200 Constitution Avenue, N.W.
Room S-2322
Washington, DC 20210
(202) 523-6871

Employment and Training Administration (ETA) programs are designed to improve training, utilize available workers, increase volunteer efforts, and enhance the basic skills of the work force.

■ Migrant and Seasonal Farmworker Employment and Training Program: Grants for projects that establish nonagricultural job opportunities and job training for farmworkers who are unemployed and their families. Contact regional ETA office.

■ Senior Community Service Employment Program: Grants for projects that utilize older, unemployed, low-income workers. Contact (202) 535-0804.

■ Summer Youth Employment and Training Program: Grants for programs that provide job training and employment opportunities for economically disadvantaged youths. Contact state ETA office.

U.S. Department of Transportation
400 7th Street, S.W.
Washington, DC 20590
(202) 366-4000

Urban Mass Transportation Administration
400 Seventh Street, S.W.
Office of Public Affairs
Room 9400
(202) 366-4043

Urban Mass Transportation Administration (UMTA) programs address public mass transit operations and related construction, extension, refurbishment, and rehabilitation activities.

■ Discretionary Capital Grant Program: Grants for projects including land acquisition, construction, and modernization for the improvement of the transit system. Contact regional UMTA office.

■ Rural Transit Assistance Program: Grants for public transportation projects in nonurban areas. Contact regional UMTA office.

INDEPENDENT AGENCIES

ACTION
Federal Domestic Volunteer Agency
1100 Vermont Avenue, N.W.
Washington, DC 20525
(202) 606-5135

ACTION is a federal organization created by Congress in 1971 to increase nationwide volunteer efforts as a solution to societal problems. ACTION programs award funding and service to those organizations and individuals who resourcefully use volunteers to carry out their projects.

■ Mini-Grants Program: Grants for projects that use volunteers to solve community problems. Contact ACTION state office.

■ Retired Senior Volunteer Program: Grants for projects that create community jobs for people over the age of 60. Contact regional ACTION program office.

■ Student Community Service Project: Grants for student volunteer community service projects that address poverty. Contact (202) 634-9424.

VISTA
1100 Vermont Avenue, N.W.
Washington, DC 20525
(202) 606-5135

VISTA's volunteer efforts address a wide variety of projects and needs. The role of VISTA volunteers must emphasize the mobilization of local human, financial, and material resources; the transference of skills to community residents; and an increase in the ability of the low-income community to solve its own problems.

■ Regular Projects: Services are provided for social projects designed to help alleviate poverty and its related problems. Contact regional ACTION office.

Commission on National and Community Service
National Press Building
529 14th Street, N.W.
Suite 428
Washington, DC 20045
(202) 724-0600

The National and Community Service Act of 1990 created the Commission on National and Community Service to provide program funds, training, and technical assistance to states and communities for developing and expanding service opportunities. The types of service activities are broadly defined: participants may perform any educational, human, environmental, or public safety service project that will benefit the community.

■ American Conservation and Youth Service Corps Programs: Grants for programs that involve teenagers and young adults who receive job and skill training, living allowances, and scholarships. Grants are awarded on a competitive basis to states, Indian tribes, and local applicants in states that do not apply.

■ Higher Education Innovative Projects for Community Service: Funds for student community service projects or teacher training in service-learning methods.

■ National and Community Service State Program: Funds to states and Indian tribes for projects that engage individuals ages 17 and older in full-time or part-time service. Participants will receive education or housing benefits upon completion of their term of service.

■ Serve America: Funds for programs sponsored by schools or

community-based agencies to involve school-aged youth in service to the community. Serve America also supports programs that involve adult volunteers in the schools. Funding will be allocated to states, Indian tribes, and local applicants in states that do not apply.

Federal Emergency Management Agency
State and Local Programs and Support Directorate
Office of Disaster Assistance Programs
500 C Street, S.W.
Washington, DC 20472
(202) 646-3615

■ Individual Assistance Program: Loans for protection, relief, and recovery from emergency disasters. Funds the repair, rehabilitation, or replacement of damaged real and personal property. Contact regional Federal Emergency Management Agency (FEMA) office.

Federal Highway Administration
Office of Engineering (HNG-12)
Room 3132
Attention: Grant Programs
(202) 366-4658

Federal Highway Administration (FHA) programs focus on construction, reconstruction, and improvement of roads, as well as landscaping, engineering, and bridge rehabilitation.

■ Construction Program: Grants for the construction of highways, roadsides, high-occupancy-vehicle facilities, and pedestrian walkways. Contact regional FHA office.

■ Highway Bridge Replacement and Rehabilitation Program: Grants for the rehabilitation or replacement of deficient highway bridges in order to enhance roadway safety. Contact regional FHA office.

■ Secondary Program: Grants for construction efforts on principal secondary and feeder roads such as farm-to-market, mail routes, and school-bus routes. Contact regional FHA office.

■ Urban Program: Grants for construction, reconstruction, and improvement of rail facilities. Contact regional FHA office.

Federal Railroad Administration
Room 8206
Attention: Grant Programs
(202) 366-0881

■ Local Freight Assistance Program: Grants for acquisition, rehabilitation, and construction of rail facilities. Contact regional Federal Railroad Administration (FRA) office.

General Services Administration
18th and F Streets, N.W.
Washington, DC 20405
(202) 708-5082

■ Surplus Real Estate Program: Hundreds of acres of federal real estate are declared surplus by various agencies each year and are sometimes donated to public or private nonprofit agencies to use for specific public purposes. Contact regional General Services Administration (GSA) office or (703) 557-1234.

■ Surplus Personal Property Program: Surplus personal property of the federal government is awarded for projects serving the public. Contact GSA regional office or state surplus property agency.

Institute of Museum Services
1100 Pennsylvania Avenue, N.W.
Washington, DC 20506
(202) 606-8539

The Institute for Museum Services is an independent federal agency, created by Congress in 1976 to increase and improve museum services. The institute has a national focus and supports all types of museums including historic houses and sites, history museums, zoological parks, aquariums, and botanical gardens.

■ General Operating Support Program: Competitive grants are awarded to museums based on the extent to which they effectively use their resources.

■ Conservation Assessment Program: Grants for an overall assessment of the conditions of a museum's environment and collections to identify conservation needs and priorities.

■ Conservation Project Support Program: Matching grants for projects geared to conserve a museum's collections.

National Endowment for the Arts
Nancy Hanks Center
1100 Pennsylvania Avenue, N.W.
Washington, D.C. 20506
(202) 682-5400

The National Endowment for the Arts (NEA) is an independent federal agency, created by Congress in 1965 to encourage and support American art and artists. The endowment awards grants to individual American artists and arts groups through a variety of artistic programs and other funding categories in order to continue excellence in the arts.

■ Challenge III Grant Programs: Grants for activities that have the potential to move the nation forward in achieving excellence in the arts. Grants are awarded to cultural organizations to open new funding sources and require a 3:1 match. Contact (202) 682-5436.

■ Design Arts Program: Grants fund activities in fields of architecture, landscape architecture, urban design and planning, historic preservation, interior design, industrial and product design, and graphic design. Contact (202) 682-5437.

■ Folk Arts Program: Grants to preserve and enhance our country's multicultural artistic heritage. Contact (202) 682-5449.

■ Locals Program: Grants to local agencies on behalf of their local government for the support and development of the arts within a city or county. Contact (202) 682-5431.

■ Museum Program: Grants for projects of artistic significance in the museum field are awarded in three categories: (1) Utilization of Museum Resources: presentation of collections, education, and catalog; (2) Care of Collections: conservation, collection, and maintenance; and (3) Museum Purchase Plan: purchase of items for museum collections. Contact (202) 682-5442.

■ Visual Arts Program: Matching grants to organizations that encourage individual artists' development and experimentation. Contact (202) 682-5448.

National Endowment for the Humanities
1100 Pennsylvania Avenue, N.W.
Room 406
Washington, D. C. 20506
(202) 786-0438

The National Endowment for the Humanities (NEH) is an independent federal agency, created by Congress in 1965 to support research, education, and public programs in the humanities. NEH offers funding programs to further the pursuit of those interested in any discipline of the humanities.

■ Challenge Grant Programs: Grants for projects designed to enhance activities in the humanities. Construction, renovation, and temporary exhibitions are funded. Funding is awarded on a 3:1 or 4:1 matching grant basis. Contact (202) 786-0361, Room 429.

■ Human Projects in Libraries and Archives: Grants for activities that increase public appreciation and understanding of the humanities through the use of books and other resources in the collections of American libraries and archives. Contact (202) 786-0271, Room 426.

■ Human Projects in Museums and Historical Organizations: Grants for the planning and implementation of exhibitions as well as the interpretation of historic sites. Contact (202) 786-0284, Room 420.

■ National Heritage Preservation Program: Grants for projects that solve problems posed by the disintegration of significant resources. Contact (202) 786-0570, Room 802.

National Historical Publications and Records Commission
National Archives Building
8th and Pennsylvania Avenue, N.W.
Washington, DC 20408
(202) 501-5600

■ Records Program: Grants for projects that preserve and make

available for use records that further an understanding and appreciation of American history.

Small Business Administration
409 Third Street, S.W.
Washington, DC 20416
(800) 827-5722

■ Certified Development Company Program: Loans help small businesses pay for construction, renovation, machinery, and architectural service costs, thereby promoting community and economic development. Contact Small Business Administration (SBA) field office.

NATIONAL TRUST FUNDING

The National Trust offers four grant and loan programs—the National Preservation Loan Fund, Inner-City Ventures Fund, Critical Issues Fund, and Preservation Services Fund—that assist with many different types of preservation projects. In addition to financial assistance, the National Trust's office of financial services offers advice, support, and technical assistance in putting together a viable financial package for a project.

NATIONAL PRESERVATION LOAN FUND

The National Preservation Loan Fund (NPLF) provides below-market-rate loans to nonprofit organizations and government agencies to acquire or rehabilitate historic properties or to establish or expand a revolving fund. NPLF loans typically are awarded in amounts ranging from $20,000 to $200,000 at an interest rate of 2 percent below prime. Loans generally are repaid within five years, although longer amortization periods may be possible. The NPLF also offers special help for threatened National Historic Landmarks. Applications to the NPLF are encouraged and are accepted at any time.

INNER-CITY VENTURES FUND

The Inner-City Ventures Fund (ICVF) helps community-based nonprofit organizations initiate historic rehabilitation projects for

the benefit of low- and moderate-income residents. ICVF awards typically include a package of grants and loans each totaling between $40,000 and $100,000. Funding rounds are held in various parts of the country as corporate and foundation support for the program becomes available.

Established with seed grants from the U.S. Department of the Interior and the National Trust, the initial $400,000 investment in the ICVF program has grown to more than $4 million in private support. As of June 1991, the ICVF program has awarded $2.5 million in funding for 40 projects in 30 cities; leveraged 23 local dollars for every ICVF dollar, resulting in $59 million in economic revitalization in project areas; created more than 1,000 housing units in 241 historic buildings; rehabilitated more than 334,000 square feet of storefront, business incubator, and community meeting space; and resulted in more than 2,000 construction and permanent jobs in project neighborhoods.

CRITICAL ISSUES FUND

The Critical Issues Fund (CIF) addresses the most widespread, pressing problems facing the historic preservation movement today. Through this fund, the National Trust works in partnership with other organizations to foster innovative research and problem-solving and to effect policy change at the local, state, and national levels. The maximum grant awards are $20,000; however, under special circumstances, a larger grant may be awarded. All grants must be matched on at least a dollar-for-dollar basis. Eligible applicants include nonprofit organizations, public agencies, universities, individuals, and for-profit entities.

PRESERVATION SERVICES FUND

The Preservation Services Fund (PSF) provides financial assistance to nonprofit organizations and public agencies to initiate preservation projects. PSF grants, which average $1,000 to $1,500, can be used for professional consultant services, preservation education programs and conferences, and rehabilitation feasibility studies. Many PSF seed grants help plan the reuse of historic buildings. Applications to this program are accepted at National Trust regional offices each February 1, June 1, and October 1.

CONTACTS

American Association of Fund Raising Counsel
25 West 43d Street
New York, NY 10036
(212) 354-5799

An association of major fund-raising consulting firms. Activities include lobbying, developing research and education programs, and monitoring ethics and standards affecting fund-raising consulting. Publishes a bimonthly newsletter, *Giving Update*, and the annual *Giving USA*, a comprehensive tabulation and analysis of the previous year's philanthropy.

American Council for the Arts
1285 6th Avenue
3d Floor
New York, NY 10019
(212) 223-2787

Provides services for artists and arts administrators including a united fund, publications, and a library.

American Prospect Research Association
1600 Wilson Boulevard
Suite 905
Arlington, VA 22209
(703) 525-1191

A membership organization focused on providing information for development researchers. Holds an annual conference and produces bibliographic materials. Publishes *Connections*, a quarterly newsletter.

Business Committee for the Arts (BCA)
1775 Broadway
Suite 510
New York, NY 10019
(212) 664-0600

Provides members with information and advice concerning their corporate giving programs to the arts. Publishes a quarterly newsletter, *BCA News*.

Center for Corporate Public Involvement
1001 Pennsylvania Avenue, N.W.
Washington, DC 20004
(202) 624-2425

Assists health and life insurance agencies involved in community service projects. Publishes *Response*, a triennial magazine, and *Social Report*, an annual bulletin.

Council for Advancement and Support of Education
11 Dupont Circle
Suite 400
Washington, DC 20036
(202) 328-5900

Publishes *Currents*, a monthly magazine, and maintains a clearinghouse for corporate matching grants through which companies match donations made by their employees to nonprofit organizations.

Council on Foundations
1828 L Street, N.W.
Washington, DC 20036
(202) 466-6512

A membership organization of grant makers. Publishes a bimonthly magazine, *Foundation News*, and a newsletter and holds conferences and seminars.

The Foundation Center
79 5th Avenue
New York, NY 10003
(212) 620-4230

Publishes extensive guides to grants from private, community, and corporate foundations. Operates several libraries (listed below) containing comprehensive fund-raising reference materials, research aids, and an orientation to fund raising. Also maintains relationships with more than 100 libraries nationwide. The center produces "Comsearch" printouts, which identify funding sources by topic, such as architecture, historic preservation, and historical societies.

312 Sutter Street
Room 312
San Francisco, CA 94108
(415) 397-0902

1001 Connecticut Avenue, N.W.
Washington, DC 20036
(202) 331-1400

Kent H. Smith Library
1422 Euclid Avenue
Suite 1442
Cleveland, OH 44115
(216) 861-1933

The Fund Raising Institute (FRI)
12300 Twinbrook Parkway
Suite 450
Rockville, MD 20852-9830
1-800-877-TAFT

A division of the Taft Group, FRI specializes in publishing texts on development in a quarterly catalog on development. Also publishes the *FRI Monthly Portfolio* newsletter.

The Grantsmanship Center
650 South Spring Street
P.O. Box 6210
Los Angeles, CA 90015
(213) 689-9222

Publishes *Grantsmanship Center News*, a bimonthly fund-raising magazine. The center also assists nonprofit groups that have not yet developed professional fund-raising skills.

KRC Development Council
431 Valley Road
New Canaan, CT 06840
(203) 972-0401

Publisher of reference books on fund-raising techniques and the *KRC Letter*.

National Assembly of Local Arts Agencies
1420 K Street, N.W.
Suite 204
Washington, DC 20005
(202) 371-2830

Assists and supports local arts agencies through advocacy, training programs, and information services. Publishes *Connections Quarterly* and *Connections Monthly*.

National Assembly of State Arts Agencies
1010 Vermont Avenue, N.W.
Suite 920
Washington, DC 20005
(202) 347-6352

Represents and protects the interests of state arts agencies through advocacy, arts policy development, training, and research services. Publishes *Art View*, a quarterly journal.

National Committee for Responsive Philanthropy
2001 S Street, N.W.
Suite 620
Washington, DC 20009
(202) 387-9177

Focuses on issues of how contributors' political positions affect their philanthropy. Publishes *Responsive Philanthropy*, a quarterly newsletter.

National Council of State Housing Agencies
444 North Capitol Street, N.W.
Suite 412
Washington, DC 20001
(202) 624-7710

Assists state housing finance agency programs that help lower- to middle-income people obtain affordable housing. Publishes *The Washington Update*, *The HFA Update*, and *The Affiliate Advisor*.

National League of Cities
1301 Pennsylvania Avenue, N.W.
Washington, DC 20004
(202) 626-3000

National organization that represents the interests of municipal governments on the federal level. The league is involved with preservation through the issues of downtown revitalization. Publishes *Nation's Cities Weekly*.

National Society of Fund Raising Executives
1101 King Street
Suite 3000
Alexandria, VA 22314-2944
(703) 684-0540

Individual membership organization consisting (usually) of senior development professionals. Publishes *The NSFRE Journal* quarterly and *The NSFRE News* eight times per year.

Taft Group
12300 Twinbrook Parkway
Suite 450
Rockville, MD 20016
(202) 966-7086

Publishes a variety of annual fund-raising reference books such as the *Taft Corporate Giving Directory* and *America's New Foundations*. Also publishes the *Corporate Giving Watch* newsletter.

United States Conference of Mayors
1620 I Street, N.W.
Washington, DC 20006
(202) 293-7330

A nonpartisan organization of mayors of American cities with populations greater than 30,000. The organization develops and formally adopts policy positions on urban issues. Also works closely with the National Park Service to assist cities on issues of open space and planning. A catalog of publications is available. Much of the organization's information is specifically tailored to the local level, where fund raising for preservation is most successful.

Washington International Arts Letter
P.O. Box 12010
Des Moines, IA 50312
(515) 255-5577

Issues a monthly newsletter with information on grants and other financial aid to groups in the arts, humanities, and education.

PERIODICALS

Affiliate Advisor
National Council of State Housing Agencies
444 North Capitol Street, N.W.
Suite 438
Washington, DC 20001
(202) 624-7710

Published bimonthly.

BCA News
Business Committee for the Arts
1775 Broadway
Suite 510
New York, NY 10019-1942
(212) 664-0600

Published quarterly.

Catalog of Federal Domestic Assistance
Superintendent of Documents
Government Printing Office
Washington, DC 20402
(202) 783-3238

Published annually. Contains eligibility guidelines and information on the type of aid offered by individual federal assistance programs.

The Chronicle of Philanthropy
The Chronicle of Higher Education
1255 23d Street, N.W.
Washington, DC 20037
(202) 466-1000

Published biweekly. One of the best overall sources of information on fund raising (including corporations, foundations, individuals, and others), nonprofit management, and related issues. Contains a useful directory of services, professional opportunities, and a listing of new books in every issue.

Connections
American Prospect Research Association
701 West Broad Street
Falls Church, VA 22046
(703) 533-3413

Published quarterly.

Connections Quarterly
National Assembly of Local Arts Agencies
927 15th Street, N.W.
12th Floor
Washington, DC 20005
(202) 371-2830

Published quarterly

Connections Monthly
National Assembly of Local Arts Agencies
927 15th Street, N.W.
12th Floor
Washington, DC 20005
(202) 371-2830

Published monthly.

Currents
Council for the Advancement and Support of Education
11 Dupont Circle, N.W.
Suite 400
Washington, DC 20036
(202) 328-5900

Published monthly. Written for higher education professionals but contains many excellent articles on fund-raising strategy and public relations.

Directory of Corporate Affiliations
National Register Publishing Company
R. R. Bowker
121 Chanon Road
New Providence, NJ 07974
(800) 323-6772

Published annually. A comprehensive guide to corporate America. Available at

larger libraries. Useful for determining subsidiary versus parent relationships of companies, as well as identifying prospects by location.

Donor Briefing
P.O. Box 830350
Birmingham, AL 35283-0350
(205) 995-1588

Biweekly. A newsletter containing digests of articles on development. Selects from a wide assortment of national and local periodicals.

Foundation News
Council on Foundations
1828 L Street, N.W.
Suite 300
Washington, DC 20036
(202) 466-6512

Published annually. A leading publication in the field.

Fund Raising Management
Hoke Communications, Inc.
224 7th Street
Garden City, NY 11530
(516) 746-6700

Published monthly. A leading publication in development, which often contains articles for small organizations.

FRI Monthly Portfolio
The Fund Raising Institute
12300 Twinbrook Parkway
Suite 450
Rockville, MD 20852-9830
(800) 877-TAFT

Published monthly.

Giving Update
American Association of Fund Raising Counsel

Published bimonthly.

Giving USA
American Association of Fund Raising Counsel
25 West 43d Street
New York, NY 10036
(212) 354-5799

Published annually. Compilation of giving statistics, including many breakdowns and cross-tabulations.

Grantsmanship Center News
The Grantsmanship Center
650 South Spring Street
P.O. Box 6210
Los Angeles, CA 90015
(213) 689-9222

Published bimonthly.

The Guide to Federal Funding for Governments and Nonprofits
Government Information Services
1611 North Kent Street
Suite 508
Arlington, VA 22209
(703) 528-1000

Published annually. Contains eligibility guidelines and information on the type of aid offered by individual federal assistance programs.

The HFA Update
National Council of State Housing Agencies
444 North Capitol Street, N.W.
Suite 438
Washington, DC 20001
(202) 624-7710

Published quarterly.

National Guide to Funding in Arts and Culture
The Foundation Center
79 5th Avenue
New York, NY 10003-3076
(212) 620-4230

Published annually. Listing by state of American sources of funding for artistic purposes, a category under which preservation is often evaluated. Provides detailed introduction, glossary, and indices.

Nation's Cities Weekly
National League of Cities
1301 Pennsylvania Avenue, N.W.
Washington, DC 20004
(202) 626-3000

Published weekly.

The NSFRE Journal
The National Society of Fundraising Executives
1101 King Street
Suite 700
Alexandria, VA 22314-2967
(703) 684-0410

Published quarterly.

The NSFRE News
The National Society of Fundraising Executives
1101 King Street
Suite 700
Alexandria, VA 22314-2967
(703) 684-0410

Published eight times a year.

Nonprofit Times
Davis Information Group, Inc.
190 Tamasack Circle
Skillman, NJ 08558
(609) 921-1251

Published monthly. A newspaper-format publication covering issues such as management, cost savings, fund raising, and legislation, often with the small organization in mind. Free to "qualified, full-time nonprofit executives."

Nonprofit World
Society for Nonprofit Organizations
6314 Odana Road
Suite 1
Madison, WI 53719

Published bimonthly. Covers issues such as board development, nonprofit management, and fund raising.

Profile
The National Society of Fundraising Executives
1101 King Street
Suite 700
Alexandria, VA 22314-2967
(703) 684-0410

Published biennially.

Response
The Center for Corporate Public Involvement
1001 Pennsylvania Avenue, N.W.
Washington, DC 20004
(202) 624-2425

Published triennially.

Responsive Philanthropy
National Committee for Responsive Philanthropy
2001 S Street, N.W.
Suite 620
Washington, DC 20009
(202) 387-9177

Published quarterly.

Social Report
The Center for Corporate Public Involvement
1001 Pennsylvania Avenue, N.W.
Washington, DC 20004
(202) 624-2425

Published annually.

Special Events Newsletter
International Events Group

Published biweekly. Focuses exclusively on event marketing and soliciting corporate sponsors for events. Each issue contains examples of recent sponsorships.

The Washington Update
National Council of State Housing Agencies
444 North Capitol Street, N.W.
Suite 438
Washington, DC 20001
(202) 624-7710

Published weekly.

REVOLVING FUNDS

Lynn Moriarity

"A revolving fund is in every sense, 'preservation-as-a-business.' Although it has a physical end product, it is a process, subject to and responsive to changing pressures and diverse needs. Lessons learned from one project will be applied to another, and a learning curve established. Future projects will be made simpler, and the energy expended for their execution will become more efficient."

The Revolving Fund Handbook, 1979

WHAT IS A REVOLVING FUND?

A revolving fund is a pool of capital created and reserved for a specific activity, such as historic preservation, with the restriction that the monies are returned to the fund to be reused for similar activities. A revolving fund can be a highly visible and aggressive tool through which a preservation organization can save endangered properties, initiate the revitalization of an entire historic neighborhood or commercial area, and demonstrate to the community the economic and social benefits of historic preservation.

Generally, there are two major types of real estate activities that funds undertake—acquisition/resale (with or without rehabilitation) and lending. Many funds do both, and most undertake various technical assistance and educational activities as well. Nonprofit preservation revolving funds often use techniques such as rehabilitation agreements, covenants, and easements to ensure appropriate rehabilitation and long-term protection of the properties they assist.

Acquisition/resale funds buy endangered properties (usually those the private sector is unwilling or unable to undertake) and resell them with protective covenants to a sympathetic buyer. The advantage of this type of fund activity is that the preservation organization can step in quickly to buy a historic property to prevent demolition or adverse development.

Lending funds use their own capital, and often borrowed funds from a bank or other lender, to make loans to individuals or organizations seeking to buy and rehabilitate historic properties. The fund acts much like a bank or any other commercial lender, except that its lending will be limited to a specific purpose, its conditions will include preservation controls, and its terms and collateral requirements often will be less stringent. Some funds buy and resell endangered properties, and then lend funds to the new owners to undertake the required rehabilitation.

Revolving funds are tools that can help create partnerships with local government, neighborhood organizations, developers, and others. Simply by having financing available for projects, a revolving fund can influence the future of a historic property or neighborhood. Often the existence of a successful fund, which has developed a track record and built credibility in the community, will be enough to get the preservation organization to the negotiating table where it can help develop a plan to save an important structure and perhaps buy needed time to implement the strategy. Other times, by virtue of the cash or development expertise it can bring to a project either as a participating lender or as a consultant, a revolving fund can ensure that preservation is a part of a larger revitalization effort (for example, a rehabilitation program run by a city agency for a low-income neighborhood).

STEPS IN ESTABLISHING A REVOLVING FUND

Although the order in which these steps are taken may vary, nonprofit organizations seeking to establish revolving funds will need to complete all of the steps in order to design an effective fund.

1. **Assess the preservation needs of the community and its support for preservation.** Evaluate whether a revolving fund is the appropriate tool to meet those needs. Is there financing available for property owners to conduct rehabilitation work? Can such financing opportunities be arranged by educating local lending institutions about the needs of owners or potential buyers of historic properties? If not, perhaps the proposed fund should lend money for that purpose. Are important historic buildings being demolished to make way for other development or to remove eyesores or social problems such as crack-cocaine houses? If zoning or other planning mechanisms have failed to halt such losses, maybe a fund is needed to buy the endangered properties and ensure their protection. The process of assessing the community's needs provides a wonderful opportunity for preservationists to educate the community about historic preservation. It may be very difficult, however, to establish a successful fund, regardless of the need for one, if the community does not understand that historic properties are assets and if local government and corporate support is lacking.

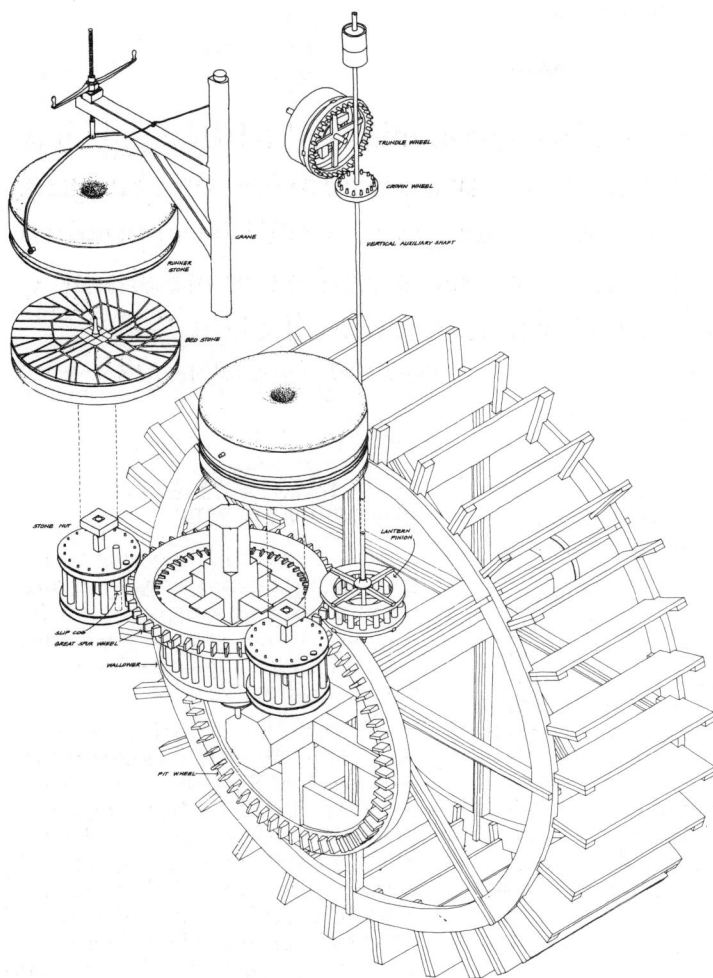

Van Wyck-Lefferts Tidemill, Huntington, Long Island, New York. (K. Hoeft, HAER)

2. **Select a governing board and determine the fund's mission, goals, and priorities.** Although the fund will need access to a variety of professionals (lawyers, accountants, developers, architects, bankers), the governing board should be composed primarily of people who have a strong interest in preservation. Nonprofit revolving funds are risky ventures by nature, and the board must be willing to define and then to accept an appropriate level of risk to accomplish its goals. A fund that targets a well-defined, limited geographic area or type of building—with flexibility to intervene on behalf of important, endangered landmarks—often works best because visible progress can be shown more

quickly. A fund that attempts to meet all the preservation needs of its community usually will not be able to meet any of them very well. Review the community's historic resource surveys and planning documents, and talk with public officials, community leaders, and neighborhood and other nonprofit organizations to choose a target area or type of property that will provide maximum positive publicity and public support. Defining the fund's goals and limits will help build its reputation and will attract the necessary financial and technical resources to ensure the fund's success. When the fund is operating successfully, it can be expanded to include other areas. If the fund will buy endangered properties, poten-

tially threatened buildings should be identified and tracked to minimize the risks associated with sudden demolition threats. If the fund will target a specific geographic area, a complete survey of the area's historic resources should be done so that potential projects can be evaluated wisely.

3. **Determine what type of fund is most appropriate.** Most funds concentrate their efforts on either lending or acquisition/resale activities, but maintain enough flexibility to meet the needs of any given situation. Decide if the fund will work principally in residential areas or in commercial areas. Lending funds usually work best in targeted areas where they are one component of a comprehensive revitalization strategy.

4. **Identify the appropriate organization to manage the fund, form the legal entity, determine staff and consultant needs, and establish accounting and management policies.** The revolving fund can be managed by a new nonprofit organization, an existing one, a government agency, or by some combination. New nonprofit organizations should apply for tax exemption under section 501(c)(3) of the Internal Revenue Code. Enlist the help of an attorney who is familiar with nonprofit organizations to draft articles of incorporation and bylaws and to incorporate the organization or amend the existing articles and bylaws as needed. Find an accountant to set up an accounting system for the fund that meets established practices and is easily understandable to the board of directors, the staff, and potential donors. Determine what the staffing requirements will be and what consulting assistance will be needed. Seek assistance from experienced lenders who can provide advice or perhaps service loans. Determine who will be responsible for negotiating deals, overseeing rehabilitation work, marketing properties, and collecting payments.

5. **Develop a fund-raising plan and timetable.** Money will need to be raised to cover start-up and administrative costs and to establish the

corpus of the revolving fund. Two budgets may be required—one for administrative costs (salaries, rent, telephone, office supplies, equipment, etc.) and one for revolving fund projects. A fund-raising consultant who knows the potential donors in the area can be useful at this stage. The consultant can help the organization's board of directors develop a concise statement about the need for the fund and what its short- and long-term goals will be. He or she can also help establish a preliminary budget. Most importantly, a good consultant can investigate informally the likelihood of donations from various sources before a capital campaign is announced. It is best to have several well-known donors already on board before announcing the organization's optimistic but not impossible fund-raising goal.

6. **Form a revolving fund committee and establish policies and guidelines for fund management and decision making.** Form a revolving fund committee unless the governing board plans to serve in this capacity as well. Membership of the committee should be similar to that of the governing board and include people with expertise in the fields of law, architecture, contracting, real estate, and finance who are also preservationists willing to take risks. The committee sets guidelines for the types of decisions to be made by the director and the kinds that will require review by the committee or by the board. Revolving fund activities often require quick action, so the decision-making process should not be too cumbersome. In addition, it is wise to keep in mind that real estate projects are not static. Changes in things like budgets, funding sources, design, and lease-up are very common throughout the development process. There should be enough flexibility in the fund's chain of command to accommodate some changes as the deal proceeds, without going back to the loan committee or the board each time.

7. **Establish eligibility criteria, type of financial and technical assistance, lending guidelines, and terms and conditions for assis-** tance. Eligibility criteria should include type of property to be assisted and its historic significance, its geographic area, and its degree of endangerment. Acceptable design standards (such as the Secretary of the Interior's Standards for Rehabilitation or Standards for Historic Preservation Projects) must be adopted for rehabilitation and restoration work financed or undertaken by the fund, and an appropriate body selected to review plans and specifications. If protective preservation easements will be required (beware that such a requirement might negate the tax deductibility of an easement donation), the form of the easement must be developed and the appropriate easement holding organization identified. Determine the degree to which technical assistance, including financial packaging, design, maintenance advice, and budgeting, will be included with the financing package offered by the fund.

For lending funds, eligible borrowers, type of collateral, term and interest rate, and other conditions must be determined. Remember that while standard lending criteria should serve as a guide to the conditions for lending capital, most preservation revolving funds are willing to take greater risks (less collateral, lower income borrowers, sweat equity in lieu of cash equity, etc.) than a typical private lending institution. In addition the fund may offer borrowers other types of assistance, including lines of credit, guarantees, interest subsidies, grants, or forgivable predevelopment loans. To ensure that they are used, lending funds must be easy to understand and the terms offered must be attractive. Acquisition/resale funds can use a variety of tools to accomplish their goals such as fee simple purchases, options, purchase agreements, rights of first refusal, and long-term leases.

8. **Establish evaluation criteria for potential projects.** Criteria should be developed to evaluate both the financial feasibility and preservation benefits of undertaking each potential project. Thought should be given to the possible public rela- tions benefits, educational value, spinoff effects, and other intangible benefits of a project in addition to its economic and preservation considerations. Each project should be evaluated against the overall goals and objectives of the fund. Specifically, consider the architectural and historical significance of the property, its contribution to the community, possible threats to the property, the appropriateness of the proposed use and rehabilitation work, the amount of staff time that will be required to manage the project, the length of time fund monies will be tied up, and the feasibility of the financing plan.

For lending programs, brief letters of intent or preapplications can be used to eliminate ineligible or inappropriate projects before too much time is spent by the applicant or the loan fund. If the project looks good, a final application can be completed. An assessment must be made as to the creditworthiness of the borrower, his or her net worth and track record, the sources identified for loan repayment, and the strength of the proposed collateral (including whether the fund would foreclose in the event of a default). An acquisition/resale fund must consider the likelihood of finding an appropriate buyer for the property, the probable time involved in and cost of marketing the property, the types of uses to which the property lends itself, and the impact of all of these factors on recovery of the fund's investment.

9. **Promote the availability of the fund's programs.** Let prospective sellers, buyers, donors, and borrowers know that the fund is in place and what it can offer. Hold informational programs for potential borrowers. Inform public officials of the availability of funds and ask for their help in locating potential projects. Keep the program in front of the public at every opportunity.

10. **Publicize the fund's accomplishments.** Always maintain good before- and after-rehabilitation slides of every property assisted by the fund to help publicize the results and educate the public about preservation. Statistics

should be kept on such facts as the number of properties bought, sold, and assisted with loans, and the total cost of each project. If funding needs are greater than available resources, keep track of the number of applications and amount of funds turned down. Donors and members will want to know what has been accomplished with their funds, and a good track record must be documented if future fund-raising efforts are to be successful. Consider a newsletter or periodic newspaper column to publicize the results of the fund's projects.

Other Things to Consider
Revolving fund activity should be closely tied to good preservation planning. No project should be accepted without first evaluating whether it meets the long-term preservation goals of the organization and whether the revolving fund is the best tool to address the problem at hand. Although many projects undertaken by revolving funds are risky (since presumably the private sector was unwilling or unable to do the job), the risk can be minimized with good planning and evaluation before the project is approved.

A fund should begin work in a targeted neighborhood or commercial area only after doing a lot of educational and outreach work to determine the needs of the area and its residents and to inform them about the benefits of preservation. Each program must be tailored specifically to the needs of the community it will serve. Revolving funds work best if the residents and local neighborhood organizations have helped design them and have a stake in their successes. If one of the goals of a fund is to prevent displacement in an area experiencing increased housing costs, the fund may need to acquire options on many properties quietly before any public strategy is announced.

A revolving fund will be in a much better position to take risks and to properly evaluate each opportunity if it has a secure source of income from which it can pay its operating costs. This will help ensure that the fund's capital is not reduced each year to cover operating deficits

and that the fund will not have to charge high interest rates on loans or reject risky projects to protect its assets. It also can assure potential donors that all their funds will be used to preserve properties, not to keep the organization alive.

Real estate projects take time. It can be several years or more between the time a fund manager first evaluates a project and the time the property finally is sold or the rehabilitation work is complete. The fund manager and the board must be patient and prepared to tie up the fund's money for long periods of time if that is what is needed to make a project work.

The organization operating the revolving fund should expect some of its loans to go into default or some of its acquisitions to take longer than anticipated to resell. Both of these conditions require that there be sufficient reserve capital so as not to deplete the revolving fund and a plan to raise additional funds as they are needed.

ACTIVE REVOLVING FUNDS

The following is a list of organizations that operate active preservation revolving funds. Loan pools and other funds that do not revolve are not included. Each entry includes a list of codes that indicate the **primary** activities of each fund. Revolving fund activity codes are as follows:

Geographic Focus of Fund
S   Statewide
CI   Citywide
CO   Countywide
N   Neighborhood
D   Downtown Area
X   Other

Type of Property Assisted
CM   Commercial
RS   Residential
XX   Other

Type of Activity
L   Loans
A/R   Acquisition/resale
A/R/R   Acquisition/rehab/resale
O   Options
Z   Other

Source of Start-up Funds
1   Federal
2   State

3   Local
4   National Trust
5   Foundations
6   Individual/membership donations
7   Fund raising
8   Sale of donated property
9   Other

Sources of start-up funds for the preservation revolving funds listed below were dependent on a number of factors, including uses of the funds (acquisition/resale, rehabilitation loans, loan guaranties); locations and types of properties targeted (commercial districts and buildings, low-income housing); lending criteria (interest rates, terms, collateral requirements); and the date the fund was established. For example, some of the preservation revolving funds established in the 1970s used Urban Development Action Grant (UDAG) repayments to local governments as start-up capital. Since UDAG funds no longer are available, the use of their repayments to start a revolving fund is highly unlikely.

ALABAMA

Mobile Historic Development Commission
P.O. Box 1827
Mobile, AL 36633
Mark McDonald, Director
(205) 438-7281
CO; RS, CM; A/R/R, L; 3

ALASKA

Anchorage Historic Properties
524 West 4th
Anchorage, AK 99501
Kerry Hoffman
(907) 274-3600
CI; CM, RS; L, A/R; 3

ARIZONA

City of Phoenix
125 East Washington
3d Floor
Phoenix, AZ 85002
Deborah Edge Abele, City Historic Preservation Officer
(602) 261-8699
CI; CM, RS; A/R, Z (grants through easement purchases); 3

ARKANSAS

Little Rock Revolving Fund
200 West Capital
Suite 1600
Little Rock, AR 72201
Mark Nichols, President
(501) 372-5659
CI; CM; A/R; 7

CALIFORNIA

Pasadena Heritage
80 West Dayton Street
Pasadena, CA 91105
Elizabeth Neaves
(818) 793-0617
CI; RS; A/R/R; 2

COLORADO

Historic Boulder
1733 Canyon Boulevard
Boulder, CO 80302
Judith Trent
(303) 444-5192
CI; CM, RS, OS; L; 8

Historic Denver
1330 17th Street
Denver, CO 80202
Kathleen Brooker, President
(303) 534-1858
D; XX; L; 3, 4

CONNECTICUT

Connecticut Trust for Historic Preservation
Connecticut Endangered Properties Fund
940 Whitney Avenue
Hamden, CT 06517
Christopher Wigren, Project Manager
(203) 562-6312
S; RS; L; 2, 5

DISTRICT OF COLUMBIA

Neighborhood Housing Services
300 I Street, N.E.
Suite 2
Washington, DC 20002
Dave Wiley
(202) 544-3298
N; RS; L; 5

FLORIDA

Dade Heritage Trust
190 S.E. 12 Terrace
Miami, FL 33131
Louise Yarbrough, Executive Director
(305) 358-9572
CO; RS; L; 3

Florida Trust for Historic Preservation
P.O. Box 11206
Tallahassee, FL 32302
(904) 224-8128
S; CM; A/R; 2, 5

Jacksonville Neighborhood Housing Services
P.O. Box 3386
Jacksonville, FL 32206
Ann Berman, Project Manager
(904) 355-1248
N; RS; A/R/R; 4, 5

Tampa Preservation, Inc
2007 North 18th Street
Tampa, FL 33605
Harriet Plyler, Revolving Fund Chairman
(813) 248-5437
CI; RS; A/R/R; 9(resale of property)

GEORGIA

Georgia Trust for Historic Preservation
1516 Peachtree Street, N.W.
Atlanta, GA 30309
Elizabeth Bryce Bell, Program Coordinator
(404) 881-9980
S; RS; O, A/R; 7

Historic Savannah Foundation
Box 1733
Savannah, GA 31402
Stephanie Churchill, Executive Director
(912) 233-7787
CI; RS; A/R; 6

Macon Heritage Foundation
P.O. Box 6092
Macon, GA 30128
Maryel Battin, Executive Director
(912) 742-5084
CO; RS; A/R, O; 1, 6, 7

ILLINOIS

Landmarks Preservation Council of Illinois Preservation Fund
53 West Jackson Boulevard
Suite 752
Chicago, IL 60604
Fred Nash, Preservation Fund and Easement Director
(312) 922-1742
S; RS; L, Z (grants); 8

INDIANA

Bloomington Restorations
P.O. Box 1522
Bloomington, IN 47402
Roy Morgan, Director
(812) 336-0909
CI; RS; L; 3, 9 (statewide preservation organization)

Historic Landmarks Foundation of Indiana Fund for Landmark Indianapolis Properties
1028 North Delaware
Indianapolis, IN 46208
David Frederick, Director
(317) 638-5264
CI; RS, CM; A/R; 9 (project by Junior League)

Historic Landmarks Foundation of Indiana Statewide Revolving Loan Fund
340 West Michigan
Indianapolis, IN 46202
Jim Connelly
(317) 639-4534
S; RS; L; 9 (endowment of HLFI)

South Bend Heritage Foundation
South Bend Heritage Fund
914 Lincoln Way West
South Bend, IN 46616
Jeff Gibney
(219) 289-1066
N; RS, CM, XX (rental); L, A/R, Z (acquisition/rental and loan guarantees); 3, 4

IOWA

Sherman Hill Association
756 16th Street
Des Moines, IA 50314
Jack Porter
(515) 284-5717
N; RS; A/R; 3

KANSAS

Lawrence Preservation Alliance
P.O. Box 1073
Lawrence, KS 66044
Nancy Shontz
CI; RS; A/R, A/R/R; 6

KENTUCKY

Blue Grass Trust for Historic Preservation
253 Market Street
Lexington, KY 40507
Patrick Lucas, Executive Director
(606) 253-0362
CO; RS; A/R/R; 4, 6

Growth
417 South 4th Street
Paducah, KY 42001
Richard Holland, Executive Director
(502) 443-9284
CI; CM; L; 2

LOUISIANA

Preservation Resource Center
604 Julia Street
New Orleans, LA 70130
Patty Gay, Director
(504) 581-7032
CI; RS; L; 6

MARYLAND

Historic Annapolis Foundation
Port of Annapolis
194 Prince George Street
Annapolis, MD 21401
Sarah Filkins
(410) 267-7619
CI; CM, RS; A/R; 5, 6

Maryland Historical Trust
100 Community Place
Crownsville, MD 21032
Charity Davidson
(410) 514-7600
S; CM, RS; L; 2

Preservation Maryland
24 West Saratoga Street
Baltimore, MD 21201
David Chase, Executive Director
(410) 685-2886
S; XX (vacant and endangered historic properties); A/R, L; 8

MASSACHUSETTS

Architectural Conservation Trust (ACT) for Massachusetts
45 School Street
Boston, MA 02108
William Steelman, Preservation Fund Coordinator
(617) 523-8678
S; CM, XX (religious); L; 1, 3, 4

Historic Boston
3 School Street
Boston, MA 02108
Henry Moss, Project Manager
(617) 227-4679
CI; CM, XX (religious); L, Z (acquisition/rental); 9 (rent from rehabilitated property)

Springfield Preservation Trust
979 Main Street
Springfield, MA 01103
Bob McCarroll
(617) 787-6020
CI; RS; A/R

Waterfront Historic Area League
13 Centre Street
New Bedford, MA 02740
Therese Kelly, Director
(617) 996-6912
CI; CM, RS; A/R

MISSISSIPPI

Historic Natchez Foundation
P.O. Box 1761
Natchez, MS 39121
Ron Miller, Director
(601) 442-2500
CI; RS; A/R; 9 (overall organization)

MISSOURI

Missouri State Historic Preservation Officer
Missouri Historic Preservation Revolving Fund
Department of Natural Resources
Historic Preservation Division
P.O. Box 176
Jefferson City, MO 65102
Jane Beetum, Director
(314) 751-5373
S; RS, CM; L, A/R; 2

NEW HAMPSHIRE

Historic Harrisville
Box 79
Harrisville, NH 03450
John J. Colony III
(603) 827-3722
X (historic district); RS; A/RR; 4, 7, 9 (sale of property)

New Hampshire Charitable Fund
P.O. Box 1335
Concord, NH 03032
Debra Cowan, Associate Director
(603) 225-6641
S; Unrestricted; L, O; 5

NEW JERSEY

Historical Society of Princeton Preservation Revolving Fund
158 Nassau Street
Princeton, NJ 08542
Emily Wallace, Director
(609) 921-6748
CI; CM, RS; L, A/R; 7

New Jersey Historic Trust
CN 404
Trenton, NJ 08625
Harriette Hawkins, Executive Director
(609) 984-0473
S; L

NEW MEXICO

State of New Mexico
Historic Preservation Division
Office of Cultural Affairs
228 East Palace Avenue
Santa Fe, NM 87503
Tom Merlan, SHPO
(505) 827-6320
S; Unrestricted; L; 2, 4

NEW YORK

Broadway-Filmore Neighborhood Housing Service
780 Filmore Avenue
Buffalo, NY 14212
Bob Siencewicz, Director
(716) 852-3130
N; RS; L; 1

City of Corning
Corning Urban Renewal Agency
7 Civic Center
Corning, NY 14830
Richard Folnsbee, Director
(607) 962-0721
CI; CM; L; 3

Geneva Historical Society
Fund for Historic Geneva
543 South Main Street
Geneva, NY 14456
Jack Mulvey, Chairman
(315) 789-5151
CI; CM, RS, XX (religious); L; 6

Historic Ithaca
Tompkins County Historic Preservation Loan Fund
120 North Cayuga Street
Ithaca, NY 14850
(607) 273-6633
CO; CM, RS, XX(religious); L; 8

Landmark Society of Western New York
133 South Fitzhugh Street
Rochester, NY 14608
Henry McCartney, Executive Director
(716) 546-7029
X(regional); CM, RS; L, A/R; 6, 7

New York Landmarks Conservancy
New York City Historic Properties Fund
141 5th Avenue
New York, NY 10010
Karen Ansis, Manager
(212) 995-5260
CI; CM, RS, XX (religious); L, Z (grants); 9 (rent from federal surplus building)

Preservation Association of Central New York
1509 Park Street
Syracuse, NY 13208
Peter Auyer, Executive Director
(315) 475-0119
CI; RS, CM; L; 9 (corporate donation)

NORTH CAROLINA

Capital Area Preservation
1 Mimosa Street
Raleigh, NC 27604
Sally Poland, Director
(919) 834-4844
CI; CM; A/R; 2

Historic Preservation Foundation of North Carolina
P.O. Box 27644
Raleigh, NC 27611
Myrick Howard, Executive Director
(919) 832-3652
S; Unrestricted; A/R, O; 5

Historic Salisbury Foundation
P.O. Box 4221
Salisbury, NC 28144
(704) 636-0103
CI; RS, CM; A/R, O; 9 (sale of optioned property)

Historic Wilmington Foundation
209 Dock Street
Wilmington, NC 28401
David Scott, Executive Director
(919) 762-2511
CO; RS; A/R; 7, 8

New Bern Preservation Foundation
P.O. Box 207
New Bern, NC 28563
Barbara Howlett, Director
(919) 633-6448
CI; RS; A/R; 5, 6

Preservation Society of Asheville and Buncombe County
P.O. Box 2806
Asheville, NC 28802
Harry Weiss, Director
(704) 253-1124
CO; CM; A/R; 8

OHIO

Miami Purchase Preservation Fund
727 Ezzard Charles Drive
Cincinnati, OH 45203
Linda Fabe
(513) 241-0504
CI; RS; A/R; 9 (overall organization)

OREGON

Baker City Economic Development Department
P.O. Box 650
Baker City, OR 97814
Brian Cole, Director
(503) 523-6541
D; CM; L, Z (grants); 2, 3, 4, 9 (bank loans)

PENNSYLVANIA

Pittsburgh History and Landmarks Foundation
400 Landmarks Building
1 Station Square
Pittsburgh, PA 15219
Stanley Lowe, Assistant Director
(412) 471-5808
CI; CM, RS; L, Z (loan guarantees); 5, 6, 8

Preservation Pennsylvania
2470 Kissel Hill Road
Lancaster, PA 17601
Grace Gary, Executive Director
(717) 569-2243
S; RS; L; 2

RHODE ISLAND

Providence Preservation Society Revolving Fund
24 Meeting Street
Providence, RI 02903
Clark Schoettle, Executive Director
(401) 272-2760
CI; Unrestricted; A/R, L; 3, 6

SOUTH CAROLINA

Historic Beaufort Foundation
P.O. Box 11
Beaufort, SC 22901
Cindy Cole
(803) 524-6334
CI; CM, RS; A/R; 7

The Palmetto Trust for Historic Preservation
P.O. Box 12547
Columbia, SC 29211
Bob Bainbridge, President
(803) 656-4094
S; CM, RS; A/R/R, O; 2, 3, 8

SOUTH DAKOTA

Historic Preservation Commission
644 Main Street
Deadwood, SD 57732
Joy McCracken
(605) 578-2082
CI; CM, RS; L; 3

Historic South Dakota Foundation
P.O. Box 2998
Rapid City, SD 57709
James Quinn
(605) 341-5820
S, CI; RS, CM; L; 2

TENNESSEE

City of Knoxville
Department of Community Development
P.O. Box 1631
Knoxville, TN 37901
Diana Gerard, Manager, Housing and Development
(615) 521-2120
X (historic districts); RS; L; 3, 4, 7

TEXAS

Galveston Historical Foundation
2016 The Strand
Galveston, TX 77550
Betty Massey
(409) 765-7834
CI; CM, RS; A/R, A/R/R; 5, 8

UTAH

Park City Historic District Incentives Program
445 Marsac Avenue
P.O. Box 148
Park City, UT 84060-1480
Suzanne McIntire, Senior Planner
(801) 645-5021
D; CM, RS; Z (grants); 9 (funds from city bond)

Utah Heritage Foundation
355 Quince Street
Salt Lake City, UT 84103
Michael Leventhal
(801) 533-0858
S, CI; RS; A/R; 3, 4, 6, 8

VIRGINIA

Historic Petersburg Foundation
215 North Market Street
Petersburg, VA 23803
(804) 732-2096
X (historic districts); Unrestricted; A/R; **6**

Historic Richmond Foundation
2407 East Grace Street
Richmond, VA 23233
Jack Zehmer
(804) 643-7407
CI; CM, RS; A/R; 8

Preservation of Historic Winchester
2 North Cameron Street
Winchester, VA 22601
Anna Thomson, Executive Director
(703) 667-3577
CI; RS, CM; A/R; 4, 7

Virginia Historic Preservation Foundation
P.O. Box 295
Charlottesville, VA 22902
David Brown, Chairman
(804) 979-3899
S; RS, CM; A/R; 2

WEST VIRGINIA

Regional Planning and Development Council
P.O. Box 1442
Princeton, WV 24740
Norman Kirkham
(304) 425-9508
X(regional); CM; L; 1, 2, 9(local banks)

WISCONSIN

Wisconsin Preservation Fund
P.O. Box 93336
Milwaukee, WI 53203-0336
Trent B. Chambers, President
(414) 374-9734
CI; CM, RS; A/R/R; 1, 5, 6, 9 (fees and income from major revenue bond projects)

FURTHER READING

Commercial Area Revolving Funds for Preservation. Peter H. Brink. Information Series, National Trust for Historic Preservation. Washington, D.C.: Preservation Press, 1976.

Handbook on Revolving Funds for Nonprofit Historic Preservation Organizations. Raleigh, N.C.: Historic Preservation Foundation of North Carolina and Young Lawyers Division of the North Carolina Bar Association, 1987.

Protecting and Preserving Communities: Preservation Revolving Funds. Susan J. Lutzker and Lyn Howell Moriarity, eds. Washington, D.C.: National Trust for Historic Preservation, 1991.

Using a Revolving Fund for Historic Preservation. J. Myrick Howard. Washington, D.C.: National Trust for Historic Preservation, 1988.

TAX INCENTIVES

Paul Edmondson and Constance Beaumont

"The power to tax involves the power to destroy."

Chief Justice John Marshall, 1819

"In preservation terms, the question is whether the potentially destructive power of taxation of which Justice Marshall spoke can be reversed to become a positive power for the protection of historic buildings."

Robert E. Stipe, *Tax Incentives for Historic Preservation*, 1981

Some states and local governments provide reductions in property and other taxes to encourage preservation of historic properties, and several federal tax incentives are available. The two most widely used federal incentives are the rehabilitation tax credit and the charitable contribution deduction.

REHABILITATION TAX CREDIT

The Tax Reform Act of 1986 permits owners and some lessees of historic buildings to take a 20 percent income tax credit on the cost of rehabilitating such buildings for industrial, commercial, or rental residential purposes. The law also permits depreciation of such improvements over 27.5 years for a rental residential property and over 31.5 years for a nonresidential property. The rehabilitated building must be a certified historic structure that is subject to depreciation, and the rehabilitation must be certified as meeting standards established by the National Park Service.

What is a certified historic structure?

A certified historic structure is one that is

- individually listed in the National Register of Historic Places, or
- certified by the National Park Service as contributing to a registered district. A registered district is one that is listed in the National Register or designated under a state or local statute that has been certified to contain criteria that will substantially achieve the purpose of preserving and rehabilitating buildings of significance to the district and that is certified as substantially meeting all of the requirements for listing of districts in the National Register.

What is a certified rehabilitation?

A certified rehabilitation is a rehabilitation certified by the National Park Service as being consistent with the historic character of the property and, where applicable, the district in which it is located. The National Park Service refers to the Secretary of the Interior's Standards for Rehabilitation in certifying rehabilitations.

How is a building or rehabilitation certified?

The National Park Service requires that owners complete a special form, the Historic Preservation Certification Application (Form-10-168), for all certification requests. The form is divided into three parts: Part 1 for evaluating the historical significance of a building, Part 2 for describing rehabilitation work, and Part 3 requesting certification of completed work. All applications are submitted to, and reviewed by, the state historic preservation office before submission to the National Park Service, which makes the final certification decision after considering the SHPO's recommendations.

Part 1 evaluations need not be prepared for buildings already individually listed in the National Register. A building within a district must be certified, based on review of a Part 1 application, as contributing to the significance of that district. Part 1 applications are also used to gain preliminary determinations of significance for individual buildings not yet listed in the register; these determinations become final when such buildings are actually listed.

Proposed rehabilitation is described in Part 2 of the application form. The National Park Service issues a preliminary approval of proposed work for projects that, as described, meet the Secretary's Standards. The preliminary approval becomes final when the work is completed and the National Park Service can certify that the standards have, in fact, been met.

Is a fee charged for certification?

The National Park Service charges a fee for reviewing rehabilitation certification requests, based on the cost of rehabilitation.

Where can application forms be obtained?

Historic Preservation Certification Application forms are available from SHPOs and from National Park Service regional offices.

How is a state law, local ordinance, or local historic district certified?

State and local governments that wish to have state laws, local ordinances, or local historic districts not included in the National Register certified so that the rehabilitation of contributing buildings may qualify for the federal tax credit should contact their SHPOs for assistance in preparing and submitting requests to the National Park Service for such certification.

REHABILITATION CREDIT PLANNING CHECKLIST

1. Determine whether the building is listed in the National Register of Historic Places or is located in a registered historic district.

2. If it is not listed in the National Register, determine whether the property is in a historic district created under state and local statutes. If so, determine whether the statute has been certified by the secretary of the interior.

3. Determine from the certificate of occupancy, county assessor's office or other authority when the building was first placed in service. Ordinarily, buildings that have been built within the past 50 years are not considered as contributing to the significance of a district.

4. Prepare plans for and estimate the cost of rehabilitation and compare it to the purchase price or other adjusted basis of the building to be certain that the rehabilitation expenditures will be "substantial." Before starting work, have the state historic preservation office review the plans to make sure that the standards and guidelines are being followed.

5. Be sure that at least 50 percent of the external walls will be retained as external walls on completion of the rehabilitation.

6. Determine that the intended use of the building will not be for residential purposes unless it is a certified historic structure.

7. To qualify for certified rehabilitation of a certified historic structure, complete a Historic Preservation Certification Application to secure certification from the secretary of the interior that the building is historic and that the rehabilitation work is in accordance with the Secretary of the Interior's Standards for Rehabilitation.

8. If the property is located in and has historical significance to a registered historic district, obtain from the Secretary of the Interior certification that the planned rehabilitation is consistent with the historic character of the building or the district.

9. If the building is located in a historic district designated by state or local statute but is not certified by the secretary of the interior, apply for certification of the statute if certified historic structure status is desired.

STATE TAX INCENTIVES

Although the definition of "historic" varies among the states, in general, these state programs require that properties be listed in the National Register of Historic Places or designated by either the state or local government.

Alabama
Property tax abatements for commercial historic properties, with properties assessed at 10 percent of their appraised values, as opposed to 20 percent (the percentage applied to all other commercial properties). No time limit is imposed on the abatement. The state also provides low-interest loans through tax-exempt bonds to finance historic property rehabilitation. No tax is levied for 20 years, and rehabilitations are exempt from sales tax.

Alaska
No special incentives.

Arizona
Reduced property tax assessments (based on 5 percent of property's cash value) for historic residential properties for up to 15 years. Non-income-producing archeological sites also eligible. Program is administered statewide rather than at local option.

Arkansas
No tax incentives. Model business grants available to historic property owners through state grant program for historic and cultural resources. Grants funded by real estate transfer tax.

California
Reduced property tax assessments available to historic property owners who sign "Mills Act" contracts foregoing future development rights.

Colorado
State income tax credit of 20 percent for rehabilitated residential or commercial historic properties.

Connecticut
Tax assessors must consider effects of any property restrictions imposed by historic easements.

Delaware
No special incentives.

District of Columbia
Tax assessments of historic properties may be based on actual, rather than "highest and best," use. To obtain reduced assessments, property owners must sign 20-year covenants guaranteeing a property's maintenance.

Bradbury Building, Los Angeles, California. (R. Giebner, HABS)

Florida
Referendum approved authorizing local governments throughout the state to provide property tax abatement to historic property owners. Up to 100 percent of the rehabilitation improvements to a historic property can be forgiven for up to 10 years. All property taxes, not merely those on the improvements, may be forgiven for properties open to the public on a regular basis. Owner-occupied residences as well as commercial structures may qualify. In addition, Florida corporations that donate to community development projects, including historic preservation, may claim state income tax credits equal to 50 percent of the donation. Credit is limited to $200,000 annually.

Georgia
Eight-year freeze on property tax assessments of rehabilitated historic structures. Rehabilitation expenses must equal 50 percent of a residential property's appraised value or 100 percent of a commercial property's appraised value. After the eighth year of a freeze, property taxes on market values are phased in; the abatement ends in the 10th year. The program is run by the state and is not a local option.

Hawaii
Property tax exemptions for up to 10 years for historic properties if owner agrees not to change land use or violate other standards.

Idaho
Tax assessors must consider reductions in property values caused by easements on historic properties.

Illinois
Eight-year freeze on property tax increases otherwise triggered by rehabilitations. Such expenses must equal 25 percent of property's value. Available to owner-occupied historic residences.

Indiana
Five-year tax exemption from increase in assessed property values attributable to rehabilitations. Limit of $2,500 per residential unit.

Iowa
Eight-year property tax abatement for historic properties if rehabilitation expenses of at least $5,000 are incurred.

Kansas
No special incentives. Rehabilitation grants available through Kansas Heritage Fund.

Kentucky
Residential properties of 25 years are eligible for five-year suspensions of property tax increases attributable to improvements. Properties must be located in "neighborhood improvement zones."

Louisiana
Up to a 10-year freeze on property tax increases for rehabilitated historic properties. Rehabilitation expenses on residential properties must equal 25 percent of property's assessed valuation; expenses on commercial properties not required to meet expenditure threshold. Program is a local option.

Maine
No special incentives.

Maryland
State income tax deductions of up to 100 percent on the costs of certified rehabilitations of historic properties. Deductions amortized over five years.

Massachusetts
No special incentives.

Michigan
No special incentives.

Minnesota
No special incentives.

Mississippi
Seven-year property tax abatement for rehabilitated historic properties. Residential and commercial structures eligible. A local option.

Missouri
Tax abatements available for rehabilitated properties in designated blighted areas.

Montana
Five-year property tax freezes on increased valuations otherwise resulting from improvements to historic properties. Limited to owner-occupied residences. A local option.

Nebraska
Restrictions on properties encumbered by easements must be considered by tax assessors.

Nevada
No special incentives.

New Hampshire
No special incentives.

New Jersey
Five-year deferrals of property tax increases due to improvements on houses at least 20 years old. Preservation grants available through Green Acres bond program.

New Mexico
State income tax credit of up to $25,000 for owners of rehabilitated historic properties. Residential and commercial structures eligible. Credit may be carried forward up to four years. Archeological site owners also eligible.

New York
Tax assessors must consider impact of easement restrictions on historic property values in calculating assessments.

North Carolina
Assessors must consider property value reductions attributable to easement restrictions placed on historic properties. Also, property taxes on historic properties may be based on 50 percent of a property's market value. Program is a local option. Program is unusual in that it does not require any rehabilitation or maintenance.

North Dakota
Three-year property tax abatement for residential structures at least 25 years old.

Ohio
Assessors must consider property value decreases attributable to restrictions on historic properties imposed by easements.

Oklahoma
A local development act approved in 1992 authorizes property tax and sales tax exemptions in designated historic districts or "reinvestment areas."

Oregon
Freeze of 15 years on property tax assessments for owner-occupied historic residences. Benefit applies

statewide. Program is currently being reexamined.

Pennsylvania
No special incentives.

Rhode Island
State income tax credit of 10 percent available to owner-occupied historic residences. Value of credit unused in one year may be carried forward indefinitely.

South Carolina
Two-year property tax freeze, followed by six-year tax abatement, for substantially rehabilitated historic properties. Residential and commercial structures eligible. A local option.

South Dakota
Five-year tax moratorium on increased valuations for rehabilitated historic properties. Also, tax assessors must consider effect of easements on property values.

Tennessee
Tax assessors must consider effect of historic easements on property values.

Texas
Constitutional amendment permits localities to devise various property tax abatement programs for historic properties.

Utah
Property owners may apply for property reappraisals to ensure that such appraisals reflect any devaluations caused by historic easement restrictions.

Vermont
Five-year property tax exemption for improvements made to residential buildings located on land unoccupied for two preceding years.

Virginia
Five-year property tax exemption available to taxpayers who improve residential structures at least 25 years old. Also, tax credits available to businesses and corporations when they contribute to neighborhood revitalization projects, including historic preservation. Credit limited to $175,000 a year but may be carried forward up to five years.

Washington
Ten-year property tax abatement for rehabilitated historic properties. A local option. Also, historic properties may be taxed according to actual, rather than "highest and best," use.

West Virginia
Income tax credit of 10 percent for the rehabilitation of historic residential and nonresidential structures. Also, tax assessors must consider impact of historic easement in making property assessments.

Wisconsin
State income tax credit of 25 percent for substantially rehabilitated, owner-occupied historic residences. Also, a 5 percent credit may be "piggy-backed" onto the federal rehabilitation tax credit of 25 percent available to depreciable properties.

Wyoming
No special incentives.

CHARITABLE CONTRIBUTIONS

Taxpayers may deduct from their federal income tax the value of historically important land areas and certified historic structures donated to governments and other appropriate recipients for preservation purposes. Donations of partial interests in such properties, e.g., easements, are also deductible.

The range of properties on which deductions for donations can be claimed is broader than that on which the rehabilitation credit can be taken. Archeological sites, rural historic districts and other land areas in or eligible for the National Register are included, as are land areas within historic districts and lands adjacent to properties individually listed in the register where such areas contribute to the historical integrity of such properties. Properties do not have to be depreciable for the charitable contribution deduction to be taken.

How does a taxpayer claim a charitable contribution deduction?
The documentation needed to support a claim for a charitable contribution deduction varies, depending on the property, the interest donated, and other factors. SHPOs and National Park Service regional offices can provide assistance in developing the necessary documentation.

Adapted from "Preservation Tax Incentives for Historic Buildings"

CONTACTS

Technical Preservation Services
Preservation Assistance Division
National Park Service
U.S. Department of the Interior
P.O. Box 37127
Washington, DC 20013-7127
(202) 343-9588
H. Ward Jandl, Chief

State Historic Preservation Offices
See listings in Part III.

FURTHER READING

A Guide to Tax-Advantaged Rehabilitation. Sally G. Oldham, Jayne F. Boyle and Stuart M. Ginsberg. Washington, D.C.: National Trust for Historic Preservation, 1986.

Historic Preservation: Law and Taxation. Tersh Boasberg, Thomas A. Coughlin and Julia H. Miller, eds. Albany, N.Y.: Matthew Bender, 1985.

Landmarks Preservation and the Property Tax. David Listokin. New Brunswick, N.J.: Center for Urban Policy and Research, Rutgers University, 1982.

"Preservation Tax Incentives for Historic Buildings." H. Ward Jandl and Patricia L. Parker. Washington, D.C.: U.S. Department of the Interior and National Conference of State Historic Preservation Officers, 1987.

State Tax Incentives for Historic Preservation. Margaret Davis. Washington, D.C.: National Trust for Historic Preservation, 1985.

See also
"Easements"

LEGISLATION

Paul Edmondson

> "It is hereby declared that it is a national policy to preserve for public use historic sites, buildings and objects of national significance for the inspiration and benefit of the people of the United States."

National Historic Sites Act of 1935

FEDERAL STATUTES

National Historic Preservation Act of 1966 (NHPA), as amended in 1980 Section 106, 16 U.S.C. § 470f. Section 106 of the NHPA requires all federal agencies to "take into account" the effects of their undertakings on historic properties, and to consult with the Advisory Council on Historic Preservation, an independent federal agency, before approving any undertaking. "Undertakings" include not only federally sponsored and funded projects, but also private activities and projects that are subject to federal licensing, permitting, or approval. For more information, contact your State Historic Preservation Officer (SHPO) or the Advisory Council on Historic Preservation. The Advisory Council has issued regulations setting forth in detail the Section 106 consultation process. 36 C.F.R. Part 800.

Section 110, 16 U.S.C. § 470h-2. Section 110 of the NHPA imposes additional preservation responsibilities on federal agencies that own or control historic properties. These duties include: assuming responsibility for preserving historic properties; using historic buildings to the maximum extent feasible; inventorying and nominating to the National Register all eligible historic properties owned or controlled by the agency; exercising caution to ensure that historic properties are not "inadvertently transferred, sold, demolished, substantially altered, or allowed to deteriorate significantly"; designating a preservation officer; and minimizing harm to National Historic Landmarks to the "maximum extent possible." The Secretary of the Interior has issued guidelines for the implementation of Section 110 of the NHPA by federal agencies. 53 Federal Register 4727 (Feb. 17, 1988).

Section 4(f) of the Department of Transportation Act of 1966, 49 U.S.C. § 303 Section 4(f) is one of our strongest federal historic preservation laws, prohibiting federal approval or funding of any transportation project that requires the "use" of any historic site, public park, recreation area, or wildlife refuge, unless (1) there is "no feasible and prudent alternative to the project," and (2) the project includes "all possible planning to minimize harm to the project." The term "use" includes not only the direct physical taking of land, but also indirect effects that would "substantially impair" the value of protected sites. *See* 23 C.F.R. § 771.135(p). Section 4(f) applies to all transportation agencies, including the Federal Highway Administration (FHWA), which funds highway and bridge projects, the Federal Aviation Administration (FAA), which approves and funds airport expansions, and the Coast Guard, which owns or operates many historic lighthouses and often has regulatory authority affecting bridges. *See also* 23 U.S.C. § 138 (highways); 49 U.S.C. § 2208(b)(5) (airports).

National Environmental Policy Act of 1969 (NEPA), 42 U.S.C. § 4332(2)(C). NEPA requires the preparation of an Environmental Impact Statement (EIS) for all "major federal actions

Lawyer's Row, Winchester, Virginia. (R. Dunnay, HABS)

significantly affecting the quality of the human environment." This environmental review must include consideration of impacts on "urban quality, historic and cultural resources, and the design of the built environment," along with alternatives to avoid these impacts. Like Section 106 of the NHPA (discussed above), NEPA is applicable to federally sponsored, funded, or licensed projects. The Council on Environmental Quality (CEQ), an independent federal agency, has issued regulations governing the implementation of the NEPA process. 40 C.F.R. Part 1500.

Archaeological Resources Protection Act of 1979 (ARPA), 16 U.S.C. §§ 470aa et seq.

ARPA prohibits the removal, excavation, or alteration of any archeological resource from federal land or Indian lands, in the absence of a permit from the Department of the Interior. Permits will be approved only for research purposes, and all artifacts remain the property of the United States. Selling, purchasing, exchanging, transporting, and trafficking in such artifacts is also prohibited. Violations are subject to civil and criminal penalties.

Native American Graves Protection and Repatriation Act (NAGPRA), 25 U.S.C. § 3001 et seq.; 28 U.S.C. § 1170.

Passed in November 1990, NAGPRA provides for the inventory and repatriation of human remains and Native American cultural items that are presently held by federal agencies and by museums, universities, and state or local agencies that receive federal funds. The law also provides a formal role for Native American tribes and individuals in making decisions about activities on federal and tribal lands that may affect archeological resources of importance to Native Americans. This process requires a determination of the tribe or descendants with the closest cultural affiliation to the site or the artifacts. NAGPRA also imposes civil and criminal penalties for illegally trafficking in Native American cultural items.

Surface Transportation and Uniform Relocation Assistance Act of 1987 (STURAA), 23 U.S.C. § 144(o).

STURAA contains two special provisions protecting historic bridges. First, STURAA requires that, prior to the approval of federal funding for the demolition of any historic bridge, the bridge must first be made available for donation to a "state, locality, or responsible private entity" that would agree to maintain the bridge and assume future responsibility for it. Second, STURAA provides that the Federal Highway Administration must make available funds to reimburse the costs of preserving or protecting a historic bridge that is no longer used for motor vehicle traffic, in an amount up to the estimated cost of demolishing the bridge.

Abandoned Shipwreck Act of 1987, 43 U.S.C. § 2101 et seq.

The Abandoned Shipwreck Act transfers ownership to the states of all historic shipwrecks within state waters (generally defined as extending three miles from the coast or from islands off the coast of the state). This transfer of ownership enables the states to regulate and protect historic shipwrecks from looters or commercial treasure salvors, who may otherwise take artifacts from these sites.

Surface Mining Control and Reclamation Act of 1977 (SMCRA), 30 U.S.C. § 1272(e).

Section 522(e) of SMCRA prohibits coal mining that will adversely affect any historic site listed in the National Register of Historic Places. In addition, the federal Office of Surface Mining (OSM) is required to comply with Section 106 of the NHPA in regulating surface coal mining.

STATE HISTORIC PRESERVATION LAWS

Many state legislatures have enacted laws that are similar to the federal laws described above—especially laws such as NEPA, Section 106 of the NHPA, and Section 4(f). These laws typically apply to projects that are funded or approved by state agencies. Therefore, historic properties that would not be protected by federal law due to a lack of federal involvement may, in many cases, be covered by these state preservation laws.

CONTACT

Center for Preservation Policy Studies Legislative Hotline
National Trust for Historic Preservation
1785 Massachusetts Avenue, N.W.
Washington, DC 20036
(800) 765–NTHP

FURTHER READING

Federal Historic Preservation Case Law: A Special Report. Charlotte R. Bell. Washington, D.C.: Advisory Council on Historic Preservation, 1985.

Federal Historic Preservation Laws. U.S. Department of the Interior, National Park Service, Cultural Resources Programs. Washington, D.C.: U.S. Government Printing Office, 1990.

LOCAL PRESERVATION ORDINANCES

Constance Epton Beaumont

"The real issue is not whether we will have change, but how great it will be, how quickly it will happen, and how shattering its impact will be."

Robert E. Stipe, *North Carolina Central Law Journal*, Vol. II, No. 1, 1980

Among the first lessons the preservationist learns is that the legal power to protect historic places lies chiefly with local government. This is a lesson often learned the hard way, for many people assume that the federal government, being the highest level of government, is the strongest guardian of historic sites. They assume that if a property is listed in the National Register of Historic Places, it must be protected automatically. This, unfortunately, is not the case. When it comes to historic preservation, the strongest protection is typically found in preservation ordinances enacted by local governments.

Local preservation ordinances in the United States date to 1931, when Charleston, South Carolina, became the first city to establish a historic district. Today, there are more than 1,700 communities with preservation ordinances in place. Big cities and small towns alike have found these laws to be an effective tool in protecting historic places from such undesirable fates as demolition or deterioration through sheer neglect.

POWER OF THE LOCAL GOVERNMENT

Preservation ordinances are local laws through which historic property owners are usually prohibited from altering or demolishing their property without local government approval. Most ordinances limit changes affecting the exterior of a structure, leaving property owners free to modify interiors as they wish. Some cities, however, have enacted ordinances that regulate changes to historic building interiors that are considered public spaces. An ordinance can protect individual landmarks, entire areas known as historic districts, or both individual landmarks and districts. Historic district ordinances typically regulate the design of new construction as well, to ensure a new building's compatibility with its older neighbors.

The power to regulate private property through land-use laws is one of the powers exercised by states under their "police-power" authority. The states have generally delegated these powers to cities and towns, and every state except Wyo-

ming has empowered local governments to regulate development affecting historic sites.

Local preservation ordinances vary widely, but they must comply with all four of the following cardinal rules of land-use law:

1. An ordinance must promote a valid public purpose; that is, it must somehow advance the public health, safety, morals, or general welfare.

2. An ordinance must not be so restrictive as to deprive a property owner of all reasonable economic use of his property.

3. An ordinance must honor a citizen's constitutional right to due process. In other words, fair hearings must be provided and rational procedures must be followed in an ordinance's administration.

4. An ordinance must comply with relevant state laws.

If an ordinance violates any one of these rules, it stands the risk of being invalidated by a court. If it violates the second rule, a court may order the local government to

Dayton Street, Cincinnati, Ohio. (R. Wyatt, W. Miller, R. Sharp, HABS)

pay a property owner *just compensation* for taking private property in violation of the Fifth Amendment to the Constitution.

The basic constitutionality of historic preservation ordinances was upheld in 1978 by the U.S. Supreme Court. In *Penn Central Transportation Co. v. City of New York* (438 U.S. 104 (1978)), the court settled two important questions.[1] First, it found historic preservation to be a valid public purpose:

"Because this Court has recognized, in a number of settings, that States and cities may enact land use restrictions or controls to enhance the quality of life by preserving the character and desirable aesthetic features of a city . . . appellants do not contest that New York City's objective of preserving structures and areas with special historic, architectural, or cultural significance is an entirely permissible government goal. . . .

The restrictions imposed (by New York's landmark ordinance) are substantially related to the promotion of the general welfare. . . ."

Second, the court held that New York's ordinance—and by inference similar ordinances enacted by other cities—had not taken private property in violation of the U.S. Constitution because the ordinance's restrictions left the Penn Central company with a "reasonable beneficial use" of its landmark property. The court punctured the oft-heard argument that property owners are entitled to make the most possible money (the "highest and best use") from their land:

". . . the submission that (property owners) may establish a 'taking' simply by showing that they have been denied the ability to exploit a property interest that they heretofore had believed was available is quite simply untenable."

But local ordinances must do more than pass muster under the federal Constitution; they must also comply with state laws and constitutions. Many states have enacted general enabling laws for local historic district or landmark statutes.

Other states have given localities broad home rule powers, permitting them to exercise all powers not expressly prohibited. Still others have conferred the power to protect historic sites on localities through statewide zoning enabling laws.

BASIC ELEMENTS OF A PRESERVATION ORDINANCE

With the legal authority for local preservation ordinances now well established, the question arises, How should an ordinance be structured? Listed below are basic elements that should be part of any ordinance.

Statement of Purpose
An ordinance should clearly state its public purpose. Although historic preservation can be justified for its own sake, many jurisdictions have found it legally and politically prudent to link historic preservation to more traditional community objectives. Some lower courts have ruled that "aesthetic regulation" is not a valid public purpose, but have sanctioned such activities as economic development as "legitimate public purposes." The ordinance in Cape May, New Jersey, includes among its purposes "to preserve and enhance the environmental quality of neighborhoods, to strengthen the township's economic base by the stimulation of the tourist industry, to establish and improve property values; to foster economic development; to manage growth. . . ."

Definitions
Technical terms such as *alterations*, *demolition by neglect*, and *environmental settings* should be clearly defined somewhere in the ordinance.

Creation of a Preservation Commission
Some entity within local government must be charged with administering the ordinance. Usually this is a preservation or design review commission composed of local citizens. Many ordinances require preservation commissioners to have special expertise in certain disciplines such as architectural history, architecture, law, or real estate to guard against claims of

arbitrary and capricious decision making. Some ordinances require that a representative of the city planning board be on the commission to ensure that local planning goals are related to historic preservation. The qualifications of commission members as well as their terms of office must be spelled out.

Commission Powers and Duties
Most commissions are charged with the duty to conduct historic surveys, maintain inventories, and keep adequate records of their actions. Their authority over the designation and regulation of historic properties varies, however. Some commissions may only make recommendations to other governmental bodies—a planning board or city council, for example—whereas others have the final word on whether and how historic properties may be altered. Obviously the more authority vested in the commission, the stronger the protection for historic sites.

Many commissions may deny proposals to demolish historic buildings; others may only delay such actions. Despite claims to the contrary, demolition denials do not constitute a *taking* in violation of the U.S. Constitution as long as a property owner has not been denied all reasonable use of his property. Reductions in property values due to regulations are not takings.[2]

Criteria for Designating Historic Properties
Objective, relevant criteria should be established for evaluating the historic or architectural worth of a structure. Appropriate criteria include such things as a building's role in national, state, or local history; its association with prominent historical figures; its architectural or engineering excellence; and its cultural significance.[3]

Designating Historic Landmarks and Districts
Ordinances must comply with basic "due process" requirements. Property owners must be given adequate notice and an opportunity to be heard before their property rights are curtailed. Otherwise, an ordinance could be invalidated by a

court. The ordinance needs to explain who can nominate properties for historic designation, how and when affected property owners are notified, how many public hearings there are, who must approve designations, and what the timetable is for these actions.

Procedures and Standards for Reviewing Alterations and Demolitions

It is important that alteration or demolition requests be acted on fairly and in a timely fashion. It is critical for commissions to review such requests according to reasonable standards clearly set forth in the ordinance. The ordinance should explain what types of changes are subject to review; often minor normal repair and maintenance are exempt. A system perceived to be rational and fair will go a long way toward avoiding legal problems.

Many cities have incorporated the Secretary of the Interior's Standards for Rehabilitation into their ordinances. Although these standards are a useful set of guiding principles for the federal programs for which they were intended, if used by local preservation commissions, they should be adapted to meet local needs and phrased in appropriate regulatory language.[4]

Economic Hardship

All historic preservation ordinances should include a process and standard for evaluating economic hardship claims. Such provisions can act as a safety valve if the ordinance is challenged in court; conversely, their absence can make an ordinance vulnerable to attack. The ordinance should explain the process for obtaining a hardship finding and spell out what information the commission needs to evaluate hardship claims. The timing for reviewing hardship claims is also important. Such claims should be considered only after an application for approval to alter or demolish a structure has been denied, not when properties are still being considered for historic designation or before applications for alterations are acted upon. In effect, economic hardship review is comparable to the variance process under zoning laws.

Penalties

Ordinances must be enforced if they are to be effective. Penalties for violating the ordinance provisions may include fines (usually levied for each day a violation continues), requirements to restore or pay for willfully damaged landmarks, denial of permission to rebuild on sites where landmarks were illegally demolished, and even jail. The stiffness of the penalty varies with each community depending on the likelihood of noncompliance.

Appeals

Even if an ordinance is silent on appeals, a citizen still has the right to challenge a commission's ruling in court. It is wise, however, to clarify the appeals process. While some ordinances make commission decisions appealable only to the courts, others find it easier and less expensive to have boards of zoning appeals or some other administrative body handle these cases. If the latter course is chosen, it is important to give such bodies clear criteria for considering appeals. Otherwise, they may use political criteria or assume unproven economic hardship on the part of the property owner. Appeal board reviews should be limited to the facts presented to the preservation commission in considering whether a decision was made arbitrarily or capriciously.[5]

Local Innovations

While most local preservation ordinances include the basic elements listed above, many have gone further to address common problems in innovative ways. Some examples:

Automobile dominance. Nothing destroys a historic area faster than subservience to the automobile. Seattle's Pioneer Square Historic District Ordinance promotes a pedestrian-friendly environment by banning gas stations, drive-in businesses, and surface parking lots. It also limits curb cuts and subjects the few parking garages that are allowed to special design review.

Environmental settings. The value of a historic structure is greatly diminished if it is surrounded by ugly, incompatible development.

The structure's setting should be protected from such development if at all possible. In Miami, Florida, the ordinance calls for drawing historic district boundaries so as to "include properties which individually do not contribute to the historic character of the district, but which require regulation in order to control potentially adverse influences on the character and integrity of the district."

Interim protection. Often the mere discussion of historic property designations will prompt property owners who are fearful of new regulations to seek demolition permits. It is important to provide interim protection for buildings nominated, but not yet officially designated, as local historic landmarks. This allows the local governing body to weigh the merits of specific nominations without witnessing a rash of demolitions. An interim control ordinance should be for a set time period and should state the public purpose for the controls. Atlanta's ordinance provides interim protection of up to 11 months.

Design guidelines. The preservation ordinance in Portland, Maine, contains unusually specific and clear guidelines for reviewing new construction in historic districts. Not only does the ordinance provide guidelines for new buildings as individual structures, but it also discusses the orientation of buildings to streets. Leesburg, Virginia, has created an overlay district to regulate the design of new construction along the highways leading up to and into the town's historic district.

Demolition by neglect. Occasionally a landowner will deliberately neglect a historic structure in the hope of obtaining a demolition permit on the ground that the building jeopardizes public safety. Many ordinances include "affirmative maintenance" provisions to prevent this. The Charlottesville, Virginia, ordinance states that a property owner shall not permit a structure to deteriorate so badly that it produces a "detrimental effect" on a historic district or landmark. The ordinance also calls for the maintenance of the "surrounding environment, e.g., fences, gates, sidewalks, steps, signs, accessory structures and landscaping."

Surface parking lots. Lest historic structures be demolished to make way for surface parking lots, Atlanta's ordinance now requires property owners to provide detailed architectural plans and evidence of financing for new building projects.[6] Salt Lake City's ordinance requires demolition permit applications to be accompanied by landscaping plans. The city planning department may obtain performance bonds to ensure that landscaping is actually done.

Uses of historic structures. Preservation ordinances typically stay out of land-use questions, yet as national chains and franchises homogenize American communities, many preservationists are searching for ways to preserve the small, locally owned businesses that give each city its unique flavor while providing ideal users for small historic structures. The guidelines of the Pike Place Market Historical District Ordinance in Seattle state that all businesses using the market are to be operated "with the owner involved in the daily management. Businesses serving local residents are preferred over those which are primarily tourism-oriented." The guidelines encourage local farmers to use the market and discourage fast-food outlets from doing so.

The question of land use raised by the guidelines of Pike Place Market in turn raises a larger issue: How do the policies embodied in a local comprehensive plan and zoning ordinance affect historic properties? While a commission is drafting a local ordinance (or soon thereafter), a preservation advocate should examine the local planning and zoning policies for their impact on preservation. Does the plan call for a road widening in the middle of the historic district? Does the zoning policy permit high-rise buildings in areas where small-scale historic structures now stand, thus putting economic pressures on property owners (or tempting them) to tear down and build anew?

Capital improvement plans and zoning policies are notorious for setting the stage for a landmark's future demolition. It is important to evaluate these plans and policies and seek appropriate changes lest they undercut the local preservation ordinance.

BENEFITS OF LOCAL PRESERVATION ORDINANCES

Because local preservation ordinances restrict what private owners can do with their property, it is critical that preservation advocates be able to articulate the broad community benefits that can flow from an ordinance. Ordinances are an important aid to our collective efforts to enhance the quality of community life. They give us a way of passing on to future generations historic sites that help explain why our society evolved as it did, while preserving architectural treasures that can never be replaced. And they help bring beauty and a sense of civility to our cities.

Here are some benefits of protecting historic resources that preservation advocates might consider as they seek to build public and political support for local ordinances:

Environmental Benefits
". . . historic conservation is but one aspect of the much larger problem, basically an environmental one, of enhancing, or perhaps providing for the first time, a quality of human life."

—Robert E. Stipe, "Why Preserve Historic Resources?" in *Legal Techniques in Historic Preservation*, 1972.

Destroying or abandoning old buildings and compact, walkable areas in favor of auto-dependent, sprawling communities contributes to air pollution while forcing taxpayers to pay higher taxes for new infrastructures—roads, sewers, utilities—and for such public services and institutions as mail delivery, schools, post offices, and libraries. The old buildings, streets, and infrastructures for which natural resources have already been extracted go underused. Compact

urban centers in which people get around using environmentally benign transportation modes, such as walking, biking, or mass transit, are abandoned for inefficient living in which everyone must rely on the automobile. Evidence suggests that such sprawl depletes natural resources, consumes farmland, and destroys trees. It also requires vast quantities of gas and oil and contributes to air and water pollution.

Demolishing existing buildings adds to waste disposal problems. According to some recycling experts, the debris from building construction and demolition can account for more than 20 percent of a municipality's solid waste stream. Because such debris is bulky and difficult to compact or burn, many landfills refuse to accept it.

Economic Benefits
A frequently heard argument is that preservation regulations will reduce property values and stifle local economic development. The experiences of many property owners and communities shows the exact opposite. A 1990 economic analysis of business activity in Denver's Lower Downtown Historic District reported that the district actually fared better economically than other parts of the city.[7] Within two years of the district's designation 114 new businesses creating 450 new jobs located in the district. While most of Denver suffered from a severe recession, the Lower Downtown district experienced substantial new business activity, according to the study.

In Fredericksburg, Virginia, properties in the historic district appreciated more in value between 1971 and 1990 than properties located elsewhere in the city.[8] Commercial property values in the historic district rose by an average of 480 percent during this period while those elsewhere rose by only 281 percent. The comparable figures for residential properties were 674 percent and 410 percent, respectively.

The ability of local communities to attract outside visitors—and the revenues and jobs they bring with them—often depends on historic preservation. As Arthur Frommer, the travel guide author, has pointed out: "Every study of travel motivations has shown that an interest in the achievements of the past is among the three major reasons why people travel. . . . Among cities with no particular recreational appeal, those that have substantially preserved their past continue to enjoy tourism. Those that haven't receive no tourism at all. It is as simple as that."

Examples of strictly regulated historic areas that boost municipal coffers by attracting tourists include Pike Place Market and Pioneer Square in Seattle; River Walk in San Antonio; Main Street in Beaufort, South Carolina; downtown San Francisco; Society Hill in Philadelphia; Historic Strand District in Galveston, Texas; and the historic center of Annapolis, Maryland. Preservation-based tourism programs are a particularly important economic development strategy in small towns and economically depressed regions of the country.

Educational Benefits
The preservation of historic buildings provides us with opportunities to give meaning and color to the lessons of history. Places like Martin Luther King's birthplace, the Alamo, and Ellis Island help us understand important events in our nation's past. The preservation of San Francisco's cable cars, New York's Brooklyn Bridge, and the Midwest's Illinois and Michigan Heritage Canal helps us appreciate the engineering and technological advances we have made as a society. And the preservation of architectural masterpieces like Frank Lloyd Wright's Fallingwater, Louis Sullivan's Carson Pirie Scott Department Store in Chicago, and Thomas Jefferson's Monticello provide inspirational models of excellence for today's youth.

Social and Psychological Benefits
One of the distinguishing characteristics of many historic areas is their walkability. The walkability of historic neighborhoods and cities facilitates important social and business contacts and helps bring

people together. This urban quality is especially important to people who do not drive.

In a world of rapid change, the presence of familiar and beautiful landmarks gives people an emotional anchor. People spend billions of dollars annually on home improvements and landscaping because they intuitively understand the relationship between pleasant surroundings and their mental health. And, most people fill their attics with old letters, furniture, and childhood toys because they value the memories these objects evoke. Through appropriate stewardship, historic preservation advocates protect these community memories by preserving valued historic structures and their surroundings.

NOTES

1. The U.S. Supreme Court has issued three major land-use rulings since 1978. While these do not focus on historic preservation, it is important to understand what they said because they may affect local preservation efforts. In *Keystone Bituminous Coal Assn. v. DeBenedictis* (480 U.S. 470 (1987)), the Supreme Court rejected a takings claim against Pennsylvania's land subsidence law. Among other things, the court observed: "Under our system of government, one of the state's primary ways of preserving the public weal is restricting the uses individuals can make of their property. While each of us is burdened somewhat by such restrictions, we, in turn, benefit greatly from the restrictions that are placed on others." In *First English Evangelical Lutheran Church v. County of Los Angeles* (482 U.S.304 (1987)) the court held that the remedy for a temporary regulatory taking is not merely the invalidation of a land-use ordinance but just compensation to the property owner for the period during which the taking occurred. And in *Nollan v. California Coastal Commission* (483 U.S. 825 (1987)), the court said there must be a nexus between the purpose of a land-use regulation and the specific regulation used to achieve that purpose. In other words, the means should further the ends. Significantly, the court did not back away from its *Penn Central* ruling in any of these decision. The court has yet to explain how compensation should be determined in a temporary regulatory case.

2. See case citations in "Court Cases," page 75.

3. Although a few ordinances require an owner's consent before a property may be officially landmarked, this is not recommended. The wishes of an individual property owner are not an objective, relevant criterion. Private individuals are not allowed to veto zoning regulations or other public laws; they should not be allowed to veto historic property designations. Owner consent provisions also raise legal questions in that they arguably represent a standardless and thus unconstitutional delegation of police powers to private individuals. To quote again from the U.S. Supreme Court's *Mugler v. Kansas* ruling:

(The power to regulate land) must exist somewhere; else society will be at the mercy of the few who, regarding only their own appetites or passions, may be willing to imperil the peace and security of the many, provided only they are permitted to do as they please. Under our system that power is lodged with the legislative branch of the government. It belongs to that department to exert what are known as the police powers of the state, and to determine primarily what measures are appropriate or needful for the protection of the public morals, the public health, or the public safety. (123 U.S. 623 (1887))

For a detailed discussion of the "owner consent" issue, see "Owner Consent Provisions in Historic Preservation Ordinances: Are They Legal?" by Julia Hatch Miller. *Preservation Law Reporter*. 10:February 1991.

4. See "The Secretary's Standards Can Be Harmful to Your Historic District," by Katherine Raub Ridley. *Preservation Forum*. Spring 1990.

5. The pros and cons of different appeals procedures are discussed in *Preparing A Historic Preservation Ordinance* and in *Responding to the Takings Challenge* (Chapter Four). See Further Reading at the end of this chapter.

6. The law in Albany, New York, which also conditions the issuance of demolition permits on the approval of new construction, was challenged but upheld in *Lemme v. Dolan*. 558 N.Y.S Appellate 2d 991 (A.D. 3 Dept. 1990)

7. "Lower Downtown: Economic Impact of Historic District Designation," by Hammer Siler George Associates. July 1990.

8. "The Economic Benefits of Community Character Preservation: A Methodology," by John E. Petersen and Susan Robinson. Government Finance Research Center of the Government Finance Officers Association. 1991.

This chapter was adapted from "A Citizen's Guide to Protecting Historic Places: Local Preservation Ordinances," a publication made possible by a grant from The Andy Warhol Foundation for the Visual Arts. The National Trust for Historic Preservation gratefully acknowledges the assistance provided by the foundation. However, the views expressed herein do not necessarily reflect the views of the foundation.

FURTHER READING

Aesthetics and Land-Use Controls: Beyond Ecology and Economics. Planning Advisory Service Report No. 399. Chicago: American Planning Association. 1986.

America's Downtowns: Growth, Politics and Preservation. Richard C. Collins, Elizabeth B. Waters, and A. Bruce Dotson. Washington, D.C.: The Preservation Press, 1991.

A Handbook on Historic Preservation Law. Christopher J. Duerksen, ed. Washington, D.C.: The Conservation Foundation, 1983. (See especially chapter two, part two, "Drafting and Administering the Ordinance.")

Maintaining Community Character: How to Establish a Local Historic District. Pratt Cassity. Information Series. Washington, D.C.: National Trust for Historic Preservation, 1992.

Preparing a Historic Preservation Ordinance. Richard J. Roddewig. Planning Advisory Service Report No. 374. Chicago: American Planning Association, 1983.

Preservation Law Reporter. Washington, D.C.: National Trust for Historic Preservation.

Preservation Law Update. Washington, D.C.: National Center for Preservation Law.

Responding to the Takings Challenge. Richard J. Roddewig and Christopher J. Duerksen. Planning Advisory Service Report No. 416. Chicago: American Planning Association, 1989. (See especially chapter four.)

"The Secretary's Standards Can Be Harmful To Your Historic District." *Preservation Forum*, spring 1990. Washington, D.C.: National Trust for Historic Preservation.

CONTACTS

National Alliance of Preservation Commissions
c/o School of Environmental Design
609 Caldwell Hall
University of Georgia
Athens, GA 30620
(404) 542–4731
Pratt Cassity

National Center for Preservation Law
1015 31st Street, N.W.
Suite 400
Washington, DC 20007
(202) 338–0392
Stephen Dennis, Director

National Trust for Historic Preservation
1785 Massachusetts Avenue, N.W.
Washington, DC 20036
(202) 673–4214
Frank B. Gilbert, Senior Field Representative

EASEMENTS

Stefan Nagel

"A preservation easement is a legal document which regulates the use of or changes to real property and may be given or sold by a property owner to a charitable organization or government body. Once recorded, an easement becomes part of the property's chain of title and usually 'runs with the land' in perpetuity, thus binding not only the present owner who conveys it but all future owners as well.

"A preservation easement gives the organization to which it is conveyed the legal authority to enforce its terms. These terms usually create negative covenants prohibiting the owner from making alterations to the property without prior review, consultation and approval by the holder. Some easements also impose positive covenants that require the owner to make certain improvements to the property or maintain it in a certain physical condition."

Charles E. Fisher III and others, *Directory of Historic Preservation Easement Organizations*

TYPES OF EASEMENTS

Scenic and open space easements
Protect open spaces, historic and scenic views, the surroundings of significant buildings, archeological sites, and ecologically significant land by restricting development rights.

Exterior and facade easements
Protect the outside appearance of buildings by controlling alterations and requiring maintenance; they may also control development and air rights of a building.

Interior easements
Protect all or part of building interiors (not so frequently used because of difficulties in reviewing privately used spaces).

EASEMENT QUESTIONS AND ANSWERS

What are the benefits of donating an easement?
Donating an easement protects a significant property even after an owner has sold or bequeathed it; provides income, gift, and estate tax advantages for the donor and, in some states, property tax advantages; and enables preservation organizations and public agencies to protect properties against adverse changes through acquisition of a partial interest rather than assumption of the full burden of property ownership.

How does an easement protect property?
The easement holder has the right to review and approve proposed alterations to a structure or its setting and to enforce the easement terms in the event of a violation.

How is an easement valued?
Valuation, made by a professional appraiser, is typically the difference between the fair market value of the property before and after the grant of easement. An easement may reduce the market value of a property because it restricts development rights.

How does a property qualify for an easement deduction under federal tax law?
The Tax Treatment Extension Act of 1980, implemented by federal regulations issued on January 14, 1986, made permanent the federal income, gift, and estate tax deductions for charitable contributions of partial interests, such as easements, in real property. Restrictions must be granted in perpetuity. Gifts of "qualified real property interests" must be made to a "qualified organization" and be "exclusively for conservation purposes," which include preservation of a "historically important" land area and a "certified historic structure." A "certified historic structure" is a building, structure, or land area, depreciable or nondepreciable, listed in the National Register of Historic Places or located in a registered historic district and certified as being of significance to the district.

How does an easement qualify as being "exclusively for conservation purposes"?
It qualifies if it preserves historically important land or a certified historic structure, if it furthers the education or outdoor recreation of the general public, if it preserves a relatively natural habitat, or if it

71

preserves publicly beneficial or scenic open space.

What are the tax consequences of an easement donation?

For federal income tax purposes, the most important benefit is that the value of the donated easement is deductible as a charitable contribution, generally not to exceed 30 percent of the taxpayer's adjusted gross income, thereby reducing the donor's taxable income; the value in excess of 30 percent may be carried over for five succeeding tax years. In some cases the ceiling on deductibility can be increased to 50 percent of adjusted gross income, with a five year carry-over. For federal estate tax purposes, the value of the estate will be reduced because of the easement's development limitations. For state income and estate taxes, state laws may authorize deductions similar to the federal provisions. An easement also may decrease a property's local tax assessment and thus its local property taxes.

Which organizations are "qualified organizations" to receive tax-deductible easement contributions?

Qualified organizations are federal, state, and local government agencies that have authority to accept property interests and private organizations that are tax-exempt, charitable, educational nonprofit groups such as state and local preservation organizations or local historical societies.

Adapted from "How to Qualify Historic Properties Under the New Federal Law Affecting Easements," National Register of Historic Places, 1981

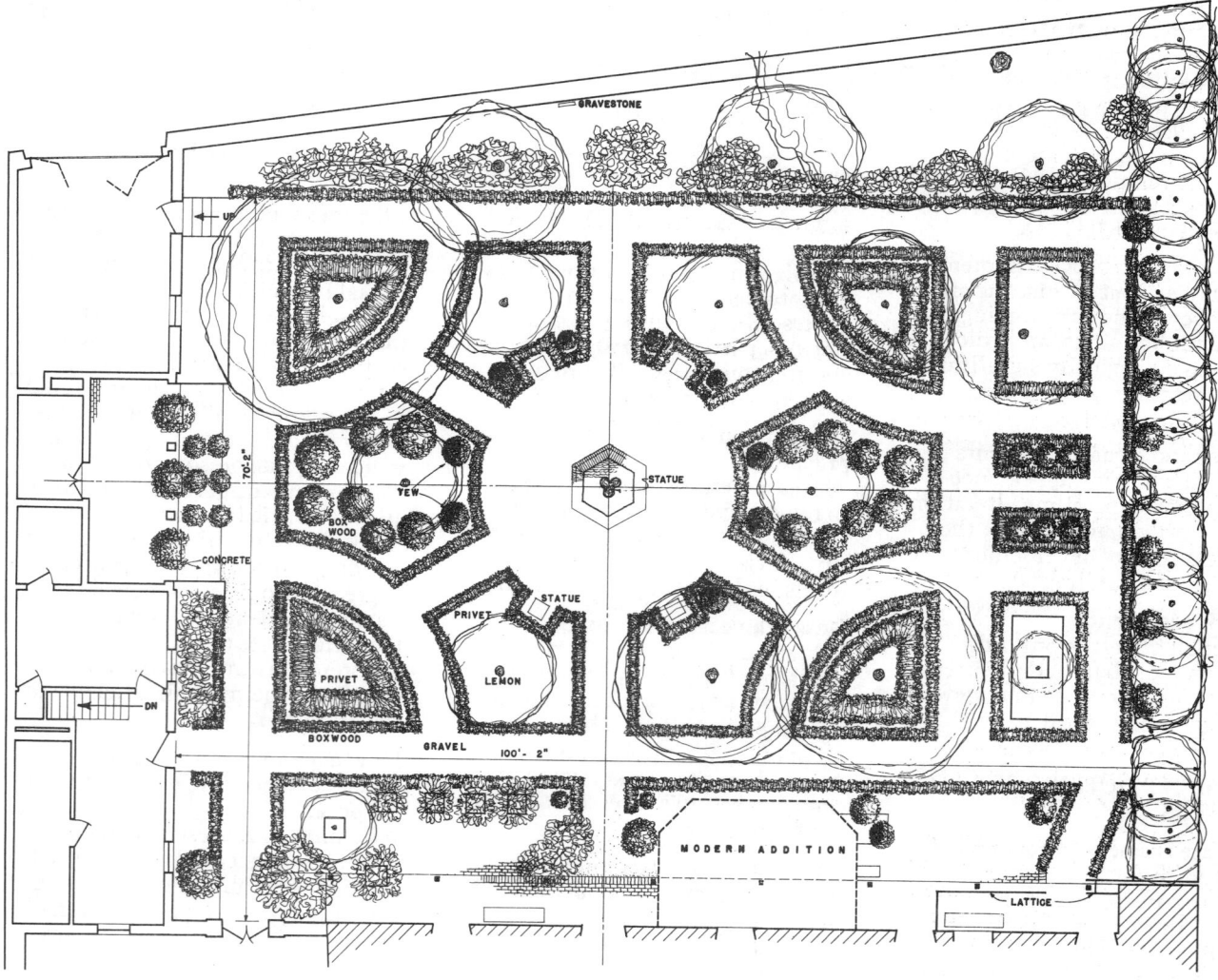

Rear Garden and Service Building, Casa Amesti, Monterey, California. (M.R. Pease, HABS)

STEPS IN OBTAINING AN EASEMENT

1. A property owner expresses an interest in donating an easement to a tax-exempt charitable organization or public agency. Less frequently, an organization may purchase an easement, sometimes stipulating that the seller use the income for preservation purposes.

2. The organization determines whether the property meets its acceptance criteria and discusses possible endowment requirements with the owner.

3. The property owner obtains legal and financial advice based on an informal estimate of the property's value.

4. Both parties decide to proceed.

5. The property owner seeks certification of the property from the U.S. Department of the Interior or listing in the National Register of Historic Places if a charitable deduction under federal income tax laws is contemplated and if it is to qualify as a historically important land area or certified historic structure, ensuring that the property is certified or listed by the time of the donation or when the tax return is filed (with extensions, generally by October 15 of the year following the donation).

6. The owner, with the assistance of the organization, documents the property, including the legal description, a location or boundary survey, photographs, and written descriptions of the property's resources that will be protected.

7. The organization, assisted by an attorney, drafts the easement deed; the property owner's attorney reviews and approves the document.

8. The property owner has an appraiser establish the value of the easement for tax purposes.

9. The deed of easement is signed by the property owner, delivered to and accepted by the organization, and recorded in the office of the local recorder of deeds and any other statutory repositories. The organization and the owner keep copies of all papers, which the owner agrees to transfer to subsequent owners.

10. The appraiser verifies the value of the easement, and the organization acknowledges the easement donation on IRS Form 8283 if a charitable deduction under federal income tax laws is contemplated.

11. The organization notifies government authorities of the easement transfer as a requirement and as a courtesy.

12. The organization monitors the property to ensure that easement provisions are observed and generally conducts an on-site inspection no less than once a year.

13. The organization continues to oversee the property's status by educating the current and future property owners, monitoring changes in property ownership, and reviewing the owner's alteration plans.

14. If a violation occurs in the easement terms, the organization should be prepared to exercise its right to compel the owner by court action to make repairs or to restore the property to its prior condition or to correct such violations itself; the organization also may seek an injunction to stop an owner's proposed action or seek monetary damages in compensation for irreversible actions. The deed of easement will set out enforcement provisions.

Adapted from *Establishing an Easement Program to Protect Historic, Scenic and Natural Resources*

SELECTED EASEMENT HOLDERS

Peninsula Open Space Trust
3000 Sand Hill Road
Building 4, Suite 135
Menlo Park, CA 94025
(415) 854-7696

Foundation for San Francisco's Architectural Heritage
2007 Franklin Street
San Francisco, CA 94109
(415) 441-3000

Historic Denver
Market Center
1330 17th Street
Denver, CO 80203
(303) 534-1858

L'Enfant Trust
2013 O Street, N.W.
Washington, DC 20036
(202) 785-9184

National Trust for Historic Preservation
1785 Massachusetts Avenue, N.W.
Washington, DC 20036
(202) 673-4035

Historic Savannah Foundation
P.O. Box 1733
Savannah, GA 31402
(912) 233-7787

Landmarks Preservation Council of Illinois
53 West Jackson Boulevard
Suite 752
Chicago, IL 60604
(312) 922-1742

Historic Faubourg St. Mary Corporation
400 Magazine Street
Room 200
New Orleans, LA 70130
(504) 524-1796

Preservation Resource Center
604 Julia Street
New Orleans, LA 70130
(504) 581-7032

Maine Coast Heritage Trust
167 Park Row
Brunswick, ME 04011
(207) 729-7366

Historic Annapolis
194 Prince George Street
Annapolis, MD 21401
(301) 267-7619

Maryland Historical Trust
100 Community Place
Crownsville, MD 21032
(410) 514-7600

Trustees of Reservations
572 Essex Street
Beverly, MA 01915
(508) 921-1944

Society for the Preservation of New England Antiquities
141 Cambridge Street
Boston, MA 02114
(617) 227-3956

Montana Land Reliance
107 West Lawrence
P.O. Box 355
Helena, MT 59624
(406) 443-7027

New York Landmarks Conservancy
141 5th Avenue
3rd Floor
New York, NY 10010
(212) 995-5260

Brandywine Conservancy
P.O. Box 141
Chadds Ford, PA 19317
(215) 388-7601

Philadelphia Historic Preservation Corporation
1616 Walnut Street
Suite 2310
Philadelphia, PA 19103
(215) 546-1146
Fax: (215) 546-1180

Historic Charleston Foundation
11 Fulton Street
Charleston, SC 29401
(803) 724-8486

Virginia Department of Historic Resources
221 Governor Street
Richmond, VA 23219
(804) 786-3143

CONTACTS

Archaeological Conservancy
415 Orchard Drive
Santa Fe, NM 87501
(505) 982-3278

Land Trust Alliance
900 17th Street, N.W.
Suite 410
Washington, DC 20006
(202) 785-1410

National Conference of Commissioners on Uniform State Laws
Uniform Conservation Easement Act
676 North St. Clair Street
Suite 1700
Chicago, IL 60611
(312) 915-0195

National Trust for Historic Preservation
Office of Law and Public Policy
1785 Massachusetts Avenue, N.W.
Washington, DC 20036
(202) 673-4035

The Nature Conservancy
1815 North Lynn Street
Arlington, VA 22209
(703) 841-5300

Trust for Public Land
116 New Montgomery Street
4th Floor
San Francisco, CA 94105
(415) 495-4014

U.S. Department of the Interior
National Park Service
Preservation Assistance Division
P.O. Box 37127
Washington, DC 20013-7127
(202) 343-9573

FURTHER READING

Appraising Easements: Guidelines for the Valuation of Historic Preservation and Land Conservation Easements. 1984. 2nd ed. Washington, D.C.: National Trust for Historic Preservation and Land Trust Exchange, 1989.

The Conservation Easement Handbook: Managing Land Conservation and Historic Preservation Easement Programs. Janet Diehl and Thomas S. Barrett. Alexandria, Va.: Land Trust Exchange and Trust for Public Land, 1988.

Conservation Easements: The Urban Setting. Thomas A. Coughlin, ed. Washington, D.C.: National Trust for Historic Preservation and Technical Preservation Services, U.S. Department of the Interior, 1981.

Directory of Historic Preservation Easement Organizations. Charles E. Fisher III, William G. MacRostie and Christopher A. Sowick. Washington, D.C.: Technical Preservation Services, U.S. Department of the Interior, 1981.

Easements and Other Legal Techniques to Protect Historic Houses in Private Ownership. Thomas A. Coughlin. Washington, D.C.: National Trust for Historic Preservation, 1988.

Establishing an Easement Program to Protect Historic, Scenic and Natural Resources. Information Series, National Trust for Historic Preservation. 1980. Rev. ed. Washington, D.C.: Preservation Press, 1982.

The Federal Tax Law of Conservation Easements. Stephen J. Small. Alexandria, Va.: Land Trust Exchange, 1986. Supplement. 1989.

Historic Preservation Law and Taxation. Tersh Boasberg, Thomas A. Coughlin and Julia H. Miller. New York: Matthew Bender, 1986.

Preservation Easements: The Legislative Framework. Steven J. Zick. Washington D.C.: National Trust for Historic Preservation, 1984.

COURT CASES

Julia Hatch Miller

"With the growth and maturity of the preservation movement comes its inevitable clash with competing interests in our pluralistic society. Attempts to save older downtown buildings will conflict with equally vociferous claims for new convention centers. Neighborhood revitalization will raise cries of low-income displacement.

"Indeed, as de Tocqueville has observed about America, 'In a modern democracy, social problems become translated into legal problems—if the democracy coheres.' This is why lawyers need to become more involved in the historic preservation movement as it gains in numbers and enlarges its objectives. Now, more than ever, there is a need to ensure that the country's laws and legal institutions keep pace with the great expectations many have for preservation in fostering economic growth, revitalizing our work places, and improving the general quality of our lives."

William K. Reilly and Tersh Boasberg, *A Handbook on Historic Preservation Law*, 1983

A number of historic properties are protected through a variety of federal and state laws and incentive programs. The level of protection afforded under these laws varies considerably, depending on their nature and scope. In general, local preservation laws provide substantive protection for historic properties while federal laws simply encourage federal agencies to consider and mitigate adverse impacts on historic properties that are within their control. Federal income tax incentives also are available under the Internal Revenue Code to encourage the preservation of historic resources through rehabilitation.

The strongest protections for historic structures are provided by local preservation ordinances that designate and regulate historic buildings as individual landmarks or as contributing structures to historic districts. Encouraged by the U.S. Supreme Court's ruling upholding New York City's landmarks law in *Penn Central Transportation Co. v. City of New York*, 438 U.S. 104 (1978), more than

1,700 local jurisdictions have adopted historic preservation ordinances. Both federal and state courts have upheld these ordinances against broad-based constitutional challenges, finding that local governments have the authority to regulate historic property and that regulation does not amount to an unlawful "taking" of property without compensation. At least 24 states have specifically considered such challenges, and in each of these decisions the reviewing court has concluded that such regulations do not unlawfully "take" private property without just compensation and/or that historic preservation is a legitimate public purpose. Only one state thus far has questioned this principle: the Pennsylvania Supreme Court in a 1991 decision, *United Artists Theater Circuit, Inc. v. City of Philadelphia*, No. J-234-1990 (1991). That decisions, however, is currently under reconsideration, and a revised decision is expected in late 1992 or early 1993. For updated information, contact the National Trust's Legal Defense Fund, (202) 673-4035.

Federal and state courts have also dismissed a number of challenges to a local government's authority to designate and regulate religious property, new construction in historic districts, and building interiors.

More limited protection is also afforded to a number of historic properties through federal legislation such as Section 106 of the National Historic Preservation Act (NHPA), 16 U.S.C. § 470f, and Section 4(f) of the Department of Transportation Act (DOTA), 49 U.S.C. § 303, and the National Environmental Protection Act (NEPA), 42 U.S.C. §4321.

Section 106 requires federal agencies to consider the effect of federally assisted projects on properties listed in or eligible for the National Register of Historic Places. If a project threatens to harm such properties, the federal Advisory Council on Historic Preservation, an independent agency created under the NHPA, must be consulted in a process designed to promote consideration of ways of avoiding or minimizing such harm.

After agencies have gone through the review and mitigation process invoked by Section 106, they may carry out a project as they see fit—even if their actions mean the destruction of historic sites. By preventing federal agencies from acting rashly, however, and by requiring them to consider alternatives to the destruction of historic resources, Section 106 can buy time and provide a forum in which preservation advocates can sometimes negotiate compromises. (If these compromises are formalized in a Memorandum of Agreement signed by the Advisory Council on Historic Preservation, the state historic preservation officer, and the federal agency involved, they may be legally enforceable.) Section 106 is rather like a stop sign. In effect, it says, "Stop and consider the options before you; once you've done that, proceed in any direction you wish."

Section 4(f) of the Department of Transportation Act is stronger than Section 106, but it only applies to transportation projects—highways, mass transit lines, and the like. Under Section 4(f), federally assisted transportation projects that could damage historic sites may not go forward unless there is no prudent or feasible alternative to using the land proposed for the project. If there is no such alternative, all possible planning to minimize harm to the historic site still must be done. Section 4(f) provides both procedural and substantive protection for historic properties and governs direct and indirect impacts on historic property, including increased noise, pollution, and traffic.

Unlike Section 106, which merely invokes procedural requirements, Section 4(f) provides substantive protection for historic resources. Section 106 can never prevent the proposed demolition of a historic structure against the federal government's will. It can never deny the proposed demolition of a landmark. However, neither Section 106 nor Section 4(f) protects historic resources from purely private development, the type of development that most often threatens such resources.

Beaver County Courthouse, Beaver, Utah. (C. Harker, HABS)

The continued use of historic properties is encouraged through tax benefits in the form of investment tax credits on rehabilitation expenses and charitable tax deductions for the donation of preservation easements. Under the Tax Reform Act of 1986, property owners may realize a 20 percent credit on rehabilitation expenditures on income-producing certified historic structures. Certification is obtained from the National Park Service. Property owners may also claim a charitable contribution for the donation of easements to qualified historic preservation organizations. This involves the relinquishment of rights to demolish or alter property in perpetuity. Cases brought pursuant to the rehabilitation tax credit have unsuccessfully challenged the National Park Service's denial of certification. Easement cases usually address questions of the proper value of the charitable donation.

Several states have adopted state preservation laws that are similar in scope to Section 106 of the NHPA or Section 4(f) of the DOTA. And many states have adopted incentive programs, usually in the form of property tax relief, designed to encourage the rehabilitation of historic properties. Individuals interested in these programs should consult the appropriate state historic preservation office.

Listed below are select decisions organized by topic. These decisions represent the types of cases that have been litigated in both federal and state courts under local preservation laws, the National Historic Preservation Act, the Department of Transportation Act, and the Internal Revenue Code. Case listing is in order of jurisdiction—U.S. Supreme Court, federal court, and state court—and by date in reverse chronological order. Decisions handed down after 1982 are discussed in full in the *Preservation Law Reporter*, a monthly publication available from the National Trust for Historic Preservation.

GOVERNMENTAL AUTHORITY

U.S. Supreme Court

Penn Central Transportation Co. v. City of New York, 438 U.S. 104 (1978) Landmark decision upholding the constitutionality of New York City's preservation ordinance. Supreme Court determined that the ordinance advanced a legitimate governmental interest and did not deny the property owner all beneficial use of its property in its application. "[The Supreme Court] has recognized in a number of settings, that States and cities may enact land-use restrictions or controls to enhance the quality of like by preserving the character and desirable aesthetic features of a city."

Berman v. Parker, 348 U.S. 26 (1954) "The concept of the public welfare is broad and inclusive. . . . The values it represents are spiritual as well as physical, aesthetic as well as monetary. It is within the power of the legislature to determine that the community should be beautiful as well as healthy, spacious as well as clean, well-balanced as well as carefully patrolled."

Mugler v. Kansas, 123 U.S. 623 (1887) The power to regulate land "must exist somewhere; else society will be at the mercy of the few who, regarding only their own appetites or passions, may be willing to imperil the peace and security of the many, provided only they are permitted to do as they please. Under our system that power is lodged with the legislative branch of the government. It belongs to that department to exert what are known as the police powers of the State, and to determine, primarily what measures are appropriate or needful for the protection of the public morals, the public health, or the public safety."

Other Federal Courts

Maher v. City of New Orleans, 516 F.2d 1051 (5th Cir. 1975), *cert denied,* 426 U.S. 905 (1976) The Vieux Carré ordinance found to be a permissible exercise of the police power and did not constitute a taking as a whole or as applied to Maher's property. Minimum maintenance provision and standards also upheld. "[P]roper state purposes may encompass not only the goal of abating undesirable conditions, but of fostering ends the community deems worthy . . . Nor need the values advanced be solely economic or directed at health and safety in their narrowest senses. The police power inhering in the lawmaker is more generous, comprehending more subtle and ephemeral societal interests."

State Courts

Faulkner v. Town of Chestertown, 428 A.2d 879 (Md. 1981) [1 PLR 3014] Regulation of nonhistoric properties within a locally designated historic district upheld. "[T]he whole concept of historic zoning would be about as futile as shoveling smoke if, *e.g.,* . . . because a building being demolished had no architectural or historical significance a historic district was powerless to prevent its demolition and the construction in its stead of a modernistic drive-in restaurant immediately adjacent to the State House in Annapolis."

A.S.P. Associates v. City of Raleigh, 258 S.E.2d 444 (N.C. 1979) Constitutionality of historic district ordinance in Raleigh that regulates both new construction and existing buildings as a legitimate use of the police power authority upheld. "While most aesthetic ordinances are concerned with good taste and beauty . . . a historic district zoning ordinance . . . is not primarily concerned with whether the subject of regulation is beautiful or tasteful, but rather with preserving it as it is, representative of what it was, for such educational, cultural, or economic values as it may have."

Figarsky v. Historic District Commission of the City of Norwich, 171 Conn. 198, 368 A.2d 163 (1976) Denial of certificate of appropriateness for demolition permit upheld against takings claim. Court found that local ordinance creating the historic district was a constitutional exercise of the police power. "The preservation of an area or cluster of buildings with exceptional historical and architectural significance may serve the public welfare."

City of Dallas v. Crownwich, 506 S.W.2d 654 (Tex. Civ. App. 1974) Authority of a municipality to use the police power to impose a moratorium on the issuance of building permits pending historic district designation upheld.

City of Santa Fe v. Gamble-Skogmo, Inc., 389 P.2d 13 (N.M. 1964) Historic district ordinance upheld under the state's general grant of zoning power to the city.

Opinion of the Justices (Nantucket), 333 Mass. 773, 128 N.E.2d 557 (1955) Proposed state legislation authorizing the operation of a historic district in Nantucket held to be a valid exercise of the police power and a proper delegation of legislation power, and to contain sufficiently definite standards for district designation and control. The proposed legislation was also held not to be a taking of property by eminent domain.

City of New Orleans v. Impastato, 3 So.2d 798 (La. 1941) Constitutionality of the Vieux Carré ordinance upheld against allegations of vague standards, denial of equal protection, and use of the police power for aesthetic purposes.

City of New Orleans v. Pergament, 198 La. 852, 5 So.2d 129 (1941) Sign regulation under preservation ordinance upheld. "The purpose of the ordinance is not only to preserve the old buildings themselves, but to preserve the antiquity of the whole French and Spanish quarter, the tout ensemble so to speak . . ."

REGULATORY TAKING

U.S. Supreme Court
Penn Central Transportation Co. v. City of New York, 438 U.S. 104 (1978). Court ruled that the designation of Grand Central Terminal and the denial of permission to construct a 55-story office tower on its site was not a taking because the "restrictions imposed are substantially related to the promotion of the general welfare." It is "quite simply untenable (to assert that property owners) may establish a 'taking' simply by showing that they have been denied the ability to exploit a property interest that

they have heretofore believed was available . . ."

Federal Courts
Rector, Warden, and Members of the Vestry of St. Bartholomew's Church v. City of New York, 914 F.2d 348 (2d Cir. 1990) [9 PLR 1103], *cert denied*, 59 U.S.L.W. 3594 (Mar. 4, 1991). Restrictions imposed by New York City's Landmarks Law did not amount to a confiscation of property without just compensation in violation of the Fifth and Fourteenth amendments.

Atlanta 10 v. City of Atlanta, No. 1:88-CU-157-HTW (N.D. Ga. Jan. 10, 1990) [9 PLR 1013]. The City of Atlanta's denial of permission to demolish the Margaret Mitchell House does not amount to an unconstitutional taking of property without compensation.

State Courts
Shubert Organization, Inc. v. Landmarks Preservation Commission of the City of New York, 570 N.Y.S. 2d 504 (App. Div. 1991) [10 PLR 1149], appeal denied, No. 107S-SSD-73 (Ct. App. Sept 23, 1991) [10 PLR 1149]. New York City's designation of the exteriors and/or interiors of 22 Broadway theaters does not effect an unlawful "taking" of historic property without compensation.

Lubelle v. Rochester Preservation Board, 551 N.Y.S.2d 127 (1990) [9 PLR 1016]. Denial of permission to demolish a designated historic structure does not amount to an unlawful taking of property without compensation.

Glisson v. Alachua County, 558 So. 1030 (Fla. Dist. Ct. App. 1990) [9 PLR 1057]. Land use law restricting development to protect the ecological and historical character of Cross Creek, Florida, does not effect a "taking" of property without compensation and is a proper use of police power authority.

Department of Natural Resources v. Indiana Coal Council, Inc., 542 N.E.2d 1000 (Ind. 1989) [8 PLR 3014]. Designation of archaeological site as unsuitable for coal mining under Indiana Surface Mining Conservation and Reclamation

Act is not an unlawful taking of property requiring compensation.

900 G Street Association v. Department of Housing and Community Development, 430 A.2d 1387 (D.C. App. 1981) [1 PLR 3001 (1982)]. Determination that an owner of a historic commercial structure had not demonstrated "unreasonable economic hardship" in a request for a demolition permit under District of Columbia preservation laws upheld. Court ruled that no taking had occurred because the owner could realize a reasonable return on his property.

Lafayette Park Baptist Church v. Board of Adjustment, 599 S.W.2d 61 (Mo. App. 1980) [2 PLR 3005]. Ordinance is not rendered unconstitutional by the mere fact that its application reduces the value of rights of some individuals.

RELIGIOUS PROPERTY

Federal
Rector, Warden, and Members of the Vestry of St. Bartholomew's Church v. City of New York, 914 F.2d 348 (2d. Cir. 1990) [9 PLR 1103], *cert denied*, 59 U.S.L.W. 3594 (Mar. 4, 1991). Application of Landmarks law to church-owned structure does not violate the First and Fifth Amendments to the U.S. Constitution.

State Courts
First Covenant Church v. City of Seattle, 14 Wash. 2d 392, 787 P.2d 1352 (1990) [9 PLR 1039], *vacated and remanded*, 59 U.S.L.W. 3593 (Mar. 4, 1991). U.S. Supreme Court vacated and remanded, for reconsideration, a Washington Supreme Court decision that had ruled that Seattle's historic preservation ordinance impinged on a church's free exercise of religion.

Society of Jesus of New England v. Boston Landmarks Commission, No. S-5415 (Mass. Dec. 31, 1990) [9 PLR 1151]. Massachusetts's highest court found designation of church interior violates state constitution's free exercise clause.

Society for Ethical Culture in City of New York v. Spatt, 415 N.E.2d 922 (N.Y. 1980) [1 PLR 3005

(1982)]. Landmark designation upheld against takings claim and interference with exercise of charitable and religious activity.

LOCAL ORDINANCES AND HISTORIC PRESERVATION COMMISSIONS

Federal Courts
Maher v. City of New Orleans, 516 F.2d 1051 (5th Cir. 1975), *cert denied,* 426 U.S. 905 (1976). Minimum maintenance requirement and standards of review within the Vieux Carré ordinance upheld.

State Courts
Historic Albany Foundation v. Coyne, 159 A.D. 73, 558 N.Y.S.2d 986 (1990) [9 PLR 1094]. Provision conditioning issuance of demolition permit on new construction plans in historic preservation ordinance found constitutional because the provision advanced a "legitimate municipal purpose."

Teachers Insurance and Annuity Association of America v. City of New York, Index No. 2583/90-001-02 (N.Y. Sup. Ct. Oct. 16, 1990) [9 PLR 1117]. Interior designation of restaurant by New York City preservation commission advanced a legitimate governmental purpose and did not amount to a "taking" of property in violation of the Fifth and Fourteenth Amendments to the U.S. Constitution.

Department of Land Conservation and Development v. Yamhill County, 99 Or. App. 441, 783 P.2d 16 (1989)[9 PLR 1001 (1990)]. The use of owner consent as a precondition to landmark designation violates Oregon's state planning law.

Parker v. Beacon Hill Architectural Commission, 27 Mass. App. Ct. 211, 536 N.E.2d 1108 (1989) [8 PLR 3054]. Denial of certificate of appropriateness to alter Beacon Hill row house affirmed. Property owner ordered to remove partially completed fifth story addition that had been constructed in violation of Boston's preservation ordinance.

Avery-Flaherty Properties, Inc. v. Montgomery County Historic Preservation Commission, Nos. 39657 & 39658 (Montgomery Cty., Md. Ct. Dec. 1, 1989) [9 PLR 1046]. Preservation commission's denial of permission to construct on side lots of Victorian house upheld.

Victorian Realty Group v. City of Nashua, 130 N.H. 60, 534 A.2d 381 (1987)[7 PLR 2042 (1988)]. Denial of application for lot line relocation in historic district upheld.

Buttnick v. City of Seattle, 105 Wash. 2d 857, 719 P.2d 93 (1986) [5 PLR 3039]. Court upheld requirement that owner of property in historic district replace a deteriorating parapet.

Weinberg v. Barry, 604 F. Supp. 390 (D.D.C. 1985) [4 PLR 3017]. Interim protection provisions under District of Columbia's historic preservation laws upheld. Interim interior designation does not deprive owner of all viable economical use of property.

A.S.P. Associates v. City of Raleigh, 258 S.E.2d 444 (N.C. 1979). Regulatory standard that precludes new construction in historic districts that "would be *incongruous* with the historic aspects of the district," despite its "subjectivity" upheld.

NATIONAL HISTORIC PRESERVATION ACT (SECTION 106)

Federal Courts
Vieux Carré Property Owners v. Brown, 875 F.2d 453 (5th Cir. 1989) *cert denied,* 100 S. Ct. 720 (1990) [8 PLR 3031]. National Historic Preservation Act creates a private cause of action against only federal agency defendants.

Boyd v. Roland, 789 F.2d 347 (5th Cir. 1986) [5 PLR 3029]. Section 106 applies to properties not officially determined eligible for listing in the National Register of Historic Places.

Benton Franklin Riverfront Trailway and Bridge Committee, 701 D.2d 7784 (9th Cir. 1983) [2 PLR 3031]. Agency should identify historic properties early in the planning process.

Neighborhood Development Corporation v. Advisory Council on Historic Preservation, 632 F.2d 21 (6th Cir. 1980) [1 PLR 3008]. Local organization has standing to challenge demolition of historically and architecturally significant buildings within an urban renewal project partially funded by a federal grant.

WATCH (Waterbury Action to Conserve Our Heritage, Inc. v. Harris, 603 F.2d 310 (2d Cir.), *cert denied,* 444 U.S. 995 (1979), *attorneys fees awarded,* 535 F. Supp. 9 (D. Conn. 1981). Department of Housing and Urban Development (HUD) held to have continuing responsibilities under the National Historic Preservation Act for federally funded urban renewal projects and to protect National Register-eligible buildings after a contract signing.

Stop H-3 Association v. Coleman, 553 F.2d 434 (9th Cir. 1976) *cert denied,* 429 U.S. 999 (1976). Determination of possible National Register eligibility triggers protection under Section 4(f) of the Department of Transportation Act.

State Court
National Trust for Historic Preservation v. U.S. Army Corps of Engineers, 552 F. Supp. 784 (S.D. Ohio 1982) [3 PLR 3001]. Power of the Advisory Council on Historic Preservation under Section 106 of the National Historic Preservation Act to review, consult, and comment on federal actions affecting properties listed or eligible for listing in the National Register confirmed.

DEPARTMENT OF TRANSPORTATION ACT (SECTION 4(F))

U.S. Supreme Court
Citizens to Preserve Overton Park v. Volpe, 401 U.S. 402 (1971). Reviewing court is to engage in a substantial inquiry and to take a "hard look" at all relevant factors to determine if the secretary satisfied his Section 4(f) obligations. If there is any feasible and prudent alternative to the use of Section 4(f) sites, that alternative must be adopted.

Other Federal Courts
Coalition Against a Raised Expressway, Inc. (CARE) v. Dole, 835 F.2d 805 (11th Cir. 1988) [7 PLR 2018]. Section 4(f) protects historic sites

from indirect adverse effects including noise, unsightliness, air pollution, and impairment of visual and physical access.

Ashwood Manor Civic Association v. Dole, 619 F. Supp. 52 (E.D. Pa.), *aff'd mem.*, 779 F.2d 41 (3d Cir. 1985), *cert denied*, 106 S. Ct. 1460 (1986) [5 PLR 3022]. Department of Transportation's practice of delegating the Secretary's Section 4(f) authority upheld.

Druid Hills Civic Association v. Federal Highway Administration, 772 F.2d 700 (11th Cir. 1985). Secretary must choose the least harmful alternative to the historic site.

Stop H-3 Association v. Coleman, 533 F.2d 4334, 445 (9th Cir.), *cert denied*, 429 U.S. 999 (1976). Construction of a highway passing within 100 to 200 feet of a historic petroglyph rock would "use" the site under Section 4(f).

Benton Franklin Riverfront Trailway and Bridge Committee v. Lewis, 701 F.2d 784 (9th Cir. 1983) [2 PLR 3031]. Section 4(f) governs structures and objects of historical significance, such as buildings and bridges.

Stop H-3 Association v. Dole, 740 F.2d 1442 (9th Cir. 1984) [3 PLR 3054]. Department of Transportation must demonstrate that a no-build alternative is "imprudent."

Louisiana Environmental Society, Inc. v. Coleman, 537 F.2d 79 (5th Cir. 1976). Alternative that minimizes harm may be rejected if it is imprudent.

REHABILITATION TAX PROJECTS

Federal Courts
Amoco Production Co. v. U.S. Department of the Interior, No. 89-C-209-B (N.D. Okla. Dec. 18, 1990) [9 PLR 1161]. National Park Service's denial of tax certification for certified rehabilitation tax project upheld.

Schneider Partnership v. Department of Interior, 693 F. Supp. 223 (D.N.J. 1988) [7 PLR 2033]. National Park Service acted within the scope of its authority in denying certification for a rehabilitation project where interior alterations did not conform to the Secretary of the Interior's Standards for Rehabilitation.

St. Charles Associates v. United States, 671 F. Supp. 1074 (D. Md. 1987) [6 PLR 3025]. National Park Service's authority to consider new construction as well as rehabilitation work in reviewing rehabilitation work upheld.

PRESERVATION EASEMENTS

Dorsey v. Commissioner, 59 T.C.M. (CCH) 592 (1990) [9 PLR 1099]. Tax court ruled that a historic preservation (facade) easement diminished the property value of a three-story commercial building by approximately 33 percent.

Griffin v. Commissioner, 56 T.C.M. (CCH) 1560 (1989) [8 PLR 1099]. Tax court ruled that facade easement on three historic build-ings in downtown New Orleans reduced market value by 20 percent.

Hilborn v. Commissioner, 85 T.C. 677 (1985) [5 PLR 3001]. Tax court upheld valuation of facade easement in Vieux Carré Historic District at 10 percent of property's market value.

CONTACTS

National Trust for Historic Preservation
Department of Law and Public Policy
1785 Massachusetts Avenue, N.W.
Washington, DC 20036
(202) 673-4035.

National Center for Preservation Law
1333 Connecticut Avenue, N.W.
Suite 300
Washington, DC 20036
(202) 338-0392

FURTHER READING

Federal Historic Preservation Case Law: A Special Report. Advisory Council on Historic Preservation. Washington, D.C.: U.S. Government Printing Office, 1985.

Handbook on Preservation Law. Christopher J. Duerksen, ed. Washington, D.C.: Conservation Foundation and National Center for Preservation Law, 1983.

Historic Preservation Law and Taxation. Tersh Boasberg, Thomas A. Coughlin and Julia H. Miller. New York: Matthew Bender, 1986.

Preservation Law Reporter. Washington, D.C.: National Trust for Historic Preservation. Monthly looseleaf service.

Preservation Law Update. Washington, D.C.: National Center for Preservation Law, 1987–present.

LEGAL ADVOCACY

Elizabeth Merritt

"If the historic preservation movement will adopt the stance of the conservationists and ecologists, they will enjoy success beyond anything encountered so far. Lawyers should be enlisted—young lawyers serving the public good—onto the staffs of every [historic preservation] organization."

Arthur Frommer, Keynote Address, Illinois Preservation Conference, June 3, 1988

Preservation law, at its most effective, has been a combination of proverbial carrots and sticks: some laws motivate and some deter. With government budgets squeezed from all sides, however, the incentives, or "carrots," such as tax benefits, have been cut back drastically, especially at the federal level. As a result, historic preservation advocates have relied increasingly on "sticks" to achieve their goals, such as restrictions applied through land-use regulation at the local level, and state and federal laws requiring agencies to consider the impacts of their projects on historic properties. In order to ensure that these mandates are followed, preservation advocates need to effectively wield the legal tools of preservation.

In controversies around the country, local and national historic preservation groups have used the courts to protect historic resources, usually with the assistance of *pro bono* legal representation. No matter how large or small a historic preservation organization is, the ability to participate in legal advocacy should be an important element of its program.

The National Trust for Historic Preservation plays a leading role in historic preservation advocacy, both in coordinating and assisting the efforts of others and in directly engaging in legal controversies. Through its Legal Defense Fund, the National Trust has taken an increasingly active role in advocating historic preservation through litigation and other legal action. A small but experienced staff of attorneys, working with the voluntary assistance of prominent lawyers from around the country, has helped many local organizations and municipalities win important preservation battles in their own communities.

Of particular importance is the National Trust's assistance in defending municipalities and preservation commissions against legal challenges to local preservation laws. For example, the National Trust, joined by local and state preservation organizations, has participated in a number of lawsuits to defend municipalities—Atlanta, New York, Dayton, Philadelphia, Chicago, and Seattle, among

Tooele County Courthouse and City Hall, Tooele, Utah. (C.D. Harker, HABS)

others—against challenges to their preservation ordinances. In many of these cases the lawsuits have involved "takings" or property rights challenges, and the National Trust has developed a nationally recognized expertise in this area.

The National Trust also works directly with federal, state, and local government agencies on disputes involving publicly funded or licensed projects. In many cases, the Trust is able to encourage cooperative solutions to these disputes without resorting to enforcement in the courts.

For example, the National Trust helped persuade the Kentucky Department of Transportation to set aside its plans to widen a 12–mile-long historic and scenic road, known as the Paris Pike, through the heart of Kentucky's historic bluegrass country. Through coalition-building, negotiation, and advocacy, an effective team of national, state, and local preservation organizations was able to stop this destructive project without the need for litigation.

In other cases, however, the National Trust—again, joined by local and state preservation organizations—has found it necessary to either bring suit or join in litigation as a "friend of the court" to protect historic properties that are threatened by publicly funded or licensed programs.

The National Trust and local preservationists successfully sued the U.S. Coast Guard for failing to comply with federal preservation laws in removing a historic Fresnel lens from the Devils Island Lighthouse in Wisconsin. As a result of the lawsuit, the Coast Guard agreed to repair the lens and return it to the lighthouse for display and interpretation. In Mobile, Alabama, the National Trust successfully participated in a lawsuit against the Federal Highway Administration, in which a federal court determined that the agency failed to comply with federal preservation law in approving a raised highway that would have passed within 43 feet of Mobile's 1858 city hall.

In numerous other cases—often involving federal highway projects, natural gas pipelines, and communications towers—litigation and administrative advocacy has been necessary to ensure compliance with federal and state preservation laws.

During the past 20 years, the National Trust has been directly involved in more than 70 litigated cases of national historic preserva-

tion significance, from state and local trial courts to the United States Supreme Court. In addition to those cases in which the Trust directly participates, the organization's staff lawyers assist local groups around the country on a daily basis in planning legal strategy and finding local counsel, and by providing research and expertise.

In addition to offering direct legal assistance to the national historic preservation constituency through the Legal Defense Fund, the National Trust's Department of Law and Public Policy also publishes the *Preservation Law Reporter*, a monthly journal that provides information on legal developments in historic preservation. This publication serves lawyers, preservation commissions, local organizations, and developers engaged in historic rehabilitation projects.

CONTACTS

Legal Defense Fund
National Trust for Historic Preservation
1785 Massachusetts Avenue, N.W.
Washington, DC 20036
(202) 673-4035

Additional legal assistance is available through many state historic preservation offices, state and local preservation organizations, and the National Trust regional offices.

FURTHER READING

A Handbook on Preservation Law. Washington, D.C.: The Conservation Foundation and the National Center for Preservation Law, 1983.

Historic Preservation Law and Taxation. Tersh Boasberg, Thomas A. Coughlin, and Julia H. Miller. New York: Matthew Bender, 1986.

Preservation Law Reporter. Washington, D.C.: National Trust for Historic Preservation. Monthly looseleaf service.

ORGANIZING LOCAL PRESERVATION

Katherine Adams

"We need civic leaders, neighborhood organizers, educators, and young people who are aware that the future, their future, is too important to leave to the work of others or to chance."

Herbert H. Smith, 1961

How many times have we heard that the key to preservation success is an organized grassroots movement? How often have we heard that an effective local organization will fix all our preservation woes?

Many local preservation organizations are born as crisis-driven coalitions of concerned citizens. The transition to a stable community-based organization—one that serves as a continuing presence in support of preservation—is never easy and never quick. Often, the group's members and board of directors alike begin to wonder, after their initial success, if their efforts to sustain commitment and resources are really having an impact on preserving the historic resources in the community. Brian O'Connell, president of Independent Sector, an organization that provides a national forum to encourage giving, volunteering, and not-for-profit initiatives, aptly describes the frequent answer to that question: "Any group as bright as we are, which has worked as long and hard as we have, must have done a lot of good."

We often get caught up in organizational development tasks and strategies and forget that they are not an end in themselves, but rather the foundation for effective local preservation activity. We confuse efficiency with effectiveness, in part because effectiveness is harder to define and often lacks a clear set of indicators to tell us when we're on the road to success.

It's much easier to know that we've been efficient—the bills are paid on time, the newsletter is mailed on schedule, the board of directors holds the prescribed number of meetings each year. In the for-profit world, efficiency and effectiveness come together on the bottom line: the result of effective business activity is profit. But preservation groups are not likely to save an old building every week. If we can measure effectiveness only by counting the number of historic resources we preserve, how do we know we're doing a good job?

Recently, National Trust field staff identified five "Traits of Effectiveness" for nonprofit preservation organizations. Of these, four are characteristic of every effective nonprofit organization. The fifth is specific to preservation groups.

The traits provide a framework for assessing preservation organization effectiveness, rather than a yardstick against which to measure the winners and losers. The traits can be applied to organizations that are just getting started, as well as those that have been in business for many years.

No organization is perfect, but the majority of these characteristics will be quickly identified in truly effective organizations. Most groups are likely to exhibit all the traits to one degree or another; those traits that exist to a greater degree are the strength to build on, while the lesser stars light up the areas that could use some improvement.

TRAITS OF EFFECTIVENESS

Mission. The mission is a clear and succinct articulation of the purpose of the organization, supported by its programs, products, and services. It is communicated equally to members, constituents, and the community.

Because many preservation groups form in response to a threat against a specific endangered landmark or historic property, their mission at the outset is crystal clear. But a decision to establish an ongoing organization focuses attention on basic issues such as developing articles of incorporation, obtaining tax-exempt status, recruiting board members, raising funds, and all the other essentials of start-up. By the time programs and services are developed, the mission has lost focus. An effective preservation group must keep its mission in mind while developing its support systems.

Leadership. Leaders have a firm grasp of the organization's mission and the ability to articulate clearly a vision and to marshal effectively human and financial resources to achieve that vision.

The people who led the fight to save an endangered landmark may or may not be the same people who can carry the organization through its developmental phases. Organizational needs change, the skills required to meet those needs vary

North Manchester Public Library, North Manchester, Indiana. (P.D. Adams, HABS)

over time, and the commitments and interests of individual leaders may shift as well.

Cultivating community preservation leadership is essential. Identifying, recruiting, and educating board members is the path most often taken to developing a pool of potential leaders, as well as to involving appropriate representatives of the community. The perfect mix of board members is often described as a combination of "work, wisdom, and wealth." Maintaining an appropriate balance on the board—from racial, ethnic, geographic, professional, skill, and talent perspectives—requires a continuous and conscious effort. But a well-balanced board of directors pays off doubly by bringing a broad community perspective to the organization as well as giving broad organizational representation to the community.

Political Savvy. Political savvy encompasses both understanding of the political process—who makes decisions, based on what information, in what time frame—and a willingness to act upon that knowledge by affirming or interceding in the process.

Many local preservation groups begin with advocacy efforts—fighting against decisions that have adverse impacts on historic resources. As these organizations mature, they take positive steps to prevent preservation crises. These steps often include active involvement in local politics.

Political involvement can take the form of electing pro-preservation candidates to local office; educating elected and appointed officials about the benefits of preservation; generating support for the creation, passage, and enforcement of local preservation ordinances; and developing coalitions with sympathetic organizations to garner broad support for preservation activities.

Resources. Adequate resources are available to support the mission, including a broad and diverse base of community support with access to the necessary human, financial, and technical resources.

Most people tend to think of resources first as financial, and most organizations think theirs are insufficient. An effective organization uses a range of fund-raising tactics including membership development campaigns, foundation and corporate donors, and special events such as historic home tours.

The objective of any combination of these strategies is to create a stable and diverse funding base. The companion objective, often overlooked, is to establish sound financial management practices for these resources.

For many groups, however, especially new ones, the primary, or only, resources are human—the citizen volunteers who came together in support of preservation. Board and committee members serve in every capacity and are the lifeblood of the organization. The members of these organizations look hopefully toward the point at which their financial resources enable them to employ a paid staff person. While this step can be a great leap forward, it also brings the temptation to undervalue and underuse a wealth of volunteer resources.

Both paid staff and volunteers can provide, directly or indirectly, access to a wealth of technical resources. If the necessary technical skills are not available through these contacts, the National Trust and numerous other service organizations can provide them on request. (See Contacts, below.)

Property. Effective preservation organizations have direct involvement with or influence on property-related issues by owning, controlling, or planning for the protection and preservation of historic resources.

Property-related activities can range from owning and maintaining house museums to operating easement programs or revolving funds, or steering decisions of local historic district and zoning commissions. The majority of preservation groups begin by emphasizing advocacy and educational programs and progress to direct involvement

in real estate activities at a more mature point in the organizational life cycle. For some groups, however, the first battle to save a historic building can result in ownership and maintenance responsibilities for the property.

For preservation groups, it is this effectiveness trait—direct involvement with historic resources—that all other traits exist to further and support. Whether the organization sponsors neighborhood walking tours, prepares National Register nominations, works to pass a preservation ordinance, or manages an active revolving fund, the clarity and communication of its mission will keep its activities focused on the preservation of historic resources. The circle is complete.

ASSISTANCE FROM THE NATIONAL TRUST

Community Preservation Organization Effectiveness Program (COEP). COEP is an assessment and planning process for local preservation organizations provided by the National Trust. It can be conducted by National Trust regional office staff or undertaken independently using the *Self-Assessment Guide for Community Preservation Organizations*, available through the National Trust's Information Series (see below). Write or call your regional office for information about COEP.

Information Series. A number of publications on organizational development and other preservation topics are available.

- *Investing in Volunteers: A Guide to Effective Management of Volunteer Programs*
- *Building on Experience: Improving Organizational Capacity to Handle Development Projects*
- *Business Ventures for Nonprofits*
- *Steering Nonprofits: Advice for Boards and Staff*
- *Legal Considerations in Establishing a Historic Preservation Organization*
- *Membership Development: A Guide for Nonprofit Preservation Organizations*

To order individual or bulk copies, write to: Information Series, National Trust for Historic Preservation, 1785 Massachusetts Avenue, N.W., Washington, DC 20036.

Preservation Leadership Training. A week-long intensive program offered twice yearly by the National Trust, Preservation Leadership Training provides leadership and organizational development skills along with the most current information on preservation practices, issues, and action strategies. Leadership training programs are also offered annually in conjunction with the National Preservation Conference. For information call or write your National Trust regional office.

CONTACTS

Accountants for the Public Interest
1012 14th Street, N.W.
Suite 906
Washington, DC 20005
(202) 347-1668

This is a national network of volunteer accountants who assist nonprofit organizations and small businesses faced with increasingly complex financial management needs to acquire better management skills. API has affiliates in 14 major cities across the country and has a directory of national volunteer accounting programs.

American Association for State and Local History
172 2d Avenue North
Suite 102
Nashville, TN 37201
(615) 255-2971

The association publishes an extensive Technical Leaflet how-to series on a wide variety of topics ranging from planning tours and using consultants to media relations. Call or write for a catalog.

American Association of Museums
P.O. Box 33399
Washington, DC 20033
(202) 289-1818

AAM offers a variety of publications on staff and volunteer management, fund raising, and other related topics. A publications list is available.

Foundation Center
79 5th Avenue
New York, NY 10003-3050
(212) 620-4230

The center conducts research for members of its Associates Program and is the source for a wide variety of general and specialized directories of corporate and foundation funders.

Funding Center
209 Madison Street
Suite 200
Alexandria, VA 22314
(800) 852-0001

The Funding Center helps nonprofit groups worldwide locate and secure public and private funds. Services include prospect research, proposal preparation, board and staff development, feasibility studies, special events, and direct mail, capital, and endowment campaigns.

Independent Sector
1828 L Street, N.W.
Washington, DC 20036
(202) 223-8100

Independent Sector is a membership organization of national nonprofits, foundations, and corporations that provides a forum to encourage giving, volunteering, and not-for-profit initiatives. In addition to information on management and fund-raising trends, legislative updates, and the latest research in the field, members get a 30 percent discount on most IS publications. Among these are IS president Brian O'Connell's popular *The Board Member's Book.*

Management Assistance Group
1835 K Street, N.W.
Room 305
Washington, DC 20006
(202) 659-1963

MAG has both Washington, D.C., and California branches offering consultation in nonprofit management and organizational and board development.

National Center for Nonprofit Boards
2001 L Street, N.W.
Suite 411
Washington, DC 20036
(202) 452-6262

This is an organization that endeavors to improve the effectiveness of nonprofit organizations by strengthening their governing boards. A request to be placed on NCNB's mailing list will provide a continuing flow of information on its training programs, information clearinghouse, new publications, and other activities.

Points of Light Foundation
736 Jackson Place, N.W.
Washington, DC 20036
(202) 408-5162

The foundation produces the *Volunteer Readership Catalog,* offering dozens of resource publications designed to increase the effectiveness of nonprofit organizations.

Support Centers of America
National Office
2001 O Street, N.W.
Washington, DC 20036
(202) 833-0300

A nonprofit organization founded in 1971, this is a national network offering reasonably priced workshops that address a wide range of nonprofit management issues, an on-site client training and consulting program, and an information and referral service. The group also offers business volunteer programs at some of its locations. Support centers are located in Boston; Chicago; Houston; Newark, New Jersey; New York City; Oklahoma City; Providence, Rhode Island; San Diego; San Francisco; and Washington, D.C.

Technical Assistance Center
1385 South Colorado Boulevard
Room A-504
Denver, CO 80222
(303) 691-9610

TAC specializes in management consulting for nonprofit organizations. Training is offered to the public at TAC's Denver location, as well as on-site to individual nonprofits. Publications include a *National Nonprofit Wage and Benefit Survey* and a *Manual for Board Members of Nonprofit Organizations.*

Voluntary Management Press
P. O. Box 9170
Downers Grove, IL 60515
(312) 964-0432

The press produces numerous monographs on management, board development, resource development, and financial and legal issues. Most publications are priced at less than $10. Call or write for a catalog.

United Way of America
National Office
701 North Fairfax Street
Alexandria, VA 22314
(703) 836-7100

Many of the 1,300 local United Way organizations offer board development training. Local volunteer centers, which are sometimes independent agencies and sometimes an arm of the United Way, generally offer training in volunteer management. For information, call or write your local United Way office or the national headquarters, listed above.

NEIGHBORHOOD CONSERVATION FOR LOW-INCOME AREAS

Jennifer L. Blake

"We are beginning to see that it is only through the healthy functioning of the neighborhoods that cities function at all."

Ada Louise Huxtable, National Conference on Neighborhood Conservation, 1975

Many of our nation's historic resources are located in low-income neighborhoods in inner cities and small towns across America. Unfortunately, low-income residents frequently are displaced when their homes deteriorate or are demolished or when their neighborhoods become the targets of outside investment.

Successful neighborhood conservation preserves these neighborhoods for the benefit of existing residents, maintaining both their historic and socioeconomic character. Herein lies the challenge. Low-income residents have a wide range of needs in their everyday lives in addition to housing—they need jobs, safe streets, good schools, affordable child care, social services, and access to other goods and services. Preserving low-income historic neighborhoods requires that these diverse needs also be addressed.

Local nonprofit organizations are effectively taking on this challenge in neighborhoods throughout the nation, achieving success by mobilizing the community around common planning objectives, by gaining access to financial and other resources, and by integrating preservation into the multiple neighborhood needs. These local groups assume a variety of roles, including community organizer, advocate, planner, owner, developer, lender, manager, and technical-assistance provider.

Because of their concern that historic preservation may displace low-income residents, many neighborhood organizations have not identified closely with the historic preservation movement. Low-income neighborhoods often are reluctant to seek formal designation as historic districts for fear of increasing property values and taxes and displacing low- and moderate-income residents.

The relationship between historic district designation and property values and taxes is complex. Some studies have found a positive correlation between the two—historic district designation contributes to increased property value and higher taxes. Other studies demonstrate that there is no direct relationship between designation and increased taxes and property values. For example, a 1989 study by Dennis Gale of the George Washington University found that historic district designation of neighborhood areas in Washington, D.C., did not increase tax assessments when compared with similar but undesignated districts in the city. In studies that have found a direct link between designation and increase in property values and taxes, contributing factors such as the dynamics of localized real estate market, assessment procedures related to improvement of dwellings, and the demand for a particular type of house in certain locations have not been incorporated. Thus, preservationists should take heart that there is no proof that historic district designation by itself will lead to higher housing values and taxes.

Other displacement threats confront residents of older neighborhoods, however. Commercial development that encroaches on residential areas can result in inappropriate adaptive use or demolition of historic structures. Other neighborhoods may suffer from disinvestment that triggers vacancies and property abandonment. From the outset, preservationists must ensure that local historic preservation projects address a myriad of displacement issues head-on and build in mechanisms to ensure that low- and moderate-income residents benefit from preservation.

THE INNER-CITY VENTURES FUND

In an effort to illustrate that historic preservation can work in partnership with neighborhood conservation, the National Trust launched its Inner-City Ventures Fund (ICVF) in 1981 to provide loans, grants, and technical assistance to nonprofit neighborhood organizations for the rehabilitation of historic buildings to be used as affordable housing and commercial properties that benefit low-income residents.

ICVF-assisted projects range from the rehabilitation of a single-room-occupancy hotel for elderly Asian immigrants in San Francisco's Chinatown Historic District to the reuse of an abandoned brewery complex as incubator space for minority-owned businesses in Boston's Jamaica Plain neighborhood. The ICVF also has helped rehabilitate historic homes, as in the Holy Cross/Westminster Historic District of Indianapolis, and apartment buildings, as in the Over-the-Rhine Historic District of Cincinnati, for moderate-income buyers and renters.

ICVF projects have demonstrated clearly that historic preservation has much to offer neighborhood

First Presbyterian Church, Springfield, New Jersey. (L. Crisson, HABS)

conservation efforts, such as financial incentives, new partnerships, job opportunities, creative adaptations of existing buildings, reduced maintenance costs through high-quality workmanship and materials, and market appeal. And, because subsidized historic buildings cannot be distinguished from others in the neighborhood, historic preservation can achieve socioeconomic diversity with little local opposition.

In addition to the economic benefits of historic preservation, neighborhoods also have received intangible benefits that improve the lives of their low-income residents. Once fears of displacement are laid to rest, preservation can offer low-income residents the emotional benefits of a beautiful, stable living environment and a heightened sense of heritage. These benefits contribute to residents' senses of security, belonging, and self worth and offer hope for a better future.

Ensuring Long-Term Affordability
A major concern often expressed by ICVF project leaders is the need to maintain the project's affordability over the long term. There are numerous techniques that can help ensure a project's long-range viability. To provide home ownership and rental opportunities for 16 low- and moderate-income households in the Holy Cross/Westminster Historic District of Indianapolis, for example, Eastside Community Investments, Inc., used creative bank mortgage financing to enable moderate-income home buyers to purchase a two-unit property; each home buyer rents one unit to low-income tenants and occupies the remaining unit. Subsidies from the U.S. Department of Housing and Urban Development tied to each rental unit guaranteed sufficient income to help qualify the home buyers for the bank mortgages.

Various cost-saving acquisition and rehabilitation techniques can also be used. For example, by obtaining site control before formally announcing project plans, project leaders can avoid paying speculative acquisition costs; in addition, the purchase of options is an inexpensive way to obtain site control while completing a project's financial packaging. Partially occupied buildings can be rehabilitated in phases so that existing tenants can be relocated within the structure during rehabilitation. The tax-exempt status and public interest goals of nonprofit organizations help attract donations of property, materials, "sweat equity" labor, and professional services to their projects.

NEIGHBORHOOD CONSERVATION MODELS

Model programs of neighborhood conservation for low-income areas exist in towns and cities all across the country. Some emphasize creating quality rental housing, while others concentrate on providing owner-occupied housing. Some of the programs rely heavily on the private and nonprofit sectors to provide the capital for the rehabilitation of existing buildings or for new construction; other programs depend on local, state, or federal subsidies and programs. Many of these programs focus exclusively on housing, while others attack a wide range of problems found in most low-income residential areas—crime, insufficient jobs, and the need to improve neighborhood commercial areas and health care.

National Models

While most neighborhood conservation models are local, two national programs that build local partnerships have proved to be very successful during the past decade. One is the Enterprise Foundation, founded in 1981 by James Rouse, the developer of Festival marketplaces and new towns. The foundation uses a comprehensive approach to neighborhood conservation. It recognizes that a community's needs go beyond providing decent housing to include technical training for local organizations involved in neighborhood conservation and providing a wide range of community services such as day-care centers, health-care services, and jobs and refocusing public policy at the local, state, and federal levels.

The Enterprise Foundation works through a network of local non-profit affiliates in 24 states and the District of Columbia. In the ten years since its founding, the foundation and its partners have provided more than 17,250 affordable housing units and more than $429 million to finance housing for low-income people. For example, in New York the foundation formed a partnership with the city, local corporations, and 37 neighborhood organizations to rehabilitate and construct 3,350 apartments; in Oakland, California, the foundation and its local partner rehabilitated 20 Victorian homes; and in Denver, the foundation and one of its local partners renovated 134 units in the city's Five Points neighborhood.

The second national organization that has had a significant effect during the past decade is Habitat for Humanity International. Founded as a housing ministry in Americus, Georgia, in 1976, Habitat is widely known because of the involvement of former President Jimmy Carter. Like the Enterprise Foundation, Habitat works through local affiliates throughout the United States and abroad. A central tenet of Habitat is that the people who are going to own the houses it builds should be directly involved in their rehabilitation and construction. Thus, Habitat requires that future owners provide "sweat equity" as well as some cash equity in the purchase of their home. Future owners are also provided with long-term, interest-free loans to assist them in acquiring the house.

In addition to these two organizations, other national organizations such as the Local Initiatives Support Corporation and the Neighborhood Reinvestment Corporation provide technical and financial assistance for neighborhood conservation in low-income areas.

The Springfield Model

Having assisted in the successful rehabilitation of more than 180 individual buildings, the National Trust's ICVF spearheaded its first neighborhood-wide planning effort in 1989 for the low-income Spring-field historic district of Jacksonville, Florida. Historic preservation was the organizing tool in the Springfield neighborhood. The National Trust worked in partnership with four local nonprofit organizations to develop a five-year (1990–1994) comprehensive revitalization strategy. The strategy addresses the preservation of the area's historic resources as well as the racial and economic diversity in the neighborhood.

The 140-square-block neighborhood of 1,800 historic structures is the largest primarily residential historic district in Florida. Adjacent to Jacksonville's central business district, it is also one of the oldest historic districts in the city and harbors most of its inner-city problems—crime, drug sales, prostitution, poverty, physical deterioration, property abandonment, and demolition—which have severely threatened the future of the neighborhood. Many of its approximately 6,850 residents are low-income African-Americans.

After an eight-month community planning effort facilitated by the National Trust, the neighborhood's nonprofit organizations, local civic and business leaders, and city officials agreed on a neighborhood revitalization strategy based on historic preservation. Yet, importantly, the Springfield Historic District Revitalization Strategy, which calls for a phased implementation of neighborhood development, affordable housing, and historic preservation programs in three stages over 10 years, is not limited to merely treating deteriorated and abandoned historic buildings. It also has an antidisplacement plan that targets 20 percent of the housing stock for nonprofit ownership or management as low- and moderate-income rental housing.

In spite of a slow start due to some early organizational problems, Springfield's major achievements have included the consensus reached on the neighborhood conservation strategy, $6.4 million raised, a 36 percent drop in violent crime, and the acquisition of 20 properties that are being rehabilitated. Implementation has been slower than expected, however, because of the organizational issues. Participation by low-income residents has not increased significantly either, despite substantial technical assistance aimed at improving outreach to this sector of the population. The ongoing challenge is to further the capacity of the area's nonprofit organizations to solve these problems and implement the remaining two phases of the project.

ORGANIZING FOR REVITALIZATION

ICVF projects have been led by a diverse range of organizational structures. Without sacrificing neighborhood control over projects, many ICVF groups have formed limited partnerships with investors who can benefit from the historic rehabilitation and low-income housing tax credits. Rather than obtain a majority interest in a project, the nonprofit organization (or, more commonly, its for-profit subsidiary) generally retains as little as 1 percent ownership as a sole or cogeneral partner. Other nonprofit organizations prefer to sell the building but retain ownership of the land, controlling the property through long-term leases.

Organizational mechanisms for preservation planning and advocacy are designed to encourage broad-based community participation and to pave the way for successful, ongoing programs. In the Springfield planning model, for example, a local advisory board was established that included key civic, business, and neighborhood leaders whose support was needed to implement the plan. A national advisory panel led by the National Trust has provided guidance to the local group. In addition, a new nonprofit organization was formed to organize low-income renters, and the resulting four nonprofit organizations in the neighborhood have been challenged by grantors to better coordinate their fund-raising and programmatic activities.

COMMITTING TO EXCELLENCE

Quality in the design, implementation, and management of real estate projects is important in neighborhood conservation. The development and management role of local nonprofit organizations ensures the project's long-term accountability to its neighborhood. These organizations and their memberships are there to stay and have a vested interest in ensuring high-quality management. Even in home-ownership programs, where the nonprofit organization no longer owns the property, neighborhood groups influence the management process by counseling first-time home buyers in budgeting, maintenance, and repairs, and by providing other support services in the event of job loss or other crises.

A commitment to excellence is essential to the long-term character of neighborhood conservation. Quality workmanship and management reduce long-term maintenance costs. In addition, successful projects whose participants—neighborhood residents, funders, contractors, insurance companies—are proud of their efforts enhance the nonprofit group's track record and make it easier to approach these and other partners for future projects.

EMPOWERING A LOCAL CONSTITUENCY TO PRESERVE ITS RESOURCES

Groups that provide funds and technical assistance and other potential partners in neighborhood conservation increasingly are looking for assurances that neighborhood-based nonprofit organizations reflect the socioeconomic make-up of project communities. The National Trust is no exception, particularly with regard to its Inner-City Ventures Fund. Community residents are acknowledging this need and expressing it in the composition of the boards of directors and membership of nonprofit organizations.

In the case of the Springfield project, lack of participation by the largest segments of its population—low-income and African-American residents—in the existing three neighborhood-based nonprofit organizations was one of the main problems detected early in the planning process.

The Springfield Historic District is a diverse community, culturally and economically. Although the majority of residents are black with low to moderate incomes, it was usually the minority white home-owners who participated in the neighborhood organizations. Out of the neighborhood planning process came a dual challenge: (1) to empower a largely disenfranchised community and (2) to make Springfield's unique heritage more relevant to its current residents.

Two approaches to empowerment have been tried with mixed success. First, a new ecumenical ministries organization was formed to identify a diverse range of social, economic, and housing needs and to organize low-income renters. Initially staffed by a social worker who was to serve as the eyes and ears of the other neighborhood groups, the organization sought to identify tenants' needs and show a connection between those needs and the other organizations' programs and activities.

Although this informational exchange has been important for building trust between the organizations and tenants, it alone has been insufficient to ensure the tenants' increased participation, which is a prerequisite to their empowerment. Two years into the model project, the existing neighborhood organizations recognized the need to make a variety of changes. For example, a conscious effort is being made to fill vacant seats on the preservation organization's board of directors with African-American residents. In addition, to attract a more diverse membership, the content of local preservation organization meetings is being updated to include more cultural topics such as "Preserving Black Craftsmanship in Springfield." Child care is being provided at community meetings to allow low-income single parents to attend.

REGULATORY MECHANISMS FOR NEIGHBORHOOD CONSERVATION

Effective regulatory mechanisms are found at the local, state, and national levels to assist neighborhood conservation efforts:

■ **Historic district designation.** Listing individual properties, thematic groups of properties, or entire districts in the National Register of Historic Places provides some protection from federally funded demolition and adverse development and gives owners access to the historic rehabilitation tax credit for income-producing properties. If stronger protection or tax benefits for owner-occupied housing are sought, local district designation should be sought.

■ **Zoning restrictions and land-use planning.** In some neighborhoods, residential disinvestment has been fueled largely by speculation stemming from inappropriate commercial zoning. Central to the residents' conservation strategies should be efforts to rezone their neighborhood back to residential use and update the land-use plan for their neighborhood.

■ **The Community Reinvestment Act (CRA) and the Home Mortgage Disclosure Act (HMDA).** The CRA and HMDA are federal regulatory tools used by low-income neighborhoods to increase the availability of credit from financial institutions. The CRA directs federally chartered or insured depository institutions to work with public-private partnerships to meet the needs of low-income persons through inner-city neighborhood revitalization, small business development, and rural economic development. HMDA data required from lenders is used by neighborhood advocates to identify credit-worthy applicants who have been denied credit due to their race, income level, or location. These neighborhood advocates then help lenders improve their record of compliance with the CRA.

In addition to HMDA data, many states have their own disclosure laws or regulations requiring more detailed data, and some cities have disclosure ordinances. Other cities and states place government deposits partly on the basis of lending records and require lenders to disclose lending information when seeking such deposits. CRA has helped motivate banks to participate in many neighborhood conservation projects.

■ **The Cranston-Gonzales National Affordable Housing Act of 1990 (NAHA).** This act is a significant piece of housing legislation for the conservation of neighborhoods. Its new Home Investment Partnership Act (HOME) initiative gives funding priority to rehabilitation over new construction, and its Community Housing Partnership (CHP) component provides a 15 percent set-aside of HOME funds for nonprofit organizations. NAHA also provides for the Comprehensive Housing Affordability Strategies (CHAS), a five-year state and local process that identifies housing needs and how they will be addressed. CHAS, which was set up to ensure that neighborhood conservation principles are applied to affordable housing strategies, is the basis on which most federal housing and community development funds will be spent in a community.

■ **Financial incentives.** Since 1976, Congress has provided important tax incentives to foster the preservation of old and historic properties. Currently, the historic rehabilitation tax credit is equal to 20 percent of the cost of rehabilitating historic buildings for rental housing or other income-producing use. A lesser rehabilitation tax credit of 10 percent is available for nonresidential, nonhistoric buildings constructed before 1936.

According to the U.S. Department of Interior's 1990 fiscal report on the preservation tax incentives program, more than 115,000 historic housing units have been rehabilitated since 1976, including more than 20,000 units for low- and moderate-income residents. Many of these projects have combined the historic rehabilitation tax credit with direct subsidies from the U.S. Department of Housing and Urban Development and other sources and, since 1986, the low-income housing tax credit. Unlike the historic rehabilitation tax credit, the low-income housing tax credit is allocated by the state government (usually the state housing finance agency) and is claimed over a 10-year period; there is a special set-aside for projects sponsored by nonprofit groups. The low-income housing tax credit is due to expire in 1992 unless renewed or permanently extended.

In some locales, abatements or alternative valuation schemes on local property taxes are also available. Low-interest loans from state or city agencies and/or nonprofit organizations are also frequently used.

State historic preservation offices are generally the best places to get information about funding sources targeted specifically to historic properties, such as revolving funds operated by nonprofit statewide historic preservation organizations. Most potential funding sources, however, are not earmarked specifically for historic preservation. State and local government offices of community and economic development, housing finance agencies, and Main Street programs are among the best sources of information on funds that can be used for preservation projects.

Sources of funds are constantly changing. Many federal programs used by early ICVF projects have since been discontinued, and nonprofit organizations are turning to new programs such as the Affordable Housing Program (direct subsidies) and Community Investment Program (below-market advances) of the Federal Home Loan Bank System. Other resources may include properties acquired by the Resolution Trust Corporation (RTC) from failed savings and loan associations as a result of the Financial Institutions Reform,

Recovery and Enforcement Act of 1989 (FIRREA). Settlement of a lawsuit brought by the Association of Community Organizations for Reform Now (ACORN) against the RTC in April 1991, has led the RTC to begin implementing an affordable housing disposition program by selling some properties in its inventory at below-market sales prices and with below-market financing, for use as low-income housing. In addition, the RTC agreed to donate to nonprofit organizations or public agencies properties that cost more to carry than RTC can recover from their sales for conversion to low-income housing and other public uses.

Many nonprofit developers look beyond financial assistance programs to opportunities to receive donations of property, materials, and services from public and private sources. For example, the Springfield project regularly receives donated real estate because of tax incentives available in the state of Florida.

IMPLEMENTATION STRATEGIES

The following strategies should be used by local preservationists when trying to implement neighborhood conservation efforts:

- Help set your local and state five-year housing agenda by participating in the Comprehensive Housing Affordability Strategies (CHAS) process required under the Cranston-Gonzales Affordable Housing Act of 1990. Ensure that historic properties are identified and viewed as resources to be rehabilitated as affordable housing. Influence local priorities for allocating financial resources from many federal programs.

- Lobby for legislation and appropriations that are favorable to neighborhood conservation efforts, and develop public-private housing partnerships that utilize these funds. On the federal level, the new Home Investment Partnership Act (HOME) program has a rehabilitation focus and a 15 percent Community Housing Partnership (CHP)

set-aside for nonprofit organizations (funds for brick and mortar and organizational capacity building).

- Work for passage of the Community Revitalization Tax Act, permanent extension of the low-income housing tax credit, and legislation that expands on the Community Reinvestment Act (CRA).

- Work to ensure that financing programs do not discriminate against historic properties. A 1990 National Trust study of federal rural housing and community development programs identified numerous instances in which historic properties were disadvantaged because of the requirements of the financing programs, such as size and design specifications and inflexible construction standards.

- Conduct CRA analyses of local financial institutions, working with lenders to address unmet credit needs. Consider Resolution Trust Corporation-owned properties, and the Federal Home Loan Bank System's Affordable Housing Program (AHP) and Community Improvement Program (CIP), and CRA opportunities when packaging rehabilitation projects. Monitor ongoing bank reform legislation.

- Influence major intermediaries such as Local Initiatives Support Corporation, The Enterprise Foundation, Habitat for Humanity, and the Neighborhood Reinvestment Corporation; secondary markets such as Federal National Mortgage Association; and public agencies such as the U.S. Department of Housing and Urban Development and the Farmers Home Administration.

CONTACTS

Association of Community Organizations for Reform Now
739 8th Street, S.E.
Washington, DC 20003
(202) 547-9292

Campaign for Human Development
3211 4th Street, N.E.
Washington, DC 20017-1194
(202) 541-3210

Center for Community Change
1000 Wisconsin Avenue, N.W.
Washington, DC 20007
(202) 342-0567

Community Information Exchange
1029 Vermont Avenue, N.W.
Suite 900
Washington, DC 20005
(202) 628-2981

Development Training Institute
4806 Seton Drive
Baltimore, MD 21215
(301) 764-0780

The Enterprise Foundation
American City Building
Suite 505
Columbia, MD 21044
(301) 964-1230

Habitat for Humanity
Habitat and Church Streets
Americus, GA 31709
(912) 924-6935

Housing Assistance Council
1025 Vermont Avenue, N.W.
Suite 606
Washington, DC 20005
(202) 842-8600

Institute for Community Economics
57 School Street
Springfield, MA 01105-1331
(413) 746-8660

Institute for Non-Profit Management Training
2025 I Street, N.W.
Suite 1111
Washington, DC 20006
(202) 646-0652

Local Initiatives Support Corporation
666 3d Avenue
New York, NY 10017
(212) 949-8560

McAuley Institute
8300 Colesville Road
Suite 310
Silver Spring, MD 20910
(301) 588-8110

National Association of Neighborhoods
1651 Fuller Street, N.W.
Washington, DC 20009
(202) 332-7766

National Community Reinvestment Coalition
1875 Connecticut Avenue, N.W.
Suite 710
Washington, DC 20009
(202) 986-7898

National Congress for Community Economic Development
1875 Connecticut Avenue, N.W.
Suite 524
Washington, DC 20009
(202) 234-5009

National Council of La Raza
810 1st Street, N.E.
Suite 300
Washington, DC 20002
(202) 289-1380

National Low-Income Housing Coalition
1012 14th Street, N.W.
Washington, DC 20005
(202) 662-1530

National Neighborhood Coalition
810 1st Street, N.E.
Suite 300
Washington, DC 20002
(202) 289-1551

National Puerto Rican Coalition
1700 K Street, N.W.
Suite 500
Washington, DC 20006
(202) 223-3915

National Trust for Historic Preservation
1785 Massachusetts Avenue, N.W.
Washington, DC 20036
(202) 673-4000

National Urban League
1111 14th Street, N.W.
Washington, DC 20005
(202) 898-1604

Neighborhood Reinvestment Corporation
1325 G Street, N.W.
Washington, DC 20005
(202) 376-3216

Planners Network
1601 Connecticut Avenue, N.W.
5th Floor
Washington, DC 20009
(202) 234-9382

FURTHER READING

Historic Preservation and Low Income Neighborhoods

"Doing Well by Doing Good," Roy Kahn. *Historic Preservation*, National Trust for Historic Preservation, May/June 1987.

Historic Preservation for Low-Income Neighborhoods. Washington, D.C.: Community Information Exchange, 1985.

"The Impacts of Historic District Designation: Planning and Policy Implications," Dennis E. Gale. *Journal of the American Planning Association*, Summer 1991.

"Keeping Properties Affordable," Susan Lutzker. *Conserve Neighborhoods*, National Trust for Historic Preservation, March 1987.

A Partnership of Pride (slide show). Office of Financial Services. Washington, D.C.: National Trust for Historic Preservation, 1987.

A Thirst for History: An Assessment of the Compatibility of Federal Rural Development Programs and Historic Preservation—Executive Summary. Rural Heritage Program. Washington, D.C.: National Trust for Historic Preservation, March 1991.

Preservation Incentives

Syndication Strategies for Community-Based Development Organizations, William E. Bivens, III, ed. Washington, D.C.: National Congress for Community Economic Development, February 1983.

"Tax Considerations for Section 501(c)(3) Organizations Engaged in Real Estate Development Activities." Thomas C. Spring. *Preservation Law Reporter*. July 1985.

Tax Credits for Low Income Housing: New Opportunities for Developers, Non-Profits and Communities Under the 1986, 1988, 1989 and 1990 Tax Acts. 7th ed. Joseph Guggenheim. Glen Echo, Md.: Simon Publications, 1992. (P.O. Box 229, Glen Echo, MD 20812, (301) 320-5771.)

Using the Community Reinvestment Act in Low-Income Historic Neighborhoods, Jennifer L. Blake and Stanley Lowe. Information Series. Washington, D.C.: National Trust for Historic Preservation, 1992.

Neighborhood Conservation Planning

Springfield Historic District Revitalization Strategy. Office of Financial Services. Washington, D.C.: National Trust for Historic Preservation, 1989.

Springfield Historic District Revitalization Strategy Phase II. Office of Financial Services. Washington, D.C.: National Trust for Historic Preservation, 1992.

Nonprofit Organizational Development

Keys to the Growth of Neighborhood Development Organizations, Neil Mayer and Jennifer L. Blake. Washington, D.C.: Urban Institute Press, 1981.

MAIN STREETS

Betsy Jackson

"Main Street is about as authentic as things get in America today. It embodies ideals that small business and the nation as a whole have long stood for. The quiet, grassroots revolution under way to revive Main Street's economic health is a real miracle in the making."

Daniel Kehrer, "Miracle on Main Street," *Independent Business*, January–February, 1992

In 1977 the National Trust for Historic Preservation, concerned about continuing threats to Main Street's historic commercial architecture, launched a pilot program—the Main Street Project—to develop a comprehensive revitalization strategy that would stimulate economic development within the context of historic preservation. It worked, and in 1980 the National Trust created the National Main Street Center to help communities throughout the nation conserve their most significant economic and civic resources, their historic commercial buildings. Today, the Main Street program is helping community leaders in hundreds of cities and towns of all sizes to find new economic options for their important historic resources.

The success of the Main Street program is based on a comprehensive strategy of work, tailored to local needs and opportunities, in four broad areas—the Main Street four-point approach:

- **Design:** enhancing the physical appearance of the commercial district by rehabilitating historic buildings, encouraging supportive new construction and developing sensitive design management systems.
- **Organization:** building consensus and cooperation among the many groups and individuals who have a role in the revitalization process.
- **Promotion:** marketing the historic commercial district's assets to customers, potential investors, new businesses, local citizens and visitors.
- **Economic Restructuring:** strengthening the district's existing economic base while finding ways to expand it to meet new opportunities.

The four-point approach is designed to improve all aspects of the downtown, producing both intangible and tangible benefits. Improving economic management, strengthening public participation, and making downtown a fun place to visit are as critical to Main Street's future as recruiting new businesses, rehabilitating buildings, and expanding parking. Building on downtown's inherent assets—rich architecture, personal service, and traditional values—the Main Street approach has rekindled entrepreneurship, downtown cooperation, and civic concern. It has earned national recognition as a practical strategy appropriately scaled to a community's local resources and conditions.

QUESTIONS AND ANSWERS ABOUT THE MAIN STREET PROGRAM

How can my community become a Main Street town?
The National Trust's National Main Street Center usually works directly with statewide and regional Main Street programs to provide technical assistance to a select number of designated communities. Most state and regional Main Street programs hold competitions through which interested communities can apply to take part in the program. The National Trust also works directly with some individual communities, depending on a town's needs and staff availability.

How does Main Street work locally?
Interest in developing a local Main Street program typically comes

Mulberry Street Block, Madison, Indiana. (H.T. Moriarity, HABS)

from merchants and property owners, city government, the chamber of commerce, historic preservationists, and other civic-minded groups. Community leaders discuss goals, establish an organization (Main Street programs are usually private nonprofits), raise money to hire a program director, and create committees and a board of directors.

What assistance is available to establish and manage a local Main Street program?

Thirty-four state and regional Main Street programs operate throughout the country, with staff members who can provide guidance to communities interested in launching a Main Street organization. The National Main Street Center provides direct technical assistance to communities, sponsors several conferences each year, and offers a variety of publications and audio-visual materials to help guide local revitalization efforts. The center also offers an organizational membership program, the National Main Street Network, through which communities receive current information on revitalization activities nationwide.

Does a Main Street program end after three years?

No! Commercial revitalization is a continuing process. Just as a shopping center has a full-time staff and works constantly to ensure proper leasing, management, and marketing, downtown and neighborhood commercial districts also need continuing attention.

STATE MAIN STREET PROGRAMS

Arizona Main Street Program
Department of Commerce and Community Development
3800 North Central
Suite 1400
Phoenix, AZ 85012
(602) 280-1350
Rod Keeling, State Coordinator

Main Street Arkansas
225 East Markham
Suite 300
Little Rock, AR 72201
(501) 324-9346
Sandra H. Hanson, State Coordinator

California Main Street Program
Department of Commerce
Office of Local Development
801 K Street
Suite 1700
Sacramento, CA 95814
(916) 322-1502
Patricia Noyes, State Coordinator

Colorado Initiatives Program
1313 Sherman Street
Room 518
Denver, CO 80203
(303) 866-4881
Nona Lee, State Coordinator

Florida Main Street Program
Bureau of Historic Preservation
Division of Historical Resources
500 South Bronough Street
Room 411
Tallahassee, FL 32399-0250
(904) 487-2333
Robert Trescott, State Coordinator

Georgia Main Street Association
c/o Downtown Cartersville
P.O. Box 1015
Cartersville, GA 30120
(404) 875-2205 (fax)
Ann Arnold, Chairperson

Hawai'i Main Street Program
Department of Land and Natural Resources
33 South King Street
6th Floor
Honolulu, HI 96813
(808) 587-0003
Tonia Moy, State Coordinator

The Main Street Partnership
220 South State Street
Suite 1880
Chicago, IL 60604
(312) 427-3688
Laurie Scott, Regional Coordinator

Indiana Main Street Program
Department of Commerce
Indiana Commerce Center
One North Capitol
Suite 700
Indianapolis, IN 46204-2243
(317) 232-8911
William Dory, State Coordinator

Iowa Main Street Program
Iowa Department of Economic Development
200 East Grand Avenue
Des Moines, IA 50309
(515) 242-4733
Thomas Guzman, State Coordinator

Kansas Main Street Program
Kansas Department of Commerce
400 West 8th Street
5th Floor
Topeka, KS 66603-3957
(913) 296-3485
Brenda Spencer, State Coordinator

Kentucky Main Street Program
Kentucky Heritage Council
300 Washington Street
Frankfort, KY 40601
(502) 564-7005
Bill Steiner, State Coordinator

Massachusetts Main Street Center
Executive Office of Communities and Development
100 Cambridge Street
Room 1803
Boston, MA 02202
(617) 727-7180
Jennifer Pardee, Assistant State Coordinator

Michigan Main Street Program
Michigan Department of Commerce
Local Development Services
525 West Ottowa Street
Law Building, 5th Floor
Lansing, MI 48911
(517) 373-8730
Bob Terry, State Coordinator

Minnesota Main Street Program
Department of Trade and Economic Development
150 East Kellog Boulevard
900 American Center Building
St. Paul, MN 55101
(612) 297-1755
Peggy Pappenfus, State Coordinator

Mississippi Main Street Program
Mississippi Department of Economic Development
P.O. Box 849
Jackson, MS 39205
(601) 359-3420
Scott Barksdale, State Coordinator

Missouri Main Street Program
Missouri Department of Economic Development
P.O. Box 118
301 West High Street
Truman Building, Room 770
Jefferson City, MO 65102
(314) 751-7939
Randy Gray, State Coordinator

Main Street New Jersey
Office of New Jersey Heritage
CN 404
Trenton, NJ 08625
(609) 292-2023
Barbara Swanda, State Coordinator

New Mexico Main Street Program
Economic Development and Tourism
1100 St. Francis Drive
Santa Fe, NM 87503
(505) 827-0272
Julie Johncox, State Coordinator

North Carolina Main Street Center
Department of Economic and Community
Development
1307 Glenwood Avenue
Suite 250
Raleigh, NC 27605
(919) 733-2850
Rodney Swink, State Coordinator

Oklahoma Main Street Program
Oklahoma Department of Commerce
PO Box 26980
6601 Broadway Extension
Oklahoma City, OK 73126-0980
(405) 841-5115
Susie Clinard, State Coordinator

Oregon Main Street Program
Oregon Downtown Development
Association
921 Southwest Morrison Street
Suite 508
Portland, OR 97205
(503) 222-2182
Kate Joncas, State Coordinator

Pennsylvania Downtown Center
301 East Second Street
Bloomsburg, PA 17815
(717) 784-0456
Beth Proper-Spokas, State Coordinator

Pennsylvania Main Street Program
Department of Community Affairs
504 Forum Building
Harrisburg, PA 17120
(717) 783-3068
Diana Kerr, State Coordinator

**South Carolina Downtown Development
Association**
P.O. Box 11637
Kittrell Center, Suite 314-C
1529 Washington Street
Columbia, SC 29211
(803) 256-3560
Ben Boozer, State Coordinator

South Dakota Main Street Program
Governor's Office of Economic
Development
Capitol Lake Plaza
Pierre, SD 57501-3369
(605) 773-5032
Kathy Lucas, State Coordinator

Tennessee Main Street Project
Department of Economic and Community
Development
320 6th Avenue, North
6th Floor
Nashville, TN 37243-0405
(615) 741-2373
Donna Emerson, State Coordinator

Texas Main Street Center
Texas Historical Commission
P.O. Box 12276
1511 Colorado Street
Austin, TX 78711
(512) 463-6092
Anice Read, State Coordinator

Virginia Main Street Program
Virginia Department of Housing and
Community Development
205 North 4th Street
Richmond, VA 23219
(804) 786-4966
Teresa Lynch, State Coordinator

Washington Downtown Revitalization
111 West 21st Avenue, MS/KL11
Olympia, WA 98504-5411
(206) 586-8977
Dick Larman, State Coordinator

West Virginia Main Street Program
Capitol Complex
Building B531
Charleston, WV 25305
(304) 348-0121
Susan Salisbury, State Coordinator

Wisconsin Main Street Program
P.O. Box 7970
123 West Washington
Madison, WI 53707
(608) 267-3855
Alicia Goehring, State Coordinator

Wyoming Main Street Project
Department of Economic Development
and Stabilization Board
Herschler Building, East Wing
3rd Floor
Cheyenne, WY 82002
(307) 777-7284
Rick Hunnicutt, State Coordinator

CONTACT

**National Main Street Center
National Trust for Historic Preservation**
1785 Massachusetts Avenue, N.W.
Washington, D.C. 20036
(202) 673-4219
Kennedy Smith, Director
R. McDuffie Nichols, Program Manager
for Technical Services
Elizabeth Jackson, Program Manager for
Information and Services Development
Linda Donavan Harper, Program Manager
for Membership and Resource
Development

FURTHER READING

Bringing Back Urban Vitality. National
Main Street Center. Washington, D.C.:
National Trust for Historic Preservation,
1988.

*The Buildings of Main Street: A Guide to
American Commercial Architecture.*
Richard Longstreth. Washington, D.C.:
Preservation Press, 1987.

Business Development for Main Street.
National Main Street Center. Washington,
D.C.: National Trust for Historic
Preservation, 1990.

Community Initiated Development.
Donovan Rypkema for the National Main
Street Center. Washington, D.C.:
National Trust for Historic Preservation,
1992.

Guiding Design on Main Street. National
Main Street Center. Washington, D.C.:
National Trust for Historic Preservation,
1988.

*Local Government and Historic
Preservation.* Richard Wagner for the
National Main Street Center. Washington,
D.C.: National Trust for Historic
Preservation, 1991.

*The Main Street Board Member's
Handbook.* National Main Street Center.
Washington, D.C.: National Trust for
Historic Preservation, 1988.

*Main Street News Guide to Building
Maintenance.* Richard Wagner for the
National Main Street Center. Washington,
D.C.: National Trust for Historic
Preservation, 1992.

Main Street: Open for Business. Linda S.
Glisson for the National Main Street
Center. Washington, D.C.: National Trust
for Historic Preservation, 1984.

Market Analysis on Main Street. National
Main Street Center. Washington, D.C.:
National Trust for Historic Preservation,
1990.

*National Main Street Center Publications
List.* Washington, D.C.: National Trust for
Historic Preservation. Includes books,
reports, technical bulletins, slide shows,
videocassettes, and computer software.

New Directions for Urban Main Streets.
Suzanne G. Dane, ed., National Main
Street Center. Washington, D.C.:
National Trust for Historic Preservation,
1989.

*The Parking Handbook for Small
Communities.* John Edwards for the
National Main Street Center and the
Institute of Transportation Engineers.
Washington, D.C.: National Trust for
Historic Preservation, 1992.

Revitalizing Downtown 1976–1986.
National Main Street Center and Urban
Institute. Washington, D.C.: National
Trust for Historic Preservation, 1988.

*The Small Business Retention, Expansion
and Recruitment Project.* National Main
Street Center. Washington, D.C.:
National Trust for Historic Preservation,
1987.

RURAL PRESERVATION

Shelley Mastran

"Every day in the United States, 12 square miles of prime farmland are lost forever. . . . In the course of a year, that adds up to more than three million acres of productive soils that are paved over to make room for urban sprawl."

Jim Riggle, American Farmland Trust, 1981

America's history rests on rural foundations. Until World War II, the vast majority of Americans lived on farms or in small towns, and agrarian values have shaped much of our nation's character. Even today, when more than 80 percent of Americans live in metropolitan areas, our rural heritage stands to remind us of our roots.

This legacy is visible on the land in the form of farmhouses, barns, fences, churches, mills, graveyards, roads, hedgerows, and fields. Yet this heritage is under threat from changes in transportation, the spread of metropolitan areas, the consolidation of farms, and the loss of population and an economic base in rural areas. Many rural resources are abandoned or neglected or are pushed aside for development.

Rural historic preservation works to promote and to mitigate the loss of the nation's agrarian heritage. This work includes land conservation—placing easements or other restrictions on land to prevent adverse development, the restoration and rehabilitation of historic structures, the revitalization of small towns and rural communities, and the promotion of commerce and tourism in areas rich in rural historic resources.

The National Trust's Rural Heritage Initiative has worked to preserve America's built and natural countryside since 1979. The program focuses on establishing a rural constituency for historic preservation, creating forums for education, publishing information on rural historic preservation issues, and influencing public policy at all levels of government.

In 1991 the Rural Heritage Initiative conducted an assessment, sponsored by the U.S. Department of Agriculture, of the compatibility of historic preservation and federal rural development programs. The study revealed many instances of compatibility and made recommendations to Congress on ways to promote the use of historic resources in rural development.

The Rural Heritage Initiative also has conducted workshops on rural community planning that were sponsored by the National Endowment for the Arts. Called "Your Town: Designing Its Future," the workshops were held in rural communities in the West and South. The initiative also has held two workshops on Low-Income Housing and Historic Preservation in North Carolina.

RESOURCES

Land Use and Natural Resources

American Farmland Trust
1920 N Street, N.W.
Suite 400
Washington, DC 20036
(202) 659-5170

The American Farmland Trust (AFT) informs Americans about the issues posed by rapid depletion of the nation's farmland, the harmful effects of soil erosion, and other threats to the agricultural viability of the land. AFT undertakes projects—with cooperating organizations and individual landowners—that demonstrate farmland protection techniques and assist farmland protection policy efforts and land-use initiatives by local, state, and federal governments.

Conservation Foundation
1250 24th Street, N.W.
Washington, DC 20037
(202) 293-4800

The Conservation Foundation, affiliated with the World Wildlife Fund-U.S., conducts policy research on emerging issues in environmental and resource management. The foundation publishes information for policy and opinion makers in government, business, academia, and the conservation community. The Conservation Foundation's Successful Communities Program helps communities and groups establish or strengthen land trusts and develop and implement plans for resource protection.

Conservation Fund
1800 North Kent Street
Suite 1120
Arlington, VA 22209
(703) 522-8008

The Conservation Fund, a national nonprofit organization, is dedicated to advancing land and water conservation through creative ideas and new resources. It provides specialized services ranging from land planning and acquisition to ecological assessment and communications support, and analyzes regulations, policies, and bills relating to natural resources and land use. The fund also publishes a bimonthly newsletter, *Common Ground*, and the bimonthly *Land Letter*, which discusses natural resource policy at the national level.

Land Trust Alliance
900 17th Street, N.W.
Suite 410
Washington, DC 20006-2596
(202) 785-1408

The Land Trust Alliance is the national organization of land trusts. The Alliance provides specialized services, publications, information, and training for land trusts and other conservation organizations, and works for public policies that advance land conservation.

Rails-to-Trails Conservancy
1400 16th Street, N.W.
Suite 300
Washington, DC 20036
(202) 797-5400

The Rails-to-Trails Conservancy (RTC), a nonprofit organization, is devoted to converting abandoned railroad rights-of-way into trails for public use. In partnership with citizen groups, public agencies, and

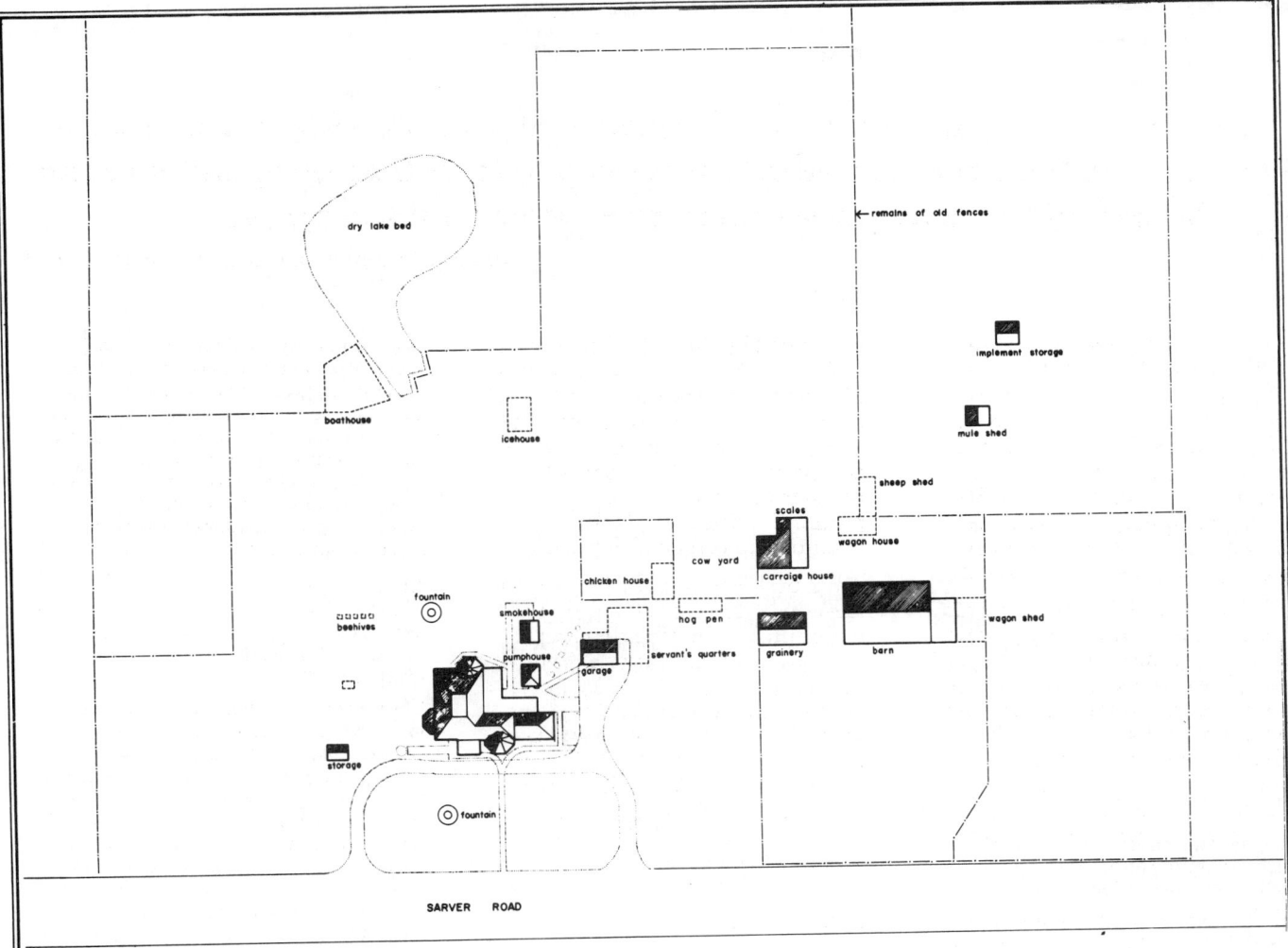

Isaac Kinsey Farm, Milton vicinity, Indiana. (R. Perlmutter, HABS)

railroads, the conservancy is working to build a transcontinental trailway network. RTC's program includes technical assistance, public education, advocacy, negotiation, legislation, and regulatory action. Publications such as the *Citizen's Manual*, *Legal Manual*, and the newsletter *Trailblazer*, in addition to conferences and statewide meetings, keep advocates up to date on rail-trail regulations and procedures.

Trust for Public Land
116 New Montgomery Street
4th Floor
San Francisco, CA 94105
(415) 495-4014

The Trust for Public Land (TPL), a nonprofit organization with 12 offices located throughout the United States, is a network of experts in real estate negotiation, finance, and law, dedicated to protecting land for the public's use and enjoyment.

Neither a membership nor an advocacy group, TPL helps public agencies, landowners, and citizens' groups protect land of recreational, historic, and scenic value. Its projects range from lot-sized neighborhood gardens to vast additions to forests, parks, and national recreation areas.

Design

Partners for Livable Places
1429 21st Street, N.W.
Washington, DC 20036
(202) 887-5990

Partners for Livable Places is an international coalition of more than 600 organizations and individuals committed to improving communities' economic health and quality of life through collaborative resource management. Partners serves as a national resource center for information on the built and natural environments and offers a research and referral service on a

broad range of community improvement projects. Partners publishes *AMENITIES* and *Livability Digest*.

Scenic America
(formerly Coalition for Scenic Beauty)
216 7th Street, S.E.
Washington, DC 20003
(202) 546-1100

Scenic America, a nonprofit membership organization, is devoted to protecting America's scenic resources and cleaning up roadside blight and visual pollution. Scenic America conducts workshops and provides information and technical assistance on sign control, tree ordinances, scenic highways, growth management, and all forms of aesthetic regulation.

Rural Development

Center for Rural Affairs
P.O. Box 405
Walthill, NE 68067-0405
(402) 846-5428

The Center for Rural Affairs works to help
low-income people. It is particularly concerned about the well-being of small,
moderate-sized, and beginning farmers,
especially in Nebraska and neighboring
states. The center publishes *Small Farm
Advocate*, a quarterly newsletter, and
Center for Rural Affairs, a monthly
newsletter.

Heartland Center for Leadership Development
941 O Street
Suite 900
Lincoln, NE 68508
(402) 474-7667
(402) 474-7672 Fax

The Heartland Center for Leadership
Development is an independent, nonprofit organization developing local leadership that responds to the challenges of
the future. A major focus of the center's
activities is the creation of practical
resources and policies for rural community survival. Programs and publications
stress the critical role played by local leadership in facing challenges, and to that
end the center's practical programs
include training communities, businesses, and organizations to develop the
capacity for locally directed strategic planning, helping policy makers clarify key
questions in the futures of communities
and states, and conducting field-based
research related to leadership and its
potential impact.

Highlander Center
Route 3, Box 370
New Market, TN 37820
(615) 933-3443

The Highlander Research and Education
Center has developed curriculum materials for adults on economic development
and the changing economy of the Appalachian region. The Highlander Economics
Education Project was developed to help
people in rural Appalachian communities
understand the changing economy and
find ways of dealing with the economy
and community economic development.
The project has created resource materials, conferences, and community
workshops.

Institute for Community Economics
151 Montague City Road
Greenfield, MA 01301
(413) 774-7956

The Institute for Community Economics
(ICE) has a range of programs that assist
the efforts of community residents to gain
control over and benefit from local economic development. ICE has developed
community land trust and loan fund
models to address the problems of
lower-income communities suffering
from limited access to land, housing, and
capital. ICE also provides organizational
and developmental assistance to community groups and public agencies around
the country. The institute is responsible
for *Community Economics*, a quarterly
newsletter, as well as other publications.

Kirkwood Rural Development Center
Kirkwood Community College
P.O. Box 2068
Cedar Rapids, IA 52406
(319) 398-5699

The Kirkwood Rural Development Center
(RDC) provides technical assistance and
training to rural families in the development of alternative enterprises that keep
rural families in their locale and add value
to the local economy. RDC facilitators
work to overcome information, marketing, or financial barriers that keep local
ideas from becoming a reality. The center
works with any rural client who wants to
start a business, the focus being on
micro-enterprises—businesses that supplement a family's income. RDC has
gained national recognition for its
grass-roots approach to rural development. It has an extensive library collection covering a wide variety of
diversification, small business, and rural
development topics and publishes a newsletter, *Tomorrow's Harvest*.

Midwest Research Institute
425 Volker Boulevard
Kansas City, MO 64110
(816) 753-7600

The Midwest Research Institute (MRI) is
an independent, nonprofit organization
that performs contract research for clients
in both the public and private sectors.
MRI's staff combine their expertise and
resources to carry out research programs
covering a broad spectrum of modern science and technology. The Economics and
Management Sciences Department provides expert assistance to clients from all
sectors of the economy. Its capabilities
include economic analyses, economic and
industrial development, management and
business studies, market and feasibility
analyses, and industry competitive
assessments.

National Association of Counties
440 1st Street, N.W.
8th Floor
Washington, DC 20001
(202) 393-6226

The National Association of Counties
(NACo) represents the more than 3,000
county governments in the United States.
Its goals are to improve county government, act as a liaison with other levels of
government, serve as a national voice for
counties, and advance public understanding of the role of counties.

National Association of Development Organizations
444 North Capitol Street, N.W.
Suite 628
Washington, DC 20001
(202) 624-7806
(202) 624-8813 Fax

The National Association of Development
Organizations (NADO), a national
grass-roots network, promotes locally
based economic development organizations in America's small cities and rural
areas. NADO members include multi-
county planning and development organizations; state, county and city agencies;
educational institutions; and private businesses. Member organizations engage in
activities designed to create and maintain
private-sector jobs. For up-to-the-minute
information, NADO publishes special
reports and *NADO News*, a weekly newsletter, and conducts an annual conference
and regional training workshops.

National Association of Regional Councils
1700 K Street, N.W.
Suite 1300
Washington, DC 20006
(202) 457-0710

The National Association of Regional
Councils (NARC) is a membership organization for regional councils across the
country, with affiliate membership for
other public and private organizations
that are interested in planning, development, and governance at the regional
level. Representing the nation's more
than 500 regional councils, NARC serves
as the national source of information for
and about regional councils. NARC holds
an annual conference that provides a
forum for information exchange and an
annual Washington Policy Conference in
Washington, D.C. Among the several publications produced by the association are
weekly and biweekly newsletters and special reports.

National Association of State Departments of Agriculture
1616 H Street, N.W.
Washington, DC 20006
(202) 628-1566
(202) 628-9461 Fax

The National Association of State Departments of Agriculture (NASDA) is a nonprofit organization composed of
commissioners, secretaries, and directors
of the departments of agriculture in the
fifty states and four trust territories.
NASDA's purpose is to provide a voluntary, nonpolitical organization to promote
unity and efficiency in administration of
agricultural statutes and regulations, to
develop cooperation between departments

of comparable agencies with the United States Department of Agriculture and with persons interested in agriculture, and to establish federal-state cooperative programs to promote agricultural interests.

National Association of Towns and Townships
1522 K Street, N.W.
Suite 730
Washington, DC 20005
(202) 737-5200

The National Association of Towns and Townships (NATaT), a membership organization, offers technical assistance, educational assistance, and public policy support to local government officials. The association publishes 10 issues per year of *NATaT's Reporter.*

National Rural Electric Cooperative Association
1800 Massachusetts Avenue, N.W.
Washington, DC 20036
(202) 857-9500

The National Rural Electric Cooperative Association (NRECA) represents the national interests of rural electric systems. NRECA provides legislative services and programs in management training, insurance, public relations, and advertising. Two major publications of NRECA are *Rural Electrification* magazine and the *Rural Electric Newsletter,* which reports mainly on legislative and governmental matters. The association and its members also support supplemental energy and environmental research and administer a program of technical advice and assistance in the development of rural electric systems in 37 countries.

North Central Regional Center for Rural Development
578 Heady Hall
Iowa State University
Ames, IA 50011
(515) 294-6257

Northeast Regional Center for Rural Development
104 Weaver Building
The Pennsylvania State University
University Park, PA 16802
(814) 863-4656

Southern Rural Development Center
P.O. Box 5446
Mississippi State, MS 39762
(601) 325-3207

Western Rural Development Center
Oregon State University
Corvallis, OR 97331
(503) 737-3621

The four Regional Centers for Rural Development assist the process of public and private decision making on rural issues by encouraging and conducting multidisciplinary research, extension, and educational programs. These activities are designed to improve the social and economic well-being of nonmetropolitan communities in the regions. Publications, conferences, interest networks, research projects, and program development assistance coordinated by the centers facilitate a strong rural development effort.

Small Towns Institute
P.O. Box 517
Ellensburg, WA 98926
(509) 925-1830

The Small Towns Institute is a nonprofit corporation dedicated to collecting and disseminating information on new and innovative ideas concerning the issues and problems facing small towns and nonurban areas. The institute publishes *Small Town,* a news journal containing perspectives from a variety of sources that share a common interest in enhancing the future of small communities.

Townscape Institute
2 Hubbard Park
Cambridge, MA 02138
(617) 491-8952

The Townscape Institute is a nonprofit public interest organization concerned with increasing the livability of cities, towns, and neighborhoods through the conservation, interpretation, and enhancement of the man-made environment. Program areas include townscape planning and interpretation, urban design and public art, preservation education, and environmental advocacy through lectures, publications, films, and exhibits.

Rural Housing

Council for Rural Housing and Development
2300 M Street, N.W.
Suite 260
Washington, DC 20037
(202) 955-9715

The Council for Rural Housing and Development (CRHD), as the only national nonprofit corporation whose sole focus is the Farmers Home Administration (FmHA) Section 515 program, advocates an effective and adequately funded rural rental housing program in a fair tax environment for private-sector participants. CRHD maintains a working relationship with FmHA and provides its members with access to a knowledgeable staff to answer questions. The Council publishes *CRHD Report,* a monthly publication; Rural Survival Alert (RSA), a newsletter; and special mailings. It holds midyear and annual meetings, plus two seminars each year on the low-income housing tax credit.

Housing Assistance Council
1025 Vermont Avenue, N.W.
Suite 606
Washington, DC 20005
(202) 842-8600

The Housing Assistance Council (HAC) is a service organization that works to increase the availability of housing for low-income people in rural areas. It administers a revolving loan fund, provides technical assistance, undertakes research and training programs, and publishes booklets on housing issues and programs. The Council also publishes *HAC News,* a biweekly newsletter.

Rural America
725 15th Street, N.W.
Suite 900
Washington, DC 20005
(202) 628-1480

Rural America is a membership organization concerned with rural transportation, especially the needs of low-income and minority groups. It publishes 10 issues per year of *Community Transportation Reporter.*

FURTHER READING

A Citizen's Guide to Conserving Land and Creating Affordable Housing. Burlington, Vermont: The Burlington Community Land Trust and the Vermont Land Trust, 1990.

Building by the Book: Pattern Book Architecture in New Jersey. Robert P. Guter and Janet W. Foster. New Brunswick, N.J.: Rutgers University Press, 1992.

Building on the Past: A Guide to Historic Preservation and Affordable Rural Housing. Washington, D.C.: Housing Assistance Council, August 1988.

Community-Based Development: Investing in Renewal. The Report of The Task Force on Community-Based Development. Washington, D.C.: National Congress for Community Economic Development, 1987.

The Conservation Easement Handbook: Managing Land Conservation and Historic Preservation Easement Programs. Janet Diehl and Thomas S. Barrett. Washington, D.C.: Trust for Public Land and Land Trust Exchange, 1988.

Creating Successful Communities. M. Mantrell, S. Harper, and L. Propst. Chicago: Island Press, 1989.

Dealing with Change in the Connecticut River Valley. Robert Yaro and Randall Arendt. Amherst, Mass.: Center for Rural Massachusetts, 1988.

From the Grassroots. Jan L. Flora and others. Profiles of 103 Rural Self-Development Projects. Washington, D.C.: USDA Economic Research Service, 1991.

Guidelines for Evaluating and Documenting Rural Historic Landscapes. Linda F. McClelland and others. National Register Bulletin 30. Washington, D.C.: U.S. Department of the Interior, National Park Service, n.d.

Landscape Architecture in the Rural Landscape. D. Coen, J. Nassauer, and R. Tuttle. Landscape Architecture Technical Information Series. Washington, D.C.: American Society of Landscape Architects, 1987.

Planning and Zoning for Farmland Protection: A Community Based Approach. American Farmland Trust. Washington, D.C.: American Farmland Trust, 1987.

Planning Small Town America. Kristina Ford with James Lopach and Dennis O'Donnell. Washington, D.C.: American Planning Association, 1990.

Plowing the Urban Fringe: An Assessment of Alternative Approaches to Farmland Preservation. Hal Hiemstra and Nancy Bushwick, eds. Fort Lauderdale: Florida Atlantic University/Florida International University Joint Center for Environmental and Urban Problems, 1989.

Recouple-Natural Resource Strategies for Rural Economic Development. Kansas City, Mo.: Midwest Research Institute, 1990.

Rural Planning and Development in the United States. Mark B. Lapping, Thomas L. Daniels, and John W. Keller. New York: The Guilford Press, 1989.

Saving America's Countryside. Samuel N. Stokes, A. Elizabeth Watson, and others. Baltimore: The Johns Hopkins University Press, 1989.

Searching for "The Way That Works." An Analysis of FmHA Rural Development Policy and Implementation. Washington, D.C.: The Aspen Institute, 1990.

The Smalltown Planning Handbook. Thomas L. Daniels and John W. Keller with Mark B. Lapping. American Planning Association, 1988.

Special Places. Breton Rouche. Boston: Little, Brown & Company, 1982.

A Thirst for History, A Study of the Compatibility of Federal Rural Development and Historic Preservation. Washington, D.C.: National Trust for Historic Preservation, 1991.

MARITIME PRESERVATION

Michael Naab

"We are all joined by a common interest, a common devotion and love for the sea."

John F. Kennedy

Our maritime heritage represents an integral part of the history of America. Native American settlements grew on lake shores and where natural harbors existed along the seacoasts and rivers. Maritime endeavors made possible the age of exploration and colonization, the exchange of cultures, and the opening of new territories and markets. Vast numbers of immigrants came to these shores by sea, and the produce of farms and factories was shipped along waterways and through seaports. American naval vessels and merchant ships, ubiquitous in every ocean, helped to establish the United States as a major economic and military power.

Yet America is losing its maritime heritage. Of the thousands of vessels that graced American harbors and waterways as recently as the turn of the century, fewer than 300 are preserved. Many of these, in precarious condition, are severely threatened by neglect, deterioration, mismanagement, and lack of

funds to properly maintain them. Historic shipwrecks are increasingly endangered by treasure hunters whose activities are made easier by advancing technologies. And historic waterfronts are being lost to opposing threats: abandonment and intensive development.

These are just some of the problems and types of maritime resources that are the focus of concern of maritime preservation. One can speak as well of abandoned lighthouses, dying skills and culture, and collapsing canals. All are part of the heritage of maritime America.

Reminders of our maritime past exist today in every region of the country. Historic vessels range from small fishing boats to aircraft carriers, and from paddlewheel steamers to the USS *Constitution*, which every schoolchild knows as "Old Ironsides." Historic lighthouses dot the nation's shores and waterways. Historic shipwrecks, lying in nearly every lake, river,

bay, and ocean, are time capsules offering unique opportunities to gather information about the past. Historic canals crisscross the landscape. Historic waterfronts, including structures associated with once-thriving maritime industries, exist in almost every town or city that has a river or harbor. Scores of museums collect and preserve the artifacts of our maritime culture, and a myriad of educational and training programs carry on maritime traditions and folkways.

Since 1976, the National Trust for Historic Preservation has maintained an Office of Maritime Preservation, which guides the Trust's involvement in maritime preservation issues. In addition to providing technical advice to individual maritime preservation projects, the Maritime Office cooperates with the regional offices and other programs within the National Trust in delivering services to maritime organizations. For example, the Maritime Office works with the Office of Law and Public Policy to

Thames Tow Boat Company, Thames Shipyard, New London, Connecticut. (B. Freemont, HAER)

foster the passage of legislation, such as the Abandoned Shipwreck Act, and to litigate maritime preservation issues in the courts. The Office of Financial Services and the regional offices, in consultation with the Maritime Office, awards grants from the Preservation Services Fund and the Critical Issues Fund and loans from the National Preservation Loan Fund to maritime projects. A host of National Trust offices rely on the maritime program as an information source for their various publications.

MARITIME PRESERVATION ORGANIZATIONS

Canals

American Canal Society
35 Towana Road
Richmond, VA 23226
(804) 288-1334
William E. Trout, III, President

General Maritime Interest

National Maritime Alliance
2000 L Street, N.W.
Suite 403
Washington, DC 20036
(410) 889-5164
Gail Shawe, Coordinator

Lighthouses

Great Lakes Lighthouse Keepers Association
P.O. Box 580
Allen Park, MI 48101
(313) 662-1200
Richard Moehl, President

Lighthouse Preservation Society
P.O. Box 736
Rockport, MA 01966
(508) 281-6336
Valerie I. Nelson, Executive Director

U.S. Lighthouse Society
244 Kearny Street
5th Floor
San Francisco, CA 94108
(415) 362-7255
Wayne C. Wheeler, President

Maritime History

National Maritime Historical Society
5 John Walsh Boulevard
Peekskill, NY 10566
(914) 737-7878
Peter Stanford, President

Maritime Museums

Council of American Maritime Museums
South Street Seaport Museum
207 Front Street
New York, NY 10038
(212) 669-9400
Peter Neill, President

Naval Ships

Historic Naval Ships Association
of North America
U.S. Naval Academy Museum
Annapolis, MD 21402-5034
(410) 267-2108
James Cheevers, Secretary

Sail Training, Sea Experience

American Sail Training Association
P.O. Box 1459
Newport, RI 02840
(401) 846-1775
Beth Bonds, Coordinator

Small Craft

Museum Small Craft Association
HCR 68, Box 196-C
Cushing, ME 04563
(207) 354-0441
Benjamin A.G. Fuller, President

Underwater Archaeology

Advisory Council on Underwater Archaeology
National Museum of American History
Room 5010
Smithsonian Institution
Washington, DC 20560
(202) 357-2025
Paul F. Johnston, Chairman

Waterfronts

The Waterfront Center
1536 44th Street, N.W.
Washington, DC 20007-2066
(202) 337-0356
Ann Breen, Codirector

FURTHER READING

Great American Lighthouses. F. Ross Holland, Jr. Washington, D.C.: National Trust for Historic Preservation, 1989.

Great American Ships. James P. Delgado and J. Candace Clifford. Washington, D.C.: National Trust for Historic Preservation, 1991.

Guidelines for Evaluating and Documenting Historic Aids to Navigation (National Register Bulletin 34). James P. Delgado and Kevin J. Foster. Washington, D.C.: Interagency Resources Division, National Park Service, U.S. Department of the Interior, n.d.

Guidelines for Recording Historic Ships. Richard K. Anderson, Jr. Washington, D.C.: Historic American Buildings Survey/Historic American Engineering Record, National Park Service, U.S. Department of the Interior, 1988.

Historic Maritime Resources: Planning for Preservation. Lynn R. Hickerson, ed. Washington, D.C.: National Trust for Historic Preservation, 1991.

Maritime America; A Legacy at Risk: Issues and Needs in Maritime Heritage Preservation. Marcia Myers. Washington, D.C.: National Trust for Historic Preservation, 1988.

The Naval Institute Guide to Maritime Museums of North America. Robert H. Smith. Annapolis, Md.: Naval Institute Press, 1990.

Nominating Historic Vessels and Shipwrecks to the National Register of Historic Places (National Register Bulletin 20). James P. Delgado. Washington, D.C.: Interagency Resources Division, National Park Service, U.S. Department of the Interior, n.d.

Sail Training Ships and Programs: An Annotated Directory. Robby Robinson, ed. Newport, R.I.: American Sail Training Association, 1990.

Sea History's Guide to American and Canadian Maritime Museums. Joseph M. Stanford. Croton-on-Hudson, N.Y.: National Maritime Historic Society, 1990.

The Secretary of the Interior's Standards for Historic Vessel Preservation Projects, With Guidelines for Applying the Standards. Michael Naab. Washington, D.C.: National Maritime Initiative, National Park Service, U.S. Department of the Interior, 1990.

HISTORIC SITES AND MUSEUM PROPERTIES

Frank Sanchis and Susan Schreiber

"Have we saved hundreds of museum-quality properties from the bulldozer and wrecking ball only to have them eventually deteriorate for want of maintenance money, effective long-term management, and adequate levels of public support?"

Gerald George, "Historic Property Museums: What Are They Preserving?" *Forum*, Summer 1989

PROFILE OF HISTORIC SITES

In 1988 the National Trust completed a study of historic sites open to the public in the United States that gathered data on the numbers, types, staffing, funding, longevity, geographical distribution, management, and other aspects of the operation of these sites.

Numbers and Location
The date collected in the study suggest that the number of historic sites open to the public is larger than previously thought—between 5,000 and 6,000 sites. As anticipated, these sites are located primarily in areas that historically have been heavily populated: the Northeast has the largest concentration of sites, the West the smallest number. Surprisingly, however, most historic sites are located in rural communities and in cities with populations of fewer than 50,000 people.

Types of Historic Sites
Of all the types of historic sites, houses tend to predominate, perhaps because individual families have committed themselves to preserving their homes. There are, however, many other types of historic sites open to the public, including Native American settlements, canals, marinas, forts, battlefields, bridges, railroad stations, barns, trails, cemeteries, apartment houses, and company towns. This diversity is perhaps one of the most positive findings of the study, indicating a willingness to preserve vernacular and ethnic architecture and to consider the historic merits of commercial and industrial structures as well as residential buildings. The trend to preserve this diversity of historic

sites is likely to continue. Historic preservationists and the general public will increasingly recognize the value of preserving all types of buildings that document the cultural, economic, and architectural history of the United States.

Historic sites open to the public are typically operated as interpretive centers and museums devoted to educating the public about some aspect of our country's history. House museums, for example, tend to be devoted to interpreting the lives and times of their prominent owners and occupants. Typically a house museum will display furnishings and artifacts associated with one or more individuals along with educational exhibits that interpret their lives.

Other historic sites are devoted to the history of an industry. For example, the Sloss Furnace Company Museum in Birmingham, Alabama, is devoted to educating the public about the steel industry at the turn of the century. Still other historic sites and museum properties are devoted to interpreting the history of America's transportation systems. The Northwest Seaport in Seattle is an example. Here, historic ships from the late 19th and early 20th centuries are used to educate the public about the maritime history of the Pacific Northwest. One of the ships in the collection, the 1889 wooden tugboat *Arthur Foss* still sails the waters of Puget Sound during waterfront festivals.

Entire communities have also been preserved as museum properties. Perhaps the best known is Williamsburg, Virginia. Begun in the 1920s, the Colonial capital of Vir-

ginia consists of dozens of buildings reconstructed as they were during the early 1770s. The foundation that oversees this museum has also taken great care to reproduce the landscape, streets, sidewalks, and other aspects of urban life during the period.

In addition to authentic museum communities such as Williamsburg, many historic villages have been created using actual historic buildings from different areas. One example is the West Virginia State Farm Museum near Point Pleasant. Here 34 historic 19th century buildings from across the state were assembled into a village that never actually existed.

Administration and Staffing
The staffing of historic sites and museum properties ranges from those run by all volunteers to those with a staff of more than 50 professionals and 500 volunteers. Nearly 80 percent of historic sites in the National Trust study operate with annual cash budgets of less than $50,000, and 65 percent have no paid staff, relying completely on volunteers for their operation.

The majority of historic sites and museum properties that maintain high standards of preservation and care and are open to the public on a regular basis employ professionally trained staff. The process of maintaining historic buildings, landscapes, and collections, researching the site's history, and interpreting the site to the public on a regular basis is labor intensive. A medium-sized historic site may have a director, curator, and education expert, with specialized personnel who contract on a fee-for-service basis. Large historic site complexes, such as Colonial Williamsburg, the Henry Ford Museum and Greenfield Village, and Mystic Seaport, employ the full range of historians, curators, conservators, educators, fund raisers, and maintenance workers. Another staffing model is found in the large regional or national organization that owns and operates a number of sites and provides technical expertise from one central location. Such organizations include the National Trust for Historic Preservation, the Society for the Preservation of New England Antiquities, and the Minnesota Historical Society.

Governance of Sites
Historic sites and museum properties vary as much in who governs them as they do in type and subject matter. Of the nearly 6,000 known historic sites, the National Park Service administers only about 4 percent. Many historic sites are governed by state or municipal authorities, receiving oversight and some, but seldom all, of their funding from these government entities. The governing agency may also be a preservation office or a university. But the majority of historic sites are privately run and are governed by a board of trustees, drawn from local communities, to

provide guidance in the development of policies, oversee the implementation of those policies, and spearhead fund-raising efforts. Whatever the governing structure, it is the responsibility of the governing body to have a clear sense of the site's mission and purpose and to remember that it holds the site in trust for the public.

How Historic Sites Support Themselves
With recent advances in the preservation and conservation of buildings and collections and a growing recognition on the part of historians and the public that such resources can play an important role in telling the diverse stories of the American experience, now is an exciting time to be working with historic sites. At the same time, however, maintaining and operating a historic site is costly and requires a long-term financial commitment.

Very seldom is a historic site supported solely by admission fees paid by visitors. More than half the sites surveyed in the National Trust study attract fewer than 5,000 visitors a year, with visitation for larger sites generally between 40,000 to

50,000 people. Only the largest and best-known sites, such as Colonial Williamsburg, attract annual visitors in the hundreds of thousands. Even sites with large numbers of visitors must attract significant dollars from other sources, which usually include fund raising, interest on an endowment if one exists, museum shop sales, and rentals of the property for private gatherings.

Renting an historic site for a private gathering or other special use is always a dilemma. While renting the property generates income and brings people to the historic sites, the educational value of having been to a wedding at a property is questionable, and the wear and tear on the site and its collections can be significant. Therefore, careful selection of events is crucial. Generally, it is best to allow fewer events at higher rental rates and to avoid large crowds. Thoughtfully planned special events can complement the site's educational program, draw people to the site, and generate income.

Even with the income from these various sources, many sites are inadequately funded. Inadequate funding often results in putting off maintenance in order to keep the doors open to the public. The result is that the amount of deferred maintenance in museum properties in the United States is extremely high. And deferred maintenance quickly turns into major capital projects as damage from leaky roofs, poorly caulked joints, and drafty windows takes its toll.

INTERPRETING HISTORIC SITES

Interpretation: "An educational activity which aims to reveal meanings and relationships through the use of original objects, by firsthand experience, and by illustrative media, rather than simply to communicate factual information."

Freeman Tilden, *Interpreting Our Heritage*, 1977

The quality of the interpretive program at an historic site or museum property is the key to its success. The program must accurately

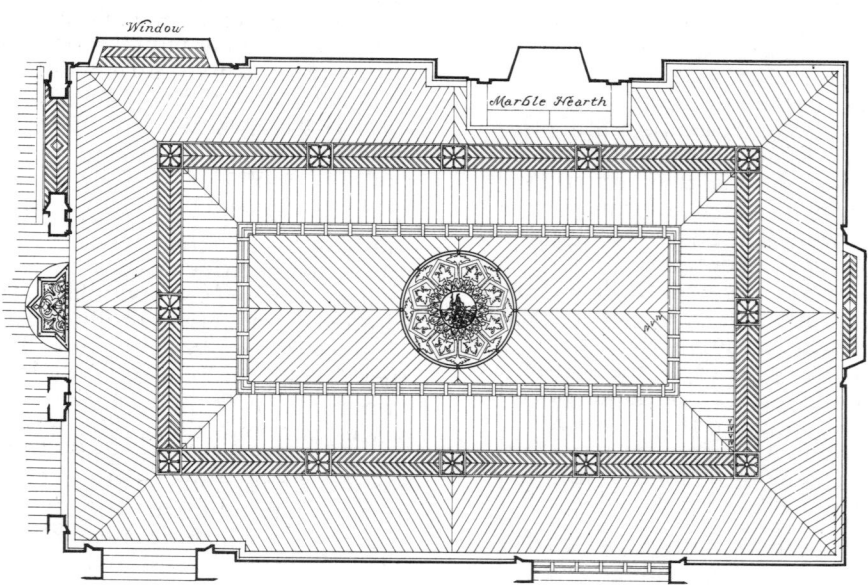

Plan of floor, Northeast Drawing Room, Decatur House, Washington, D.C. (H.R.J. Thompson, HABS)

reflect the history of the property, its use and inhabitants, and effectively communicate this information to the public. Even with a tight budget, a portion of the site's funding should be set aside for research and interpretation. If an interpretive program is not possible for the foreseeable future, the research should be conducted anyway. Much useful historical evidence is lost to the passage of time: papers disintegrate, artifacts are destroyed, and knowledge dies with people.

Planning an Interpretive Program
An interpretive program should illustrate the importance of the historic site to the individual and to the community. Some sites are chosen for the beauty of the buildings and the surrounding landscape, others are chosen because they represent a particular type of building, and still others for the people who have lived there or the events that have occurred there. Thus, the foundation of the interpretive program will vary based largely on the reason the museum property was created.

Ideally, the interpretive program should be considered before the site is preserved or restored. In fact, the interpretation of a site can be important in determining the course of its preservation, especially for a site that has experienced several significant periods in its history or represents multiple social and cultural events. The National Trust property Montpelier, in Orange, Virginia, is a good example of such a site. Montpelier was home to three generations of the Madison family, including President James Madison. A later owner, William Du Pont, added significantly to the size of the house. A strict restoration to the time of Madison would destroy evidence of the Du Ponts, a family of note in the history of business and industry in the United States.

Determining what at Montpelier should be restored to President Madison's time and what of the Du Pont's tenure should remain is central to the site's interpretation. The solution required much research and many discussions among preeminent preserva-

tionists. The decision finally was made *not* to restore the entire house to its condition when President Madison lived there but to keep parts of it as they had been during the residence of the Du Ponts, interpreting Madison's within the entire history of the house.

Principles of Interpretation
Before planning the specifics of an interpretive program, several basic principles should be addressed:

■ Base interpretation on well-founded research. If the information is inaccurate, the interpretation is useless.

■ Make the site relevant to visitors by using modern comparisons to emphasize historical facts and figures.

■ Explain why the site and the people or events associated with it are important to the community, the state, the nation, and perhaps even the world.

■ Use imagination in translating the site to the audience. Interpretation is more than information, it is an art. Tell a coherent story, leaving out extraneous information that would detract from the audience's understanding and enjoyment.

■ Illustrate the connections between the many aspects of what is being observed, while putting the site in its historical context.

■ Sensitize visitors to the importance of saving the historic site and tell them how they can help.

Methods of Interpretation
The method used to interpret an historic site or museum property is critical to its understanding by the public. Literally hundreds of different methods have been successfully used in sites across the country. Some of the most innovative include:

■ *Historical layers.* Present different periods, or layers of time, in the same building to allow the public to experience the complexity of the site's history.

■ *Time lines.* Isolate the sequential change in the life of an

individual or building and give focus to the process of change.

■ *Environmental immersion.* Surround observers with a total environment, causing them to step or peer into a different era, or to experience the sensation of another scale or viewpoint.

■ *Leading questions.* Pose questions designed to lead the public along a pathway where many aspects of a site's significance may be discovered. For example, questions about architectural style can lead to an exploration of the social status of the original owners and their role in the community.

■ *Place making (trails).* Use the history of a site to challenge artists and artisans to create objects that reflect the site and its significance.

It is important to remember that people learn not just through dialogue and thinking, but through all five senses. People often visit historic sites to experience an environment. For example, visitors to Alcatraz have difficulty in understanding actual prison conditions without the presence of inmates and guards. To overcome this problem, an audio program that simulates the sounds made by prisoners and guards has been developed as part of the interpretive program. Also included on tape are interviews with former inmates describing conditions in the prison. Other historic sites used odors and dramatic lighting to enhance the experience of the visitor.

OTHER OPTIONS FOR HISTORIC SITES

Although the increasing number of historic sites open to the public is an encouraging sign that the historic preservation movement has gained momentum, it is disheartening that so many of these sites are poorly funded and understaffed. Below are several strategies that may help ameliorate some of the financial problems faced by historic sites.

■ Find ways other than creating additional museums to save old buildings. Ideally, many historic buildings can continue to be used—if not for their original purpose, then for an adapted use. Train stations have been turned into hotels and shopping malls; schools have been adapted for use as artists' studios; apartment buildings have become offices. If the growth in the number of museums is limited, there is a better chance that we can maintain those already in existence.

■ Use easements to protect historic buildings. A property owner donates an easement to a tax-exempt charitable organization or public agency and in return the owner receives income, gift, and estate tax advantages and, in some states, property tax advantages. An easement will typically protect the property from major redevelopment, protect the outside appearance of a building, and sometimes protect all or part of the building's interior.

■ Develop methods to integrate historic preservation into local planning and development policies. Many historic sites are developed as museum properties to save them from redevelopment or removal. By integrating their preservation and continued use into local planning and development procedures, many of these properties will continue to survive without being turned into museums.

Diversifying the types of historic sites open to the public and preserving other historic buildings as part of the economic fabric of our communities will demonstrate the full range of America's social, industrial, and commercial history. We need to increase the number of industrial sites, commercial buildings, ethnic structures, and vernacular buildings in the collection of historic museum properties. We also need to develop other methods to protect our historic sites and to integrate them into our communities.

CONTACTS

American Association of Museums
1225 I Street, N.W.
Suite 200
Washington, DC 20005
(202) 289-1818

American Association for State and Local History
172 2nd Avenue, North
Suite 202
Nashville, TN 37201
(615) 255-2971

National Trust for Historic Preservation Stewardship of Historic Properties
1785 Massachusetts Avenue, N.W.
Washington, DC 20036
(202) 673-4146

Funding Organizations

Some of the most significant of the organizations that make grants to historic sites are listed below:

Institute of Museum Services
1100 Pennsylvania Avenue, N.W.
Suite 510
Washington, DC 20506
(202) 786-0539

National Endowment for the Arts
1100 Pennsylvania Avenue, N.W.
Room 624
Washington, DC 20506
(202) 682-5442

National Endowment for the Humanities
1100 Pennsylvania Avenue, N.W.
Room 503
Washington, DC 20506
(202) 786-0310

FURTHER READING

Directory of Historical Agencies in North America. Betty Pease Smith, ed. 13th ed. Nashville: American Association for State and Local History, 1986.

The Good Guide: Sourcebook for Interpreters, Docents and Tour Guides. Alison L. Grinder and E. Sue McCoy. Scottsdale, Ariz.: Ironwood Press, 1985.

History Museums in the United States: A Critical Assessment. Warren Leon and Roy Rosenzweig, eds. Urbana: University of Illinois Press, 1989.

If Walls Could Talk: Telling the Story of a Historic Building to Create a Market Edge. Ronald Lee Fleming and Judith Hadden Edington. Washington, D.C.: National Trust for Historic Preservation, 1989.

Interpretation of Historic Sites. William T. Alderson and Shirley Payne Low. 2nd ed. Nashville: American Association for State and Local History, 1985.

Interpreting Our Heritage. Freeman Tilden. Chapel Hill: The University of North Carolina Press, 1977.

The Living History Sourcebook. Jay Anderson, ed. Nashville: American Association for State and Local History, 1985.

National Register of Historic Places 1966–1991. National Park Service. Nashville: American Association for State and Local History, 1992. Distributed by the National Trust for Historic Preservation.

The Official Museum Directory 1992. American Association of Museums. Wilmette, Ill.: National Register Publishing Company, 1991.

Starting Right: A Basic Guide to Museum Planning. Gerald George and Cindy Sherrell-Leo. Nashville: American Association for State and Local History, 1986.

The Wages of History: The AASLH Employment Trends and Salary Survey. Charles Phillips and Patricia Hogan. Nashville: American Association for State and Local History, 1984.

EDUCATION

Kathleen Hunter and Richard Wagner

"Through education there is the potential of engendering a national preservation ethic . . . that will nurture a new national value system based on environmental maintenance rather than environmental disposal."

Chester H. Liebs, *Preservation News*, January 1978

PROGRAMS FOR ELEMENTARY AND SECONDARY SCHOOLS

"Heritage education is democratizing the historic landscape, making it accessible to large numbers of young people for a widely supported public good: learning. It is framing historic environments in much broader terms for a large segment of the population, using them imaginatively to encourage learning that is both memorable and enjoyable. Heritage education is turning historic areas into lexicons of information, testing grounds for theories, laboratories where students practice a whole range of new skills—places where young people are actively engaged and involved, habitually taking and making meaning from these environments in very personal ways."

National Council for Preservation Education, *A Heritage at Risk: A Report on Heritage Education*

Heritage education is an approach to teaching and learning about history and culture that uses the natural and built environments, historic objects, oral histories, community practices, music, dance, and written documents to help students understand their local heritage and connections to other regions of the country and to the nation as a whole.

Heritage education supports the goals of preservationists. When a community understands and values its environment as a positive reflection of its heritage, it protects and preserves its surroundings. Individuals, neighborhoods, communities and their schools can work together to research, document, and interpret their historical and cultural environment.

Heritage education also supports the goals of educators. The heritage education approach offers an interdisciplinary perspective on many subjects in the school curriculum, such as history, geography, social studies, and the arts and sciences. Students learn to link their local history and culture to broader themes, issues, events, and people that have shaped the nation and the world. Teachers can use heritage education activities in the classroom to strengthen students' basic skills in reading, writing, and mathematics, and their critical thinking skills in investigation, analysis, and interpretation.

Heritage Education in Action

The heritage education approach encourages school-community partnerships that foster positive civic attitudes and a respect for continuity and diversity within the community. Following are some examples of these partnerships at work through heritage education programs and activities.

■ A local preservation group invites teachers, students, a librarian, a museum curator, and business leaders to restore an old school building as a community heritage interpretation center. Teachers adapt the project to the school curriculum. As a lesson in language arts, students gather oral histories about the area. As a geography project, students research the route of the Underground Railroad through their region and the culture of the Quaker farmers and merchants who settled in the area. Students in a civics class testify at a county commissioner's hearing to protect the deteriorating old school and provide the documen-

tation needed to nominate the building to the National Register of Historic Places. Over the summer, students in scouting and 4-H groups work with preservation craftspersons on restoration projects at the old school that are sponsored by local businesses.

■ In a western hamlet, fourth-graders research the heritage of their homes as a part of a local history curriculum. Students talk to parents, relatives, and neighbors, and they visit the county courthouse and library to do research. A preservation architect from the state university helps them understand the design and structure of their homes. A business woman donates small file boxes in which students keep a journal, drawings, and copies of documents about their homes. The teacher arranges for the materials to be displayed at the county historical society, and the society awards each student with a Heritage Home certificate. The story of the fourth-graders' Heritage Home project appears in a local newspaper and finds its way to a regional television station.

■ A school district encourages students to volunteer for community projects. High school history teachers work with the staff at a historic plantation to develop a student intern program. Over the summer, the interns help the interpreter and curator research and document the plantation's history. During the next school year, the interns are released from their American history course one day per month to guide elementary school groups through the plantation. The area's businesses donate a $100 savings bond for each student who completes the one-year program.

- A middle school principal wants to help Southeast Asian students who have recently arrived in the United States better understand how they fit into their new neighborhood's heritage. With the help of a local preservation organization and a senior citizens group, the students study the cultures of the many immigrant groups who settled in their urban neighborhood. The stu-

dents identify the buildings and sites that reflect the layering of diverse cultures in their community. They design a traveling display for use at schools, libraries, and city hall.

- A local preservation group and a historical society help Head Start teachers in two inner-city neighborhoods design a "walk around our neighborhood" program that introduces youngsters to their own surroundings and to the other Head Start neighborhood. The students meet their neighbors and explore the area's architectural features, parks and

gardens, and roads and bridges. The children visit businesses, the library, the police station, the post office, and places of worship, and they talk with community elders about neighborhood landmarks. Then the students visit their sister neighborhood, comparing it with their own, drawing pictures of favorite sites, and exchanging these with students from the other school.

Heritage education is a combination of accessible history, imaginative education, and effective community action. It is a program that can find success in any community by following these guidelines:

- Base the program on sound research, scholarship, and accepted preservation practices.

- Tell the whole story of the neighborhood or community—clearly, accurately, and respectfully. Show how rich and diverse local experiences link the local community to the region, the state, the nation, and the world.

- Engage students in an experiential learning process that moves from ideas to action. Start young, integrating the heritage education approach into classroom subjects.

- Involve the whole community: preservationists, historians, educators, librarians, public agencies, civic and religious groups, professionals, lay persons, and retirees.

Programs and Resources
The National Trust for Historic Preservation is particularly interested in preserving and teaching about those reflections of our heritage that remain in objects, structures, buildings, and sites. Through its six regional offices and one field office, the National Trust provides technical assistance and small start-up grants to support local heritage education activities. At each of its 17 properties, National Trust employees provide heritage education programs for children, teens, and adults through tours, workshops, seminars, field studies, and interpretive displays. Contact the Heritage Education Center for more

Gymnasium (now Fine Arts Building), Vanderbilt University, Nashville, Tennessee. (W.H. Edwards, HABS)

information on the National Trust's Heritage Education efforts.

PRESERVATION PROGRAMS AT COLLEGES AND UNIVERSITIES

More than 45 colleges and universities in the United States offer programs in historic preservation. Many are interdisciplinary, reflecting the diverse nature of the field. Some of the programs offer undergraduate degrees in historic preservation or a B.A. or B.S. with a specialization in preservation. Other programs offer graduate degrees. In addition, some preservation programs are based in allied disciplines, such as archaeology, anthropology, architecture, history, museum studies, and the like. Such programs offer degrees in those disciplines with an emphasis on historic preservation. Some of the college and university based programs require internships; others offer foreign study and summer institutes.

Most of the historic preservation programs at colleges and universities, as well as those offering crafts training programs, belong to the National Council for Preservation Education. Founded in 1980 with the support of the National Trust for Historic Preservation, the National Council works to establish and improve educational standards and serves as a voice for educators involved in historic preservation. The National Council publishes *The Directory of College, University, Craft and Trade Programs in Cultural Resource Management*, a guide to post-secondary education in historic preservation.

UNDERGRADUATE PROGRAMS IN PRESERVATION

Maryland

Goucher College
Historic Preservation Program
Baltimore, MD 21204
(410) 337-6267
R. Kent Lancaster, Director

Michigan

Michigan State University
Department of Art
East Lansing, MI 48824-1119
(517) 355-7612
Sadayoshi Omoto, Interim Chairman

Missouri

Southeastern Missouri State University
Historic Preservation Program
Department of History
One University Plaza
Cape Girardeau, MO 63701-4799
(314) 651-2808
Art H. Manningly, Coordinator

Ohio

Belmont Technical College
Science and Engineering
120 Fox Shannon Place
St. Clairsville, OH 43950
(614) 695-9500 x46
David R. Mertz, Coordinator

Pennsylvania

Bucks County Community College
Social Sciences Department
Historic Preservation Program
Swamp Road
Newtown, PA 18940
(215) 968-8286
Lyle L. Rosenberger, Director

Rhode Island

Roger Williams College
Historic Preservation Program
Bristol, RI 02809
(401) 254-3396
Kevin E. Jordan, Director

Virginia

Mary Washington College
Department of Historic Preservation
1301 College Avenue
Trinkle Hall, Room 40B
Fredericksburg, VA 22401
(703) 899-4122
Carter L. Hudgins, Chairman

GRADUATE PROGRAMS IN PRESERVATION

Georgia

Georgia State University
Heritage Preservation Program
Atlanta, GA 30303
(404) 651-2250
Timothy J. Crimmins, Director

University of Georgia
Graduate Studies in Historic Preservation
Caldwell Hall
Athens, GA 30602
John Waters, Coordinator

Indiana

Ball State University
Department of Architecture
Muncie, IN 47306
(317) 285-1909
David R. Hermansen, Director

Massachusetts

Boston University
Preservation Studies Program
226 Bay State Road
Boston, MA 02215
(617) 353-2948
Richard M. Candee, Director

Michigan

Eastern Michigan University
Historic Preservation Program
Department of Geography
Ypsilanti, MI 48197
(313) 487-0218
Marshall McLennan, Director

New York

Columbia University
Division of Historic Preservation
400 Avery Hall
New York City, NY 10027
(212) 854-3518
Robert A. M. Stern, Director

Cornell University
Graduate Program in Historic Preservation Planning
106 West Sibley Hall
Ithaca, NY 14853
(607) 255-7261
Michael Tomlan, Director

Oregon

University of Oregon
Historic Preservation Program
105 Lawrence Hall
Eugene, OR 97403
(503) 346-1419
Michael Shelenbarger, Director

Pennsylvania

University of Pennsylvania
Graduate Program in Historic Preservation
115 Myerson Hall
Philadelphia, PA 19104-6311
(215) 898-3169
David G. DeLong, Chairman

Tennessee

Middle Tennessee State University
Center for Historic Preservation
Vaughn House
Box 80
Murfreesboro, TN 37132
(615) 898-2947
James Huhta, Director

Vermont

University of Vermont
Historic Preservation Program
Department of History
Burlington, VT 05405
(802) 656-4006
Chester H. Liebs, Director

GRADUATE PROGRAMS WITH A SPECIALIZATION IN PRESERVATION

Arizona

Northern Arizona University
Department of Anthropology
Flagstaff, AZ 86011
(602) 523-6575
Charles Hoffman, Associate Professor

California

California State University, Dominquez Hills
History Department
1000 East Victoria Street
Room SBS E 310
Carson, CA 90747
(310) 516-3494
John Auld

University of California, Riverside
History Department
Riverside, CA 92521
(714) 787-5403
Ronald Tobey, Director

University of California, Santa Barbara
Department of History
Santa Barbara, CA 93106
(805) 893-8439
Otis Graham, Professor

Colorado

Colorado State University
Department of History
Fort Collins, CO 80523
(303) 491-6334
Liston Leyendecker, Professor

District of Columbia

George Washington University
Graduate Program in Historic Preservation
American Studies Program
Washington, DC 20052
(202) 994-6098
Richard Longstreth, Director

Florida

University of Florida
Graduate Program in Architectural Preservation
Department of Architecture
Gainesville, FL 32611
(904) 392-0215
Herschel Shepard, Director

Georgia

Armstrong State College
Public History Program
History Department
11935 Abercorn Street
Savannah, GA 31419-1997
(912) 927-5283
Sarah Brown, Coordinator

Savannah College of Art and Design
Historic Preservation Department
201 W. Charlton Street
Savannah, GA 31401
(912) 238-2487
Maggie O'Connor, Chairman

Hawaii

University of Hawaii, Manoa
American Studies Department
1890 East-West Road
Moore Hall 324
Honolulu, HI 96822-2318
William Murtagh, Director of Historic Preservation
(808) 956-6599

Illinois

University of Illinois at Urbana-Champaign
Department of Urban and Regional Planning
907 1/2 West Nevada Street
Urbana, IL 61801
(217) 244-5404
Lewis D. Hopkins, Head

Kansas

Kansas State University
Graduate Studies in Historic Preservation
College of Architecture and Design
212 Seaton Hall
Manhattan, KS 66506
(913) 532-5953
Wayne Charney, Coordinator

Kentucky

Western Kentucky University
Program in Folk Studies
Department of Modern Languages
IWFAC Room 251
Bowling Green, KY 42101
(502) 745-5898
Michael Ann Williams, Director

Louisiana

University of New Orleans
Historic Preservation Concentration
College of Urban Studies and Public Affairs
New Orleans, LA 70148
(504) 286-6514
Jane Brooks, Coordinator

Maryland

University of Maryland
Department of History
College Park, MD 20742-7315
(301) 405-4313
J. Kirkpatrick Flack, Chairman, Committee on Historic Preservation
(310) 405-4313

Nevada

University of Nevada
Historic Preservation Program
College of Arts and Sciences
Reno, NV 89557
(702) 784-6851
Don Fowler, Director

Ohio

Kent State University
Historic Preservation Graduate Program
School of Architecture and Environmental Design
Kent, OH 44242
(216) 672-2869
Daniel Vieyra, Coordinator

Ohio State University
Department of Architecture
189 Brown Hall
190 West 17th Avenue
Columbus, OH 43210
(614) 292-9063
Paul Young, Jr., Professor

University of Cincinnati
Historic Preservation Certification Program
Department of Anthropology
Cincinnati, OH 45221-0180
(513) 556-5787
Kent Vickery, Acting Chairman

Oklahoma

Oklahoma State University
History Department
Stillwater, OK 74078-0611
(405) 744-8183
Bill Bryans, Coordinator

South Carolina

University of South Carolina
Applied History Program
Department of History
Gambrell Hall
Columbia, SC 29208
(803) 777-4854
Constance Schultz, Director

Tennessee

University of Tennessee
College of Architecture and Planning
School of Architecture
1715 Volunteer Boulevard
Knoxville, TN 37996-2400
(615) 974-5265
J. S. Rabun, Associate Professor

Texas

Texas A & M University
College of Architecture
College Station, TX 77843-3137
(409) 845-7850
David Woodcock, Professor

Texas Tech University
Department of Architecture
P.O. Box 42091
Lubbock, TX 79409-2091
(806) 742-2790
John P. White, Professor

University of Texas
Department of Architecture
P.O. Box B
Austin, TX 78713
(512) 471-0159
Wayne Bell, Director

Virginia

University of Virginia
Preservation Program
School of Architecture
Campbell Hall
Charlottesville, VA 22903
(804) 924-3976
Mario di Valmarana, Director

Washington

University of Washington
Department of Architecture
J0-20
Seattle, WA 98195
(206) 543-4180
Jeffrey K. Ochsner, Professor

Western Washington University
Historic Preservation Program
History Department
Bellingham, WA 98225
(206) 676-3439
Donald Whisenhunt, Chairman

Wisconsin

University of Wisconsin
Department of Landscape Architecture
25 Agricultural Hall
Madison, WI 53706
(608) 263-8973
William H. Tishler, Professor

CONTINUING EDUCATION, SEMINARS, SHORT COURSES, AND WORKSHOPS

Preservation organizations, professional associations, colleges, universities, and others offer a wide range of post-secondary educational programs in historic preservation. Some target professionals in various fields wishing to acquire specific preservation skills and knowledge. Others focus on individuals wanting to become acquainted with the field. Some of these programs offer continuing education unit (CEU) credits applicable to certificates in different disciplines. Many of the programs listed below are conducted on a regular basis, while others are offered only occasionally by the sponsoring agency.

Connecticut

University of Connecticut
Courses in American Maritime Studies
Summer Sessions Office
Storrs, CT 06269-4056
(203) 486-4631
W. Matthew McLoughlin, Director

Delaware

Winterthur Museum and Gardens
Route 52, Kennett Pike
Winterthur, DE 19735
(302) 888-4645
Peter Hammell, Director of Education

District of Columbia

Bureau of National Affairs
1231 25th Street, N.W.
Washington, DC 20037
(202) 833-7296
Gretchen Zekiel, Marketing Specialist

International Downtown Executives Association
915 15th Street, N.W.
Suite 900
Washington, DC 20005
(202) 783-4963
Richard H. Bradley, President

National Main Street Center
National Trust for Historic Preservation
1785 Massachusetts Avenue, N.W.
Washington, DC 20036
(202) 673-4219
Elizabeth Jackson, Senior Program Manager

National Preservation Institute
401 F Street, N.W.
Pension Building
Washington, DC 20001
(202) 393-0038
Peggy Boucher, Administrator

Partners for Livable Places
1429 21st Street, N.W.
Washington, DC 20036
(202) 887-5990
Penny Cuff, Senior Program Officer

Smithsonian Institution
Office of Museum Programs
900 Jefferson Drive, S.W.
Room 2235
Washington, DC 20560
(202) 357-3148
Bruce Craig, Professional Services Manager

Florida

Preservation Institute: Nantucket
September–May:
c/o Department of Architecture
331 Arch
University of Florida
Gainesville, FL 32611
(904) 392-7003
June–August:
P.O. Box 1139
Nantucket, MA 02554
(508) 228-2429
Susan Tate, Director

Illinois

American Planning Association
1313 East 60th Street
Chicago, IL 60637
(312) 955-9100
Carolyn Torma, Continuing Education Director

Campbell Center for Historic Preservation Studies
P.O. Box 66
Mount Carroll, IL 61053
(815) 244-1173
Mary Wood Lee, Director

Maryland

Goucher College
Center for Continuing Studies
Baltimore, MD 21204

Preservation Resource Group
P.O. Box 1768
Rockville, MD 20849-1768
(301) 309-2222
Carol Blundell, Sales and Marketing

Massachusetts

Boston University
Community Service Center
730 Commonwealth Avenue
Room 253
Boston, MA 02215
(617) 353-4710
Stephen McMahon, Director

Harvard Graduate School of Design
Office of Special Programs
Continuing Education
48 Quincy Street
Cambridge, MA 02138
(617) 495-1680
Jennifer Leahy, Staff Assistant

Historic Deerfield
Fall Forums
P.O. Box 321
Deerfield, MA 01342
(413) 774-5581
Norma Woods, Assistant to Director of Education

Society for the Preservation of New England Antiquities
141 Cambridge Street
Boston, MA 02114
(617) 227-3956
Barbara Levy, Director of Education and Interpretive Programs

New York

Practising Law Institute
810 7th Avenue
New York, NY 10019
(212) 765-5700
Sandra Geller, Director of Programming

Royal Oak Foundation
285 West Broadway
Suite 400
New York, NY 10013
(212) 966-6565
Damaris S. Horan, Executive Director

Seminars in American Culture
New York State Historical Association
P.O. Box 800
Cooperstown, NY 13326
(607) 547-2533
Milo Stewart, Associate Director

North Carolina

North Carolina Department of Cultural Resources
Archeology and Historic Preservation
109 East Jones Street
Room 214
Raleigh, NC 27611
(919) 733-4763
David Brook, Administrator

Pennsylvania

The Pennsbury Society
Pennsbury Manor
400 Pennsbury Memorial Road
Morrisville, PA 19067
(215) 946-0400
Stacy Roth, Education Director

The Victorian Society in America
219 South 6th Street
Philadelphia, PA 19106
(215) 627-4252
Judith van Buskirk, Summer School Director

Wharton School
Fundamentals of Finance and Accounting
for Nonprofit Organizations
University of Pennsylvania
Steinberg Hall
3620 Locust Walk
Philadelphia, PA 19104
(215) 898-7640
Mike Baltes, News Officer

Tennessee

American Association for State and Local History
172 2nd Avenue, North
Suite 202
Nashville, TN 37201-1925
(615) 255-2971
Rob Meeks, Publications Manager

Vermont

University of Vermont
Historic Preservation Summer Institute
Department of History
Burlington, VT 05405
(802) 656-4006
Chester H. Liebs, Director

Virginia

Association for Preservation Technology International
904 Princess Anne Street
P.O. Box 8178
Fredericksburg, VA 22404
(703) 373-1621
Susan Ford Johnson, Executive Director

Williamsburg Seminar on Historical Administration
Sponsors:
American Association of Museums
1225 I Street, N.W.
Suite 200
Washington, DC 20005
(202) 289-1818
Meg McCarthy, Director of Continuing Education

National Trust for Historic Preservation
1785 Massachusetts Avenue, N.W.
Washington, DC 20036
(202) 673-4000
Gigi Barrett, Forum Membership Assistant

American Association for State and Local History
172 Second Avenue, North
Suite 202
Nashville, TN 37201-1925
(615) 255-2971
Rob Meeks, Publication Manager

Colonial Williamsburg Foundation
P.O. Box B
Williamsburg, VA 23185
(804) 220-7215
Velva Henegar, Secretary

RESTORATION CRAFTS TRAINING PROGRAMS

The increasing interest in rehabilitating historic buildings in the past quarter century has resulted in a demand for skilled crafts and trades people. The programs listed in this section offer a variety of short seminars and hands-on work-shops and long-term apprenticeship programs covering preservation technology, materials, and methods of construction.

District of Columbia

U.S. Committee, International Council on Monuments and Sites
Summer Intern Program
1600 H Street, N.W.
Washington, DC 20006
(202) 842-1862
Ellen Delage, Program Officer

New York

RESTORE
41 East 11th Street
New York, NY 10036
(212) 477-0114
Jan C. K. Anderson, Executive Director

North Carolina

Durham Technical Institute
Residential Carpentry
1637 Lawson Street
Durham Tech Community College
Durham, NC 27703
(919) 598-9397
Russell Pratt, Associate Dean

Stagville Preservation Center
P.O. Box 71217
Durham, NC 27722-1217
(919) 620-0120
Kenneth McFarland, Site Manager

West Virginia

National Park Service
Stephen T. Mather Training Center
P.O. Box 77
Harpers Ferry, WV 25425-0077
(304) 535-6404

Canada

St. Lawrence College
Restoration Technology Workshops
Technology Department
2288 Parkdale Avenue
Brockville, Ontario K6V 5X3, Canada
(613) 345-0660
Donn Atchison, Chairman

CONTACTS

American Association of Museums
1225 I Street, N.W.
Suite 200
Washington, DC 20005
(202) 289-1818

Information on museum and site interpretation.

American Association of State and Local History
172 2d Street North
Suite 200
Nashville, TN 37201
(615) 225-2971

Information on volunteers, professionals, and organizations related to state and local history.

National Council for Preservation Education
Cornell University
106 West Sibley Hall
Ithaca, NY 14853
(617) 255-7261
Michael A. Tomlan

Information about undergraduate, graduate, and training programs in historic preservation.

National Register of Historic Places
National Park Service
P.O Box 37127
Washington, DC 20012-7127
(202) 343-9536

Maintains information on all sites and districts listed on the National Register of Historic Places.

National Trust for Historic Preservation Heritage Education Center
Old School
P.O. Box 202
Waterford, VA 22190
(202) 673-4040

The National Trust's Heritage Education Center offers a number of materials on heritage education that are helpful to communities and their schools:

■ **Forum**
A journal of background papers on the educational concepts and principles of the heritage education approach.

■ **Heritage Education Resource Database**
A growing database of community and school programs, training programs, and educational materials related to heritage education.

■ **Information: Heritage Education**
A practical guide to starting a heritage education program or improving an existing program. Examples, resources, and strategies are included.

■ **Old School**
A quarterly newsletter on heritage education that provides updates on programs and shows how historic places deepen our knowledge of history and culture.

■ **Teaching With Historic Places**
A series of lesson plans that use sites from the National Register of Historic Places to enrich social studies and art curricula. Appropriate for upper-elementary and secondary-school students. Forthcoming from The Preservation Press, National Trust for Historic Preservation.

In addition, National Trust properties and regional offices offer a variety of technical assistance and educational materials for heritage education. They can also refer readers to other education programs.

EMPLOYMENT

Greg Coble and Karen Peil

"The preservation movement needs engineers, lawyers, economists and other technicians of contemporary society but not to the exclusion of generalists. . . ."

Roderick S. French, "On Preserving America: Some Philosophical Observations," in
Preservation: Toward an Ethic in the 1980s

EMPLOYMENT OUTLOOK

Careers in historic preservation are a relatively recent employment opportunity. They exist at all levels of government and nonprofit organizations and with private firms.

Employment opportunities exist in government agencies, nonprofit organizations, and private firms whose main mission or service is historic preservation. In the public sector, the National Park Service, state historic preservation offices, and local government preservation divisions are examples of agencies dedicated to the preservation of America's heritage. Examples of nonprofit organizations include the National Trust, statewide preservation organizations, and hundreds of local preservation groups. Private firms whose main service is providing preservation expertise include architects, planners, archaeologists, historical research firms, and a host of others specializing in historic preservation.

In addition to employers whose main mission is historic preservation, thousands of others hire people with preservation skills. For example, all federal government departments are required to have a preservation officer. Thus, opportunities exist with such diverse departments as the Department of Defense and Health and Human Services. Many state agencies involved in housing, community development, transportation, and other services hire people with preservation backgrounds. In addition, thousands of private firms engaged in design, development, financing, and other aspects of creating and changing the physical environment have positions that require someone with knowledge and skills in preservation.

Preservation Agencies, Organizations, and Firms

During the past quarter century, thousands of jobs have been created in government agencies, nonprofit organizations, and private firms specializing in historic preservation. For example:

- In 1970 the National Trust had fewer than 80 paid employees; today, the number exceeds 300.

- Eighty percent of state and local nonprofit preservation organizations hired their first paid employee after 1970 (according to a 1990 National Trust survey).

- In 1980 fewer than 10 people were employed in downtown revitalization positions in communities associated with the National Trust's Main Street program. Today the number is nearly 1,000, with hundreds of similar positions in communities practicing preservation based commercial district revitalization.

- In 1966 the state historic preservation office system was established, creating hundreds of preservation jobs.

- Hundreds of private firms specializing in the many aspects of historic preservation have been founded in the past two and one half decades, creating thousands of jobs for people with particular skills and knowledge.

Competition for available jobs, however, has increased as public interest in historic preservation has broadened. For example, the National Trust Office of Human Resources estimates that it receives 25 inquiries weekly from individuals requesting information on how to pursue a career in historic

preservation. In 1990 more than 800 students were enrolled in undergraduate or graduate programs in historic preservation. In the past few years, this competition has been especially acute as the country's economy has slowed.

A 1990 National Trust survey of nonprofit preservation groups indicates that nonprofit organizations do not depend on job advertisements for recruitment: 78 percent of state organizations and 65 percent of local organizations use personal contacts as their primary recruitment strategy. Job ads can be easily located, however. The National Trust's *Historic Preservation News* publishes more than 100 preservation job ads annually, and *Main Street News* publishes between 35 and 40 ads each year for jobs in Main Street communities.

MAJOR AREAS OF EMPLOYMENT CONCENTRATION

While the majority of persons employed in historic preservation work in local communities, employment options also exist at the state and national levels. The types and magnitude of jobs for all levels are listed below.

Nonprofit State and Local Preservation Organizations

Where: More than 150 local and more than 30 state nonprofit organizations employ paid staff

Magnitude: 70 percent of these organizations have less than five paid employees

Types of jobs: Organizational/ program managers, planners, general field service providers, fund raisers, specialized service providers (real estate, financial assistance, and historic sites), educators, administrative support

Local Government
Where: Planning, economic, and community development divisions of county and city governments

Magnitude: Most counties and cities administer housing rehabilitation, economic development, and community development programs

Types of jobs: Preservation planners, economic development specialists, housing rehabilitation specialists, urban and community planners, community development specialists

Local Historic District Commissions
Where: Communities throughout the nation

Magnitude: 1,500 local commissions; about 72 percent employ staff

Types of jobs: Preservation planners, urban and community planners

Main Street Programs
Where: 700 communities throughout the United States; 32 state governments with programs (usually found in the state historic preserva-

tion office [SHPO], commerce, or economic development departments)

Magnitude: 750 to 800 jobs in community nonprofit organizations (about 15 percent of these jobs are in city government); most of the 32 state programs employ a staff of three people

Types of jobs: Local program managers; 10 percent of local programs have assistant managers; states employ coordinators and assistant coordinators; half of state programs employ architects or designers

Historic Sites, House Museums, Local Historical Societies
Where: Communities throughout the nation

Magnitude: Estimated at 10,000 jobs

Types of jobs: Site managers, curators, interpretation specialists, historians, museum educators, archeologists, restoration artisans, administrative support

State Government
Where: State historic preservation offices

Magnitude: Size of staff varies widely from state to state (range of 5 to 30 jobs per office)

Types of jobs: Managers, architects, historians, planners, archeologists, site managers, preservation education specialists, administrative support

National Nonprofit
Where: National Trust for Historic Preservation, Washington, D.C.; 7 regional locations and 17 museum properties

Magnitude: 300 jobs

Types of jobs: Organizational/program managers, site managers, planners, field service providers, fund raisers, attorneys, lobbyists, marketing and communications specialists, educators, administrative support

Federal Government
Where: Cultural Resources Program of the National Park Service, Washington, D.C.; 10 regional locations and 357 parks and sites

Magnitude: 500 jobs

Types of jobs: Archeologists, architects, landscape architects, planners, historians, curators, interpreters, managers

Where: Advisory Council for Historic Preservation, Washington, D.C.

Prentis Store, Williamsburg, Virginia. (E. Phillups, HABS)

Magnitude: 30 to 35 jobs

Types of jobs: Historic preservation specialists, archaeologists, historians

Where: Federal agencies

Magnitude: estimated at approximately 100

Types of jobs: Agency preservation officers

Rehabilitation Crafts and Trades
Where: Communities throughout the country

Magnitude: Estimated to be thousands nationwide

Types of jobs: Contractors, craftspersons, and artisans

Education
Where: Colleges and universities (departments such as history, architecture, landscape architecture, planning, archaeology, museum studies)

Magnitude: 45 degree programs in preservation and allied disciplines

Types of jobs: Professors, instructors, lecturers

Where: Preservation crafts and trades training programs nationwide

Magnitude: 55 training programs

Types of jobs: Training coordinators, program directors, instructors

Professional Firms
Where: Firms and self-employed individuals providing specialized preservation services

Magnitude: Several thousand

Types of jobs: Architects, planners, attorneys, real estate developers, market and business analysts, organizational development specialists, landscape architects, archaeologists, historians, urban designers, engineers

EXPERIENCE, EDUCATION, AND SKILLS REQUIRED

A 1990 National Trust survey of state and local nonprofit preservation groups shows that 73 percent of top or second-level professionals have fewer than 10 years experience in preservation. Many of these professionals held nonpreservation positions prior to employment with state and local nonprofit organizations. This reflects, in part, the relatively recent founding of degree programs in preservation and in part the high rate of growth of preservation jobs in the past two decades.

The survey does not show an overwhelmingly dominant educational background among respondents: while preservation is the leader at 27 percent, history, planning, and art history are other common backgrounds. Main Street jobs are filled by people from preservation, business, and planning backgrounds. The current trend is to hire Main Street managers with business experience who have worked as community group volunteers or advocates.

Success in small organizations requires a combination of technical and general skills. While a proficiency in historic preservation must be exhibited, skills in management, communication, fund raising, board relations, marketing, and event planning also are needed. Large organizations are more likely to have a number of very specialized positions requiring specific training and experience.

Preservation employers look for applicants with commitment and experience, which can be obtained in a number of ways: working as an intern, volunteering for the summer teams of the Historic American Buildings Survey (HABS) or the Historic American Engineer Record (HAER), being active in community and neighborhood advocacy, and attending preservation conferences. Employees who look for ways in a current job to demonstrate a commitment to or an accomplishment in historic preservation are likely to find advancement. Establishing a track record as a volunteer leader with a preservation organization may be the best way to obtain experience, demonstrate commitment, and become acquainted with people currently working in historic preservation. Community preservation organizations and historic sites offer a significant number and range of volunteer opportunities.

Private firms, nonprofit organizations, and government agencies whose primary mission is not preservation nonetheless hire people with preservation skills and knowledge. For example, an architectural firm specializing in housing will have a position for someone who understands the National Register and the Investment Tax Credit processes. Nonprofit housing organizations will need people with the same skills. And federal agencies and some state agencies are required to have someone on staff who understands historic preservation and how it can assist in fulfilling the mission of the agency.

Careers Chart
A chart entitled the "Historic Preservation Family Tree" was produced in 1991 by the National Center for the Study of History as part of its series illuminating the value of history. The chart presents a comprehensive view of historic preservation and highlights steps to clarify the process for selecting a career or avocation in historic preservation. A listing of 495 sample occupations are included. Charts for walls and notebooks are available.

National Center for the Study of History
R.R. #1
Box 679
Cornish, Maine 04020
(207) 637-2873
Robert W. Pomeroy, Director

CONTACTS
Local preservation organizations, Main Street projects, your state's historic preservation office, and college-level preservation programs can provide invaluable advice about historic preservation careers. The National Trust Office of Human Resources can also respond to questions. National Trust regional offices can respond to questions about opportunities in specific geographical regions. In addition, preservation job openings can be found in the following publications:

AVISO
American Association of Museums
1225 I Street, N.W.
Suite 200
Washington, DC 20005
(202) 289-1818

Center City Report
International Downtown Association
915 15th Street, N.W.
Suite 900
Washington, DC 20005
(202) 783-4963

Economic Developments
National Council for Urban Economic
Development
1730 K Street, N.W.
Washington, DC 20006
(202) 2234735

Historic Preservation News
National Trust for Historic Preservation
1785 Massachusetts Avenue, N.W.
Washington, DC 20036
(202) 673-4000

History News Dispatch
American Association for State and Local
History
172 2d Avenue, North
Nashville, TN 37201
(615) 255-2971

Job Mart
American Planning Association
1313 East 60th Street
Chicago, IL 60637
(312) 955-9100

Main Street News
National Main Street Center
National Trust for Historic Preservation
1785 Massachusetts Avenue, N.W.
Washington, DC 20036
(202) 673-4219

CHRONOLOGY

Dwight Young

"The longer you can look back, the farther you can look forward."

Winston S. Churchill, 1944

1812
First national historical organization, the American Antiquarian Society, is founded in Worcester, Massachusetts.

1816
Philadelphia purchases Independence Hall to save it from demolition.

1828
Touro Synagogue in Newport, Rhode Island, is the nation's first restoration.

1850
New York legislature purchases the Hasbrouck House, George Washington's headquarters in Newburgh, and opens it to the public as the nation's first historic house museum.

1853
Ann Pamela Cunningham initiates efforts to save Mount Vernon; her Mount Vernon Ladies' Association of the Union is chartered in 1856, and restoration of Mount Vernon begins in 1859.

1856
Tennessee legislature authorizes purchase of The Hermitage, Andrew Jackson's home near Nashville.

1872
Yellowstone is named America's—and the world's—first National Park.

1876
Old South Meeting House in Boston is rescued from demolition.

1888
Association for the Preservation of Virginia Antiquities is formed as the nation's first statewide preservation organization.

1889
America's first National Monument designation is awarded to Casa Grande, near Coolidge, Arizona.

1890
Chickamauga Battlefield in Georgia becomes the first National Military Park.

1891
Trustees of Public Reservations is incorporated in Massachusetts, and becomes a model for the English (1894) and American (1949) national trusts.

1896
In *United States v. Gettysburg Electric Railway Company*, the U.S. Supreme Court hears its first case involving preservation.

1906
Antiquities Act, the first major federal preservation legislation, is passed to preserve archeological sites.

1909
Essex Institute opens America's first outdoor museum of historic buildings in Salem, Massachusetts.

1910
Society for the Preservation of New England Antiquities is incorporated as America's first regional preservation organization.

1916
National Park Service is created, taking over nine existing National Monuments.

1926
Greenfield Village, Henry Ford's collection of historic buildings and artifacts in Dearborn, Michigan, opens to the public.

Restoration of Williamsburg, Virginia, begins under a plan formulated by W. A. R. Goodwin, with funding from John D. Rockefeller, Jr.

1931
America's first municipal preservation ordinance is passed in Charleston, South Carolina.

Moore House at Yorktown, Virginia, is the first National Park Service restoration.

1933
Responsibility for operation and maintenance of battlefields and other historic federal property is transferred to the National Park Service.

Historic American Buildings Survey is begun.

1935
National Historic Sites Act is passed, authorizing the U.S. Department of the Interior to survey historic sites under a National Historic Landmarks Program and to acquire historic properties for public use.

1936
Louisiana state constitution is amended to create a commission to preserve the Vieux Carré in New Orleans.

1938
First National Historic Site designation is awarded to Salem Maritime National Historic Site in Massachusetts.

1940
Society of Architectural Historians and the American Association for State and Local History are founded.

1944
This Is Charleston is published in Charleston, South Carolina, representing the nation's first attempt at a citywide inventory of historic buildings.

1947
National Council for Historic Sites and Buildings, the first nationwide private preservation organization and predecessor of the National Trust, is formed.

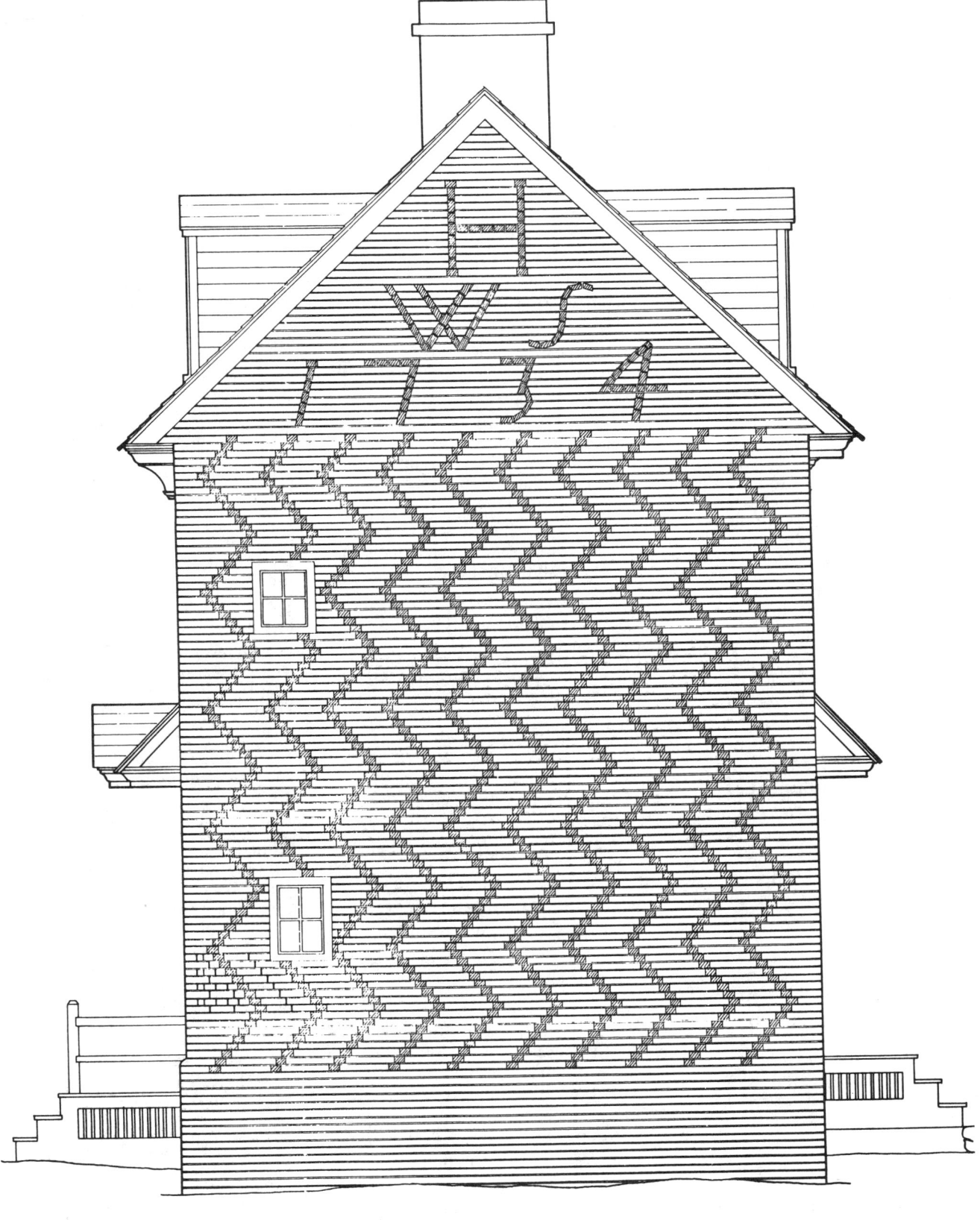

Hancock House, Hancock Bridge, New Jersey. (E.R. Coutch, HABS)

1949
National Trust for Historic Preservation in the United States is chartered by Congress.

1951
National Trust acquires Woodlawn Plantation in Virginia as its first museum property.

1952
First issue of *Historic Preservation* magazine appears.

1954
National Council for Historic Sites and Buildings merges with the National Trust.

1959
College Hill Demonstration Study in Providence, Rhode Island, is the first urban renewal study to address preservation concerns.

1960
A limited register of nationally significant landmarks, predecessor of the National Register of Historic Places, is initiated by the National Park Service.

Mount Vernon Ladies' Association of the Union is named the first winner of the National Trust's Crowninshield Award.

1961
First issue of *Preservation News* appears.

1963
Despite widespread public outcry and celebrity picket lines, the demolition of New York's Pennsylvania Station begins.

1964
Columbia University School of Architecture offers the first graduate-level course in historic preservation.

1966
With Heritage So Rich is published.

National Historic Preservation Act is passed, establishing an expanded National Register of Historic Places, an Advisory Council on Historic Preservation, and matching grants-in-aid to the states and the National Trust.

Department of Transportation Act declares a national policy of preservation of natural and historic sites on highway routes.

Demonstration Cities and Metropolitan Development Act redirects urban renewal to recognize and fund preservation projects.

1967
First state historic preservation officers and the first keeper of the National Register are appointed.

1968
Association for Preservation Technology is founded.

New York City enacts the nation's first ordinance allowing transfer of development rights.

1969
U.S. Department of the Interior makes the first preservation grants to the states.

National Environmental Policy Act stresses federal preservation responsibility and creates environmental impact statements.

Historic American Engineering Record is established, and its first site-recording project is carried out at Great Falls Canal and Locks in Virginia.

Proposed Vieux Carré elevated expressway in New Orleans becomes the first interstate highway stopped for environmental reasons.

1970
First Earth Day is celebrated.

1971
Executive Order 11593 directs federal agencies to preserve, restore, and maintain their cultural properties.

National Trust opens a Western Regional Office in San Francisco as the first of six regional offices.

Society for Industrial Archeology is established.

1972
Congress authorizes transfer of surplus significant federal property to local public agencies for preservation.

1973
First National Historic Preservation Week is celebrated.

New York City's Landmarks Preservation Commission is authorized to designate historic interiors.

Nation's first urban homesteading program is launched in Wilmington, Delaware.

1974
Preservation Action is formed as a national citizens' lobby.

1976
Celebration of the Bicentennial of the American Revolution strengthens interest in preservation.

Tax Reform Act of 1976 provides the first major preservation tax incentives for the rehabilitation of certified historic income-producing properties.

Public Buildings Cooperative Use Act encourages restoration and adaptive use of historic buildings for federal use.

1977
National Trust's Main Street Project, forerunner of the National Main Street Center, is launched in Galesburg, Illinois; Hot Springs, South Dakota; and Madison, Indiana.

1978
U.S. Supreme Court rules in its first major preservation case, involving Grand Central Terminal, that New York City's preservation ordinance is constitutional.

1979
Miami Beach becomes the site of the first National Register historic district made up entirely of 20th-century buildings.

1980
Amendments to the National Historic Preservation Act are passed directing federal agencies to nominate and protect historic federal properties, broadening participation of local governments, and requiring owner consent for National Register listing.

1981
Economic Recovery Tax Act provides significant new investment tax credits for rehabilitation.

1982
Zero preservation funding proposed by the Reagan administration is fought, and funding is restored after an intensive nationwide campaign.

1983
Congress approves restoration rather than extension of the West Front of the U.S. Capitol.

1984
National Trust acquires Montpelier, the 2,677-acre Virginia home of President James Madison.

1985
McDonald's announces plans to restore the first roadside stand built by Ray Kroc in 1955 in Des Plaines, Illinois.

San Francisco adopts a downtown master plan with the strongest design controls yet devised in an American city, including the protection of more than 250 landmark buildings.

1986
Nationwide campaign helps save the rehabilitation tax credits in the Tax Reform Act of 1986.

Centennial and restoration of the Statue of Liberty are celebrated on July 4.

1987
International Council on Monuments and Sites (ICOMOS) holds its 8th General Assembly in Washington, D.C., its first such meeting in the Western Hemisphere.

1988
Federal Abandoned Shipwrecks Act protects significant shipwrecks and authorizes state management of them.

Manassas National Battlefield Park in Virginia is saved from adjacent shopping mall development.

National Trust issues its first "11 Most Endangered Historic Places" list.

1989
Jobbers Canyon warehouse area of Omaha is demolished, the largest National Register historic district lost to date.

1990
President Bush's 1991 budget proposal reinstates preservation funding, ending a decade in which preservation had been "zeroed out" every year. The budget document states that "the preservation, understanding and passing on of the best of [our] local heritage is essential if Americans are to know what it means to be 'American.' "

1991
In two separate cases, the U.S. Supreme Court affirms the right of localities to designate religious properties as landmarks.

Pennsylvania Supreme Court, stating that landmark designation without owner consent represents a "taking," rules Philadelphia's preservation ordinance unconstitutional.

GLOSSARY

**"'Then you should say what you mean,' the March Hare went on.
"'I do,' Alice hastily replied; 'at least—at least I mean what I say—that's the same thing, you know.'
"'Not the same thing a bit!' said the Hatter."**

Lewis Carroll, *Alice's Adventures in Wonderland*, 1865

Adaptive use
The process of converting a building to a use other than that for which it was designed, e.g., changing a factory into housing. Such conversions are accomplished with varying alterations to the building.

Amenity
A building, object, area, or landscape feature that makes an aesthetic contribution to the environment, rather than one that is purely utilitarian.

Amicus curiae
Friend of the court (Latin). A party that may be allowed to present a brief on an issue before a court, frequently one with relevant special expertise.

Architect
Generally, one who designs and supervises the construction of buildings; legally, only professionals who are licensed by a state, territory, or the District of Columbia to practice architecture.

Architecture
The art and science of design and construction of buildings; also refers to particular periods, styles, or methods of construction, as in Victorian architecture.

Assessed value
Economic value of land or buildings determined by location, use, improvements, and other factors for tax purposes.

Background buildings
Buildings that may lack exemplary character or significance but that are nonetheless essential to maintain a sense of place.

Building code
Standards adopted by local governments regulating methods and materials of construction, egress, and accessibility and other factors affecting the construction of safe and sanitary buildings. Federal, state, territorial, and the District of Columbia governments also issue building codes used for government or government-funded buildings.

Building occupancy
Classification of buildings by use in building codes; used to determine methods and materials of construction, egress, and other requirements.

Certificate of appropriateness
A document awarded by a preservation commission or architectural review board allowing an applicant to proceed with a proposed alteration, demolition, or new construction in a designated area or site, following a determination of the proposal's suitability according to applicable criteria.

Certified historic structure
For the purposes of the federal preservation tax incentives, any structure subject to depreciation as defined by the Internal Revenue Code that is listed individually in the National Register of Historic Places or located in a registered historic district and certified by the secretary of the interior as being of historical significance to the district.

Certified rehabilitation
Any rehabilitation of a certified historic structure that the secretary of the interior has determined is consistent with the historical character of the property or the district in which the property is located.

Code compliance
Ensuring that a building is built in agreement with the appropriate building code.

Code enforcement
Local regulation of building practices and enforcement of safety and housing code provisions; a principal tool to ensure neighborhood upkeep.

Commercial archeology
The study of structures and artifacts created in connection with popular commercial activity, such as diners, motels, gasoline stations, and signs.

Conservation archeology
A field of archeology concerned with limiting excavations to a minimum consistent with research objectives and with preserving archeological sites for future scientific investigation.

Contingency budget
Unassigned construction funds set aside to pay for unexpected expenses or changes to the contract documents. A typical restoration or rehabilitation project should have a contingency budget of 10 percent.

Contract documents
Detailed plans, known as working drawings, and instructions and materials lists, known as specifications, concerned with the construction of a building.

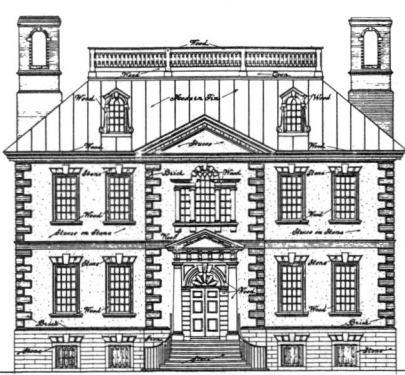

Mt. Pleasant Mansion, Philadelphia, Pennsylvania. (H. Stevenson, HABS)

Cultural resource
A building, structure, district, site, object, or document that is of significance in American history, architecture, archeology, or culture.

Demolition by neglect
The destruction of a building through abandonment or lack of maintenance.

Demolition delay
A temporary halt or stay in the planned razing of a property, usually resulting from a court injunction obtained by preservationists to allow a period of negotiation.

Design development
Phase of the design process when plans, elevations, and sections of a building are fully developed, when materials and methods of construction are determined, and when preliminary cost estimates are generated.

Design guidelines
Criteria developed by preservation commissions and architectural review boards to identify design concerns in an area and to help property owners ensure that rehabilitation and new construction respect the character of designated buildings and districts.

Design process
Method by which buildings are designed. The process typically has the following phases: determining the requirements of the owner and the proposed use; researching applicable codes and other legal requirements; analyzing the existing building, in the case of preservation projects; schematic design; design development; contract documents; bidding; site observation; and post-occupancy evaluation.

Design review
The process of ascertaining whether modifications to historic and other structures, settings, and districts meet standards of appropriateness established by a governing or advisory review board.

Dismantling
Taking apart a structure piece by piece, often with the intention of reconstructing it elsewhere.

Displacement
The movement of individuals, businesses, or industries from property or neighborhoods because of real estate activities.

Easement
A less-than-fee interest in real property acquired through donation or purchase and carried as a deed restriction or covenant to protect important open spaces, building facades, and interiors.

Egress
Term used in building codes for exiting a building; egress requirements are designed to allow occupants to exit safely in the event of fire or other disaster.

Eminent domain
The power of a government to acquire private property for public benefit after payment of just compensation to the owner.

Enabling legislation
Federal or state laws that authorize governing bodies within their jurisdictions to enact particular measures or delegate powers such as enactment of local landmarks and historic district ordinances, zoning, and taxation.

Energy audit
Analysis of a building's energy consumption for heating, cooling, lighting, etc. The purpose of an audit is to determine means to reduce energy consumption by improving the thermal performance of foundations, walls, doors, windows, roofs, and other exterior surfaces and the efficiency of heating, cooling, and lighting systems.

Extended use
Any process that increases the useful life of an old building, e.g., adaptive use or continued use.

Fabric
The physical material of a building, structure, or city, connoting an interweaving of component parts.

Facadism
The retention of only the facade of a historic building during conversion while the remainder is severely altered or destroyed to accept the new use.

Fenestration
Openings in an external wall such as doors and windows.

Financial incentives
Grants, low-interest loans, tax deductions and credits, easements, and other monetary inducements that improve the financial feasibility of a building project.

Found space
Old buildings or spaces within them that have been retrieved from near oblivion for rehabilitation or adaptive use after having been abandoned or "lost."

Gentrification
British term for the process by which young professionals or "gentry" buy into inner-city areas as part of a neighborhood preservation trend.

Historical archeology
The study of the cultural remains of literate societies, including excavated material as well as above-ground resources such as buildings, pottery, weapons, tools, glassware, cutlery, and textiles.

Historic district
A geographically definable area with a significant concentration of buildings, structures, sites, spaces, or objects unified by past events, physical development, design, setting, materials, workmanship, sense of cohesiveness, or related historical and aesthetic associations. The significance of a district may be recognized through listing in a local, state, or national landmarks register and may be protected legally through enactment of a local historic district ordinance administered by a historic district board or commission.

Homesteading
Programs under which abandoned buildings are made available at little or no cost in return for an agreement to rehabilitate and occupy them for a specified period of time. Similar programs to recycle commercial structures may be called shopsteading.

House museum
A museum whose structure itself is of historical or architectural signifi-

cance and whose interpretation relates primarily to the building's architecture, furnishings, and history.

Human scale
A combination of qualities in architecture or the landscape that provides an appropriate relationship to human size, enhancing rather than diminishing the importance of people.

Industrial archeology
The study of the history and development of industry as revealed by industrial buildings and artifacts such as bridges, transportation systems, and other engineering landmarks.

Infiltration
Movement of air between the interior of a building and the exterior. Eliminating gaps around windows, doors, and their joints in a building's exterior surfaces reduces infiltration and increases the building's thermal performance.

Landmarks register
A listing of buildings, districts, and objects designated for historical, architectural, or other special significance that may carry protection for listed properties.

Landscape
The totality of the built or human-influenced habitat experienced at any one place. Dominant features are topography, plant cover, buildings or other structures, and their patterns.

Leverage
The use of a small amount of funding to attract additional money to provide project capital; also, the use of fixed-cost funds to acquire a property that is expected to produce a higher rate of return through income or appreciation.

Life-safety code
Sections of a building code dealing with egress, fire separation, and other aspects of occupant safety; also a specialized code dealing with safety in buildings.

Maintenance program
Specific methods, procedures, and schedule to maintain and repair a building.

Massing
Composition of a building's volumes and surfaces that contribute to its appearance; for example, many classical-style buildings have a central mass or pavilion, flanked by subordinate masses or wings.

Material culture
Tangible objects used by people to cope with the physical world, such as utensils, structures, and furnishings, all of which provide evidence of culturally determined behavior.

Methods of construction
Systems and types of construction usually related to the types of structural materials used. For example, the balloon frame method of construction used in most single-family homes employs light structural wood members (2x4s, 2x6s, etc.) nailed together; a poured-in-place method of construction uses concrete poured into forms, reinforced with steel bars.

Mixed use
A variety of authorized activities in an area or a building, as distinguished from the isolated uses and planned separatism prescribed by many zoning ordinances.

Outdoor museum
A restored, re-created or replica village site in which several or many structures have been restored, rebuilt, or moved and whose purpose is to interpret a historical or cultural setting, period, or activity.

Police power
The inherent right of a government to restrict individual conduct or use of property to protect the public health, safety, and welfare; it must follow due processes of the law but, unlike eminent domain, does not carry the requirement of compensation for any alleged losses. Police power is the basis for such regulations as zoning, building codes, and preservation ordinances.

Preservation
Generally, saving from destruction or deterioration old and historic buildings, sites, structures, and objects and providing for their continued use by means of restoration, rehabilitation, or adaptive use. Specifically, "the act or process of applying measures to sustain the existing form, integrity, and material of a building or structure, and the existing form and vegetative cover of a site. It may include stabilization work, where necessary, as well as ongoing maintenance of the historic building materials" (Secretary of the Interior's Standards).

Preservation commission
A generic term for an appointed municipal or county board that recommends the designation of and regulates changes to historic districts and landmarks. It may be called a historic district review board or commission, architectural or design review board, or landmarks commission; the latter's authority may be limited to individual buildings.

Proportions
The relative size of two or more dimensions of a building; many architectural styles use highly developed mathematical proportions to determine the composition of facades and volumes of interior spaces.

Reconstruction
"The act or process of reproducing by new construction the exact form and detail of a vanished building, structure, or object, or a part thereof, as it appeared at a specific period of time" (Secretary of the Interior's Standards).

Redlining
A practice among financial institutions and insurance companies of refusing to provide services to certain supposedly high-risk geographical areas, regardless of the merits of individual applicants; derived from the red line that the institutions may draw around the area on a map.

Redundant building
British term for a building or site no longer in demand for its original or current use. In the United States the terms "endangered property" and "surplus property" are most often used.

Rehabilitation
"The act or process of returning a property to a state of utility through repair or alteration which makes possible an efficient contemporary use while preserving

those portions or features of the property which are significant to its historical, architectural, and cultural values" (Secretary of the Interior's Standards).

Reinvestment
The channeling of public and private resources into declining neighborhoods in a coordinated manner to combat disinvestment.

Renovation
Modernization of an old or historic building that may produce inappropriate alterations or eliminate important features and details.

Restoration
"The act or process of accurately recovering the form and details of a property and its setting as it appeared at a particular period of time by means of the removal of later work or by the replacement of missing earlier work" (Secretary of the Interior's Standards).

Revolving fund
A funding source that makes loans to accomplish some preservation purpose, e.g., purchase and rehabilitation of an endangered property. The loans are repaid to maintain the fund for other projects.

Salvage archeology
Rescue of archeological materials and data threatened by damage or destruction.

Schematic design
Preliminary plans, elevations, and sections showing the approximate location and size of rooms, composition of the facade, and other aspects of a building project. Typically, a number of schematic designs are developed for each project to study alternative solutions.

Section 106
The provision of the National Historic Preservation Act of 1966 that requires the head of a federal agency financing or licensing a project to make a determination of the effect of the project on property in or eligible for the National Register of Historic Places.

Seismic code
Building code governing structural design and methods and materials of construction to prevent the collapse of a building during an earthquake.

Sense of place
The sum of attributes of a locality, neighborhood, or property that give it a unique and distinctive character.

Setback
Distance between the facade of a building and the property line. Setback requirements are usually found in zoning codes.

Site coverage
Percentage of a site covered by buildings.

Site observation
Services provided by an architect during construction to ensure that the building is being constructed according to the contract documents.

Stabilization
"The act or process of applying measures designed to re-establish a weather resistant enclosure and the structural stability of unsafe or deteriorated property while maintaining the essential form as it exists at present" (Secretary of the Interior's Standards).

Standing to sue
The doctrine that cases presented to a court must be concrete controversies between parties with a real stake in the dispute, such as a financial injury. In environmental matters, other parties may gain the right to sue.

Street furniture
Municipal equipment placed along streets, including light fixtures, fire hydrants, police and fire call boxes, signs, benches, and kiosks.

Streetscape
The distinguishing character of a particular street as created by its width, degree of curvature, paving materials, design of the street furniture, and forms of surrounding buildings.

Style
A type of architecture distinguished by special characteristics of structure and ornament and often related in time; also, a general quality of distinctive character.

Sweat equity
The investment of property owners' or occupants' own labor in rehabilitation work as a form of payment.

Taking
The appropriation by government of private property, e.g., condemnation through eminent domain for public use with just compensation. A "taking issue" arises when the use of the police power appears to diminish the value of affected property, such as a decision under a preservation ordinance.

Tax incentive
A tax reduction designed to encourage private investment in historic preservation and rehabilitation projects.

Thermal performance
The ability of wall, roof, foundation, and other external building surfaces to regulate the exchange of heat or cold between the interior and exterior. Thermal performance is determined by the materials used as well as methods of construction.

Townscape
The relationship of buildings, shapes, spaces, and textures that gives a town or area its distinctive visual character or image.

Underwater archeology
A field of archeology concerned with the identification, analysis, and documentation of sites and properties submerged under water, e.g., shipwrecks.

Vernacular buildings
Buildings designed and built without the aid of an architect or trained designer; buildings whose design is based on ethnic, social, or cultural traditions rather than on an architectural philosophy.

Visual pollution
Anything that, because of its placement or intrinsic nature, is offensive to the sense of sight, e.g., garbage dumps and billboards.

Zoning code
Local government regulation governing the use of land and buildings, setbacks, site coverage, and other aspects of land development.

PERIODICALS

Julie Griffith

"What happens in print goes on happening indefinitely. Print is the medium of continuity—the medium of history."

"The Talk of the Town," *The New Yorker*, October 2, 1978

AIC Newsletter and **Journal**
American Institute for Conservation of Historic and Artistic Works
1400 16th Street, N.W.
Suite 340
Washington, DC 20036
(202) 364-1036
Marcia Anderson, Editor

American Heritage
60 5th Avenue
New York, NY 10011
(212) 206-5500
Byron Dobell, Editor

Americana
29 West 38th Street
New York, NY 10018
(212) 398-1550
Sandra Wilmot, Editor

APT Bulletin and **Communique**
Association for Preservation Technology
P.O. Box 8178
Fredericksburg, VA 22404
(703) 373-1621, 373-1622
Marylee MacDonald, Editor, *Bulletin*
Connie Garnett, Editor, *Communique*

Architectural Record
McGraw-Hill Information Systems Company
1221 Avenue of the Americas
New York, NY 10020
(212) 512-2000
Mildred F. Schmertz, Editor

Architecture
American Institute of Architects
1130 Connecticut Avenue, N.W.
Washington, DC 20036
(202) 828-0993
Deborah Deitsch, Editor

Art and Archaeology Technical Abstracts
Getty Conservation Institute
4503 Glencoe Avenue
Marina del Rey, CA 90292
(310) 822-2299
Jessica S. Brown, Editor

Blueprints
National Building Museum
401 F Street, N.W.
Washington, DC 20001
(202) 272-7784
Joyce Elliot, Editor

Commercial Renovation
20 East Jackson Boulevard
Suite 700
Chicago, IL 60604
(312) 922-5402
Craig Shutt, Editor

Conservation Foundation Letter
The Conservation Foundation/World Wildlife Foundation
1250 24th Street, N.W.
Suite 500
Washington, DC 20037
(202) 293-4800
Jonathan Adams, Editor

Cultural Resource Management Bulletin
National Park Service
U.S. Department of the Interior
P.O. Box 37127
Washington, DC 20013-7127
(202) 343-9500
Carol Shull, Manager

Environmental Law Reporter
Environmental Law Institute
1616 P Street, N.W.
Suite 200
Washington, DC 20036
(202) 328-5150
Barry Breen, Editor

Fine Homebuilding
The Taunton Press
63 South Main Street
P.O. Box 5506
Newtown, CT 06470
(203) 426-8171
Mark Feirer, Managing Editor

Historic Preservation and **Preservation News**
National Trust for Historic Preservation
1785 Massachusetts Avenue, N.W.
Washington, DC 20036
(202) 673-4070
Anne Elizabeth Powell, Editor-in-Chief

History News
American Association for State and Local History
172 2nd Avenue, North
Suite 202
Nashville, TN 37201
(615) 255-2971
Candace Floyd, Editor

Housing and Development Reporter
2300 M Street, N.W.
Suite 100
Washington, DC 20037
(202) 955-9610
Barry G. Jacobs, Editor

IA Journal and **Newsletter**
Society for Industrial Archeology
National Museum of American History
Room 5014
Smithsonian Institution
Washington, DC 20560
(202) 357-2228
David R. Starbuck, Editor, *Journal*
Robert M. Frames, Editor, *Newsletter*

Journal of Architectural Education
Association of Collegiate Schools of Architecture
1735 New York Avenue, N.W.
Washington, DC 20006
(202) 785-2324
David Bell, Editor

Landscape
P.O. Box 7107
Berkeley, CA 94707
(510) 549-3233
Bonnie Lloyd, Editor

Landscape Architecture
American Society of Landscape Architects
4401 Connecticut Avenue, N.W.
5th Floor
Washington, DC 20008-2302
(202) 686-2752
James Trulove, Editor

LHAT Bulletin
League of Historic American Theatres
1511 K Street, N.W.
Suite 923
Washington, DC 20005
(202) 783-6966
Tara Dicken, Editor

Livability Digest and **PLACE**
Partners for Livable Places
1429 21st Street, N.W.
Washington, DC 20036
(202) 887-5990
Catherine Young, Editor

The Magazine Antiques
575 Broadway
New York, NY 10012
(212) 941-2800
Wendell Garrett, Editor

Main Street News
National Main Street Center
National Trust for Historic Preservation
1785 Massachusetts Avenue, N.W.
Washington, DC 20036
(202) 673-4219
Linda S. Glisson, Editor

Marquee
Theatre Historical Society of America
152 North York Road
2d Floor
Elmhurst, IL 60126
(708) 782-1800
John Fischer, Editor

Material Culture
Pioneer America Society
c/o Department of Earth Sciences
Southeast Missouri State University
Cape Girardeau, MO 63701
(314) 651-2354
Charles Martin, Editor

Metropolis: The Architecture and Design Magazine of New York
177 East 87th Street
New York, NY 10128
(212) 722-5050
Susan S. Szenasy, Editor

Museum News
American Association of Museums
1225 I Street, N.W.
Suite 200
Washington, DC 20005
(202) 289-1818
Bill Anderson, Editor

National Parks
National Parks and Conservation
Association
1776 Massachusetts Avenue, N.W.
Washington, DC 20036
(202) 223-6722
Michele Strutin, Editor

Nineteenth Century
The Victorian Society in America
219 South 6th Street
Philadelphia, PA 19106
(215) 627-4252
Judith Snyder, Editor

The Non Profit Times
190 Tamarack Circle
Skillman, NJ 08558
(609) 921-1251
Larry Sterne, Editor

Old-House Journal
435 Ninth Street
Brooklyn, NY 11215
(718) 788-1700
Patricia Poore, Editor

Planning
American Planning Association
1313 East 60th Street
Chicago, IL 60637
(312) 955-9100
Sylvia Lewis, Editor

Preservation Action Alert
Preservation Action
1350 Connecticut Avenue, N.W.
Suite 401
Washington, DC 20036
(202) 659-0915
Nellie L. Longsworth, President

Preservation Forum
National Trust for Historic Preservation
1785 Massachusetts Avenue, N.W.
Washington, DC 20036
(202) 673-4037
Ann Elizabeth Powell, Editor

Preservation Law Reporter
National Trust for Historic Preservation
1785 Massachusetts Avenue, N.W.
Washington, DC 20036
(202) 673-4035
Julia H. Miller, Editor

Progressive Architecture
600 Summer Street
P.O. Box 1361
Stamford, CT 06904
(203) 348-7531
John Morris Dixon, Editor

The Public Historian
University of California Press
Journals Division
2120 Berkeley Way
Berkeley, CA 94720
(510) 642-4191
G. Wesley Johnson, Jr., Editor

SCA News Journal
Society for Commercial Archeology
Smithsonian Institution
National Museum of American History
Room 5010
Washington, DC 20560
(202) 882-5424

Oswego City Library, Oswego, New York. (H. McKee, HABS)

Sea History
National Maritime Historical Society
P.O. Box 68
Peekskill, NY 10566
(914) 737-7878
Lincoln P. Paine, Editor

Small Town
Small Towns Institute
P.O. Box 517
Ellensburg, WA 98926
(509) 925-1830
Anne S. Denman, Editorial Director

**Society of Architectural Historians
Newsletter, Journal**, and **Forum**
1232 Pine Street
Philadelphia, PA 19107-5944
(215) 735-0246
Marjorie Pearson, Editor, *Newsletter*
(212) 553-1126
Patricia Waddy, Editor, *Journal*
(315) 443-5099
Michael Tomlan, Editor, *Forum*
(607) 255-7261

Technology and Conservation
One Emerson Place
Boston, MA 02114
(617) 227-8581
Susan E. Schur, Editor

Urban Outlook
215 Park Avenue, South
Suite 1301
New York, NY 10003
(212) 228-0246
Frank C. Frantz, Editor

Urban Design International
Institute for Urban Design
390 Riverside Drive
Suite 4F
New York, NY 10025
(619) 455-1251
Ann Ferebee, Editor

Urban Land
Urban Land Institute
625 Indiana Avenue, N.W.
Suite 400
Washington, DC 20004-2930
(202) 624-7000
Libby Howland, Editor

US/ICOMOS Newsletter
1600 H Street, N.W.
Washington, DC 20006
(202) 842-1866
Paul Hallam, Editor

Vernacular Architecture Newsletter
Programs in Folk Studies/MLIS
Western Kentucky University
Bowling Green, KY 42101
(502) 745-2401
Michael Ann Williams, Editor

Victorian Homes
P.O. Box 61
Millers Falls, MA 01349
(413) 659-3785
Carolyn Flaherty, Editor

Waterfront World
The Waterfront Center
1536 44th Street, N.W.
Washington, DC 20007
(202) 337-0356
Ann Breen and Dick Rigby, Coeditors

**Winterthur Portfolio: A Journal of
American Material Culture**
Henry Francis du Pont Winterthur
Museum
Winterthur, DE 19735
(302) 888-4600
Ian M. G. Quimby, Editor

FURTHER READING

**Ulrich's International Periodicals
Directory, 1988–89.** New York: R. R.
Bowker, 1988. 2 vols.

**Index to Historical Preservation
Periodicals.** Boston: G. K. Hall and
Company, 1988.

REHABILITATION STANDARDS

Richard Wagner

"The standards are intended to create a strong framework for responsible preservation practices, to be used in conjunction with the accompanying guidelines for applying the standards. The underlying concern of the standards is the preservation of the significant historic and architectural characteristics of a structure that is being rehabilitated.

"In addition to providing guidance to individual property owners as they plan their rehabilitation work, the standards are used by the Secretary of the Interior to determine whether a rehabilitation project qualifies as a 'certified rehabilitation' for federal tax benefits under the Economic Recovery Tax Act of 1981. They are also used to assist state and local governments and individuals in planning and carrying out rehabilitation work on historic buildings."

Technical Preservation Services, National Park Service,
Respectful Rehabilitation: Answers to Your Questions About Old Buildings, 1982.

The Secretary of the Interior's Standards for Rehabilitation were originally published in 1977 and revised in 1990. The standards compose one section in the Secretary of the Interior's Standards for Historic Preservation Projects and appear in Title 36 of the Code of Federal Regulations, Part 68, which governs alterations to buildings listed in the National Register of Historic Places. The standards, which pertain to the exterior and interior of historic buildings, deal with design, methods of construction, and materials. They also apply to sites, landscapes, and archaeological resources listed in the register.

In addition to the standards, the Department of the Interior also issues "Guidelines for Rehabilitating Historic Buildings," which contains recommended (and not recommended) treatments commonly used in restoration and rehabilitation projects.

Since each proposed design and historic building and landscape is unique, the standards are applied on a case-by-case basis, taking into account specific economic and technical considerations. The standards and the guidelines, in addition to preserving and enhancing the architectural character of a building and its site, present sound technical advice to help reduce future maintenance problems.

THE SECRETARY OF THE INTERIOR'S STANDARDS FOR REHABILITATION

1. A property shall be used for its historic purpose or be placed in a new use that requires minimal change to the defining characteristics of the building and its site and environment.

2. The historic character of a property shall be retained and preserved. The removal of historic material or alteration of features and spaces that characterize a property shall be avoided.

3. Each property shall be recognized as a physical record of its time, place, and use. Changes that create a false sense of historic development, such as adding conjectural features or architectural elements from other buildings, shall not be undertaken.

4. Most properties change over time; those changes that have acquired historic significance in their own right shall be retained and preserved.

5. Distinctive features, finishes, and construction techniques or examples of craftsmanship that characterize a property shall be preserved.

6. Deteriorated historic features shall be repaired rather than replaced. Where the severity of deterioration requires replacement of a distinctive feature, the new feature shall match the old in design, color, texture, and other visual qualities and, where possible, materials. Replacement of missing features shall be substantiated by documentary, physical, or pictorial evidence.

7. Chemical or physical treatments, such as sandblasting, that cause damage to historic materials shall not be used. The surface cleaning of structures, if appropriate, shall be undertaken using the gentlest means possible.

8. Significant archaeological resources affected by a project shall

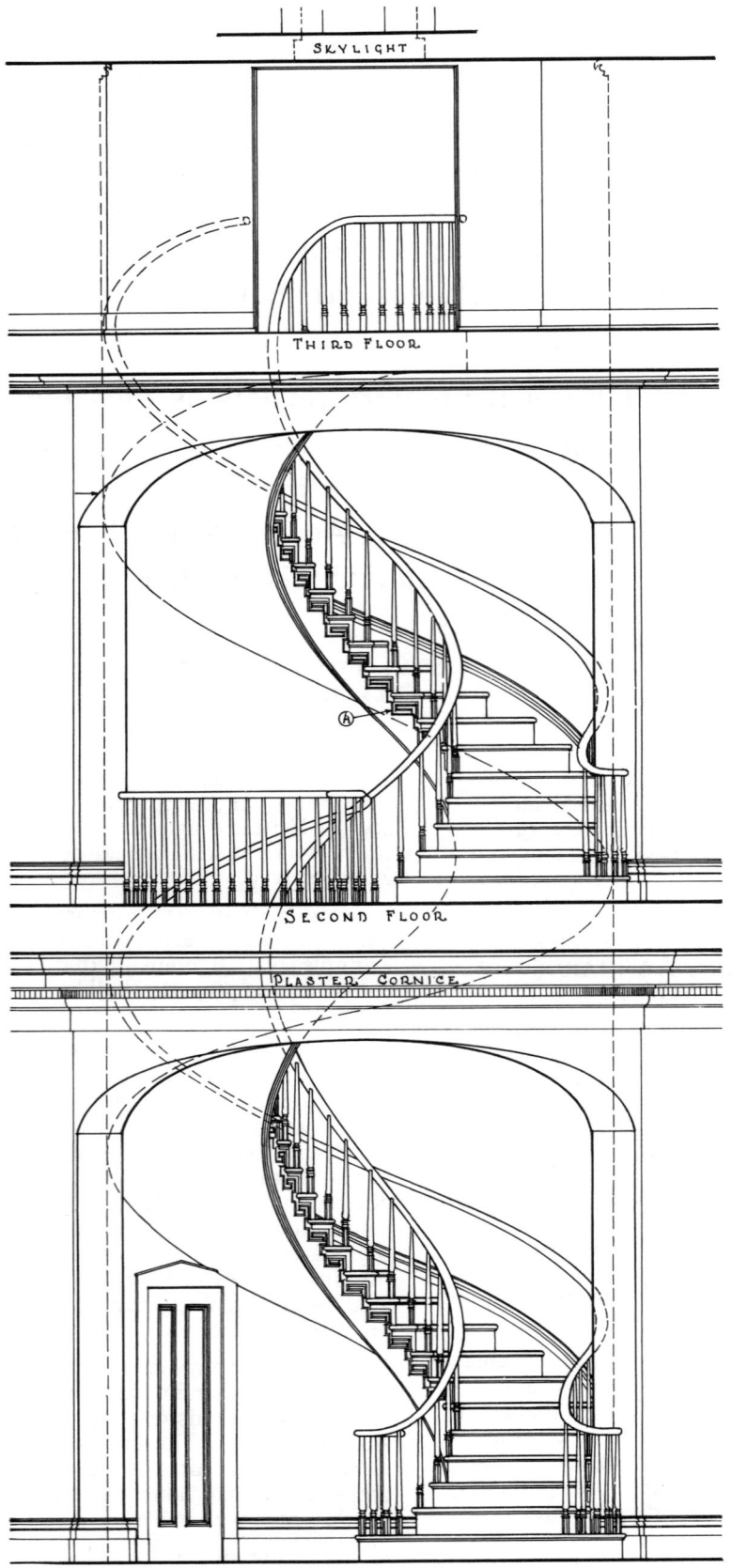

SKYLIGHT

THIRD FLOOR

SECOND FLOOR

PLASTER CORNICE

Stairway, James F.D. Lanier Home, Madison, Indiana. (J.F. Ehlert, HABS)

be protected and preserved. If such resources must be disturbed, mitigation measures shall be undertaken.

9. New additions, exterior alterations, or related new construction shall not destroy historic materials that characterize the property. The new work shall be differentiated from the old and shall be compatible with the massing, size, scale, and architectural features to protect the historic integrity of the property and its environment.

10. New additions and adjacent or related new construction shall be undertaken in such a manner that, if removed in the future, the essential form and integrity of the historic property and its environment would be unimpaired.

FURTHER READING

The Secretary of the Interior's Standards for Rehabilitation. Rev. ed. Washington, D.C.: Preservation Assistance Division, U.S. Department of the Interior, 1990.

The Secretary of the Interior's Standards for Rehabilitation and Guidelines for Rehabilitation Historic Buildings. W. Brown Morton III and Gary L. Hume. 1979. Rev. ed. Washington, D.C.: Technical Preservation Services, U.S Department of the Interior, 1983.

Interpreting the Secretary of the Interior's Standards for Rehabilitation. Washington, D.C.: Technical Preservation Services, U.S. Department of the Interior, 1980–. Sets and occasional bulletins.

Preservation Briefs Series. Technical Preservation Services, U.S. Department of the Interior. Washington, D.C.: U.S. Government Printing Office, 1975–.

Preservation Tech Notes. Technical Preservation Services, U.S. Department of the Interior. Springfield, Va.: National Technical Information Service, 1985–.

II. PRODUCTS AND SERVICES

ARCHITECTS

Arnouts Associates
87 Cannon Street
Poughkeepsie, NY 12601
(914) 473-5245
(914) 473-5314 (Fax)
Robert A. Arnouts, AIA

Beyer Blinder Belle

Architects & Planners

Beyer Blinder Belle
41 East 11th Street
New York, NY 10003
(212) 777-7800
(212) 475-7424 (Fax)
John Belle, FAIA, RIBA

The focus of Beyer Blinder Belle's 23 years of practice has been the pursuit of excellence in the field of historic preservation. Projects range from the restoration and interpretation of small-scale historic sites—such as the Alice Austen House and Grounds on Staten Island and the analysis of 19th-century structures at High Breeze Farm in New Jersey's Wawayanda State Park—to large-scale technically complex restorations such as Ellis Island and the Cathedral of the Madeleine in Salt Lake City and the preservation of Grand Central Terminal in New York City.

Buttrick White & Burtis
475 10th Avenue
New York, NY 10018
(212) 967-3333
(212) 629-3749 (Fax)
Jean C. Parker, AIA

Buttrick White & Burtis specializes in historic preservation and has won six preservation awards. New York City projects include several buildings in Central Park, St. James's Church, the Fordham University chapel, a warehouse at the South Street Seaport, and several Upper East Side town houses. The firm is also restoring the Hawkins-Mount Homestead, a National Historic Landmark, in Stony Brook on Long Island, New York.

Carey & Co. Architecture
300 Brannan Street
Suite 402
San Francisco, CA 94107
(415) 957-0100
(415) 957-1199 (Fax)
Alice Ross Carey

Crawford & Stearns
Architects and Preservation Planners
134 Walton Street
Syracuse, NY 13202
(315) 471-2162
(315) 471-2965 (Fax)
Randall T. Crawford and Carl D. Stearns

Crawford & Stearns, founded in 1978, is a medium-sized professional firm specializing in the restoration, preservation, expansion, and adaptive use of historic properties. Services include documentation, conditions assessment, historic structures reports, facade revitalization programs, preservation planning, historic district commission consulting, and alterations for accessibility.

Ehrenkrantz & Eckstut Architects
23 East 4th Street
New York, NY 10003
(212) 353-0400
(212) 228-3928 (Fax)
Kate Burns Ottavino

Ehrenkrantz & Eckstut Architects is a rare alliance of preservationists, architects, and urbanists with extensive experience in the restoration, preservation, and adaptive use of historic buildings, landmarks, and landscapes. Other areas of expertise include the design of buildings, interiors, and public spaces.

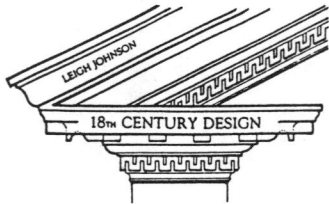

18th Century Design Associates
397 Massey Road
Springfield, VT 05156
(802) 885-1122
Leigh Johnson

Founded in 1979, 18th Century offers complete architectural services. The firm's expertise in preserving historic houses and in designing reproductions and additions is based on many years of preservation woodwork/contracting and countless hours of historical research. State and national registrations written. Consultations with owner, architect, and contractor.

GAEDE SERNE ARCHITECTS INC.

Gaede Serne Architects, Inc.
820 Superior Avenue, N.W.
Suite 100
Cleveland, OH 44113
(216) 241-3339
(216) 241-1823 (Fax)
Robert C. Gaede

Gaede Serne Architects, Inc., has long been affiliated with historic preservation. Founded in 1956, the firm has fulfilled a variety of assignments including master plan studies, feasibility studies, and restoration plans or adaptive use programs for historic structures. Experience with public buildings, churches, collegiate structures, and commercial buildings predominate.

Jamie Gibbs & Associates
Landscape Architects and Interior Designers
340 East 93rd Street
Suite 14C
New York, NY 10128
(212) 722-7508
(212) 722-7508 (Fax)
Jamie Gibbs

Thirteen years of experience, 11 of them as the head of his own practice, have taken Jamie Gibbs across three continents. His specialty is the creation of traditional interiors, and his work in this field has encompassed a wide range of architectural styles. Gibbs has also written several books and frequently gives lectures and conducts study tours through famous houses, many of which are not usually open to the public.

David H. Gleason Associates, Inc.
520A North Eutaw Street
Baltimore, MD 21201
(410) 728-1810
David H. Gleason, AIA

Award-winning architectural office with more than 30 years of experience in commercial and residential restoration, rehabilitation, and adaptive use as well as new construction. The firm is also a leader in developing design review programs and historic district design guidelines for local governments.

Douglas S. Heckrotte, AIA
Architect
P.O. Box 3674
Greenville, DE 19807
(302) 656-0242
Douglas S. Heckrotte

**David H. Gleason Associates
Architects**

ARCHITECTURE
URBAN DESIGN
PRESERVATION
DESIGN GUIDELINES

**520 N. Eutaw St.,
Baltimore, Md. 21201
(410) 728-1810**

The Office of Allen Charles Hill, AIA
Historic Preservation and Architecture
25 Englewood Road
Winchester, MA 01890
(617) 729-0748
Allen C. Hill

Architectural and preservation problem
solving for old buildings and their owners.
Architectural services for restoration, conservation, replication, and extension.
Building evaluation, analysis, technical
assistance, and troubleshooting. Research,
consulting, and historic structures
reports. Lectures and workshops.
Award-winning specialists in residences,
churches, and history museums. Please
call for further information.

Robert E. Meadows, P.C., Architects
40 Dover Street
New York, NY 10038
(212) 962-2645
(212) 233-8568 (Fax)
Robert E. Meadows, AIA

Robert E. Meadows, P.C., Architects provides reports, analyses, construction documents, and innovative solutions for
problems unique to the restoration of historic structures. Restoration projects
range from a thatched-roof cottage to the
18th-century residence of a New York
City mayor. Other projects have included
the relocation of four town houses in
Brooklyn and the preservation of decorative house museums and church interiors.

JOHN MILNER ASSOCIATES
ARCHITECTS • ARCHEOLOGISTS • PLANNERS
(JMA)

John Milner Associates (JMA)
309 North Matlack Street
West Chester, PA 19380
(215) 436-9000
(215) 436-9000 (Fax)
Allan Steenhusen

John Milner Associates (JMA) is a professional firm specializing in architectural
design, historic building restoration and
rehabilitation, archeology, and building
materials conservation. Offices are
located in West Chester, Pennsylvania,
Philadelphia, and Alexandria, Virginia.

MINEWEASER & ASSOCIATES
preservation architecture / construction

Mineweaser & Associates
1775 Junction Avenue, Suite 100
San Jose, CA 95112-1010
(408) 441-1755
(408) 441-1757 (Fax)
Craig Mineweaser, A.I.A.

Craig Morrison/Platt, Wyckoff & Coles, Architects
305 East 46th Street
New York, NY 10017
(212) 753-5910
(215) 423-3595
(212) 826-9163 (Fax)
(215) 423-3595 (Fax)
Craig Morrison, AIA

Founded in 1897 by noted architect Charles Adams Platt, the firm continues his practice of classical architecture and interiors, maintaining a high standard of integrity, responsiveness, scholarship, and respect for the needs of its clients and their buildings. The firm has a wide resume of outstanding traditional and historic restoration projects.

Dan Peterson AIA & Associates, Inc.
146 Bishop Alley
Point Richmond, CA 94801
(510) 235-2753 (Northern California)
(818) 796-8454 (Southern California)
Dan Peterson

Dan Peterson, AIA, is a nationally recognized historic architect with extensive experience in all aspects of architectural preservation, restoration, conservation, relocation, and assessments. The firm specializes in sensitive reuse of historic buildings, solving special technical, code, and EIR problems in both northern and southern California.

Short and Ford
864 Mapleton Road
Princeton, NJ 08540
(609) 452-1777

Thirtieth Street Architects, Inc.
2821 Newport Boulevard
Newport Beach, CA 92663
(714) 673-2643
(714) 673-8547 (Fax)
Leonard Klinwinski

Thirtieth Street Architects, Inc., has more than 15 years of experience in historic preservation projects including the restoration and adaptive use of historic structures (many of them certified tax credit projects), National Register nominations, HABS documentation, historic resource inventories, historic district identification, and design guidelines for historic structures and districts.

Watson and Henry Associates
12 North Pearl Street
Bridgeton, NJ 08302
(609) 451-1779
(609) 451-0471 (Fax)
Michael C. Henry, PE, AIA

Watson and Henry Associates provides professional services in architecture, engineering, and historic preservation. The firm specializes in preservation studies, architectural/engineering assessments, and museum-quality preservation and restoration of historic structures. Public and private projects include lighthouses, museums, mills, agricultural and industrial structures, churches, residences, and commercial and public buildings.

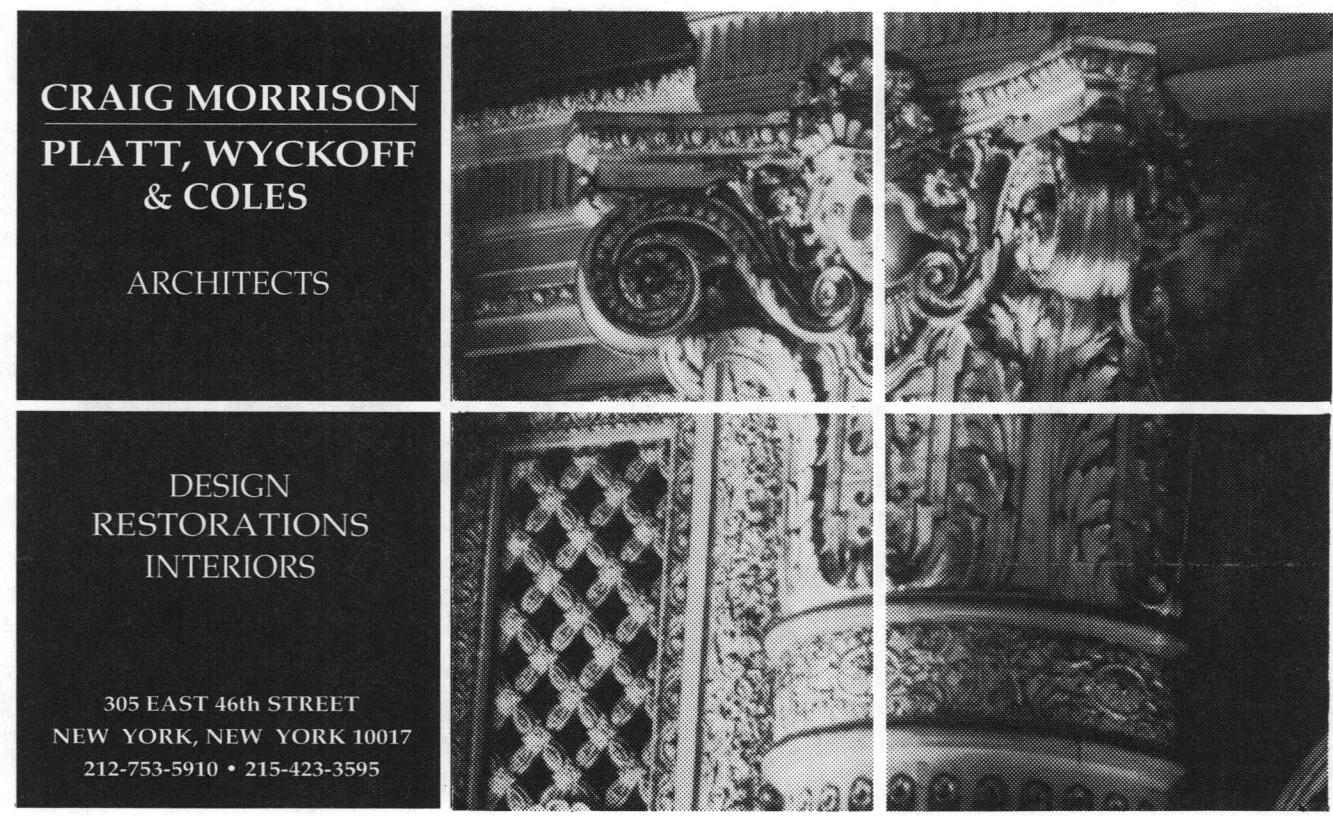

SHORT AND FORD AND PARTNERS
ARCHITECTS

Preservation Planning
Site Analysis
Design Standards
Existing Condition Surveys
Feasibility Studies
Preservation Plans
Facility Studies
Maintenance Plans

Documentation
Measured Drawings
Photography
Photogrammetry Coordination
Historic American Buildings Survey

Historic Structures Reports
Historic Research
Physical Investigation
Graphic Documentation
Development Analysis
Restoration Recommendations
Cost Estimates
Guide Specifications

Conservation Services
Structural Evaluations
Paint Analysis Coordination
Decorative Finishes Restoration
Materials Testing Analysis

Restoration Services
Research and Design
Financial Feasibility
Grant Writing
Code Consulting
Tax Act Certification
Design Development
Construction Documents
Construction Administration

Consultation
Historic District Commissions
Main Street Preservation
Historic Color Schemes

864 Mapleton Road Princeton, New Jersey 08540 609-452-1777 fax 609-452-7192

ARCHITECTURAL DETAILS

Amherst Woodworking & Supply, Inc.
P.O. Box 718
Northampton, MA 01061
(413) 585-3003
(413) 585-0288 (Fax)
David Short

(See description under Millwork)

Anything Fiberglass
8510 Lava Hill Road
Austin, TX 78744
(512) 243-3303
(512) 243-1782
Jerry Post

Anything Fiberglass reconstructs, restores, or replicates ornate moldings, carvings, and castings through the use of composites. Techniques yield visual and structural results that are faithful to original materials such as cast iron, bronze, aluminum, marble, wood, plaster, pressed tin, and ceramic. Artist on duty to custom design original details. Nationwide service.

Creative Woodworking, Ltd.
26 Friendship Street
Westerly, RI 02891
(401) 596-4463
(401) 596-3418 (Fax)
Ray Matteson

(See description under Millwork)

Denninger Cupolas & Weather Vanes
RD 1, Box 447 L
Middletown, NY 10940
(914) 343-2229
Alfred H. Denninger

Finely crafted cupolas and weather vanes for historic homes, carriage houses, barns, and stables. Cupolas with standing-seam copper roofs and clear redwood carpentry in traditional styles. Copper weather vanes in period designs: full and swell-bodied, silhouettes, church scrolls, Victorian banners. Replication and custom commissions are welcome. Free brochure. Catalog $4.00.

Historical Arts & Casting, Inc.
4130 West 1939 South, Unit F
Salt Lake City, UT 84104
(801) 974-0242
(801) 974-5832 (Fax)
Robert A. Baird

Specializing in the restoration and replication of traditional architectural cast-metal ornamentation such as canopies, facades, grilles, fountains, gates, and furniture. Manufacturing materials include cast iron, brass, bronze, and aluminum. Design, research, documentation, estimating, manufacturing, and installation services are provided.

ARTISTIC METAL WORK
EUROPEAN QUALITY CRAFTSMANSHIP

Railings — Gates — Doors — Marquees
Fine Interior and Exterior Metalwork

Specialist in Historic Monument Restoration
• Our References: Statue of Liberty, New York, U.S.A. •
• Château de Versailles, France • Place Stanislas, Nancy, France •
• Pont Alexandre III, Paris, France •
• Monument de Maisonneuve, Montreal, Canada •

and Private Customers in Paris, Montreal, New York,
Monaco, Switzerland, Japan . . .

LMC Corporation
(Les Metalliers Champenois USA)
118 Second Avenue
Paterson, NJ 07514
(201) 279-3573
(201) 881-0235 (Fax)
Jean Wiart, General Manager

(See description under Metalwork)

New England Woodturners
75 Daggett Street
New Haven, CT 06519
(203) 776-1880
(203) 776-1880 (Fax)
Richard Swartwout

(See description under Millwork)

Robinson Iron Corporation
P.O. Box 1119
Robinson Road
Alexander City, AL 35010
(205) 329-8486
(205) 329-9860
J. Scott Howell

Robinson specializes in cast-metal restoration and the custom casting of iron, aluminum, or bronze architectural details. Expert craftsmen evaluate projects in situ, generate required shop drawings and patternwork, and coordinate installation of the material for proper fit and appearance. A color brochure illustrating authentic reproductions of 19th-century lawn and garden ornaments is available.

A. F. Schwerd Manufacturing Company
3215 McClure Avenue
Pittsburgh, PA 15212
(412) 766-6322

Specialists in production of wood columns up to 35 feet in height and from 4 to 50 inches in diameter with a choice of 14 styles of capitals. Matching pilasters are also available. Maintenance-free aluminum bases and ventilated plinths are available for exterior columns. Custom work done. Write or call for a free catalog.

(See display ad, page 138)

Traditional Line, Ltd.
143 West 21st Street
New York, NY 10011
(212) 627-3555
(212) 645-8158 (Fax)
James Boorstein or Anthony Lefeber

(See description under Restoration)

Frederick Wilbur
Carver
P.O. Box 425
Lovingston, VA 22949
(804) 263-4827
(804) 263-5958 (Fax)

Hand-carved traditional moldings, mantelpieces, ecclesiastical decoration, brackets, garlands, coats of arms, as well as contemporary polychromed or gilded logos and lettering for architects, designers, and individuals. Clients

include the Library of Congress; Grace and Trinity Cathedral, Kansas City; University of Virginia; Locsin-York Design, San Diego; and John W. Kluge, Charlottesville, Virginia. Please request brochure and resume.

BIRD CONTROL

Bird-X, Inc.
730 West Lake Street
Chicago, IL 60661
(312) 648-2191
(312) 648-0319 (Fax)
Joyce Dall

BRIDGES

Bridgemasters
6130 Mill Road
Somerset, CA 95684
(209) 245-3014
(209) 245-6124 (Fax)
Peter M. Newton

Prevents destruction of metal-truss highway bridges not deemed worthy of preservation in place by acquiring them from public agencies. Remarkets and relocates them nationally for private-sector use on farms, ranches, rural estates, golf courses, equestrian parks, and other uses. Consults on disassembly, acquiring permits at new site, and re-erection.

CARPENTERS

Marion H. Campbell
Cabinetmaker
39 Wall Street
Bethlehem, PA 18018
(215) 837-7775

All manner of architectural woodwork to restore, extend, or recreate period work. Design, fabrication, finishing, and installation services. Mantel, paneling, molding, doors, and cabinetry executed in hardwoods and paint-grade material. Brochure $1.00.

Muckle & Associates, Inc.
354 Andover Street
Lowell, MA 01852
(508) 937-2747
Sue Muckle

(See description under Contractors)

Pacific States Carpentry
3249 Briggs Avenue
Alameda, CA 94501
(415) 523-0906
James Spaulding

Carpentry and woodworking repairs for the old-building owner. Pacific States concentrates on repairing or replacing interior and exterior millwork.

CHEMICALS

Diedrich Chemicals
7373 South 6th Street
Oak Creek, WI 53154
(800) 323-3565
(414) 764-6993 (Fax)
Larry Kotke

Professional restoration chemicals for building exteriors: masonry restorer-cleaner; paint-removal system for lead abatement for wood, masonry, and metal. Chemicals sold only to distributors/contractors nationwide. Products specified and used on more than 300 buildings in the National Register of Historic Places. Video demonstrating products is available. Write for brochure.

CHIMNEYS

Homestead Chimney
P.O. Box 5182
Clinton, NJ 08809
(800) 242-7668
(908) 537-7642 (Fax)
Robb Stasyshyn

Homestead Chimney, a chimney and fireplace restoration company, is experienced in all types of masonry work including brick, block, stone, and concrete. All types of fireplace and chimney work are performed from cleaning, inspecting, and evaluating to relining and rebuilding. Specialize in Golden Flue cast-in-place liners and Z-Flex stainless steel liners.

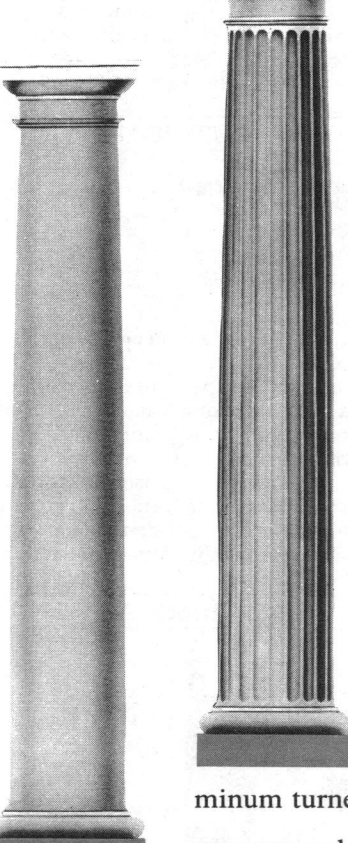
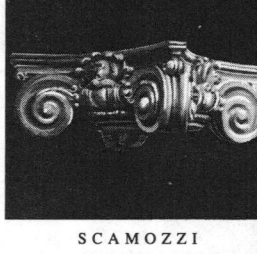

CLEANING

THOMANN-HANRY®

Thomann-Hanry, Inc.
555 Madison Avenue
Suite 250
New York, NY 10022
(212) 755-5550
(212) 755-6385 (Fax)
Eric J. Saine

Developed in France, Thomann-Hanry's Powder Cleaning System softly cleans historic masonry without using water, chemicals, or impactive abrasion. The system is based on a low-pressure, air-blown glass powder which circulates against a surface like an eraser on a sheet of paper. Ideal for delicate and architecturally detailed surfaces. Safe for the environment due to an enclosed cabin which traps both powder and dirt.

CLOCKS

Electric Time Company, Inc.
45 West Street
P.O. Box 466
Medfield, MA 02052
(508) 359-4396
(508) 359-4482 (Fax)

Electric Time has been manufacturing quality street and tower clocks since 1928. Their most popular products are 15-foot-high two- and four-dial post clocks cast in aluminum from original 1890 patterns. Please call or write for literature or a specific quotation.

VERDIN
V
SINCE 1842

The Verdin Company
444 Reading Road
Cincinnati, OH 45202
1-800-543-0488
(513) 241-4010
(513) 241-1855 (Fax)
Emily Embertson

The Verdin Company, in business since 1842, manufactures a wide variety of street clocks, tower clocks, and clock chiming systems. Verdin also provides bells and bell-ringing equipment and clock and bell renovation and servicing. Call toll free for information.

CONSULTANTS

William J. Baxter &
Associated Preservationists
Rural Route 2
Titusville, NJ 08560
(609) 737-1801
William J. Baxter

A research and redevelopment firm specializing in historic architectural work. Consulting services are available to home owners, contractors, and developers on contracting, design, restoration, preservation, finishing techniques, architectural antiques sources, period furnishings, and Old World art work. Educational presentations are also available. Projects include downtown and neighborhood redevelopment and commercial, residential, and estate planning.

Joseph Chillino
3 Emerald Lane
Suffern, NY 10901
(914) 357-6772
(914) 357-2348 (Fax)
Joseph Chillino

(See description under Contractors)

The Color People
1546 Williams Street, Suite 201
Denver, CO 80218
(800) 541-7174
(303) 388-8686 (Fax)
James Martin

Architectural color consultants specializing in building exteriors, The Color People work nationwide in person and through a unique mail-order system. Specialties include historic residential and commercial buildings, townscapes, and Main Streets. Their seminar on creating appropriate color schemes for historic buildings has been presented at the National Trust's annual conference.

Crawford & Stearns
Architects and Preservation Planners
134 Walton Street
Syracuse, NY 13202
(315) 471-2162
(315) 471-2965 (Fax)
Randall T. Crawford and Carl D. Stearns

(See description under Architects)

Facade Consultants
27 Pine Brook Road
Chestnut Ridge, NY 10977
(914) 426-6022
(914) 426-2644 (Fax)
Maurice Schickler

Facade Consultants provides detailed surveys of the exterior conditions of buildings, including written and photographic reports. The firm is familiar with restoration techniques used on all types of materials including brick, natural stone, concrete, cast iron, and terra cotta. Ser-

vices in architecture, geology, engineering, and construction management are provided.

Alan M. Farancz Painting Conservation Studio, Inc.
361 West 36th Street
New York, NY 10018
(212) 563-5550
(212) 947-1186

The studio conserves easel paintings, removes post-historic paint from murals and frescos, consults on the materials and methods of replicating historic stenciling, analyzes and identifies paint materials, conserves sculpture and ornamental plaster, gives lectures on conservation, and writes specifications for conservation and preservation projects. Clients include architects, government agencies, and building owners.

David H. Gleason Associates, Inc.
520A North Eutaw Street
Baltimore, MD 21201
(410) 728-1810
David H. Gleason, AIA

(See description under Architects)

The Office of Allen Charles Hill, AIA
Historic Preservation and Architecture
25 Englewood Road
Winchester, MA 01890
(617) 729-0748

(See description under Architects)

Historic Property Associates
P.O. Box 1002
St. Augustine, FL 32085-1002
(904) 824-5178
(904) 824-5857
W. R. Adams

Historic preservation consulting firm offering regional and statewide services for National Register and Section 106 activity, surveys, tax certifications, phase-one environmental audits, and preservation planning. Historical and architectural tours of St. Augustine on a scheduled basis.

H. Cliff Ivester, ASID
104 North 4th Street
P.O. Box 260
Sayre, OK 73662
(405) 928-5739

The Real Estate Services Group
1785 Massachusetts Avenue, N.W.
Washington, DC 20036
(202) 673-4258
Donovan D. Rypkema

A Washington, D.C.-based consulting firm specializing in the economic revitalization of downtowns and the development of historic properties. The firm has a national clientele of nonprofit organizations and state and local governments. Services to these clients include: feasibility analyses for real estate acquisition and development, demographics and market evaluation, economic revitalization strategies for business districts, development plans for the rehabilitation of historic properties, training in real estate and economic development, and research and publications on related issues.

ReUseIt Institute
The Cultivator's Organization
P.O. Box 440
Grist Mill Square
Lebanon, NJ 08833
(908) 236-7030
R. C. "Bob" Congdon

Practical, specialized consulting in reuse of buildings as theme restaurants, general stores and delicatessens, fitness centers, business technology centers, and housing—daily uses that people need and want. The institute analyzes and evaluates feasibility, profitability, and construction economics from real, hands-on experience. Rural and small-city environs only.

Svendsen Tyler, Inc.
215 Main, Suite 901
Davenport, IA 52801
(319) 324-1910
Marlys Svendsen

TEXANA Development and Design Company
526 11th Street at Church
Galveston Island, TX 77550
(409) 762-4042
(409) 762-9141 (Fax)
Lennie Brown

Established in 1980, TEXANA Development and Design Company provides a variety of services ranging from planning, fund raising/project financing, and historic nominations to programs for rural, commercial, and residential redevelopment, design guidelines, and exterior and interior restorations. Based in Texas, this company of associates serves communities across the country.

Traditional Line, Ltd.
143 West 21st Street
New York, NY 10011
(212) 627-3555
(212) 645-8158 (Fax)
James Boorstein or Anthony Lefeber

(See description under Restoration)

The URBANA Group
P.O. Box 1028
Urbana, IL 61801-9028
(217) 344-7526
(217) 344-7535 (Fax)
Alice Edwards

Comprehensive preservation planning for states, local governments, areas, and organizations. Surveys, building evaluations, feasibility studies, strategic planning, and action programs. Historic and architectural research for National Register documentation, thematic or district planning, urban design guidelines, workshop training, and consultation. Highly experienced staff. Expert testimony in urban planning, zoning, and preservation planning.

Watson and Henry Associates
12 North Pearl Street
Bridgeton, NJ 08302
(609) 451-1779
(609) 451-0471 (Fax)
Michael C. Henry, PE, AIA

(See description under Architects)

CONTRACTORS

Bellvale Construction
Box 127
Bellvale, NY 10912
(914) 651-4552
Ingrid Lange

Marcus H. Brandt: Restorations
35 Curley Mill Road
Chalfont, PA 18914
(215) 997-2117
(215) 822-7293 (Fax)
Marcus H. Brandt

Restoration contractor specializing in vernacular and industrial architecture. Structural and detail work from cornerstones to lightning rods. Services include period carpentry, timber framing, plasterwork, stone masonry, slating, tinwork, and consulting. Previous projects include barns, covered bridges, canal locks, springhouses, summer kitchens, and farmhouses. References by request. Inquiries welcome.

Joseph Chillino
3 Emerald Lane
Suffern, NY 10901
(914) 357-6772
(914) 357-2348 (Fax)

Restoration contractor and consultant Joseph Chillino, Jr., does all phases of reconstruction work. Son of Joseph Chillino, Sr., formerly master restorationist and director of the Restoration Workshop, National Trust for Historic Preservation. Interior and exterior restorations of old buildings as well as period-conforming additions, renovations, and alterations.

The Cultivator's Organization
P.O. Box 440
Grist Mill Square
Lebanon, NJ 08833
(908) 236-7030
R. C. "Bob" Congdon

Construction management, consulting, interior decoration and design, and public-private participation in specialized reuse of buildings for people. Destination-driven end uses involving theme restaurants, general stores and delicatessens, housing, health and fitness centers, and business technology centers. Real hands-on practical construction and ownership experience. Rural and small-city environs only.

The House Carpenters
P.O. Box 281
Leverett, MA 01054
(413) 367-2189
James Mizaur

Located in western Massachusetts, The House Carpenters are licensed general contractors in both Massachusetts and California. Historic New England, Victorian, and Edwardian homes restored and reproduction homes designed and constructed. Can provide hand-finished and decorated timber frames, plastering, millwork, and cabinetry. Brochure available.

Muckle & Associates
Fine Building and Restoration

Muckle & Associates, Inc.
354 Andover Street
Lowell, MA 01852
(508) 937-2747
Sue Muckle

Exterior and interior restoration of historic properties. National Trust Award, 1990. Fine craftsmanship and expertise in preservation technology. Specialties include windows, doors and stairs, epoxy consolidation, and custom architectural millwork to match existing work. General contractor, licensed in Massachusetts.

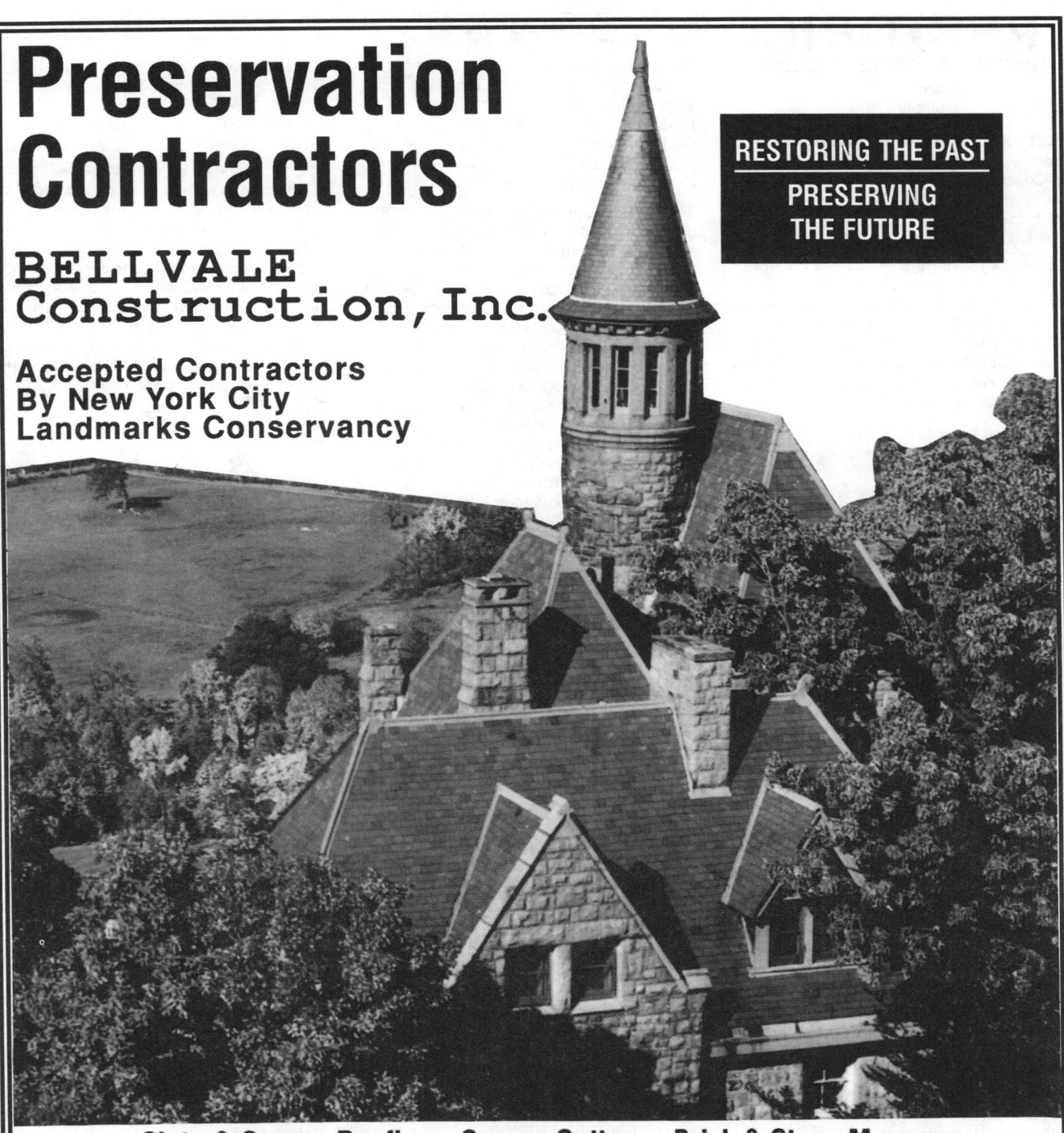

RESTORIC

Historical Restoration
Company, Inc.

RESTORIC Historical Restoration Co., Inc.
69 Coventry Loop
Staten Island, NY 10312
(718) 356-0077
Vincent Usuriello

Landmark restoration specialists for windows, doors, architectural woodwork, porches, porticos, roofs, and cornices. Can restore or replicate existing woodwork. Window projects involve everything from basic double-hung window replacement to elaborate restoration of arched and curved stained-glass windows.

Traditional Line, Ltd.
143 West 21st Street
New York, NY 10011
(212) 627-3555
(212) 645-8158 (Fax)
James Boorstein or Anthony Lefeber

(See description under Restoration)

ALBERT J. WAGNER & SON
INCORPORATED
ESTABLISHED 1894
SHEET METAL CONTRACTORS

Albert J. Wagner & Son
3762 North Clark Street
Chicago, IL 60613
(312) 935-1414
(312) 935-4127 (Fax)
Albert J. Wagner III

Architectural sheet metal contractor specializing in the repair and replacement of copper and galvanized metal cornices, inlaid and other custom gutters, copper decking, bay facade, and gable and hip-style skylights. Also repair and install slate and tile roofing, as well as all metal flashings associated with these roofs.

DOORS

Touchstone Woodworks
P.O. Box 112
Department LM
Ravenna, OH 44266
(216) 297-1313
Tina R. Walters

Touchstone Woodworks is a custom woodworking shop specializing in Victorian and country wooden screen-storm doors. Every door is custom made to order and size from Honduras mahogany. Brochure illustrates a variety of styles from which to choose. Custom designs are also welcome. Brochure $2.00.

EDUCATION

Mary Washington College
Department of Historic Preservation
1301 College Avenue
Fredericksburg, VA 22041-5358
(703) 899-4037
(703) 899-4123 (Fax)
Carter L. Hudgins

The Department of Historic Preservation offers courses that lead to an undergraduate degree in historic preservation. Graduates of this program enter preservation careers with a broad range of technical and intellectual skills with concentrations in architectural conservation, archeology, material culture studies, history museums, or preservation planning. Students may complete the Bachelor of Arts or the Bachelor of Liberal Studies degrees; work at the graduate level is also available. The academic department is also linked to the Center for the Study of the Architecture of Bahia at the Federal University of Bahia, Salvador, Brazil.

The Preservation Institute for the Building Crafts

AT WINDSOR HOUSE

The Preservation Institute for the Building Crafts (PIBC)
P.O. Box 1777
Windsor, VT 05089
(802) 674-6752
Judy L. Hayward

PIBC, a division of Historic Windsor, Inc., offers workshops on the history, theory, and practical application of traditional building skills, and the latest in preservation technology. On-site training at historic properties in preservation skills is a growing part of the program.

The Wool Bureau, Inc.
240 Peachtree Street, N.W.
Merchandise Mart Suite 6-F-11
Atlanta, GA 30303
(404) 524-0512
(404) 659-6974 (Fax)
Laura Diffenderffer

(See description under Fabrics)

ENGINEERS

Cline Contec South
157 Stevens Avenue
Oldsmar, FL 34677
(813) 855-6645
(813) 855-6392 (Fax)
Rocco Capabianco

Dedicated to serving your varied and complex needs for rehabilitating historic facilities. Professional engineering, technical, and management services offered, beginning with prepurchase site assessments, geotechnical explorations, building condition surveys and evaluations, construction materials and nondestructive testing, threshold inspections, roofing, and sealants, and ending with performance evaluations of completed facilities.

Degenkolb
STRUCTURAL
ENGINEERS

H. J. Degenkolb Associates
350 Sansome Street
Suite 900
San Francisco, CA 94104
(415) 392-6952
(415) 981-3157 (Fax)
Susan L. Kennedy

H. J. Degenkolb has been practicing structural/earthquake engineering for more than 50 years. The firm has expertise in evaluating historic structures for their earthquake resistant capacities and in designing strengthening measures that are sensitive to a building's architectural character. Recent seismic strengthening projects for historic buildings include the Ferry Building in San Francisco and Stanford University's Memorial Church.

Robert Silman Associates, P.C.
88 University Place
New York, NY 10003
(212) 620-7970
(212) 620-8157 (Fax)
Robert Silman

Structural engineers with 25 years of experience in restoration of landmarks, both large and small. Specialists in facades, evaluations, upgrades, and stabilization while maintaining sensitivity to historic fabric. Projects include Ellis Island, Carnegie Hall, Brooklyn Bridge, Mount Vernon, New York State Capitol, and many historic houses. Familiar with all official preservation guidelines.

Smith and Faass
12800 Middlebrook Road
Germantown, MD 20874
(301) 428-3222
(301) 428-0706 (Fax)
Cheryl Fleet

Smith and Faass provides sensitive engineering design to integrate modern heating, ventilation, air conditioning, electrical, plumbing, and fire protection systems into historic structures. Specialized experience includes design of precise environmental controls to eliminate moisture problems and to accommodate artifact storage and display. Projects include the White House, Monticello, and the Hermitage.

Watson and Henry Associates
12 North Pearl Street
Bridgeton, NJ 08302
(609) 451-1779
(609) 451-0471 (Fax)
Michael C. Henry, PE, AIA

(See description under Architects)

FABRICS

Classic Revivals, Inc.
1 Design Center Place
Suite 545
Boston, MA 02210
(617) 574-9030
(617) 574-9027 (Fax)
J. Buscemi or Paul Tomaso

Suppliers of authentic fabrics (custom-woven and stocked goods); block-printed French papier peint; documented 18th-, 19th-, and 20th-century papers; friezes and borders; historic Brussels, Wilton, and ingrain carpets and trimmings. To the trade only. We do not do adaptations.

Family Heir-Loom Weavers
R.D. 3, Box 59E
Red Lion, PA 17356
(717) 246-2431
(717) 246-2216 (Fax)
David Kline

(See description under Floorcoverings)

The Wool Bureau, Inc.
240 Peachtree Street, N.W.
Merchandise Mart Suite 6-F-11
Atlanta, GA 30303
(404) 524-0512
(404) 659-6974 (Fax)
Laura Diffenderffer

Resource and product information on wool and wool-blend carpet and upholstery designed to be historically accurate, durable, and easily maintained is available free from regional contract consultants of The Wool Bureau. Also available at no charge is WICS, an exclusive, complete carpet specifying service.

(See display ad, page 145)

FENCES

Custom Ironwork, Inc.
P.O. Box 180
Union, KY 41091
(606) 384-4122
(606) 384-4848 (Fax)
Roger Scott

Historical Fence and Ironworks
P.O. Box 141459
Cincinnati, OH 45250
(513) 244-1442
(513) 244-1494 (Fax)
David Bouwkamp

(See description under Ironworks)

Texas Standard Picket Co.
606 West 17th Street
Suite 304
Austin, TX 78701
(512) 472-1101
(512) 473-2419 (Fax)
Frances Morey

FLOORCOVERINGS

J. R. Burrows and Company
393 Union Street
P.O. Box 522
Rockland, MA 02370
(617) 982-1812
John Burrows

Historical design merchants serving as America's only fully authorized agent for Woodward Grosvenor & Co. of Kidderminster, England, a carpet mill founded in 1790 with an archive of 10,000 historic Wilton and Brussels designs from the Federal, Empire, and Victorian periods. Other specialties are William Morris Wilton and Axminster carpets, Aesthetic Movement period wallpaper, printed fabric designs by Candace Wheeler, and documented Victorian Nottingham Lace Curtains.

Designs In Tile
Box 358, Department Y
Mount Shasta, CA 96067
(916) 926-2629

(See description under Tile)

Family Heir-Loom Weavers
R.D. 3, Box 59E
Red Lion, PA 17356
(717) 246-2431
(717) 246-2216 (Fax)
David Kline

Eight late 18th-century through early 20th-century designs of ingrain carpet are offered in 36-inch width and two designs are available in $22^{1}/_{2}$ inch width for hall and stairway use. Personalized jacquard coverlets and other historic fabrics are also offered.

(See display ad, page 144)

K.V.T. International, Ltd.
240 Peachtree Street
Suite 4G5
Atlanta, GA 30303
(404) 525-5643
(404) 525-0941 (Fax)
Teresa Beard

Langhorne Carpet Company, Inc.
201 West Lincoln Highway
Pendel, PA 19047-0824
(215) 757-5155
(215) 757-2212 (Fax)
William H. Morrow

Edward Molina Designs
196 Selleck Street
Stamford, CT 06902
(203) 967-9445
(203) 967-9551 (Fax)
Catch Molina

For 35 years Edward Molina Designs has been associated with better wool rugs and carpeting for hotel public spaces, high-end commercial uses, and grand homes. Their purpose is to listen carefully, understand thoroughly, and expedite efficiently special needs of architects, specifiers, and designers. Call and let them solve your carpeting needs.

Pennsylvania Woven Carpet Mills, Inc.
401 East Allegheny Avenue
Philadelphia, PA 19134
(215) 425-5833
(215) 634-2543 (Fax)
Marianne L. Embiscuso

The Pennsylvania Woven Carpet Mills have manufactured quality carpets for many prestigious historical buildings. Their Jacquard Wilton carpet looms manufacture patterned carpet in the same manner as was done hundreds of years ago. To create a design of the past, call or write to the mill facility, which has woven carpet for more than 100 years.

The Wool Bureau, Inc.
240 Peachtree Street, N.W.
Merchandise Mart Suite 6-F-11
Atlanta, GA 30303
(404) 524-0512
(404) 659-6974 (Fax)
Laura Diffenderffer

(See description under Fabrics)

FURNISHINGS

G. R. Clidence
18th Century Woodworks
Box 272 James Trail
West Kingston, RI 02892
(401) 539-2558
Ray Clidence

Late 18th-century bed and table reproductions crafted using period tools and techniques. Featuring pencil post, Sheraton field, low post, and cannonball beds with traditional roping or adapted for today's foundation styles. All with hand-rubbed finish of milk paint, oil, or varnish on your choice of wood. Custom orders. Brochure $3.00.

North Woods Chair Shop
237 Old Tilton Road
Canterbury, NH 03224
(603) 783-4595
Lenore Howe

Fine handmade dining chairs and tables, rocking chairs, children's chairs, and occasional tables in the Shaker style. Custom work in cherry, bird's eye maple, tiger maple, and walnut. All pieces signed, dated, and registered to original owner. Reseating of antique Shaker chairs. Canterbury showroom, ship nationwide. Catalog $3.00.

GARDENING

Select Seeds-Antique Flowers
180 Stickney Hill Road
Union, CT 06076
(203) 684-5655
Marilyn Barlow

Offering 120 documented antique species flower seeds for garden restoration and recreation. Owner Marilyn Barlow writes, "The names themselves can send one into romantic reveries; Cupid's Dart, Meadow Sweet, Dame's Violet. . . ." Facsimile reproductions of 19th-century gardening books and collections of Vining, Fragrant, Victorian, and Immortelle flowers complete this most extensive selection. Catalog $2.00.

There is always a good reason to specify wool...

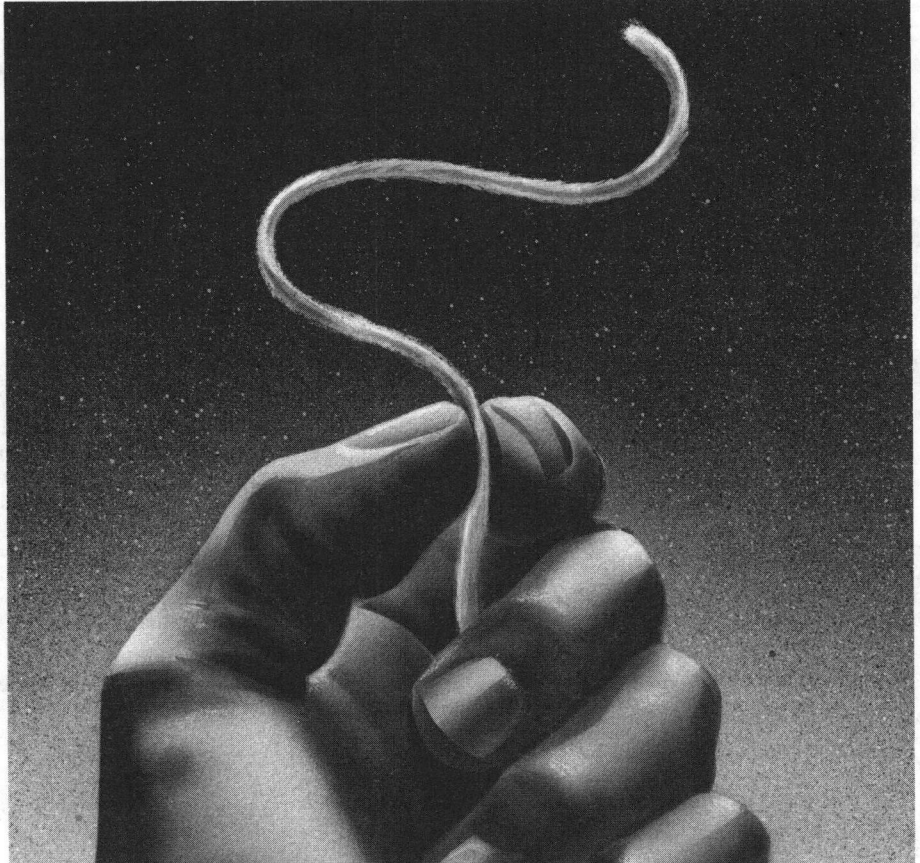

and The Wool Bureau is always there.

We are part of the International Wool Secretariat (IWS), a global non-profit network devoted to the use of 100% wool and wool blends in Interior textiles. We provide free assistance to the manufacturing and design communities in the United States and Canada.

Look to us for:

Domestic and International Wool Manufacturing Resources for Carpets and Interior Textiles. Selection and specification guidance for proper commercial applications of wool carpet and upholstery. Technical information and education seminars.

Contract Consultants

LINDA CAPPER
West/Los Angeles
(213) 659-9981

MOIRA MURRY
Northeast/New York
(212) 986-6222

GEORGE FORTE
South/Atlanta
(404) 524-0512

STACY KRONER
Central/Kansas City
(816) 361-3118

CAROL INNES
Mid-Atlantic/Chevy Chase, MD
(301) 907-0314

PURE WOOL ®

The Wool Bureau, Inc.

240 Peachtree Street, NW, Suite 6F11, Atlanta, GA 30303 • (404) 524-0512
360 Lexington Avenue, New York, NY 10017-6572 • (212) 986-6222

GLASS

Flickinger Glassworks, Inc.
204 Van Dyke, Pier 41
Brooklyn, NY 11231
(718) 875-1531
(718) 875-4264
Charles Flickinger
Steve Baxter

Glass bending for lighting restoration, brownstone windows, fine cabinetry, custom lighting, antique furniture restoration, and department store display cases. Other glass specialty work includes stained glass design and restoration, sand blasting, cast glass, and commercial glass blowing.

HARDWARE

Antique Hardware
509 Tangle Drive
Jamestown, NC 27282
(919) 454-3583
Jim Leonard

Specializing in 18th- and 19th-century wrought-iron door hardware and fireplace equipment. Great thumb latches, square and keyhole plate latches, elbow latches, box locks, strap hinges, slide bolts, andirons, cranes, trammels, utensils, broilers, toaster hooks, utensil racks, Betty lamps, wood/glass lanterns, and sawtooth trammel lighting device holders. Send $2.00 for photos and descriptions.

Arden Forge Company
301 Brintons Bridge Road
West Chester, PA 19382
(215) 399-1530
Peter A. Renzetti

Blaine Window Hardware, Inc.
12430 McDade Road
Hagerstown, MD 21740
(800) 678-1919
(301) 797-2510 (Fax)
G. Patrick Howard

(See description under Windows)

Cirecast, Inc.
380 7th Street
San Francisco, CA 94103
(415) 863-8319
(415) 863-7721 (Fax)
Peter S. Morenstein

Conant Custom Brass, Inc.
270 Pine Street
Burlington, VT 05401
(802) 658-4482
(802) 864-5914 (Fax)
Steve Conant

(See description under Metalwork)

D.E.A. Bathroom Machineries
495 Main Street
P.O. Box 1020
Murphys, CA 95247
(209) 728-3860
(800) 255-4426
(209) 728-2320 (Fax)

(See description under Plumbing)

Gunnebo Fastening Corporation
180 South Hartman Street
York, PA 17403
(717) 846-2200
(717) 852-0809
Dale Sventeck

Paxton Hardware, Ltd.
7818 Bradshaw Road
Upper Falls, MD 21156
(410) 592-8505
Ledley Clarke Boyce

Authentic, hard-to-find, solid brass cabinet hardware available for immediate shipment. These pulls, locks, casters, hinges, and lamp hardware have enhanced fine antiques and handcrafted furniture for more than 20 years. To receive a copy of their 75-page catalog, please send $4.00 to Paxton Hardware, P.O. Box 256, Dept. HTG, Upper Falls, MD 21156.

Windy Hill Forge
3824 Schroeder Avenue
Perry Hall, MD 21128-9783
(410) 256-5890
Ray Zeleny

Custom-forged hardware, old locks repaired, and iron keys cut. Museum-quality restoration work. Wall washers, shutter hold-backs, snow irons. Cast metals catalog, $3.00 (refundable).

HISTORIC INNS AND HOTELS

The Brown Palace Hotel

The Brown Palace Hotel
321 17th Street
Denver, CO 80202
(303) 297-3111
(303) 297-3928 (Fax)
Jane Andrade

Copper King Mansion and Restaurant
219 West Granite Street
Butte, MT 59701
(406) 782-7580
Clancy Stockham

Tour the home of one of America's wealthiest mining magnates daily 9 a.m. to 5 p.m. Dine in Victorian elegance on gourmet dinners served from 6 p.m. to 9 p.m. Sleep like a copper king! Bed and breakfast year-round. Reservations suggested for dinner and B&B. A national historic place.

Elmwood House
2609 College Avenue
Berkeley, CA 94704
(510) 540-5123
John Ekdahl

Built in 1902 as the residence of University of California professor William Augustus Merrill and his family, Elmwood House is a brown-shingled, redwood-trimmed building epitomizing the Berkeley Bay tradition of turn-of-the-century architecture. Conveniently located between the fashionable Elmwood shopping district and the U.C. Berkeley campus. Continental breakfast included.

The Gastonian
220 East Gaston Street
Savannah, GA 31401
(912) 232-2869
(800) 322-6603
(912) 234-0006 (Fax)
Hugh Lineberger

The Gastonian—an 1868 historic inn completely furnished with English antiques. Central air conditioning, elevator, in-room jacuzzi baths with showers, cable/color/remote television, fruit and wine on arrival, nightly turndown with sweets and cordials. Full, hot, Southern-style breakfast, beautiful gardens, sun deck with hot tub, and off-street parking. The only inn in Georgia to receive the Mobil Travel Guide four-star and AAA four-diamond ratings.

Inn at Narrow Passage
U.S. 11 South
Woodstock, VA 22664
(703) 459-8000
Ed Markel

This historic log inn, surrounded by five acres on the Shenandoah River, has been welcoming travelers since the 1740s. Furnished in antiques and colonial reproductions, the fully air-conditioned inn features a large living room, dining room, and 12 comfortable guest rooms, most with private baths and working fireplaces. Executive conference room available.

RetroTie

THE ADVANCED REMEDIAL WALL TIE WITH A "TWIST"

ADVANTAGES

- Designed specifically for cavity walls
- Has lateral flexibility to accommodate normal movements
- Quick and easy installation
- Hammer or power driven
- Easily proof tested
- Multiple drip points reduce water flow
- Low installation costs

- No joints or moving parts
- Minimal defacing of exterior wall
- Self-anchors in inner wythe, thus eliminating excessive stress on base material
- Anchorage into inner wythe is easily tested by using a test key and/or Helifix load tester
- Hi fin design swirls resin ensuring complete bond to exterior wythe.

GUNNEBO
FASTENING

P.O. Box 1589 • 180 S. Hartman St., York, PA 17405 • 717-846-2200 • FAX 717-852-0809

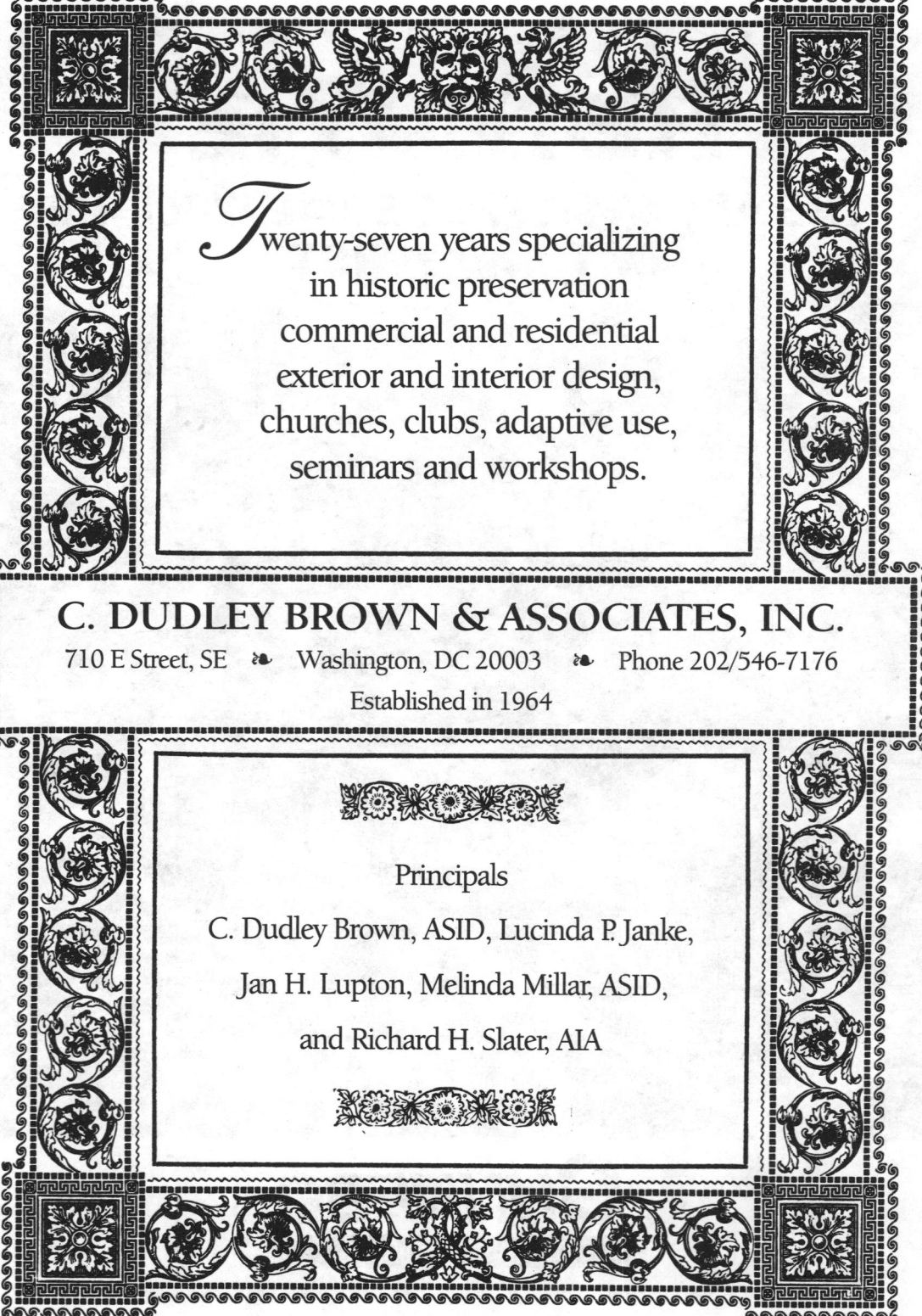

The Newel Post
3428 Uniontown Road
Uniontown, MD 21157
(301) 775-2655
(800) 732-2655
Janet Michael

This Victorian bed and breakfast in historic Uniontown is convenient to Baltimore and Washington and is only minutes from Gettysburg. The spacious rooms with private or shared bath are comfortably furnished to pamper and delight guests. Original stained-glass windows, lavish period wallcoverings, and decorative carved woods are all unique features.

Todd House Bed and Breakfast
Todd's Head
Eastport, ME 04631
(207) 853-2328
Ruth McInnis

Westchester House Bed and Breakfast
102 Lincoln Avenue
P.O. Box 944
Saratoga Springs, NY 12866
(518) 587-7613
Stephanie Melvin

Gracious, in-town, high Victorian Gothic inn combines elegant Old World ambience with up-to-date comforts. Enjoy double parlors, wraparound porch, old-fashioned gardens, extensive library, and baby grand piano. Private baths and air conditioning. Explore the romance of Saratoga's gilded past and glamorous present . . . more than a bed and breakfast.

HOUSE MOVERS

The House Relocater
631 Cross Avenue
Los Angeles, CA 90065-4018
(213) 749-2600
Jim Dunham

Consulting firm that assists clients in finding, evaluating, and executing the relocation of historic buildings. Services include finding free buildings and arranging the financing for moving them. Jim Dunham, consultant, has published *Moving Buildings for Profit*, a definitive step-by-step investment guide for the entire relocation process. He also holds workshops on investing in relocation projects.

INTERIOR DESIGNERS

C. Dudley Brown & Associates, Inc.
710 E Street, S.E.
Washington, DC 20003
(202) 546-7176
(202) 546-4234 (Fax)

Award-winning interior design firm with nationwide experience in residential, commercial, and institutional design. Expertise in historic preservation/restoration and adaptive use. Specializing in antique and traditional furniture and furnishings including carpet and lighting fixtures and period window, wall, and color treatments.

Allyn R. Gardner
P.O. Box 8851
Naples, FL 33940-8852
(813) 793-4810

Design associate, member of the National Trust for Historic Preservation, and allied member of American Society of Interior Designers (ASID). Twenty-five years of experience in interior design and decoration, restoration and renovations. Listed in *Who's Who in Interior Design*, International, 1991.

Jamie Gibbs & Associates
340 East 93rd Street
Suite 14C
New York, NY 10129
(212) 722-7508
(212) 722-7508 (Fax)
Jamie Gibbs

(See description under Architects)

Hughes Group Ltd.
P.O. Box 84
North Salem, NY 10560
(212) 420-0220
(212) 420-0289 (Fax)
Nina Hughes, ASID

Space planners and interior designers specializing in period interiors with in-depth experience in commercial projects.

(See display ad, page 151)

H. Cliff Ivester, ASID
104 North 4th Street
Box 260
Sayre, OK 73662
(405) 928-5739

Terry Peterson, ASID
Period Restoration and Design
302 Arlington Avenue
Natchez, MS 39120
(601) 442-4515

Richardson Design Studio
496 Ferry Point Road
Annapolis, MD 21403
(410) 269-6186
(410) 269-6187 (Fax)
Janet K. Richardson, ASID

Offering award-winning interior design services for historic preservation and adaptive use commercial and residential projects. Past projects include: The Imperial Hotel, a Victorian renovation; Mt. Airy Plantation, Lord Calvert's Georgian home converted to a restaurant; Oakland Manor, a Federal-style manor house now used for receptions; and numerous historic homes.

Robert W. Smith Interiors
212 North Quincy Street
Hinsdale, IL 60521
(708) 986-0238
Robert W. Smith, ASID

Edwin Turrell Associates
202 Fair Oaks Street
San Francisco, CA 94110
(415) 282-2000
(415) 282-1236 (Fax)
Ted Turrell, ASID

IRONWORKS

Custom Ironwork, Inc.
P.O. Box 180
Union, KY 41091
(606) 384-4122
(606) 384-4848 (Fax)
Roger Scott

Historical Fence and Ironworks
P.O. Box 141459
Cincinnati, OH 45250
(513) 244-1442
(513) 244-1494 (Fax)
David Bouwkamp

The craftsmen at Historical Fence and Ironworks believe in time-honored designs and workmanship. Authentic designs are rich in both European and early American cultures. More than 25 styles ranging from $12.50 to $30.00 per linear foot are shown in a catalog, which is available for $2.00.

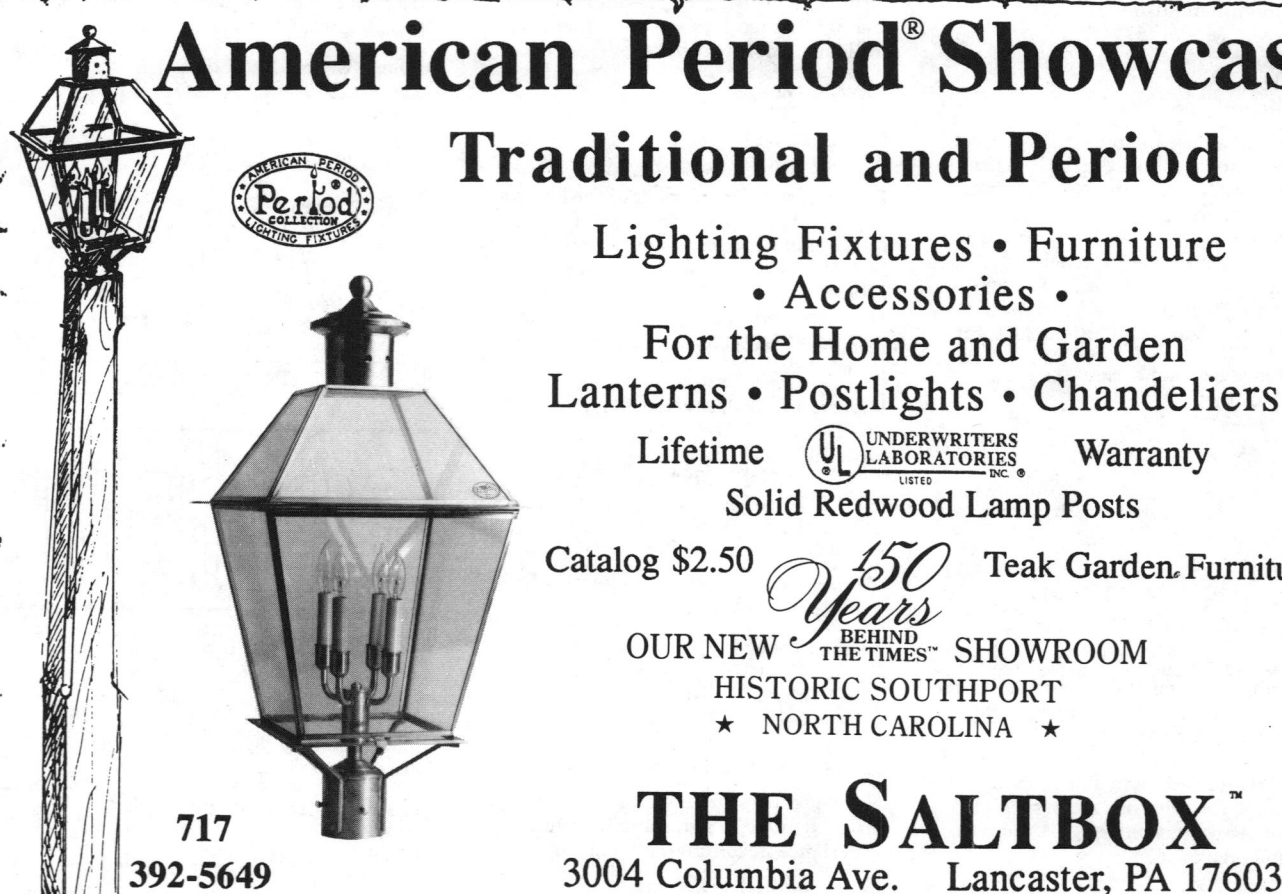

LADDERS

JOMY Safety Ladder Company
1728 16th Street
Suite 201
Boulder, CO 80302
(800) 255-2591
(303) 440-9091 (Fax)
Paul F. Enright

The JOMY Safety Ladder is a discreet, collapsible, industrial-strength ladder with a safety rail that looks like a drainpipe. Used for access or supplementary egress, the JOMY Safety Ladder is virtually invisible and burglar proof. It is an ideal solution where space or aesthetics are of primary importance. UL listed.

LIGHTING

American Period Lighting
3004 Columbia Avenue
Lancaster, PA 17603
(717) 392-5649

Dutch Products and Supply Company
166 Lincoln Avenue
Yardley, PA 19067
(215) 493-4873
(215) 493-4873 (Fax)
Marianne Leisinger or Martin Int Hout

Colonial chandeliers sand cast in solid brass plus a complete line of Delft chande-liers, Delft tiles for kitchen and fireplace, and old-country street and garden lights.

Elcanco, Ltd.
P.O. Box 682
Westford, MA 01886
(508) 392-0830
800-423-3836
Marcia Linton

Specializing in custom, handcrafted electric wax candles and beeswax candle covers, 6-volt candlewick bulbs, and decorative candelabra bulbs for outfitting fine design lighting fixtures. Services, crafts, and skills for the restoration of antique lighting. Known for installations.

Herwig Lighting, Inc.
P.O. Box 768
Russellville, AR 72801
(501) 968-2621
(501) 968-6762 (Fax)
Donald F. Wynn

Hurley Patentee Lighting
464 Old Route 209
Hurley, NY 12443
(914) 331-5414
Carolyn M. Waligurski

Hurley Patentee Lighting offers reproductions of outstanding quality including a large and varied collection of handcrafted colonial chandeliers, lamps, sconces, and lanterns aged to resemble the antique originals. Each part of every fixture is created with great care and attention to authenticity. The only modifications are to accommodate electrical wiring. Catalog $3.00.

(See display ad, page 152)

Gates Moore Lighting
River Road Silvermine
Norwalk, CT 06850
(203) 847-3231
Patricia Moore

Paxton Hardware, Ltd.
7818 Bradshaw Road
Upper Falls, MD 21156
(410) 592-8505
Ledley Clarke Boyce

Reproduction fittings and shades for antique lamps available for immediate shipment. Glass, fabric, and parchment shades. Replacement burners, sockets, wire, shade holders, and vase caps. To receive a copy of the 75-page catalog, please send $4.00 to Paxton Hardware, P.O. Box 256, Dept. HTG, Upper Falls, MD. 21156.

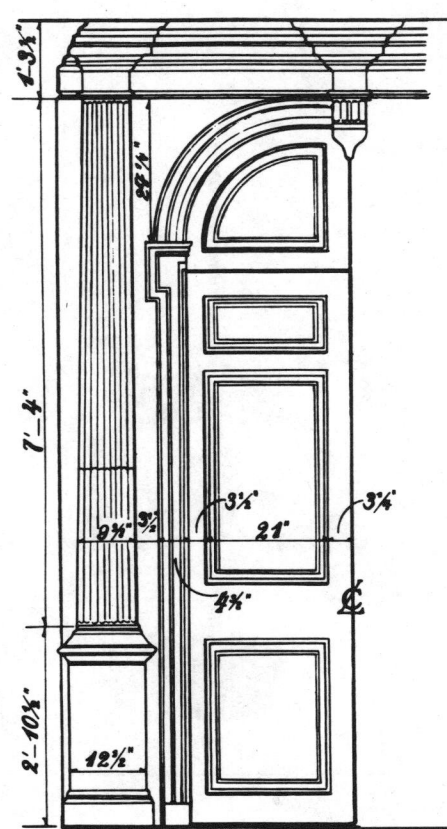

Victorian Lighting/Watertower Pines
P.O. Box 1067
Kennebunk, ME 04043
(207) 985-6868
Judy Oppert

Victorian Lighting specializes in antique kerosene, gas, and early electric lighting from 1850 to 1930. Chandeliers, wall sconces, table lamps, floor lamps, and some outdoor fixtures are available. Some pieces are in original condition, while others have been discreetly wired for modern use. Usually open daily.

LITTER RECEPTACLES

Best Litter Receptacles
P.O. Box 5038
Florence, SC 29502
(803) 667-8188
(803) 665-5963 (Fax)
Carol Doty

Receptacles are constructed of reinforced, exposed aggregate and weigh 350 pounds, making them a permanent and attractive method of litter control. Colors are river-rock brown with dark brown lids or gray with black lids. Lids are chained to the inside of the concrete units, and liners have rocker-shaped bottoms to prevent use outside of receptacles.

LUMBER

Bear Creek Lumber, Inc.
P.O. Box 669
Winthrop, WA 98862
(509) 997-3110
(509) 997-2040 (Fax)
Dick Garing

Western red cedar lumber products milled to a variety of patterns including channel and bevel siding and decking and tongue and groove paneling. Clear pine and fir flooring, edge and center-bead paneling and trim. California redwood siding and paneling. Appalachian hardwood flooring, stainless steel nails and fasteners. Delivery direct to your job site nationwide. Call for quotation.

Palmer Creek Hand-Hewn Wood Products
1826 King Street
Santa Rosa, CA 95404
(707) 578-0870
Larry McCanse

The woodwrights of Palmer Creek produce authentic hewn timbers using axe, broad-axe, and adze. Restorations include 1991 California Preservation Foundation award-winning Reed Mill, circa 1834, and Fort Ross, circa 1812. Products include

large, hard-to-find timbers, redwood, Douglas fir to 40 feet, mortise-and-tenon joinery, mantels, wainscoting and doors for today's eclectic designs.

The Woods Company
2357 Boteler Road
Brownsville, MD 21715
(301) 432-8419
(301) 432-8439 (Fax)
Barry Stup

Specialty supplier of antique wood flooring, molding, millwork, and beams. Inventory includes heart pine, oak, chestnut, white pine, and hemlock as well as new hardwoods. Hand-hewn and old sawn beams are also available for restoration needs. Most products are custom milled to order. Free literature.

(See display ad, page 154)

MASONRY

Buckley Rumford Fireplace Company
P.O. Box 21131
Columbus, OH 43221
(614) 486-1725
Jim Buckley

Cathedral Stoneworks
1047 Amsterdam Avenue
New York, NY 10025
(212) 316-7457
(212) 316-7404 (Fax)
David Teitelbaum, G.P.

Marshalltown Trowel Company
P.O. Box 738
Marshalltown, IA 50158
(515) 753-0127
(515) 753-6341 (Fax)
Steve Lyons

Preserving a better tomorrow by building the highest quality line of masonry tools today, Marshalltown manufactures in the most efficient way possible—yet never compromises on quality. Included in their complete line are Marshalltown tuck-pointers, and pointing, and margin trowels. Put them to work on your next restoration project. The professional's choice—since 1890.

Old Carolina Brick Company
P.O. Box 786
Anniston, AL 36208
(205) 237-2890
(205) 237-0667 (Fax)
Art Burkhart

Genuine handmade brick, architectural shapes, arches, pavers, and matching res-

toration brick. Can authentically match almost any color, shape, or size brick for restoration work. World's largest producer of genuine handmade brick.

A. Ottavino Corporation
80-60 Pitkin Avenue
Ozone Park, NY 11417
(718) 848-9404
(718) 848-7156 (Fax)
A. G. Ottavino

Stone restoration: cutting, carving, and sculpting all kinds of natural stone as well as furnishing and setting stone and inserting dutchman repairs. Work performed in complete shop with all cutting machinery. Among their restorations are the Temple of Dendur, Maine Monument, Firemen's Monument, Brooklyn Borough Hall, and Brooklyn Appellate Court.

Watertrol, Inc.
P.O. Box 163
Cranford, NJ 07016
(908) 245-6622
(908) 298-8718 (Fax)
S. James Papandrea

Watertrol has 30 years of experience specializing in exterior restoration, preservation, and reconstruction of masonry and stonework projects. Services include concrete restoration, stone and brick restoration and repair, lintel repair and replacement, pointing, mask grouting, masonry cleaning, and waterproofing applications.

METALWORK

Conant Custom Brass, Inc.
270 Pine Street
Burlington, VT 05401
(802) 658-4482
(802) 864-5914 (Fax)
Steve Conant

Conant Custom Brass offers the widest variety of brass restoration services available anywhere. Capabilities include everything from basic polishing to the repair and replacement of broken or missing parts. Custom fabricating facilities enable Conant to provide services far beyond most refinishing shops. Custom new fixtures make up a large portion of their work.

Historical Arts & Casting, Inc.
4130 West 1939 South
Unit F
Salt Lake City, UT 84104
(801) 974-0242
(801) 974-5832 (Fax)
Robert A. Baird

(See description under Architectural Details)

LMC Corporation
(Les Metalliers Champenois USA)
118 Second Avenue
Paterson, NJ 07514
(201) 279-3573
(201) 881-0235 (Fax)
Jean Wiart, General Manager

LMC Corporation, an architectural metal-working firm, blends quality craftsmanship and scrupulous attention to historic, architectural, and technical detail to preserve, restore, and reproduce original fine metal details on landmark buildings and historic monuments. Products include domes, marquees, stair rails, doors, gates, balconies, winter gardens, and other metal detail work. Technical consulting services available.

Robinson Iron Corporation
P.O. Box 1119
Robinson Road
Alexander City, AL 35010
(205) 329-8486
(205) 329-9860
J. Scott Howell

(See description under Architectural Details)

Salem Cross Inn, West Brookfield, Massachusetts

Museum Quality

Architectural Components, Inc. specializes in custom made, period reproduction windows, doors, doorways and interiors for commercial and residential restorations and rehabilitations. Traditional joinery and authentic detailing, along with competitive pricing and professional service, guarantee customer satisfaction.

Send $3.00 for a complete brochure of our products and services.

Architectural Components, Inc.

Dept. LYP, 26 North Leverett Rd., Montague, MA 01351
(413)367-9441   FAX (413)367-9461

AMHERST WOOD WORKING

Amherst Woodworking and Supply, Inc.
P.O. Box 718
Northampton, MA 01061
(413) 584-3003
(413) 585-0288 (Fax)
David Short

Manufacturers of commercial and residential architectural woodwork including moldings, doors, paneling, cabinets, and flooring. Fourteen species of lumber are stocked. Descriptive literature is available including a catalog of molding patterns and door and paneling design guides. Jobs, including installation, can be bid from plans.

Architectural Components, Inc.
26 N. Leverett Road
Montague, MA 01351
(413) 367-9441
(413) 367-9461 (Fax)
C. Bellinger

Architectural Components is a full-service millwork company specializing in the manufacture of high-quality interior and exterior woodwork for historic and reproduction residential and commercial buildings. Products include reproduction windows and sash, doors and doorways, paneling, molding, mantels, shutters, turnings, and more.

Beech River Mill Corporation
Old Route 16, Box 168
Centre Ossipee, NH 03814
(603) 539-2636
Greg Dales

(See description under Shutters)

Creative Woodworking, Ltd.
26 Friendship Street
Westerly, RI 02891
(401) 596-4463
(401) 596-3418 (Fax)
Ray Matteson

Creative Woodworking, now in its 13th year of operation, is an extensive millwork, architectural woodwork, and supply source for southern New England. Stock and custom molding, profile grinder on premises, doors and windows, shutters to your specification, flooring, curves, and much more. Inquiries invited.

Millwork Specialties
189 Prospect Avenue
Brooklyn, NY 11215
(718) 768-7112
(718) 965-3974 (Fax)
Cosmo Cotroneo

Established in 1966. Specializing in wood windows, doors, and molding for older buildings. Recent projects include the replication of 103 windows for the old Saturday Evening Post building in New York and doors, windows, and woodwork for the early 19th-century Weeksville Museum in Brooklyn, New York.

NEW ENGLAND WOODTURNERS

New England Woodturners
75 Daggett Street
New Haven, CT 06519
(203) 776-1880
(203) 776-1880 (Fax)
Richard Swartwout

New England Woodturners is a full-service studio offering made-to-order architectural turnings at competitive prices: balusters, newels, porch posts, columns, and finials. Twist, fluting, and reeding available in lengths up to 20 feet. Short runs are a specialty. "Since 1976 we've been turning your world just a bit nicer."

Pagliacco Turning and Milling
P.O. Box 225
Woodacre, CA 94973
(415) 488-4333
(415) 488-9372 (Fax)
Steve Evans

Manufacturer of a complete line of custom turnings. Over 150 stock designs for balusters, newel posts, railings, and classic columns. All products are based on authentic designs and can be custom ordered in decay-resistent all-heart redwood, Douglas fir, or hardwoods. Call or write for 24-page catalog.

Touchstone Woodworks
P.O. Box 112
Department LM
Ravenna, OH 44266
(216) 297-1313
Tina R. Walters

(See description under Doors)

MOSAIC ART

Palanza Mosaic Art Company
85 Harrison Street
Staten Island, NY 10304
(718) 816-7147
William Palanza

William Palanza is a mosaic artist who specializes in the restoration and reproduction of all types of mosaic styles, tesserae, smalti venetian glass, pietra dura, and inlay. He is recognized by the New York City Landmarks Commission and has worked for the Metropolitan Museum of Art. Portfolio available upon request.

MUSEUMS

Hancock Shaker Village
Route 20
P.O. Box 898
Pittsfield, MA 01202
(413) 443-0188
Lawrence J. Yerdon, Director

Twenty restored buildings housing the largest collection of Shaker furniture and artifacts in an original Shaker site (1790–1960) and a farm on 1,200 acres. Resident craftspeople produce furniture, oval boxes, baskets, and textiles. Museum shop has custom or kit Shaker furniture, accessories, and a large book selection. Open April–November.

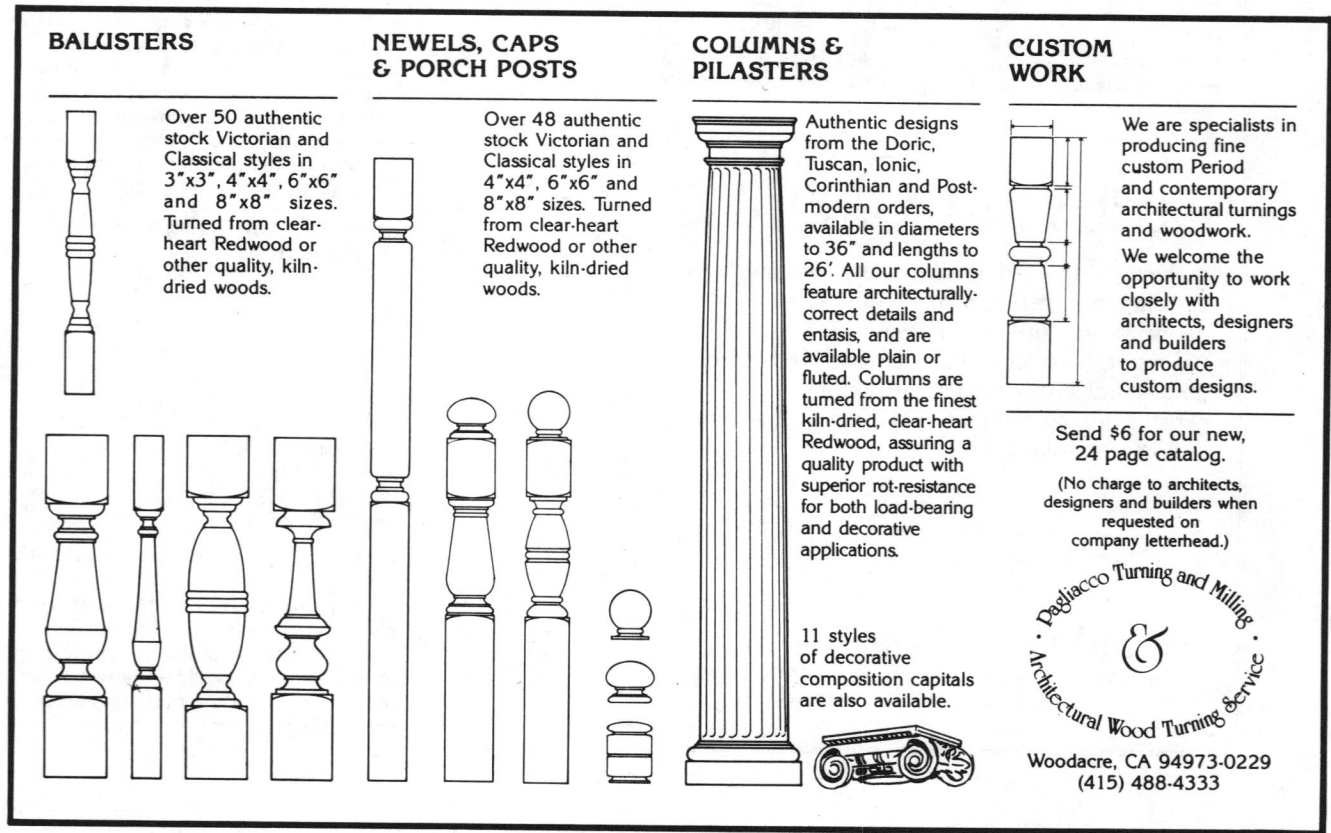

HISTORIC DEERFIELD *offers eleven early 19th century wallpapers accurately reproduced through the hand screen process from samples discovered in and around Deerfield, Mass. All papers are printed in the document colorways but can be custom colored as desired.*

For further information contact Director of Merchandising, Historic Deerfield Museum Store, Deerfield Massachusetts 01342. (413) 774-5581.

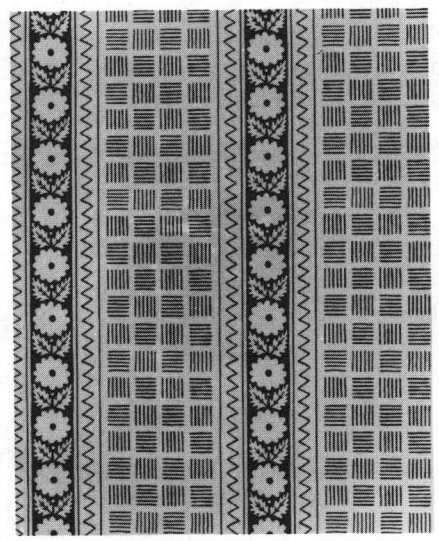

Historic
DEERFIELD
MASSACHUSETTS

JOIN PRESERVATION FORUM
A *Valuable Partner*

PUBLICATIONS

Historic Preservation Forum, a bimonthly professional journal including case studies, debates on current issues, and the practical, technical side of preservation developments. *Information Series* on organizational development, preservation of special building types, funding, and other topics. *Historic Preservation* and *Historic Preservation News,* the Trust's acclaimed four-color bimonthly magazine and monthly newspaper.

AN ESSENTIAL NETWORK

Preservation Forum provides a vital network among those who are undertaking projects and programs and those who can provide technical advice and professional services.

TECHNICAL ASSISTANCE

The Trust's network of regional offices helps communities protect historic resources, organize and establish group goals, influence key decision makers, and helps you find the information and experts you need.

FINANCIAL ASSISTANCE

Grants and loans from the Trust help nonprofits, public agencies, and owners of National Historic Landmarks fund their projects. Forum members are eligible to apply.

GROUP INSURANCE

Available exclusively to Forum members, options include property and casualty, employee benefits, directors and officers, and maritime insurance.

OTHER BENEFITS

Other benefits include discounts on conferences, training programs, and materials to promote Preservation Week. Forum members also receive a 20% discount on Preservation Press publications.

ANNUAL DUES

Annual dues in Preservation Forum are only $75. If you are already an individual member of the Trust, we will upgrade your membership for $60. Simply compete the card below; we will activate your Forum membership and bill you for dues. You will continue receiving present benefits and enjoy the extended benefits of Preservation Forum.

YES, I'LL JOIN...

Member Name _____

Street Address _____

City _____ State _____ Zip _____

Principal Contact _____

Phone _____

FEEL FREE TO PHOTOCOPY THIS FORM.

Please Check One

____I am currently an individual member of the National Trust. Please upgrade my membership to Forum and bill me $60.00

____I am not currently a member of the National Trust. Please bill me $75 for annual dues.

____Enclosed is a check in the amount of $_____.

____I'd like to know more about membership in the Preservation Forum. Please send me a brochure.

Historic Deerfield, Inc.
The Street
Deerfield, MA 01342
(413) 774-5581
Grace Friary

Historic Deerfield, Inc., a museum village complex of 53 buildings on 100 acres in rural western Massachusetts, celebrated its 40th anniversary of incorporation in 1992. Located within the Old Deerfield National Historic Landmark, Historic Deerfield maintains 13 historic house museums with outstanding collections of American decorative arts along a mile-long thoroughfare called, simply, The Street. Open daily all year, Historic Deerfield owns and operates the 1884 Deerfield Inn and the Museum Store at the Sign of J.G. Pratt.

Surratt House Museum
9118 Brandywine Road
Clinton, MD 20735
(301) 868-1121
Laurie Verge, Park Historian

A page in American history: This 1852 middle-class farmhouse served as a family home, a tavern and hostelry, a post office, and a polling place during the crucial decades before the Civil War. During the war, it became a safe house in the Confederate espionage system. In 1864, the Surratt family became involved in a scheme to kidnap Abraham Lincoln—a scheme that ultimately turned to assassination.

ORGANIZATIONS

American Association for State and Local History
172 2nd Avenue, North
Suite 202
Nashville, TN 37201
(615) 255-2971
(615) 255-2979 (Fax)
David Rector

The American Association of State and Local History (AASLH) provides leadership and service to members worldwide. Services benefit history professionals and volunteers working in historical societies, museums, historic sites, parks, libraries, archives, historic preservation organizations, schools, and colleges.

Association for Preservation Technology International
P.O. Box 8178
Fredericksburg, VA 22404
(703) 373-1621
(703) 373-6050 (Fax)
Susan Ford Johnson

The Association for Preservation Technology International (APT) is an interdisciplinary membership organization dedicated to the practical application of the principles and techniques necessary for the care and wise use of the built environment.

Historic Hawai'i Foundation
P.O. Box 1658
Honolulu, HI 96806
(808) 537-9564
(808) 526-3989 (Fax)
Phyllis G. Fox

Historic Hawai'i Foundation's mission is to preserve and to encourage the preservation of historic buildings, objects, and sites relating to the history of Hawai'i. The following services are provided: a monthly magazine; workshops; elementary, secondary, and high school educational programs; undergraduate and graduate university programs; and information and resources.

Historic Lexington Visitors Bureau
102 East Washington Street
Lexington, VA 24450
(703) 463-3777
(703) 463-1105 (Fax)
Martha Doss

Mary Washington College
Center for Historic Preservation
1301 College Avenue
Fredericksburg, VA 22401-5358
(703) 899-4037
(703) 899-4123 (Fax)
Carter L. Hudgins

Established in 1980 to support the undergraduate program in Historic Preservation at Mary Washington College, the Center for Historic Preservation supports a wide range of preservation activities in the mid-Atlantic region. Through lectures, seminars, symposiums, workshops, and conferences, the center disseminates information to a wide range of preservation constituents. The center also offers a range of professional services, particularly in the areas of archaeology, architectural conservation, museum administration, and preservation planning. In addition to an active research program, the center administers the James Monroe Museum and the Germanna Archaeological District.

Preservation Forum
The National Trust for Historic Preservation
1785 Massachusetts Avenue, N.W.
Washington, DC 20036
(202) 673-4296
(202) 673-4223 (Fax)

Preservation Forum provides a vital network among those who are undertaking preservation projects and programs and those who can provide technical advice and professional services. Publishes *Historic Preservation Forum*, a bimonthly professional journal and the *Information Series*, booklets on preservation topics. Forum members are eligible to apply for National Trust-sponsored grants and loans and group insurance.

THE
PRESERVATION SOCIETY
OF CHARLESTON
ORGANIZED 1920
INCORPORATED 1928

Preservation Society of Charleston
147 King Street
Charleston, SC 29401
(803) 722-4630
John Meffert, Director

Victorian Alliance
824 Grove Street
San Francisco, CA 94117
(415) 566-6630
Vikki-Marie Powers

The Victorian Alliance of San Francisco invites you to attend their annual house tour in October. Monthly meetings are held in pre-1906 homes. Membership is open to anyone who cares to join this all-volunteer organization dedicated to preserving the architecture of the Victorian era.

INFORMATION

BACK TO BASICS

The National Trust for Historic Preservation's *Information* booklets provide concise information on basic and frequently used preservation techniques. The National Trust has assembled two customized packages of *Information* publications for *Landmark Yellow Page* readers - addressing the "basics" of preservation:

Organizational Development Package $20.00
Steering Nonprofits: Advice for Boards and Staffs
Membership Development: A Guide for Nonprofit Preservation Organizations
Investing in Volunteers: A Guide to Effective Volunteer Management
Legal Considerations in Establishing a HistoricPreservation Organization
A Self-Assessment Guide for Community Preservation Organzations

Essential Preservation Issues Package $20.00
Basic Preservation Procedures
Rescuing Historic Resources: How to Respond to a Preservation Emergency
The Economics of Rehabilitation
Maintaining Community Character: How to Establish a Local Historic District
Establishing an Easement Program to Protect Historic, Scenic, and Natural Resources

Each package of five *Information* booklets is specially priced at $20, including postage and handling. (Individual booklets cost $5 -a $25 value.) Orders must be prepaid. To order *Information* packages, send payment and order form to:
Information Series
National Trust for Historic Preservation
1785 Massachusetts Ave., N.W.
Washington, D.C. 20036

Please send:
_____ Both packages @ $40
_____ Organizational Development Package @ $20
_____ Essential Preservation Issues Package @ $20
_____ Free *Information* Order Form - listing all titles available (heritage education, infill housing, fundraising, building codes, rural preservation, revolving funds, and many more)

Name _____

Address _____

CAL CREW PAINTING

★ *The Finest Work* ★

Lic. #480102

Cal Crew Painting
1495 East Francisco Boulevard
San Rafael, CA 94901
(415) 459-2621
(415) 459-0919 (Fax)
Anna Shubeau

Leaders in Victorian restoration and artistic finishes. European craftsmanship. Latest technologies applied to restoration of wood and masonry buildings. Color design. Have restored over 200 landmark buildings in the San Francisco Bay area.

Fine Paints of Europe
P.O. Box 104
Blooming Grove, NY 10914
(914) 496-8989
(800) 332-1556
John Lahey

Fine Paints of Europe is the exclusive importer and distributer of Hasco oil house paints and enamels, the legendary finishes from Holland. These heavily pigmented coatings based on traditional formulations offer exceptionally beautiful, durable finishes and are ideal for restoration work. They are available through select retailers or directly from Fine Paints of Europe. (Free nationwide delivery)

John Canning & Co., Ltd.
50 Center Street
P.O. Box 822
Southington, CT 06489
(203) 621-2188
(203) 621-2188 (Fax)
John Canning

Services include field research, paint analysis, consultation and assessment, documentation of original interior decoration, and the creation of new designs suitable for period interiors. Skilled in the execution of gold-leafing, stenciling, marbling, woodgraining, trompe l'oeil, murals, conservation, and painting.

(See display ad, page 162)

The Color People
1546 Williams Street, Suite 201
Denver, CO. 80218
(800) 541-7174
(303) 388-8686
(303) 388-8686 (Fax)
James Martin

(See description under Consultants)

Matthew John Mosca
Historic Paint Research
2513 Queen Anne Road
Baltimore, MD. 21216
(410) 466-5325
(410) 466-5715 (Fax)

Historic paint finishes determined by microscopic and chemical analyses that indicate original pigments used. Examination of painted finishes and wallpaper from the 18th to the 20th centuries. Special preparation of early paint types. Significant restorations include: Mount Vernon, the Hermitage, U.S. Treasury, Hope Lodge, George Read House, and 1925 Warner Theater.

Othmar Carli/Studio Carli
100 Springdale Road
York, PA. 17403
(717) 846-3341
Othmar Carli

Specializing in historic painted decorative finishes and murals for 35 years in Europe, the United States, and Asia. Research, conservation, restoration, replication, and new interiors. Award-winning projects: courthouses, palaces, churches, theaters, museums, temples, and private homes. Call for comprehensive assessment of your project. Lectures, seminars, hands-on training. Member AIC, IIC.

PLANNERS

The URBANA Group
P.O. Box 1028
Urbana, IL 61801-9028
(217) 344-7526
(217) 344-7535 (Fax)
Alice Edwards

(See description under Consultants)

PLAQUES

Erie Landmark Company
4449 Brookfield Corporation Drive
Chantilly, VA 22021
(800) 874-7848
(703) 818-2157 (Fax)
Barbara Bamberger

Erie Landmark Company offers bronze and aluminum markers for indoor and outdoor use. Bronze 10-inch by 7-inch house plaques start at $95.00. National Register plaques start at $50.00. The company casts all sizes from medallions to roadside markers and can reproduce logos or graphics. Erie Landmark offers discounts to historical societies. Ask about time capsules. Free brochure.

(See display ad, page 164)

PLASTERING

Frank J. Mangione
21 John Street
Saugerties, NY 12477
(914) 246-9863

Can repair, restore, or produce new ornamental plaster including the reproduction and/or restoration of medallions, capitals, and other ornamental work. Flat-work restoration also done.

Morell-Brown Plastering Corporation
2112 Broadway
New York, NY 10023
(212) 873-4014
Murray Hartstein

PLUMBING

D.E.A. Bathroom Machineries
495 Main Street
P.O. Box 1020
Murphys, CA 95247
(209) 728-2031
(800) 255-4426
(209) 728-2320 (Fax)

Antique, old, and one-of-a-kind toilets, tubs, sinks, faucets, and accessories are treasures to the people at Bathroom Machineries. Specializing in unique Victorian bathroom fixtures and brass hardware, Bathroom Machineries carries both reproductions and antiques that are in such high demand they can barely keep them in stock. Color catalogs, one featuring fixtures and accessories and another showing hardware, are available for $3.00 each.

PUBLICATIONS

Traditional Building
69A 7th Avenue
Brooklyn, NY. 11217
(718) 636-0788
(718) 636-0750 (Fax)
Magnolia Shepherd

Traditional Building is a bimonthly publication for historical architects, contractors, and other building professionals. It provides up-to-date listings and reviews of sources for hard-to-find historical products and services for restoration, renovation, and historically styled new construction. Subscriptions $18 per year.

REAL ESTATE

IsaBell K. Horsley, R.E.
P.O. Box 194
Urbanna, VA 23175
(804) 758-0740
(804) 758-2355 (Fax)
Dave Johnston

Dave Johnston, "The Old-House Man," specializes in period and historic properties. He covers a 15-county area known as Tidewater Virginia, the first part of the commonwealth to be settled. Call for information on *all* old-house listings in this region in any price range.

John Petraglia Real Estate Services
P.O. Box 1046
East Douglas, MA 01516
(508) 476-7745
(508) 476-7271 (Residence)
John Petraglia, Broker

Petraglia Real Estate Services specializes in the sale and brokerage of older, antique, and historic properties throughout central and southeastern Massachusetts, from 18th-century Capes to 20th-century Georgian Revival. Professional personal service including renovation, preservation, and restoration consulting and historic research.

The Real Estate Services Group
1785 Massachusetts Avenue, N.W.
Washington, DC 20036
(202) 673-4258

(See display ad, page 165)

REPRODUCTIONS

Anything Fiberglass
8510 Lava Hill Road
Austin, TX 78744
(512) 243-3303
(512) 243-1782 (Fax)
Jerry Post

(See description under Architectural Details)

Stephen P. Bedard
Durrell Mountain Farm
P.O. Box 2
Gilmanton Iron Works, NH 03837
(603) 528-1896

Windsor chair reproductions using the same procedures and locally grown woods as those utilized by 18th-century chair-makers. Authentic to their period in every detail including line, form, and technique. Established 1980. Send $3.00 for 12-page catalog and paint samples. Hours by appointment.

Denninger Cupolas & Weather Vanes
Rural Route 1, Box 447 L
Middletown, NY 10940
(914) 343-2229
Alfred H. Denninger

(See description under Architectural Details)

Design in Tile
Box 358, Department Y
Mount Shasta, CA 96067
(916) 926-2629
Selene Seltzer

(See description under Tile)

RESEARCH

Architectural Researches, Inc.
W5722 Sherwood Drive
La Crosse, WI 54601
(608) 788-5932
Joan Rausch

Traceries
1606 20th Street, N.W.
Washington, DC 20009
(202) 232-6870
(202) 232-7106 (Fax)
Emily Hotaling Eig

Traceries is a research and consulting firm specializing in the study of architectural history and preservation. The firm provides a wide range of services related to the identification, survey, evaluation, preservation, and understanding of historic architecture, interiors, and neighborhoods. Traceries, a woman-owned company, is the largest firm of its kind in the District of Columbia.

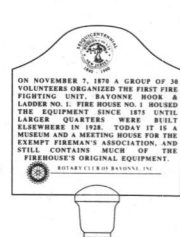

RESTORATION

Architectural Reclamation
312 South River Street
Franklin, OH 45005
(513) 746-8964
Susan Stewart

Specialists in historic restoration and rehabilitation for 15 years. Sensitive, responsible, and creative in their work. Experienced in log and timber frame work, custom woodworking, plastering, masonry, sheet metal, structural repairs, and more. Bondable to $500,000. Cincinnati-Dayton vicinity. References and literature available.

R. W. Bolling Company, Inc.
2128 Florida Avenue, N.W.
Washington, DC 20008-1925
(202) 462-3100
(202) 462-2442 (Fax)
Sterling R. Bolling, Jr.

R.W. Bolling Company specializes in restoration, renovation, and practical repair. Founded in 1913, the firm built many fine homes in Washington's desirable Kalorama neighborhood and other locations. Relying on skilled personnel and quality materials, the company provides renovation services that include finely crafted kitchens, baths, and additions.

Problem-solving repairs and restoration projects are traditional specialties.

Chereskin Restorations Company
80 Morris Avenue
Farmingville, NY 11738
(516) 736-6376
Robert Chereskin

Listed in *The Restorations Directory*, a publication of the New York Landmarks Conservancy. Member of Preservation Forum of the National Trust for Historic Preservation and Advisory Council on Historic Preservation, New York Division. Providing interior and exterior restoration services since 1962. Specializing in paint removal/refinishing, replicative carpentry, wood gutters, and roofing. Structural analyses, project management, and procedural specifications services available.

Fifty Three Restorations
Box 337, Cooper Station
New York, NY 10276
(212) 505-1586
(212) 228-4123 (Fax)
Vincent Lepre

Vincent Lepre began his career in historic preservation as an apprentice at the National Trust's Restoration Workshop in Tarrytown, New York. Since then he has

installed period rooms and been the general contractor for many exterior building restorations. He has the skill and the organization to properly execute your grand or modest project.

Hladun-Goldman, Ltd.
4 Sutton Square
New York, NY 10022
(212) 355-0885
(212) 545-4555 (Fax)
Vira L.M.H. Goldman

Restoration of pre-World War II lobbies in New York City buildings.

Henry Page House Restoration Consultants
P.O. Box T419
Gilmanton, NH 03237
(603) 524-0963
Henry Page or Stephen Bedard

Over 30 years of experience in all aspects of restoration/reproduction ranging from structural problems to detail replacement. All levels of projects accepted, from one room to entire estates and from initial project analysis to complete project supervision. Video of projects $10.00.

Rockworks, Ltd.
114 Linden Street
Oakland, CA 94607
(510) 763-3113
(510) 763-3464 (Fax)
Jay Dunham and Charles Kibby

Rockworks, Ltd. serves the western United States and specializes in the restoration of natural and cast stone, terra cotta, tile, mosaic and decorative plaster surfaces, and architectural elements. In conjunction with Professional Stone Care, which serves eastern states, Rockworks, Ltd., offers professional services throughout the country in project planning and/or execution.

Conrad Schmitt Studios
2405 South 162nd Street
New Berlin, WI. 53151
(414) 786-3030
(800) 969-3033
(414) 786-9036 (Fax)
Heidi Gruenke

Since 1889 the Conrad Schmitt Studios has demonstrated its ability to provide solutions in a variety of media. Experienced in the preservation and restoration of historic interiors, the studio will become an effective part of the restoration team. Whether working directly with the client or through the architect or designer, the studio will continue to strive for excellence through the quality of its art.

STRASSER & ASSOCIATES, INC.
Architectural Restoration • Period Interiors

Strasser & Associates, Inc.
35 Hillside Avenue
Monsey, NY 10952
(914) 425-0650
(914) 425-1842 (Fax)
Peter J. Strasser

Specialists in top-of-the-line restoration, conservation, and reproduction of period interiors, exteriors, and furniture. Strasser personally was involved in restoring period rooms for the Metropolitan Museum of Art. Since then, he has been a principal in many restoration projects throughout the New York metropolitan area, including several large residences at the Dakota apartment house.

Traditional Line, Ltd.
143 West 21st Street
New York, NY 10011
(212) 627-3555
(212) 645-8158 (Fax)
James Boorstein or Anthony Lefeber

Full restoration services: fine woodworking and comprehensive finishing techniques, furniture, consultation, research, condition reports, drawings, carvings, project management (including coordinating mechanical trades). Large or small projects. Past work includes mantel reproductions, restorations of entire Landmark Apartments, wood-paneled library, and American Wing period rooms at the Metropolitan Museum of Art. Information available on request.

RESTORATION PRODUCTS

ABATRON, INC.
33 Center Drive
Gilberts, IL 60136
(708) 426-2200
(708) 426-5066 (Fax)
Marsha Caporaso

Manufacturers of adhesives and wood consolidants, patching and resurfacing compounds for structural and decorative restoration, structural wood substitutes and patching compounds, floor coatings and resurfacers, laminating resins for fiberglass, casting resins for tooling, patterns, and molds, architectural coatings and resurfacing materials, maintenance

and marine repair kits, industrial primers and coatings, porcelain refinishing materials, sealants, and protective coatings. Free brochure.

(See display ad, page 168)

Allied Resin Corporation
Weymouth Industrial Park
East Weymouth, MA 02189
(617) 337-6070
(617) 340-1341 (Fax)
Charles P. Howland

ROOFING

Bellvale Construction, Inc.
P. O. Box 127
Bellvale, NY 10912
(914) 986-5107
(914) 986-6709 (Fax)
Mike Forman

For 13 years Bellvale Construction has taken pride in the preservation of historic buildings with an emphasis on slate and copper roofing. Their crew of experienced craftsmen, many of whom have been with the company since its founding, ensure quality workmanship and prompt completion of projects. Accepted by the New York Landmarks Conservancy in 1990.

Northern Roof Tile Sales Company
P.O. Box 275
Millgrove, Ontario, LOR 1VO
(416) 627-9648
Stuart Matthews

We offer a portfolio of quality clay roofing tiles from around the world, a full range of tiles in all shapes and finishes, including ornamental ridge tiles, finials, decorative shingle tiles for vertical panels, turret tiles, and natural and man-made slates, snow guards, and other roofing accessories.

Schnabels Roofing Corporation
59 Remington Boulevard
Ronkonkoma, NY 11779
(516) 585-7073
(516) 585-7077 (Fax)
Henry A. Schnabel

Schnabels Roofing Corporation is a long-standing professional contractor skilled in every aspect of roofing. A family business that protects its name by taking pride in its work and employing only skilled and experienced craftsmen. Serving New York City and Long Island.

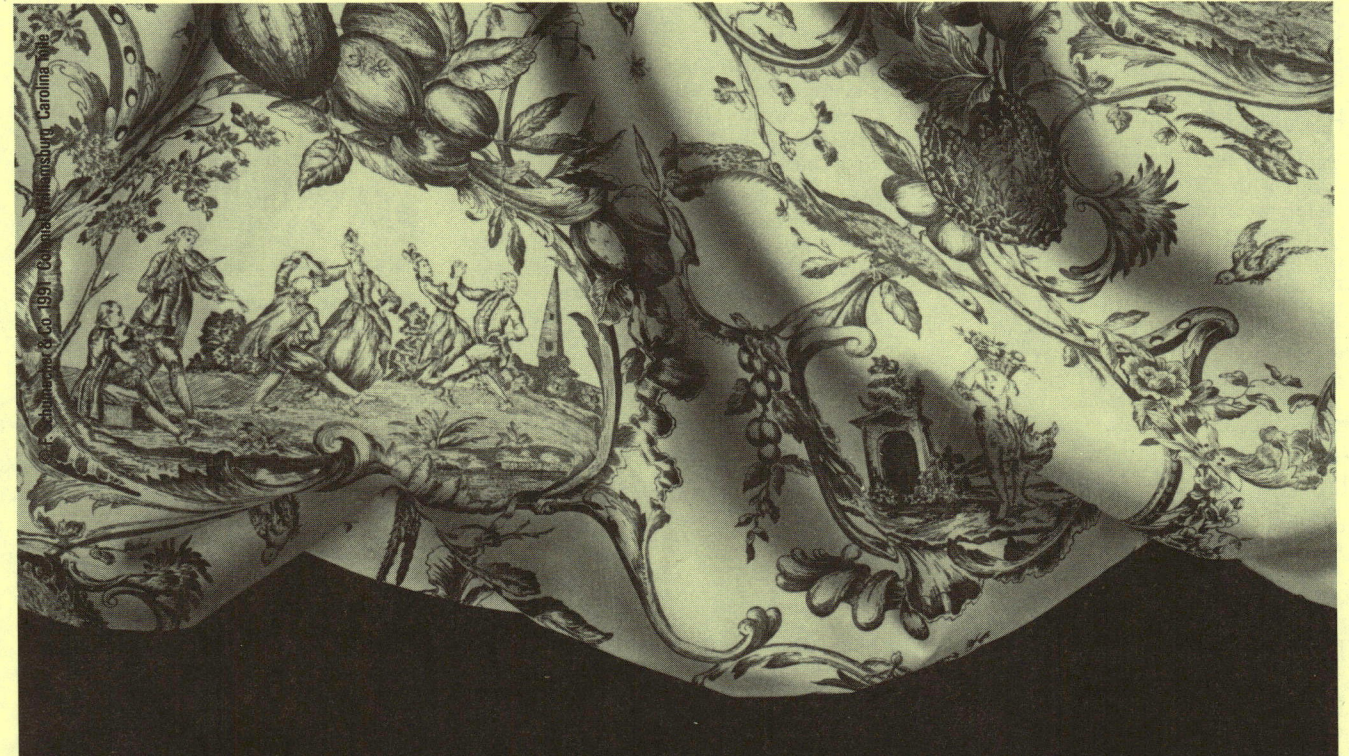

SCHUMACHER®

C L A S S I C   D E S I G N

Vande Hey-Raleigh Manufacturing, Inc.
1565 Bohn Drive
Little Chute, WI 54140
(800) 236-8453
(414) 766-0776 (Fax)
Don Vande Hey

Vande Hey-Raleigh architectural roof tile comes with a 50-year warranty. Completely weatherproof, it withstands tropical summers or frigid winters with equal ease. This termite-proof, rodent-proof, rot-proof concrete roof tile will offer protection from the elements for as long as the structure stands, thus eliminating reroofing costs. Practical beauty never before possible. Costs less than fire-treated cedar shakes, clay, or slate. Custom colors and surfaces offered.

SCULPTURE

New York Conservation Center, Inc.
519 West 26th Street
New York, NY 10007
(212) 714-0620
(212) 714-0149 (Fax)
John Scott, MA, MA-CAS

Top professional-quality consulting, specifications writing, and full services in the conservation of sculptural features of architectural landmarks, monuments, interiors, etc. All sculptural media and periods.

SHUTTERS

Beech River Mill Corporation
Old Route 16, Box 168
Centre Ossipee, NH 03814
(603) 539-2636
Greg Dales

Shutters and hardware that are history! Established in 1856, this historic mill manufactures custom exterior or interior louvered or paneled shutters of superior quality. Duplication of any shutter is possible. Also manufacture and carry the most complete line of shutter hardware available. References and samples upon request. Brochure $5.00.

STAINED GLASS

Cummings Stained Glass Studio, Inc.
P.O. Box 427
North Adams, MA 01247-0427
(413) 664-6578
(413) 664-6570 (Fax)
Nigel D. Johnson

Founded in 1922, Cummings is a high-quality leaded art glass restoration, design, and fabrication studio specializing in complex project management and craftsmanship. Good location has allowed Cummings to successfully complete projects in significant historic areas throughout the country. Please call or write for further information.

Rohlf's Stained and Leaded Glass Studio, Inc.
783 South Third Avenue
Mt. Vernon, NY 10550
(914) 699-4848
(212) 823-4545
(212) 823-4717 (Fax)
Peter A. Rohlf

STAINED GLASS

Devoted to excellence, whether in skilled restoration or in new glass design and installation, Lamb Studios has a worldwide reputation. Over the years, Lamb has created approximately 10,000 original stained glass windows in America and abroad, and more than 7,000 windows have been restored, some in the most important architectural works ever built in this country. Some of our better known projects have been at Cornell University - Sage and Anabelle Taylor Chapels, The Breakers Mansion, Newport, Rhode Island, Plymouth Church of the Pilgrims - Brooklyn Heights, (A New York City Landmark), Veterans Administration Chapel, Bronx, (New York State Landmark) and Bay Ridge High School - Brooklyn, New York. These and many other projects are available for in depth review by writing to our office.

EXCELLENCE THROUGH EXPERIENCE

P.O. Box 291, Philmont, NY 12565   518-672-7267/7597

STONE CARVING

DMS Studios
550 51st Avenue
Long Island City, NY 11101
(718) 937-5648
(718) 937-2609 (Fax)
Daniel Sinclair

Specializing in stonecutting on 36-inch-blade gantry saw. Can reproduce in marble, bronze, or sandstone complex moldings, returns, fireplaces, ornamentation, statues, and lettering. Custom work of all types. Projects have included the Cloisters at the Metropolitan Museum of Art, the Morgan Library, the Museum of Stony Brook, and the Vanderbilt Museum restoration.

TILE

Antique Delft Tiles
12643 Hortense Street
Studio City, CA 91604
(818) 761-2756
Helen Williams

Designs in Tile
Box 358, Department Y
Mount Shasta, CA 96067
(916) 926-2629
Selene Seltzer

Custom historical hand-decorated ceramic tiles for fireplaces, kitchens, baths, fountains, and floors. Residential and commercial installations. Full-service art-tile studio specializing in consultation, design development, material/installation specification. Art styles and techniques suitable for interior and exterior applications. Coordinated patterns in historical sizes. Specializing in English and American Arts and Crafts. Color brochure $3.00.

WALLCOVERINGS

Bentley Brothers
2709 South Park Road
Louisville, KY 40219
800-824-4777
(502) 969-1702 (Fax)
Whitney Lewis

Bentley Brothers is the East Coast distributor for Anaglypta and Lincrusta embossed, paintable wallcoverings from England. This type of wallcovering dates back to 1877. National distributor for Carton Pierre embossed, paintable cardboard ornaments that date back to 1924.

Flexi-Wall Systems
P.O. Box 88
Liberty, SC 29657
(800) 843-5394
(803) 843-9318 (Fax)
Hank Levy

Specializing in $1/32$-inch-thick gypsum-impregnated flexible fabric repair system for walls and ceilings. Installs like wallpaper over cracked or damaged plaster. Class A flame-spread rating. Lead-based paint barrier system. Can be painted, plastered, or covered with wallcovering. Fabric reinforcing prevents cracking. Breathes at 40 perms to prevent mildew.

Schumacher
79 Madison Avenue
New York, NY 10016-7878
(212) 213-7900

Schumacher provides the international design community with an exclusive selection of the finest quality fabrics, wallpapers, trimmings, and floorcoverings for both contract and residential use. Schumacher's dedication to keeping the best historical designs alive while simultaneously recognizing the classics of the future has enabled the company to bring the finest in timeless design to its clientele for more than a century.

(See display ad, page 169)

Victorian Collectibles, Ltd.
845 East Glenbrook Road
Milwaukee, WI 53217
(414) 352-6971
(414) 352-7290 (Fax)
Wendye Schroeder-Girard

The Brillion Collection offers the finest reproductions of wallpapers, borders, and ceiling patterns from the largest and most pristine cache of Victorian wallpapers ever found in America. Available in a wide array of styles and colors appropriate for commercial or residential installation.

WINDOWS

A&S Window Associates
88-19 76th Avenue
Glendale, NY 11385
(718) 275-7900
(718) 997-7683 (Fax)
Alan Herman

Allied Window, Inc.
2724 West McMicken Avenue
Cincinnati, OH 45214
(513) 559-1212
(800) 445-5411
(513) 559-1883 (Fax)
David Martin

Specialists in the manufacture of custom aluminum storm windows for historic homes, museums, and commercial build-ings. Storm windows can be mounted inside or outside and fit flush with existing windows. Screen or glass panels are available in standard or custom colors and in arched, eyebrow, circle, gothic, or custom shapes. Fixed, magnetic, sliding, or lift-out designs can be ordered. Call or write for free 8-page catalog.

(See display ad, page 174)

Blaine Window Hardware, Inc.
12430 McDade Road
Hagerstown, MD 21740
(800) 678-1919
(301) 797-2510
G. Patrick Howard

Blaine has been the nation's leading supplier of current and obsolete replacement hardware since 1954, offering an extensive inventory of more than 20,000 parts for windows, closets, doors, and more. High quality sand castings are produced in Blaine's foundry. Antique reproductions are a specialty. The foundry has proven to be of significant importance to historic preservationists, private contractors, and anyone restoring old buildings. Call toll free for more information or a current catalog.

Courtaulds Performance Films, Inc.
P.O. Box 5068
Martinsville, VA 24115
(703) 629-1711
(703) 629-8333 (Fax)
V. Kubler

Llumar window films block 98 percent of
incoming ultraviolet rays, reducing fading
and heat gain while maintaining a high
level of visible light and neutral appearance. Films are applied by professional
dealer network on the interior of any type
of glass without disruption of activities or
setting.

Lamb Studios
Box 291
Philmont, NY 12565
(518) 672-7267
(518) 672-7597
Donald Samick

Lamb Studios has restored over 7,000 windows throughout the United States and
continues to be a leader in this specialized
field. For 125 years master craftsmen have
handed down quality techniques in the
restoration and fabrication of stained-glass
windows. Any of the following signs may
indicate that your windows need the kind
of attention available at Lamb Studios: an
increase in heating and cooling costs;
cracked, broken, or loose glass; bulging
panels; loose support bars; cracked or
loose glazing.

(See display ad, page 171)

Marvin Windows
Warroad, MN 56763
(800) 346-5128
(800) 263-6161 (Canada)

Marvin Windows is the nation's largest
manufacturer of made-to-order wood and
clad-wood windows and doors. Marvin
offers ideal products for historic renovation and replacement. Unique and
unusual designs in round tops and authentic divided-light patterns, state-of-the-art
glazings, and a nearly unlimited selection
of options and sizes.

Millwork Specialties
189 Prospect Avenue
Brooklyn, NY 11215
(718) 768-7112
(718) 965-3974 (Fax)
Cosmo Cotroneo

(See description under Millwork)

Thermo-Press Corporation
5406 Distributor Drive
Richmond, VA 23225
(804) 231-2964
(804) 232-0454 (Fax)
John Watkins

Preserve your building's exterior appearance. Thermo-Press makes
interior-mounted acrylic insulating panels
that attach with Velcro. Improve energy
efficiency by 60 percent. Custom made to
fit any opening. Virtually invisible once
installed. Patented system used by
National Park Service, University of Virginia, and Association for the Preservation
of Virginia Antiquities.

The Woodstone Company
P.O Box 223
Patch Road
Westminster, VT 05158
(802) 722-9217
(802) 722-9528 (Fax)
Sandy Furlong

From the Rockingham Meeting House in
Vermont to the Harvard Club in New York
City, Woodstone provides historic landmark reproduction windows and doors.
Narrow muntins to $7/8$-inches-wide with
true divided-light insulating glass, grade
60 performance weight and pulley-hung
sash, and matching molding profiles in
various woods.

(See display ad, page 173)

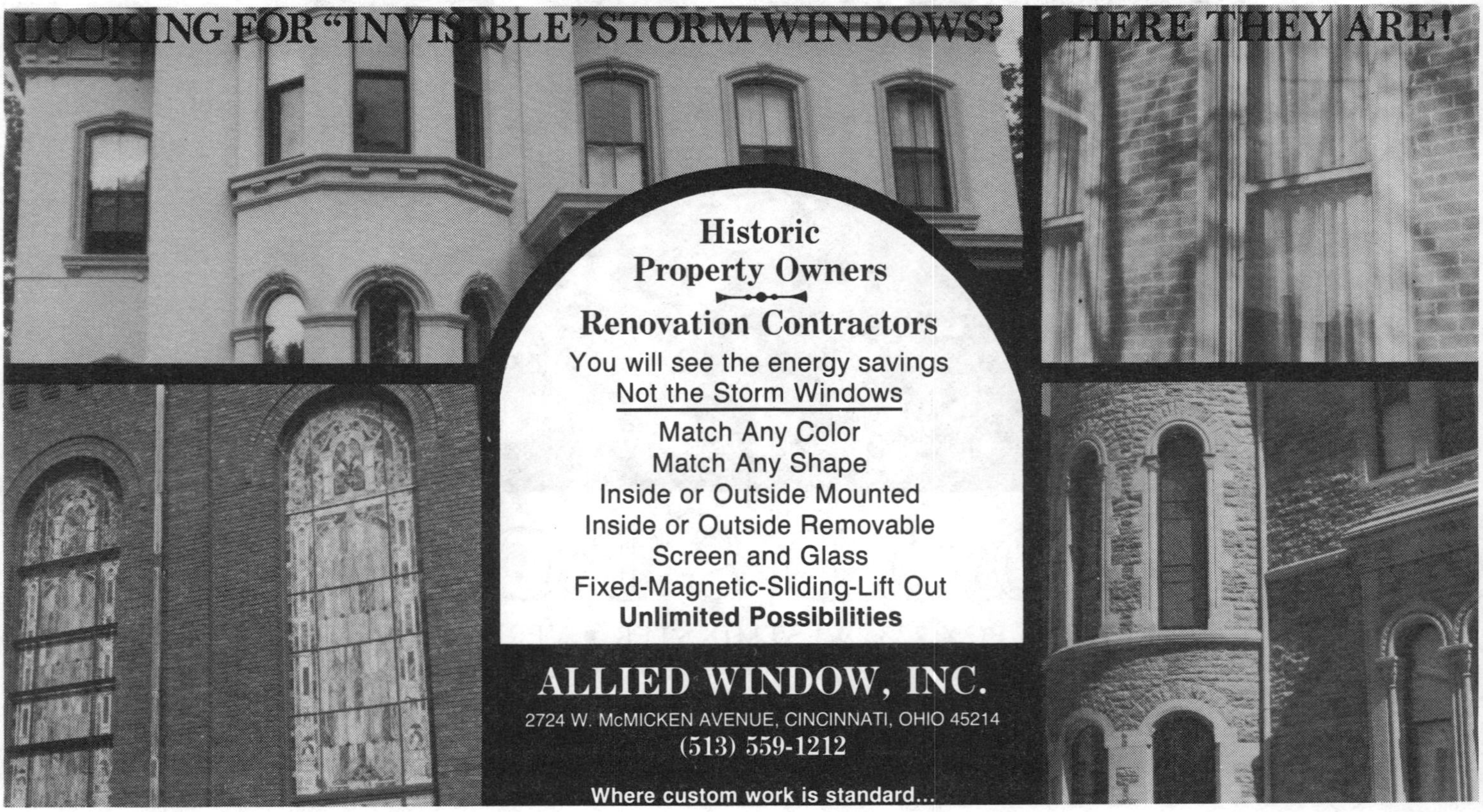

III. THE PRESERVATION NETWORK

FEDERAL PROGRAMS

Vicki Andrews

"Although the major burdens of historic preservation have been borne and major efforts initiated by private agencies and individuals, and both should continue to play a vital role, it is nevertheless necessary and appropriate for the Federal Government to accelerate its historic preservation programs and activities, to give maximum encouragement to agencies and individuals undertaking preservation by private means, and to assist State and local governments and the National Trust for Historic Preservation in the United States to expand and accelerate their historic preservation programs and activities."

Section 1(b)(7), National Historic Preservation Act of 1966, as amended

In addition to the National Park Service (see separate listing), a number of federal programs and agencies affect historic preservation in the United States. The principal programs and agencies are listed below.

EXECUTIVE BRANCH

U.S. Department of Agriculture
Agricultural Stabilization and
Conservation Service
Planning and Evaluation Branch
P.O. Box 2415
Washington, DC 20013
(202) 447-3264

Forest Service
Recreation Division
P.O. Box 96090
Washington, DC 20090-6090
(202) 205-1754

Rural Electrification Administration
Electric Staff Division
14th Street and Independence Avenue,
S.W.
Room 1246
Washington, DC 20240
(202) 382-9093

Soil Conservation Service
Economic and Social Sciences Division
P.O. Box 2890
Washington, DC 20013-2890
(202) 382-1510

U.S. Department of Commerce
Center for Building Technology
National Institute of Standards and
Technology
Building 226, Room B-216
Gaithersburg, MD 20899
(301) 975-5900

Economic Development Administration
14th Street and Constitution Avenue,
N.W.
Room 7804
Washington, DC 20230
(202) 377-5113

National Oceanic and Atmospheric
Administration
Office of Ocean and Coastal Resource
Management
Coastal Programs Division
1825 Connecticut Avenue, N.W.
Room 20036
Washington, DC 20235
(202) 606-4122

U.S. Department of Defense
U.S. Air Force
Office of Air Force History
HQ USAF/CHO
Building 5681
Bolling Air Force Base
Washington, DC 20332-6098
(202) 767-5764

U.S. Army
Chief of Military History
Museum Branch
Washington, DC 20314
(202) 287-9000

U.S. Army
Corps of Engineers
20th Street and Massachusetts Avenue,
N.W.
Washington, DC 20314
(202) 272-0660

U.S. Marine Corps
History and Museums Division
HD, Building 58
Washington Navy Yard
Washington, DC 20374-0580
(202) 433-2273

U.S. Navy
Navy Memorial Museum
Building 76
Washington Navy Yard
Washington, DC 20374
(202) 433-2651

U.S. Department of Education
400 Maryland Avenue, S.W.
Washington, DC 20202
(202) 708-5366

U.S. Department of Energy
1000 Independence Avenue, S.W.
Washington, DC 20585
(202) 586-5000

**U.S. Department of Housing and Urban
Development**
451 7th Street, S.W.
Washington, DC 20410
(202) 708-1422

Community Development Block Grants
(202) 755-6587

Historic Preservation Information
(202) 755-6610

Section 202—Housing for the Elderly
and Assisted Housing
(202) 755-5720

Section 221—Public Housing Program
(202) 426-7212

Section 312—Rehabilitation Loans
(202) 755-0367

Title I—Property Improvement Loan
Program
(202) 755-5740

Urban Homesteading Program
(202) 755-5327

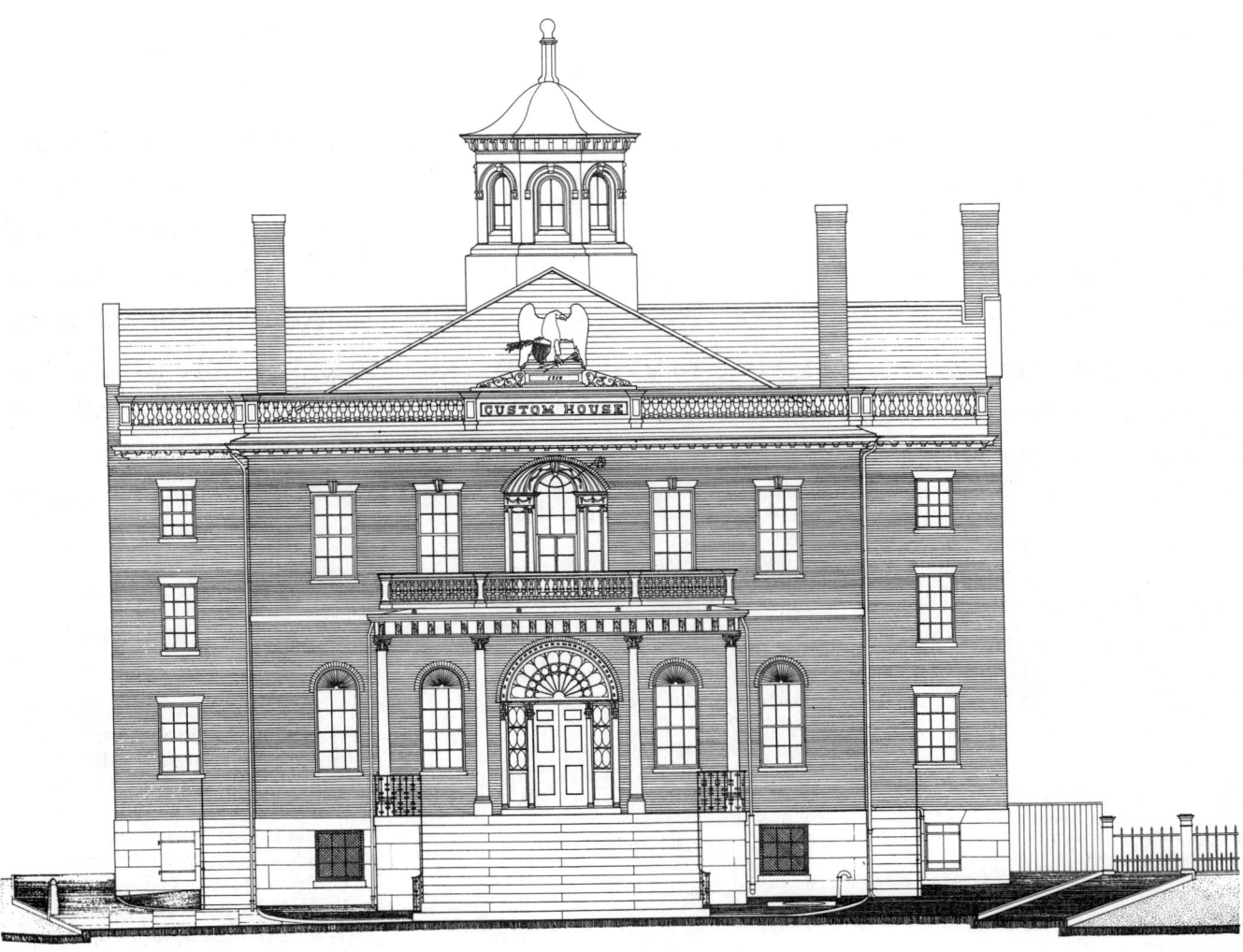

U.S. Customs House and Public Stores, Salem, Massachusetts. (L.D. Nichols, HABS)

U.S. Department of the Interior
Assistant Secretary for Fish and Wildlife
and Parks
(202) 208-4416
John Michael Hayden

National Park Service
18th and C Streets, N.W.
Washington, DC 20840
(202) 208-4621
James M. Ridenour, Director

Associate Director for Cultural Resources
(202) 208-7625
Jerry L. Rogers

Deputy Associate Director for Cultural
Resources
(202) 343-9596
Roland T. Bowers

Assistant to the Associate Director
(202) 343-3411
Ronald M. Greenberg

Office of International Affairs
(202) 343-7063
Robert C. Milne, Chief

International Liaison Officer for Cultural
Resources
(202) 343-7069
John Poppeliers
Also see separate listing

U.S. Department of Labor
Employment and Training Administration
200 Constitution Avenue, N.W.
Washington, DC 20210
(202) 523-6871

U.S. Department of Transportation
Coast Guard
Aids to Navigation Office
2100 2nd Street, S.W.
Washington, D.C. 20593
(202) 267-0349

Federal Highway Administration
Environmental Analysis Division
400 7th Street, S.W.
HEP-42
Washington, DC 20590
(202) 366-2067

Federal Railroad Administration
Office of Policy
400 7th Street, S.W.
Room 8300
Washington, DC 20590
(202) 366-0881

Urban Mass Transportation
Administration
400 7th Street, S.W.
Room 9301
Washington, DC 20590
(202) 366-0096

U.S. Department of the Treasury
Bureau of the Mint
10001 Aerospace Road
Lanham, MD 20706
(301) 436-7400

Comptroller of the Currency
250 E Street, S.W.
Washington, DC 20219
(202) 874-5000

Internal Revenue Service
500 North Capitol Street
Washington, DC 20001
(800) 829-1040

INDEPENDENT AGENCIES

Advisory Council on Historic Preservation
1100 Pennsylvania Avenue, N.W.
Suite 809
Washington, DC 20004
(202) 786-0503
John F. W. Rogers, Chairman
Robert D. Bush, Executive Director

American Battle Monuments Commission
20 Massachusetts Avenue, N.W.
Room 5127, Pulaski Building
Washington, DC 20314-0300
(202) 272-0533
Maj. Gen. A. J. Adams, Secretary

Amtrak
National Railroad Passenger Corporation
60 Massachusetts Avenue, N.E.
Washington, DC 20002
(202) 906-3000
W. Graham Claytor, Chairman and
President

Bicentennial Commission on the U.S. Constitution
808 17th Street, N.W.
Washington, D.C. 20006
(202) 653-2486
Warren Burger, Chairman
Herbert Atherton, Director

Commission of Fine Arts
441 F Street, N.W.
Room 312
Washington, DC 20001
(202) 504-2200
J. Carter Brown, Chairman
Charles H. Atherton, Secretary

Council on Environmental Quality
722 Jackson Place, N.W.
Washington, DC 20503
(202) 395-5750
Michael Deland, Chairman

Federal Emergency Management Agency
500 C Street, S.W.
Washington, DC 20472
(202) 646-2641
Alfred E. Warren, Management Division
Chief

General Services Administration
Public Buildings Service
7th and D Streets, S.W.
Washington, DC 20407
(202) 708-5082
Dale M. Lanzone, Director, Art and
Historic Preservation

National Archives and Records Administration
National Historical Publications and
Records Commission
8th and Pennsylvania Avenue, N.W.
Washington, D.C. 20408
(202) 501-5600
Gerald George, Executive Director
Don W. Wilson, Archivist of the United
States

National Foundation on the Arts and Humanities
Institute of Museum Services
1100 Pennsylvania Avenue, N.W.
Suite 510
Washington, DC 20506
(202) 786-0536
Susannah Simpson Kent, Director

National Endowment for the Arts
1100 Pennsylvania Avenue, N.W.
Room 624
Washington, DC 20506
(202) 682-5442
John E. Frohnmayer, Chairman

National Endowment for the Humanities
1100 Pennsylvania Avenue, N.W.
Room 503
Washington, DC 20506
(202) 786-0310
Lynn V. Cheney, Chairman

Neighborhood Reinvestment Corporation
1325 G Street, N.W.
Suite 800
Washington, DC 20005
(202) 376-2400
George Knight, Executive Director

Pennsylvania Avenue Development Corporation
1331 Pennsylvania Avenue, N.W.
Suite 1220 N
Washington, DC 20004
(202) 724-9091
Jay Brodie, Executive Director

Small Business Administration
409 3rd Street, S.W.
Washington, DC 20416
(202) 205-6605
Patricia Saiki, Administrator

Smithsonian Institution
National Institute for Conservation of
Cultural Property
3299 K Street, N.W.
Suite 403
Washington, DC 20007
(202) 625-1496
Larry Reiger, President

Tennessee Valley Authority
Cultural Resources Program
2C, Natural Resources Building
Norris, TN 37828
(615) 632-1585
Maxwell D. Ramsey, Manager

Quincentennial Commission
810 1st Street, N.E.
Washington, DC 20002
(202) 289-1661
Pablo Sedillo, Executive Director

LEGISLATIVE BRANCH

Library of Congress
101 Independence Avenue, S.E.
Washington, DC 20540
(202) 287-5000
James H. Billington, Librarian

Architect of the Capitol
U.S. Capitol Building
Washington, DC 20515
(202) 225-1200
George M. White, Architect
William L. Ensign, Assistant Architect

FEDERAL PRESERVATION OFFICERS

"The head of each Federal agency shall, unless exempted under section 214, designate a qualified official to be known as the agency's 'preservation officer' who shall be responsible for coordinating that agency's activities under this Act. Each Preservation Officer may, in order to be considered qualified, satisfactorily complete an appropriate training program established by the Secretary under section 101(g)."

Section 11(c), National Historic Preservation Act of 1966, as amended

U.S. Department of Agriculture
Agricultural Stabilization and
Conservation Service
P.O. Box 2415
Washington, DC 20013
(202) 447-6221
James R. McMullen, Director

Farmers Home Administration
14th Street and Independence Avenue,
S.W.
Room 6303
Washington, DC 20250
(202) 382-9619
Susan Wieferich, Senior Environmental
Specialist

Forest Service
P.O. Box 96090
Washington, DC 20090-6090
(202) 205-1427
Evan De Bloois, Preservation Officer

Rural Electrification Administration
Environmental Compliance Branch
Electric Staff Division
14th Street and Independence Avenue,
S.W.
Room 1269
Washington, DC 20250
(202) 720-5093
Lawrence R. Wolfe, Chief

Soil Conservation Service
Economic and Social Sciences Division
P.O. Box 2890
Washington, DC 20013-2890
(202) 720-2171
Diane Gelburd, Assistant Director

U.S. Department of Commerce
Office of Federal Property Programs
National Programs Division
14th Street and Constitution Avenue,
N.W.
Room 1037
Washington, DC 20230
(202) 377-3580
James T. McCombs, Chief

Economic Development Administration
14th Street and Constitution Avenue,
N.W.
Room 7019
Washington, DC 20230
(202) 377-4208
Frank Monteferrante, Senior
Environmental Officer

National Oceanic and Atmospheric
Administration
Office of Ocean and Coastal Resource
Management
1825 Connecticut Avenue, N.W.
Suite 714
Washington, DC 20235
(202) 606-4111
Trudy Coxe, Director

Office of Federal Property Programs
National Programs Division
14th Street and Constitution Avenue,
N.W.
Room 1037
Washington, DC 20230
(202) 377-3580
James T. McCombs, Chief

U.S. Department of Defense
U.S. Air Force
Environmental Safety and Occupational
Health
The Pentagon
Room 4C916
Washington, DC 20330-1000
(703) 697-9297
Gary D. Vest, Deputy Assistant Secretary

U.S. Army
Attention CEHSC-FN
Fort Belvoir, VA 22060-5516
(703) 704-1631
Constance Werner Ramirez, Historic
Preservation Officer

U.S. Army
Office of Environmental Policy
20 Massachusetts Avenue, N.W.
CECW-PO
Washington, DC 20314-1000
(202) 272-0166
William Klesch, Chief

U.S. Marine Corps
Installations and Facilities
OASN (I&E)
Crystal Plaza
Building 5, Room 218
Washington, DC 20360-5000
(703) 602-2686
Frederick S. Sterns, Deputy Assistant
Secretary of the Navy (Installations and
Facilities)

U.S. Navy
Installations and Facilities
OASN (I&E)
Crystal Plaza
Building 5, Room 218
Washington, DC 20360-5000
(703) 602-2686
Frederick S. Sterns, Deputy Assistant
Secretary of the Navy (Installations and
Facilities)

U.S. Department of Energy
Federal Energy Regulatory Commission
825 North Capitol Street, N.E.
Washington, DC 20426
(202) 208-0400
Lois D. Cashell, Secretary

**U.S. Department of Health and Human
Services**
330 Independence Avenue, S.W.
Room 4714
Washington, DC 20201
(202) 619-0426
Scott Waldman, Historic Preservation
Officer

**U.S. Department of Housing and Urban
Development**
Office of Environment and Energy
451 7th Street, S.W.
Room 7240
Washington, DC 20410
(202) 708-2894
Richard H. Broun, Director

U.S. Department of the Interior
Bureau of Indian Affairs
Office of Trust and Economic
Development
1849 C Street, N.W.
Room 4513
Washington, DC 20240
(202) 208-5831
George Farris, Chief, Environmental
Services

Bureau of Land Management
1849 C Street, N.W.
Room 3360
Washington, DC 20240
(202) 208-3353
John G. Douglas, Senior Archeologist

Fish and Wildlife Service
Refuges and Wildlife
18th and C Streets, N.W.
Washington, DC 20240
(202) 208-5333
David L. Olsen, Assistant Director

Minerals Management Service
Environmental Operations and Analysis
381 Elden Street
Herndon, VA 22070
(703) 787-1736
Melanie Stright, Archeologist

National Park Service
History Division
P.O. Box 37127
Washington, DC 20013-7127
(202) 343-8167
Edwin C. Bearss, Chief

Office of Surface Mining
Office of the Director
1951 Constitution Avenue, N.W.
Washington, DC 20240
(202) 208-2700
Suzanne Hudak, Policy Analyst

U.S. Geological Survey
Environmental Affairs Program
12201 Sunrise Valley Drive
Mail Stop 423
Reston, VA 22092
(703) 648-6828
Norman Wingard

U.S. Department of Justice
Real Property and Space Management
1425 K Street, N.W.
Suite 700
Washington, DC 20530
(202) 633-1598
Cynthia D'Agostino, Assistant Director

U.S. Department of Labor
Division of Administrative Services
200 Constitution Avenue, N.W.
Room C-4513
Washington, DC 20210
(202) 535-8710
Michael O'Malley, Architect

U.S. Department of Transportation
Office of Transportation Regulatory
Affairs
P-14
400 Seventh Street, S.W.
Washington, DC 20591
(202) 366-4866
Robert F. Crecco, Historic Preservation
Officer

Federal Aviation Administration
Noise Abatement Division
Office of Environment and Energy
Room 432
800 Independence Avenue, S.W.
Washington, DC 20591
(202) 267-3561
Laurette Fisher, Environmental Specialist

Federal Highway Administration
Environmental Quality Branch
400 7th Street, S.W.
HEP-42
Washington, DC 20590
(202) 366-9173
Charles Des Jardins, Chief

Federal Railroad Administration
Office of Policy
400 7th Street, S.W.
RRP-32
Washington, DC 20590
(202) 366-0358
Marilyn Klein

U.S. Department of the Treasury
Treasury Annex Building
Room 6140
Washington, DC 20220
(202) 377-9165
William M. McGovern, Environmental
Programs Officer

U.S. Department of Veterans Affairs
Historic Preservation Office
086B
810 Vermont Avenue, N.W.
Washington, DC 20420
(202) 233-3447
Karen Tupek, Historic Preservation
Officer

INDEPENDENT AGENCIES

Environmental Protection Agency
Office of Enforcement
Office of Federal Activities (A-104)
Waterside Mall, S.W.
Washington, DC 20460
(202) 382-5910
John Gerba, Environmental Protection
Specialist

Federal Communications Commission
1919 M Street, N.W.
Room 621
Washington, DC 20554
(202) 632-6410
Donna R. Searcy, Secretary

Federal Deposit Insurance Corporation
Supervision and Resolutions
550 17th Street, N.W.
Washington, DC 20429
(202) 898-6946
Paul G. Fritts, Executive Director

General Services Administration
Public Buildings Service
18th and F Streets, N.W.
Room 1300
Washington, DC 20405
(202) 501-1256
Dale Lanzone, Director, Arts and Historic
Preservation

**National Aeronautics and Space
Administration**
Facilities and Operations Maintenance
Division
NXG
400 Maryland Avenue, S.W.
Washington, DC 20546
(202) 453-1956
Kenneth Kumor

National Capital Planning Commission
Operations Office
801 Pennsylvania Avenue, N.W.
Washington, DC 20576
(202) 724-0210
Margot Stephenson, Assistant Director

Nuclear Regulatory Commission
1717 H Street, N.W.
Mail Stop 12-H-5
Washington, DC 20555
(301) 492-1219
Roberta Ingram

**Pennsylvania Avenue Development
Corporation**
1331 Pennsylvania Avenue, N.W.
Suite 1220 North
Washington, DC 20004
(202) 724-9068
Jan Frankina, Historic Preservation
Officer

Small Business Administration
Office of Portfolio Management
409 3rd Street, S.W.
8th Floor
Washington, DC 20416
(202) 205-6481
Annie McCluney, Analyst Specialist

Tennessee Valley Authority
River Basin Operations
400 West Summit Hill Drive
OCH Room E-K
Knoxville, TN 37902
(615) 632-6367
Billy J. Bond, Vice President

FEDERAL AGENCY AREA OFFICES

Listed in local telephone directories by
agency under "United States
Government."

FEDERAL INFORMATION CENTERS

Listed in local telephone directories under
"Federal Information Center, United
States Government" and connected to
other agencies by toll-free lines.

FURTHER READING

*The Catalog of Federal Domestic
Assistance.* Office of Management and
Budget. Washington, D.C.: U.S.
Government Printing Office. Annual.

Federal Staff Directory. Ann L. Brownson,
ed. Mount Vernon, Va.: Congressional
Staff Directory, 1989.

*Guide to Federal Funding for
Governments and Non-Profits* Arlington,
Va.: Government Information Services.
Annual.

United States Government Manual.
Office of the Federal Register, National
Archives and Records Service, U.S.
General Services Administration.
Washington, D.C.: U.S. Government
Printing Office. Annual.

Washington Information Directory.
Washington, D.C.: Congressional
Quarterly. Annual.

NATIONAL AND REGIONAL CONTACTS

Julie Griffith and Stephanie Reeves

"Because governments respond most readily to organized constituencies, preservationists working in concert and in alliance with other groups are far more influential than individuals working autonomously."

Robert M. Utley, Barry Mackintosh, and others, *Monumentum*, 1976

Alliance for Historic Landscape Preservation
82 Wall Street
Suite 1105
New York, NY 10005
(617) 491-3727
Shary Page Berg, President

American Association for State and Local History
172 2nd Avenue, North
Suite 202
Nashville, TN 37201
(615) 255-2971
Patricia Gordon Michael, Executive Director

American Association of Museums
1225 I Street, N.W.
Suite 200
Washington, DC 20005
(202) 289-1818
Edward Able, Director

American Council for the Arts
1285 Avenue of the Americas
New York, NY 10019
(212) 245-4510
Milton Rhodes, President

1213 29th Street, N.W.
Washington, DC 20007
(202) 333-5841
Jack Duncan, Special Counsel

American Historical Association
400 A Street, S.E.
Washington, DC 20003
(202) 544-2422
Samuel R. Gammon, Executive Director

American Institute for Conservation of Historic and Artistic Works
1400 16th Street, N.W.
Suite 340
Washington, DC 20036
(202) 232-6636
Sarah Rosenburg, Executive Director

American Institute of Architects
1735 New York Avenue, N.W.
Washington, DC 20006
(202) 626-7300
James P. Cramer, Executive Vice President

Committee on Historic Resources
Regional Urban Design Assistance
Teams
(202) 626-7452

American Planning Association
1313 East 60th Street
Chicago, IL 60637
(312) 955-9100
Israel Stollman, Executive Director

Historic Preservation Division
1776 Massachusetts Avenue, N.W.
Washington, DC 20036
(202) 872-0611
Stephen Cochran

Urban Design and Preservation
Division
(303) 556-2755
Hamid Shirvani, Chairman

American Society for Conservation Archeology
c/o U.S. Department of the Interior
Fish and Wildlife Service
1849 C Street, N.W.
Room 2343
Washington, DC 20240
(202) 208-5634
Kevin Kilcullen, Archeologist

American Society of Civil Engineers
Committee on History and Heritage
of American Civil Engineering
1015 15th Street, N.W.
Suite 600
Washington, DC 20005
(202) 789-2200
Kelly Cunningham, Coordinator

American Society of Interior Designers
Historic Preservation Committee
608 Massachusetts Avenue, N.E.
Washington, DC 20002
(202) 546-3480
Robert H. Angle, Executive Director

American Society of Landscape Architects
4401 Connecticut Avenue, N.W.
5th Floor
Washington, DC 20008-2302
(202) 686-2752
David Bohardt, Executive Vice President

American Society of Mechanical Engineers
History and Heritage Program
345 East 47th Street
New York, NY 10017-2392
(212) 705-7740
David Belden, Executive Director

America the Beautiful Fund
219 Shoreham Building
Washington, DC 20005
(202) 638-1649
Paul Bruce Dowling, Director

Archaeological Conservancy
415 Orchard Drive
Santa Fe, NM 87501
(505) 982-3278
Mark Michel, President

Archeological Institute of America
675 Commonwealth Avenue
Boston, MA 02215
(617) 353-9361
Mark Meister, Executive Director

Association for Living Historical Farms and Agricultural Museums
c/o Smithsonian Institution
National Museum of American History
Room 5035
Washington, DC 20560
(202) 357-2095
Terry Scharer, Curator

Association for Preservation Technology
904 Princess Anne Street
P.O. Box 8178
Fredericksburg, VA 22404
(703) 373-1621, 373-1622
Susan Ford Johnson, Executive Director

Association of American Geographers
1710 16th Street, N.W.
Washington, DC 20009
(202) 234-1450
Ronald Abler, Executive Director

Association of Junior Leagues
660 1st Avenue
2nd Floor
New York, NY 10016-3241
(212) 683-1515
Holly Sloan, Executive Director

Public Policy Office
1319 F Street, N.W.
Suite 604
Washington, DC 20004
(202) 393-3364

Classical America
227 East 50th Street
New York, NY 10022
(212) 753-4376
Henry Hope Reed, President

The Conservation Foundation/World Wildlife Foundation
1250 24th Street, N.W.
Suite 500
Washington, DC 20037
(202) 293-4800
Kathryn Fuller, President

Council of American Maritime Museums
c/o Chesapeake Bay Maritime Museum
P.O. Box 636
St. Michaels, MD 21663
(301) 745-2916
John R. Viallont, Chairman, Membership Committee

Council of Planning Librarians
1313 East 60th Street
Chicago, IL 60637
(312) 955-9100
Dennis Janiks, Editor

Council of Preservation Executives
Historic Landmark Foundation of Indiana
1028 North Delaware Street
Indianapolis, IN 46202
(317) 638-5264
David Frederick, Director

Council on America's Military Past—U.S.A.
P.O. Box 1151
Fort Myer, VA 22211
(202) 479-2258
Col. Herbert M. Hart, Secretary

Decorative Arts Society
c/o Brooklyn Museum
200 Eastern Parkway
Brooklyn, NY 11238
(718) 638-5000
Kevin Stayton

Early American Society
6405 Frank Drive
Harrisburg, PA 17105
(717) 657-9555
Frances Carnahan, Editor

Environmental Action Foundation
6930 Carroll Avenue
Suite 600
Takoma Park, MD 20912
(301) 891-1100
Ruth Caplan, Director

Environmental Defense Fund
257 Park Avenue, South
New York, NY 10010
(212) 505-2100
Frederick Krupp, Executive Director

Environmental Law Institute
1616 P Street, N.W.
Suite 200
Washington, DC 20036
(202) 328-5150
J. William Futrell, President

Pigeon Point Lighthouse, Pescadero vicinity, California. (S. Farneth, HABS)

Friends of Cast-Iron Architecture
235 East 87th Street
Room 6C
New York, NY 10128
(212) 369-6004
Margot Gayle, President

Friends of the Earth
218 D Street, S.E.
Washington, DC 20003
(202) 544-2600
Jane Perkins, President

Great Lakes Lighthouse Keepers Association
24200 West Ater Drive
P.O. Box 580
Allen Park, MI 48101

Institute for Urban Design
2453 Karensue Avenue
San Diego, CA 92122
(619) 455-1215
Ann Ferebee, Director

Institute of Early American History and Culture
P.O. Box 220
Williamsburg, VA 23187
(804) 221-1110
Ronald Hoffman, Director

International Downtown Association
915 15th Street, N.W.
Suite 900
Washington, DC 20005
(202) 783-4963
Richard Bradley, President

League of Historic American Theatres
1511 K Street, N.W.
Suite 923
Washington, DC 20005
(202) 783-6966
Mary Margaret Schoenfield, Executive Director

Lighthouse Preservation Society
P.O. Box 736
Rockport, MA 01966
(508) 281-6336
Valerie Nelson, Executive Director
James Hyland, President

Metropolitan Museum of Art
1000 5th Avenue
New York, NY 10028
(212) 879-5500
Philippe de Montebello, Director

National Alliance of Preservation Commissions
c/o School of Environmental Design
609 Caldwell Hall
University of Georgia
Athens, GA 30620
(706) 542-4731
Pratt Cassity, Director

National Alliance of Statewide Preservation Organizations
c/o Historic Massachusetts
Old City Hall
Boston, MA 02108
(617) 723-3383
Alan G. Schwartz, Executive Director

National Association for Olmsted Parks
5010 Wisconsin Avenue, N.W.
Room 308
Washington, DC 20016
(202) 362-9511
Phyllis Knowles, Administrator

National Association of Conservation Districts
509 Capitol Court, N.E.
Washington, DC 20002
(202) 547-6223
Ernest Shea, Executive Vice President

National Association of Counties
440 1st Street, N.W.
Washington, DC 20001
(202) 393-6226
Larry Naake, Executive Director

National Association of Housing and Redevelopment Officials
1320 18th Street, N.W.
Washington, DC 20036
(202) 429-2960
Richard Y. Nelson. Jr., Executive Director

National Building Museum
Pension Building
401 F Street, N.W.
Washington, DC 20001
(202) 272-2448
Robert W. Duemling, President
and Director

National Center for Preservation Law
1333 Connecticut Avenue, N.W.
Suite 300
Washington, DC 20036
(202) 338-0392
Stephen Dennis, Executive Director

National Conference of State Historic Preservation Officers
444 North Capitol Street, N.W.
Suite 342
Washington, DC 20001-1512
(202) 624-5465
Eric Hertfelder, Executive Director

National Council for Preservation Education
Heritage Preservation Program
Georgia State University
Atlanta, GA 30303
(404) 651-3255
Timothy J. Crimmins, Rector

National Institute for the Conservation of Cultural Property
c/o Smithsonian Institution
900 Jefferson Drive, S.W.
Room 2225
Washington, DC 20560
(202) 357-2295
Lawrence L. Reger, President

National Parks and Conservation Association
1226 Massachusetts Avenue, N.W.
Suite 200
Washington, DC 20036
(202) 223-6722
Paul C. Pritchard, President

National Recreation and Park Association
2775 South Quincy Street
Suite 300
Arlington, VA 22206
(703) 820-4940
Dean Tice, Executive Director

National Trust for Historic Preservation
1785 Massachusetts Avenue, N.W.
Washington, DC 20036
(202) 673-4000
Richard Moe, President

National Wildlife Federation
1400 16th Street, N.W.
Washington, DC 20036
(202) 797-6800
Jay D. Hair, President

Natural Resources Defense Council
40 West 20th Street
New York, NY 10011
(212) 727-2700
John Adams, Executive Director

The Nature Conservancy
1815 North Lynn Street
Arlington, VA 22209
(703) 841-5300
John Sawhill, President

Organ Historical Society
National Office
P.O. Box 26811
Richmond, VA 23261
(804) 353-9226
William Van Pelt, Executive Director

Organization of American Historians
112 North Bryan Street
Bloomington, IN 47408
(812) 855-7311
Arnita Jones, Executive Secretary

Partners for Livable Places
1429 21st Street, N.W.
Washington, DC 20036
(202) 887-5990
Robert H. McNulty, President

Partners for Sacred Places
1616 Walnut Street
Suite 2310
Philadelphia, PA 19103
(215) 546-1288
Diane Cohen, Codirector

Pioneer America Society
c/o Department of Earth Sciences
Southeast Missouri State University
Cape Girardeau, MO 63701
(314) 651-2354
Michael Roark, Director

Popular Culture Association
Popular Culture Center
Bowling Green University
Bowling Green, OH 43403
(419) 372-2981
Ray B. Browne, Secretary-Treasurer

Preservation Action
1350 Connecticut Avenue, N.W.
Suite 401
Washington, DC 20036
(202) 659-0915
Nellie L. Longsworth, President

Project for Public Spaces
153 Waverly Place
4th Floor
New York, NY 10014
(212) 620-5660
Fred Kent, President

Public Works Historical Society
1313 East 60th Street
Chicago, IL 60637
(312) 667-2200
Richard Sullivan, Executive Secretary

Railroad Station Historical Society
430 Ivy Avenue
Crete, NE 68333
(402) 826-3356
William F. Rapp, Editor

Scenic America
21 Dupont Circle, NW
Washington, DC 20036
(202) 833-4300
Sally Oldham, Executive Director

Sierra Club
730 Polk Street
San Francisco, CA 94109
(415) 776-2211
Michael Fischer, Executive Director

Small Towns Institute
P.O. Box 517
Ellensburg, WA 98926
(509) 925-1830
Kenneth Munsell, Director

Smithsonian Institution
Museum Reference Center
Arts and Industries
Room 2235
900 Jefferson Drive, S.W.
Washington, DC 20560
(202) 357-3101
Sylvia Churgin, Librarian

Society for American Archaeology
808 17th Street, N.W.
Suite 200
Washington, DC 20006
(202) 223-9774
Jerome A. Miller, Executive Director

Society for Commercial Archeology
Room 5010
National Museum of American History
Washington, DC 20560
(202) 882-5424
Rebecca Shiffer, President

Society for Historical Archaeology
P.O. Box 704
Egg Harbor, NJ 08215
(609) 965-9073
Bud Walker, President

Society for Industrial Archeology
c/o Smithsonian Institution
National Museum of American History
Room 5014
Washington, DC 20560
(202) 357-2228
Helena Wright, Curator

Society for the Preservation of Mills
Gardiner Road
Wiscasset, ME 04578
(207) 882-7260
Donald W. Martin, President

Society for the Preservation of New England Antiquities
141 Cambridge Street
Boston, MA 02114
(617) 227-3956
Nancy Coolidge, Director

Society of American Historians
603 Fayerweather Hall
Columbia University
New York, NY 10027
(212) 854-2555
Kenneth T. Jackson, Executive Secretary

Society of Architectural Historians
1232 Pine Street
Philadelphia, PA 19107-5944
(215) 735-0246
David Bahlman, Executive Director

Society of Professional Archeologists
P.O. Box 7807
Wake Forest University
Winston-Salem, NC 27109
(919) 761-5497
J. Ned Woodall, President

Trust for Public Land
116 New Montgomery Street
4th Floor
San Francisco, CA 94105
(415) 495-4014
Martin J. Rosen, President

Urban Land Institute
625 Indiana Avenue, N.W.
Suite 400
Washington, DC 20004-2930
(202) 624-7000
Rick Slosan, Executive Vice President

U.S. Lighthouse Society
244 Kearney Street
5th Floor
San Francisco, CA 94108
(415) 362-7255
Wayne C. Wheeler, President

The Victorian Society in America
219 East 6th Street
Philadelphia, PA 19106
(215) 627-4252
Judith Snyder, Executive Director

The Wilderness Society
1400 I Street, N.W.
Washington, DC 20005
(202) 842-3400
George Frampton, President

ADVISORY COUNCIL ON HISTORIC PRESERVATION

Marcia Smith

"The head of any Federal agency having direct or indirect jurisdiction over a proposed Federal or federally assisted undertaking in any State and the head of any Federal department or independent agency having authority to license any undertaking shall prior to the approval of the expenditure of any Federal funds on the undertaking or prior to the issuance of any license...take into account the effect of the undertaking on any district, site, building, structure or object that is included in or eligible for inclusion in the National Register. The head of any such Federal agency shall afford the Advisory Council on Historic Preservation established under Title II of this Act a reasonable opportunity to comment with regard to such undertaking."

Section 106, National Historic Preservation Act of 1966, as amended

Advisory Council on Historic Preservation
1100 Pennsylvania Avenue, N.W.
Suite 809
Washington, DC 20004
(202) 606-8503

An independent federal agency, the council is the primary policy adviser to the president and Congress on historic preservation. The council's main function is to review and comment on federal and federally assisted and licensed projects that affect properties listed in or eligible for the National Register of Historic Places, as provided under Section 106 of the National Historic Preservation Act of 1966. The council also makes annual recommendations on how to improve the national preservation program; publishes special reports and studies on topics of interest to preservationists; and provides advice and technical assistance to Congress on legislative proposals relating to historic preservation.

In addition, the council coordinates U.S. membership in the International Centre for the Study of the Preservation and Restoration of Cultural Property. A three-day training program, "Introduction to Federal Projects and Historic Preservation Law," is offered by the council throughout the country for federal, state and local officials who encounter preservation-related law in their jobs. The council has also cosponsored with the U.S. Department of the Interior an awards program recognizing achievements in privately funded and federally assisted preservation projects.

MEMBERS

John C. Harper, Chairman

Joan Wellhouse Stein, Vice Chairman

Katherine Elkins Boyd

Lucille Clarke Dumbrill

Avery Faulkner

Lynn Kartavich

Dennis F. Mullins

John Reynolds, III

Gov. Michael N. Castle, Delaware

Mayor William J. Althaus, York, Pennsylvania

Chairman, National Trust for Historic Preservation

President, National Conference of State Historic Preservation Officers

Secretary of Agriculture

Secretary of Housing and Urban Development

Secretary of the Interior

Secretary of Transportation

Secretary of the Treasury

Architect of the Capitol

Director, Office of Administration

STAFF

Executive Director: Robert D. Bush

General Counsel and Deputy Executive Director: John M. Fowler

Director, Office of Communications and Publications: Marcia A. Smith

Director, Office of Program Review and Education: Ronald D. Anzalone

Director, Eastern Office of Project Review: Don L. Klima

Director, Western Office of Project Review: Claudia Nissley (303) 231-5320 (730 Simms Street, Room 401, Golden, CO 80401)

A BRIEF LOOK AT SECTION 106 REVIEW

How does Section 106 review work? Federal regulations spell out the specific process by which an agency affords the council an opportunity to comment on the agency's proposed activity. The council's regulations, "Protection of Historic Properties," appear in the U.S. Code of Federal Regulations at 36 CFR Part 800. These regulations were revised and reissued on September 2, 1986 (51 FR 31115). A simplified look at the process follows.

Before beginning the specific steps of Section 106 review, the agency determines that its proposed activity is an "undertaking" as defined by council regulations and establishes the area that the undertaking may potentially affect. The agency then proceeds with the review process:

1. Identification and evaluation of historic properties

First, the agency reviews all the available information that could help determine whether historic properties may be located in the area of the proposed activity. Based on this review, the agency decides whether any additional survey work is needed to locate possible historic properties.

Next, the agency identifies all properties listed in the National Register that might be affected by the proposed activity. The agency also identifies properties not actually listed in the Register but that appear to meet eligibility criteria. The agency and the state historic preservation officer (SHPO) then apply the National Register criteria (found at 36 CFR 60.4) to decide whether the properties are eligible for listing and thus subject to Section 106 review. When eligibility is in question, the agency may obtain a final determination of eligibility from the keeper of the National Register. The agency documents the results of its identification and evaluation findings.

2. Assessment of effects

Once historic properties have been identified and found to meet National Register criteria, the federal agency determines whether its proposed activity will affect them in any way. The agency again

works with the SHPO, making judgments based on criteria found in the council's regulations. There are three possible findings:

■ No effect: If there will be no effect of any kind on the historic properties, the agency notifies the SHPO and interested parties of its determination of no effect. If the SHPO does not object, the agency proceeds with the project.

■ No adverse effect: If there could be an effect but the effect would not be harmful to any historic property, the agency obtains SHPO concurrence and submits to the council a determination of no adverse effect. Or, the agency can submit its determination of no adverse effect directly to the council for review and notify the SHPO of its determination. Unless the council objects, the agency proceeds with its project or activity.

■ Adverse effect: If there could be a harmful effect to a historic property, the agency begins the consultation process.

3. Consultation

During this step, an effort is made to find acceptable ways to reduce the harm to the historic properties. The consulting parties are the agency and the SHPO; council involvement in consultation is optional. Other interested parties (such as a local government, Indian tribe or federal applicant for a grant, license or permit) may also be invited to join the consultation and must be invited under certain circumstances.

The agency gathers needed documentation, informs the public that consultation is under way, and works with the consulting parties to find a solution. When the consulting parties have agreed on steps to avoid or reduce harm to historic properties, they sign a memorandum of agreement.

In some instances, the consulting parties cannot agree on a solution, in which case the consultation is terminated. The agency may then submit documentation to the council and request the issuance of written council comments.

4. Council comment

Unless the council has already signed the agreement (by virtue of being a consulting party), the agency submits the signed agreement to the council for review. The council can accept the agreement, request changes to it, or opt to issue written comments on the proposed activity.

If the consulting parties have terminated consultation, the council issues written comments about the proposed action directly to the head of the agency.

5. Proceed

If the Section 106 review process has resulted in a council-accepted memorandum of agreement, the agency proceeds with its proposed activity according to the terms of the agreement. In the absence of an agreement, the agency must take into account the council's written comments, after which the agency makes the final decision about how (or whether) to proceed with its proposed activity. The agency notifies the council of its decision.

Either outcome concludes the Section 106 review process and satisfies the agency's statutory responsibilities under Section 106 of the National Historic Preservation Act of 1966.

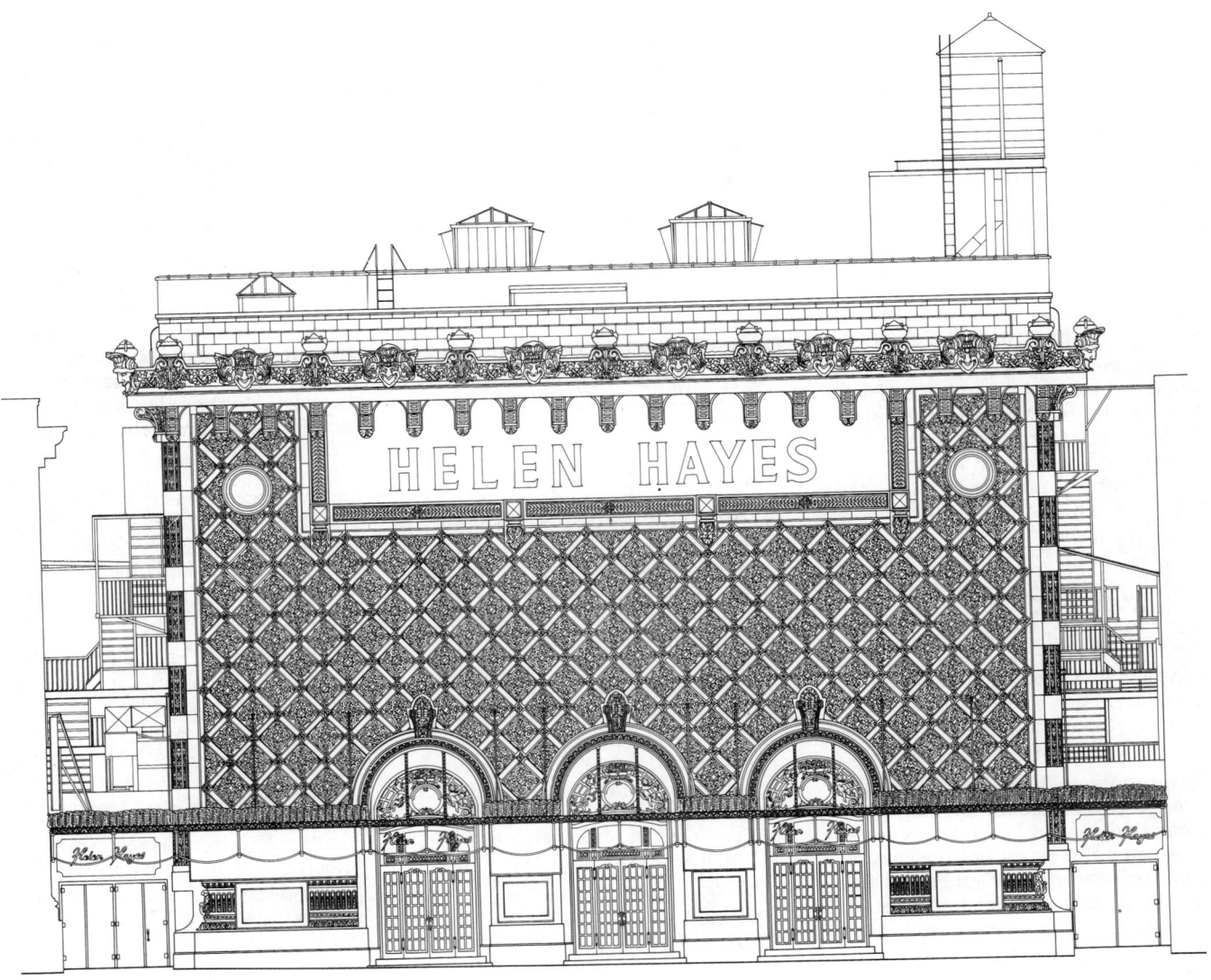

Helen Hayes Theater, New York, New York. (R. Hartman, Meadows/Woll Architects, HABS)

NATIONAL PARK SERVICE

Richard Wagner

"...the historical and cultural foundations of the nation should be preserved as a living part of our community life and development in order to give a sense of orientation to the American people."

National Historic Preservation Act of 1966, Section 1, 16 USC 470

The National Park Service (NPS) is the principal federal agency responsible for historic preservation. Part of the Department of the Interior, NPS administers the national park system and is responsible for a number of programs that assist privately held historic resources.

National Park Service
18th and C Streets, NW
Washington, DC 20840
(202) 208-4621
James M. Ridenour, Director

Associate Director for Cultural Resources
(202) 208-7625
Jerry L. Rogers

Anthropology Division. This division is responsible for developing archaeological and ethnographic program policies, guidelines, and standards for the NPS and monitoring their operations in the parks and centers. A principal goal of the division is to develop effective partnerships with Native Americans and other ethnic communities to promote the conservation of natural and cultural resources.

Anthropology Division
(202) 343-8161
Douglas H. Scovill, Chief

Archaeological Assistance Division. This division provides federal and state agencies with technical assistance on the identification, evaluation, and preservation of archaeological sites and properties. It can provide limited technical assistance on special projects for the recovery of important archaeological data threatened by federal undertakings; it plans meetings and programs to coordinate federal archaeological activities; it maintains the National Archaeological Database; and it prepares publications on archaeological subjects.

Archaeological Assistance Division
(202) 343-4101
Frank McManamon, Chief

Certified Local Government Program. Local governments that have established an historic preservation commission and program that meets certain federal and state standards are eligible to participate in the Certified Local Government (CLG) program. Participation in the program provides recognition for local government preservation efforts and allows CLGs to apply for earmarked grants, to participate in the National Register nomination process, and to receive technical assistance and training.

Certified Local Government Program
(202) 343-9505
Steve Morris, Coordinator

Curatorial Services Division. The Curatorial Services Division develops and coordinates policies, standards, and procedures for the management of NPS museum collections. To accomplish its mission, the division maintains the Automated National Catalog Systems, develops and conducts curatorial training programs, produces the technical information series *Conserve O Grams*, and provides NPS parks and centers with technical assistance on the care and maintenance of museum collections.

Curatorial Services Division
(202) 343-8138
Ann Hitchcock, Chief Curator

Historic American Buildings Survey/Historic American Engineering Record Division. The Historic American Buildings Survey (HABS) and Historic American Engineering Record (HAER) Division produces measured drawings, large-format photographs, and written histories of historic sites, buildings, structures, and objects that are significant to the architectural, engineering, and industrial heritage of the United States. Since 1935 the division has documented more than 22,000 historic resources. The collection is housed in the Prints and Photographs Division of the Library of Congress in Washington, DC.

Historic American Buildings Survey/Historic American Engineering Record Division
(202) 343-9606
Robert J. Kapsch, Division Chief

Historic American Buildings Survey
(202) 343-9611
Paul Dolinsky, Chief HABS

Historic American Engineering Record
(202) 343-9603
Eric N. Delony, Chief HAER

History Division. This division administers the National Historic Landmarks program, which identifies and designates the most significant historic sites, buildings, structures, and objects in the country. The division also conducts programs to document and interpret the history of the NPS, manages a maritime preservation program, and participates in developing policies and guidelines related to the management of cultural resources within the National Park system.

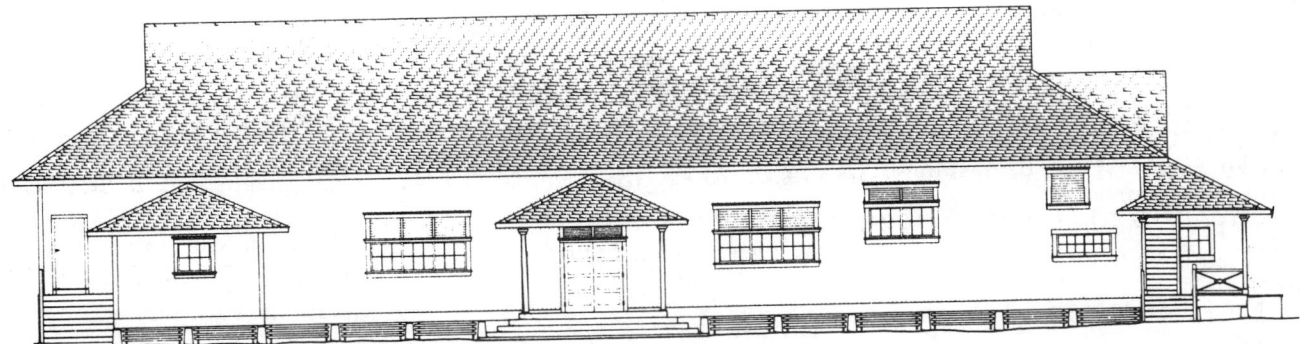

Elevation of Paschoal Hall, Kalaupapa National Historic Park, Molokai, Hawaii. (Angela Hasenyager and Katharine Slocumb, HABS.)

History Division
(202) 348-8167
Edwin C. Bearss, Chief

National Historic Landmarks
Program
(202) 343-8164
Benjamin Levy, Senior Historian

Interagency Resources Division.
The two principal branches within this division are the National Register of Historic Places program and the Preservation Planning Branch. The Interagency Resources Division assists federal, state, and local governments, private groups and organizations, and individuals with a wide range of preservation activities. Among its responsibilities are: establishing standards and guidelines for the identification, evaluation, and protection of historic properties; administering the National Register program; administering the Historic Preservation Fund (a grant-in-aid program); establishing preservation planning standards and guidelines; and providing technical assistance and sponsoring workshops on preservation issues and topics.

Interagency Resources Division
(202) 343-9500
Lawrence E. Aten, Chief
Arthur Stewart, Deputy Chief

National Register of Historic
Places
(202) 343-9536
Carol D. Shull, Chief

Preservation Planning Branch
(202) 343-9595
De Teel Patterson Tiller, Chief

Park Historic Architecture Division. This division conducts activities related to the preservation of historic and prehistoric structures, buildings, sites, objects, and cultural landscapes located within the National Park system. It develops and monitors the application of policies, guidelines, and standards; develops preservation techniques and methods; conducts training programs for NPS employees; maintains an inventory of historic resources in parks and centers; and provides oversight for the Historic Properties Leasing Program.

Park Historic Architecture
Division
(202) 343-8146
Randall Biallas, Chief

Preservation Assistance Division.
The Preservation Assistance Division conducts a variety of activities and programs for federal, state, and local governments, the NPS, and the general public. Among them are developing and publishing technical information on preservation; providing training on preservation techniques and methods; administering the federal Investment Tax Credit (Preservation Tax Incentives) program; and managing the Historic Preservation Fund.

Preservation Assistance Division
(202) 343-9573
E. Blaine Cliver, Chief

Technical Preservation Services
Branch
(202) 343-9584
H. Ward Jandl, Chief

Grants Administration Branch
(202) 343-9564
Joseph Wallis, Acting Chief

National Park Service Regional Offices
The following NPS regional offices manage the Preservation Tax Incentives program for the rehabilitation of income-producing National Register property and can provide information on the other programs and divisions listed above.

Mid-Atlantic Regional Office
2nd and Chestnut Streets
2nd Floor
Philadelphia, PA 19106
(215) 597-5129
Connecticut, Delaware, District of Columbia, Indiana, Maine, Maryland, Massachusetts, Michigan, New Hampshire, New Jersey, New York, Ohio, Pennsylvania, Rhode Island, Vermont, Virginia, West Virginia.

Southeast Regional Office
75 Spring Street, NW
Atlanta, GA 30303
(404) 331-2632
Alabama, Arkansas, Florida, Georgia, Kentucky, Louisiana, Mississippi, North Carolina, South Carolina, Tennessee, Puerto Rico, Virgin Islands.

Rocky Mountain Regional Office
12795 West Alameda Parkway
P.O. Box 25287
Denver, CO 80225-0287
(303) 969-2875
Colorado, Illinois, Iowa, Kansas, Minnesota, Missouri, Nebraska, New Mexico, North Dakota, Oklahoma, South Dakota, Texas, Utah, Wisconsin, Wyoming.

Western Regional Office
600 Harrison, Suite 600
P.O. Box 36063
San Francisco, CA 84107
(415) 774-3988
   Arizona, California, Hawai'i,
   Idaho, Nevada, Oregon, Washing-
   ton, American Samoa, Guam.

Alaska Regional Office
2525 Gambell Street
Room 107
Anchorage, AS 99503
(907) 257-2687
   Alaska.

Additional Regional Centers
In addition to the regional centers
listed above, the NPS has the following centers dealing primarily
with NPS-owned properties.

National Capital Regional Office
1100 Ohio Drive, SW
Washington, DC 20242
(202) 619-7222

North Atlantic Regional Office
15 State Street
Boston, MA 02109
(617) 835-8800

Midwest Regional Office
1709 Jackson Street
Omaha, NE 68102
(402) 864-3431

Southwest Regional Office
Box 728
Santa Fe, NM 87501
(505) 476-6388

Pacific Northwest Regional
Office
83 South King Street, Suite 212
Seattle, WA 98104
(206) 399-5565

Harpers Ferry Center
National Park Service
Harpers Ferry, WV 25425
(304) 925-6588, ext. 621

Denver Service Center
National Park Service
P.O. Box 25287
Denver, CO 80225
(303) 969-2100

NATIONAL TRUST FOR HISTORIC PRESERVATION

Byrd Wood

"The National Trust will not only preserve property, but it will also contribute to the perpetuation of the historical fabric of our national life in a way that only the mobilized participation of all interested societies and individual citizens can do."

Report of the Committee on Organization of the National Trust, 1948

National Trust for Historic Preservation
1785 Massachusetts Avenue, N.W.
Washington, DC 20036
(202) 673-4000

The National Trust, a private, non-profit organization, is the only national preservation organization chartered by Congress (in 1949) to encourage public participation in the preservation of sites, buildings, and objects significant in American history and culture. The National Trust acts as a clearinghouse of information on all aspects of preservation, assists in coordinating efforts of preservation groups, provides professional advice on preservation, conducts conferences and seminars, maintains 17 historic properties as museums, administers grant and loan programs, and issues a variety of publications. Seven regional and field offices provide localized preservation advisory services and represent the National Trust in their regions. A board of trustees directs the organization's policies, and a board of advisors representing the states helps implement programs. Financial support for National Trust programs comes from membership dues, endowment funds, contributions, and matching grants from federal agencies, including the U.S. Department of the Interior, National Park Service, under the National Historic Preservation Act of 1966. National Trust headquarters are in the McCormick Apartments—Andrew Mellon Building (1915–17, Jules Henri de Sibour) in Washington, D.C.

Lyndhurst, Tarrytown, New York. (T. Price, R. Ericson, HABS)

OFFICES, PROGRAMS, AND SERVICES

During the past 40 years, the scope of programs at the National Trust has expanded dramatically to include downtown revitalization, support for state and local preservation organizations, litigation, advocacy, tourism, growth management, and rural and maritime preservation. We have rallied to save Civil War battlefields; testified on Capitol Hill to protect underwater shipwrecks; marched in South Pasadena, California, to protest construction of a freeway; sent emergency assistance to disaster-stricken Charleston, St. Croix, and northern California; charted the future of James Madison's home, Montpelier; funded inner-city housing; and argued in court to halt a proposed elevated highway along the historic waterfront of Mobile, Alabama.

With such a diversity of activities, it can be difficult to determine which office to call with your preservation questions. The listing of offices, programs, and services, below, will help direct you to the right place. Each entry includes a phone number at headquarters in Washington, D.C.. The best place to start in most instances, however, is with your regional office. Staff members there can help direct your inquiry to the appropriate office or program. Regional offices advise state and local preservation organizations on preservation issues; organize field visits, workshops, and seminars of interest to the region; and conduct special projects focusing on regional historic preservation resources and issues.

Advertising: To place an ad in *Historic Preservation News* or *Historic Preservation* magazine call: (202) 673-4068. To advertise in the National Preservation Conference brochure, call (800) YES-NTHP.

Advisors: See Board of Advisors.

Advocacy: See Center for Preservation Policy Studies.

Award Programs: See BARN AGAIN!, Crowninshield Award, Honor Awards, Great American Home Awards.

BARN AGAIN!: Award program cosponsored by the National Trust and *Successful Farming* magazine to encourage farmers to restore and adapt their barns to new agricultural uses. For information, call (303) 623-1504.

Bed and Breakfasts: *The National Trust Guide to Historic Bed & Breakfasts, Inns, and Small Hotels*, published in 1992 by the National Trust, lists more than 500 historic bed-and-breakfast inns. To order copies of the book, call Mail Order at (202) 673-4200 or (800) 274-3694. To order bulk quantities wholesale, call Preservation Press at (202) 673-4200. To be listed in the next edition, call (202) 673-4169.

Bequests: See Planned Giving.

Board of Advisors: Network of volunteer preservation leaders in each state who support the work of the National Trust. Each state, the District of Columbia, Puerto Rico, and the Virgin Islands has two advisors. Call your National Trust regional office or (202) 673-4120 for more information.

Board of Trustees: The National Trust is governed by a Board of Trustees elected by the membership at the annual membership meeting. For information, call (202) 673-4105.

Books: See Preservation Press.

Bookshop: See Decatur House Shop.

Center for Historic Houses: Provides information and sponsors educational programs for National Trust members who are private owners of old or historic houses. The Center publishes the Old House Starter Kit, a guide to restoring or rehabilitating your old house. For information, call (202) 673-4021. See Old House Starter Kit.

Center for Preservation Policy Studies: Provides research and testimony on pending federal legislation, advocacy of preservation issues, coordination of national lobbying network, and research on preservation policy issues. For information, call (202) 673-4254.

See Legislative Information Hotline.

Christmas Cards: The National Trust's holiday cards are available exclusively by mail. To order cards, call (800) 632-4200. To have your name added to the mailing list, call (202) 673-4267.

Churches: See Inspired Partnerships

COEP: See Community Preservation Organization Effectiveness Program.

Commercial Buildings: See National Main Street Center.

Communications: The National Trust's Office of Communications carries out media-relations programs to promote National Trust activities, coordinates Preservation Week activities, and provides information to National Trust members. For information, call (202) 673-4141.

Community Preservation Organization Effectiveness Program (COEP): National Trust program that helps communities increase their organizational effectiveness to better achieve their preservation goals. Call your National Trust regional office for more information.

Computer Network: See Preserve Link.

Conferences: The National Trust offers a variety of educational conferences, seminars, workshops, and outreach programs including the annual National Preservation Conference and ReHabitat. For information, call (800) YES-NTHP. For information on regional and state preservation conferences, call your National Trust regional office.

Conference Room Rental: The National Trust headquarters building in Washington, D.C., has two conference rooms, the Boardroom (40-person conference seating, 90-person theater seating) and the Executive Conference Room (15-person conference seating), available to outside groups for a reasonable fee. For information, call (202) 673-4018.

Congressional Liaison: See Center for Preservation Policy Studies.

Critical Issues Fund (CIF): The Critical Issues Fund supports innovative research and model projects that examine the most serious issues facing the historic preservation movement today. The program's goal is to integrate preservation values into public policy at the local, state, and national levels. Call your National Trust regional office or (202) 673-4255 for information.

Crowninshield Award: Named for Louise DuPont Crowninshield, an avid preservationist, this is the National Trust's highest recognition for superlative lifetime achievement in the preservation movement, including interpretation of our historic, architectural, and maritime heritage. Winners are selected by the preservation committee of the National Trust's board of trustees. Although there is no formal nomination process, the committee chairman welcomes candidate suggestions. For information, call (202) 673-4162.

Decatur House Shop: The National Trust's primary book and gift shop is located at Decatur House, 1600 H Street, N.W., in downtown Washington. A full selection of National Trust and Preservation Press publications is available along with an extensive selection of books and gifts relating to preservation, architecture, and history. For information on shop hours, call (202) 842-1856. To order books or gifts by mail, call Mail Order at (800) 274-3694. In the Washington area, call (202) 673-4200.

Deferred Gifts: See Planned Giving.

Design Competition: See Student Competitions.

Disaster Relief: The National Trust has provided technical assistance and emergency funding when natural disasters threaten or destroy historic sites. Call your National Trust regional office for information.

Downtown Revitalization: See National Main Street Center.

Easements: The National Trust helps organizations initiate and administer preservation easement programs along with other protection tools. The National Trust also maintains its own active easement program and holds preservation easements on more than 70 properties. For information, call (202) 673-4035.

Education: See Educational Supplement and Heritage Education Initiative.

Educational Supplement: A list of historic preservation degree programs in colleges, universities, and preservation training centers is published annually in *Historic Preservation News*. To receive a copy, call (202) 673-4296.

Employment: See Human Resources.

Endangered Places: The National Trust maintains a list of the Eleven Most Endangered Historic Places in the nation. The list is updated periodically. Call your National Trust regional office for more information.

Fax Numbers: Fax numbers most often requested at headquarters are: Washington office main number, (202) 673-4038; Publications, (202) 673-4172; President's Office, (202) 673-4082; and Programs, Services, and Information, (202) 673-4223. Contact your regional office for its fax number.

Funding: See Preservation Services Fund, Critical Issues Fund, National Preservation Loan Fund, Inner-City Ventures Fund, or call your National Trust regional office.

Gift Shop: The National Trust operates gift shops at each of its historic properties. The shops carry gifts and books relating to the history of the specific site and region. See Decatur House Shop.

Grants: See Preservation Services Fund, Critical Issues Fund, Inner-City Ventures Fund, or call your National Trust regional office.

Great American Home Awards: Annual recognition program for outstanding residential rehabilitation projects in the United States cosponsored by the National Trust's Center for Historic Houses and *Historic Preservation* magazine. Winners are announced in *Historic Preservation*. For information and nomination forms, call (202) 673-4283.

Heritage Education Initiative: The National Trust maintains an information database on heritage education resources, supports model programs, and develops educational materials related to heritage education. For information, call (202) 673-4040.

Heritage Society: National Trust membership program for members who contribute $1,000 or more annually. For information, call (202) 673-4188.

Historic District Commissions: The National Trust's state and local preservation program works to strengthen state preservation legislation and local preservation ordinances. For information, call (202) 673-4255.

Historic Hotels of America: Membership program for hotels that are distinguished by their historic character and architectural quality. To inquire about membership or obtain a hotel directory call (202) 673-4267. To make reservations at participating hotels, call (800) 678-8946.

Historic Houses: See Center for Historic Houses.

Historic Preservation: Bimonthly magazine published by the National Trust. Calendar listings are published without charge, space permitting. Send items to *Historic Preservation*, Calendar, National Trust for Historic Preservation, 1785 Massachusetts Avenue, N.W., Washington, DC 20036. For back issues, call (202) 673-4069. To place an ad, call (202) 673-4068.

Historic Preservation Forum: Bimonthly journal published by the National Trust for professional and organizational preservation constituents. Preservation Forum members receive copies of the journal among other membership benefits. To join or to order back issues, call (202) 673-4296.

Historic Preservation News:
Monthly newspaper published by the National Trust. To place an ad, call (202) 673-4068. For back issues, call (202) 673-4164.

Historic Properties: The National Trust owns and operates 17 historic properties across the country. For information, call (202) 673-4151 or call the properties directly for hours of operation.

Historic Real Estate Program:
National Trust membership program for real estate companies that provides realtors with information and training to assist owners and buyers of historic buildings. For information, call (202) 673-4169.

Historic Rehabilitation Tax Credit:
Federal income tax incentives are available for preservation projects designated by the Secretary of the Interior as rehabilitations of certified historic structures. For information call your state historic preservation office or the National Park Service at (202) 343-9573. A National Trust publication, *A Guide to Tax-Advantaged Rehabilitation*, explains how the historic rehabilitation tax credit works. To order a copy, call (202) 673-4189.

Honor Awards: The National Trust's annual recognition program for organizations and individuals whose projects demonstrate outstanding dedication and commitment to excellence in historic preservation. Awards are presented at the National Preservation Conference. For information and nomination forms, call (202) 673-4162.

Hotels: See Historic Hotels of America.

Human Resources: For information on volunteer opportunities, job openings, careers in preservation, and internships, call (202) 673-4120.

Information Series: National Trust publications that provide concise information on basic and frequently used preservation and organizational development techniques. To order a copy or request a list of titles, call (202) 673-4189.

Inner-City Ventures Fund: Matching grants and low-interest loans to nonprofit community organizations to help revitalize older historic neighborhoods for the benefit of low-and moderate-income residents. Call your National Trust regional office for more information.

Insurance: Insurance programs available to Preservation Forum members include property, employee benefits, health, directors and officers, and maritime insurance. For information call (202) 673-4296. For information on insurance for private homeowners, call 1-800-899-6004.

Internships: See Human Resources.

Inspired Partnerships: A three-year demonstration program to provide a wide range of training, technical services, and information designed to promote effective stewardship of older religious properties in the Chicago area. Inspired Partnerships is now a separate nonprofit organization and a grantee of the National Trust. For information, call (312) 294-0077.

Investment Tax Credit: See Historic Rehabilitation Tax Credit.

Jobs: See Human Resources.

Leadership Training: The National Trust offers periodic workshops on organizational development and preservation leadership training. For information on upcoming workshops, call your National Trust regional office or (202) 673-4162.

Legal Services: The legal department provides education and advice on preservation law and historic districts; zoning; the monitoring of preservation litigation; intervention as an *amicus curiae* (friend of the court) in litigation; and advice on tax incentives and easements. For information, call your National Trust regional office or (202) 673-4035.

Legislative Information Hotline: To obtain updates on preservation legislation or to place an order for copies of legislative bills, testimony, fact sheets, and other public-policy information, call 1-800-765-NTHP.

Library: The National Trust's library is located at the University of Maryland's College Park campus and is maintained as a separate collection in the Architecture Library. For information, call (301) 405-6319.

Licensing: The National Trust has a licensing program that offers historic reproductions of gifts, accessories, and home furnishings. For information on the licensees and where to find their products, call (202) 673-4267. For information on how to become a licensee, call (202) 673-4096.

Lighthouses: See Maritime.

Litigation: The National Trust regularly enters court cases throughout the United States in support of local preservation organizations when an issue of national significance is involved. For information, call your National Trust regional office or (202) 673-4035.

Loan Programs: See Inner-City Ventures Fund, National Preservation Loan Fund.

Lobbying: See Center for Public Policy Studies.

Mail Order: To order books and gifts seen in advertisements or catalogs, call (800) 274-3694 (in the Washington area, call (202) 673-4200)

Main Street: See National Main Street Center.

Main Street Certification Institute: Advanced, professional certification program for Main Street managers. For information, call (202) 673-4219.

Main Street Network: See National Main Street Network.

Main Street News: Monthly newsletter of the National Main Street Center. For information, call 202-673-4219.

Mailing Lists: To exchange or purchase mailing lists, contact the following departments: Preservation Forum, (202) 673-4296; National

Trust Membership, (202) 673-4184; National Main Street Network, (202) 673-4219.

Maritime Program: Offers technical assistance and field services on maritime preservation issues and awareness of maritime heritage. For information, call (202) 673-4127.

Membership Inquiries: For questions concerning new membership, renewals or change of address call, (202) 673-4129; for questions concerning membership in Preservation Forum, call (202) 673-4296; for questions concerning National Main Street Network membership, call (202) 673-4219.

Media Relations: See Communications.

Museum Properties: See Historic Properties.

National Main Street Center: National Trust program to stimulate economic revitalization of business districts in the context of historic preservation. The center conducts training courses, provides technical assistance to states, towns, and cities and helps build business and government partnerships. For information, call (202) 673-4219.

National Main Street Network: Organizational membership program through which communities receive current information on business district revitalization techniques and activities. Members receive the monthly newsletter, *Main Street News*, and technical advice and referrals on issues of local concern. For information, call (202) 673-4219.

National Preservation Conference: Annual conference organized by the National Trust that includes educational sessions, affinity group meetings, Rehabitat, tours, annual meeting of the National Trust, and special events. Future National Preservation Conferences will be held in St. Louis, Missouri, September 22–26, 1993; Boston, October 26–30, 1994; and Fort Worth, Texas, October 11–15, 1995. For information, call (800) YES-NTHP.

National Preservation Loan Fund: Below-market-rate loans to nonprofit organizations and public agencies to help preserve properties listed, or eligible for listing, in the National Register of Historic places. For information, call your National Trust regional office.

National Preservation Week: See Preservation Week.

National Town Meeting: Annual conference of Main Street revitalization professionals and volunteers. For information, call (202) 673-4219.

Neighborhood Business Districts: See National Main Street Center.

Old House Starter Kit: See Center for Historic Houses.

Personnel: See Human Resources.

Planned Giving: The National Trust welcomes donations in the form of real estate, cash, securities, personal property, and bequests. For information, call (202) 673-4188.

Preservation Forum: National Trust membership program designed specifically for preservation professionals and organizations. Benefits include subscriptions to *Historic Preservation Forum*, *Historic Preservation*, *Historic Preservation News*, participation in financial/insurance assistance programs, technical advice, and substantial discounts on professional conferences and publications. For membership information, call (202) 673-4296.

Preservation Law Reporter: Legal periodical on federal, state, and local preservation legislation and administrative decisions. Includes reports on recent legislative developments, comprehensive model ordinances, and easement provisions. Reduced rates for National Trust Forum members. For information, call (202) 673-4035.

Preservation Press: The book-publishing division of the National Trust. For general information, call (202) 673-4057; for trade marketing, call (202) 673-4066; for wholesale customer service, call (202) 673-4058; for individual orders, call Mail Order at (800) 274-3694. (In the Washington area, call (202) 673-4200.)

Preservation Services Fund: Matching grants to nonprofit organizations, universities, and public agencies to initiate preservation projects. Call your regional office for more information.

Preservation Shops: See Decatur House Shop.

Preservation Week: Usually celebrated the second full week in May, Preservation Week is a nationwide celebration of our local, state, and national heritage and efforts to preserve them. For a Preservation Week Kit, call (202) 673-4141.

PRESERVE LINK: Computer-based communication system designed by the National Trust for the preservation community. It provides computer access to information and allows local, state, and national preservation organizations to network together. For information call (202) 673-4067.

Public Relations: See Communications.

Railroad Depots: The National Trust's Mountains/Plains regional office offers technical assistance for organizations or individuals interested in preserving historic railroad depots. For information, call (303) 623-1504.

Real Estate: See Historic Real Estate Program.

ReHabitat: A national exposition of products, service, skills, and crafts for restoring, renovating, and maintaining old and historic buildings and sites. Rehabitat takes place during the National Preservation Conference. For information, call (800) YES-NTHP.

Regional Offices: See list below.

Religious Properties: See Inspired Partnerships

Reproductions: See Licensing.

Rural Preservation: The National Trust's rural heritage initiative offers support to rural communities on preservation issues and public policy. Call your National Trust regional office for information or (202) 673-4037.

Seminars: See Conferences.

Student Competitions: The National Trust sponsors a design competition and a papers competition for students enrolled in college and university historic preservation programs. For more information call (202) 673-4162.

Study Tours: The National Trust hosts study tours throughout the world focusing on architecture and historic preservation. Open to members and nonmembers. See also Work/Study Projects. For information, call (202) 673-4138.

Tax Credits: See Historic Rehabilitation Tax Credit.

Tourism: The National Trust's Tourism Initiative serves 16 pilot projects in Indiana, Tennessee, Texas, and Wisconsin and conducts a national awareness program in the responsible use of historic resources in tourism development. For information, call (303) 623-1504.

Tours: See Study Tours.

Volunteers: See Human Resources or call individual National Trust historic properties.

Workshops: See Conferences.

Work/Study Projects: Work/study projects take place at National Trust properties during the summer months. Participants have the opportunity to join the work of National Trust properties and learn firsthand the process of preservation. Each project is led by an expert director and includes lectures, tours, meetings, and instruction in the project at hand. For information, call (202) 673-4138.

Zoning: See Legal Services.

EXECUTIVE STAFF

President: Richard Moe
Finance and Administration: Senior Vice President, Bonnie R. Cohen
Law and Public Policy: Vice President and General Counsel, David A. Doheny
Membership and Communication: Vice President, Gerald F. Dunaway

Programs, Services, and Information: Vice President, Peter H. Brink
Resources Development: Vice President, Jane A. Couch
Stewardship of Historic Properties: Vice President, Frank E. Sanchis III

REGIONAL OFFICES

Northeast Regional Office
Seven Faneuil Hall Marketplace
5th Floor
Boston, MA 02109
(617) 523-0885
Vicki Sandstead, Director
Connecticut, Maine, Massachusetts, New Hampshire, New York, Rhode Island, Vermont

Mid-Atlantic Regional Office
6401 Germantown Avenue
Philadelphia, PA 19144
(215) 438-2886
Patricia Wilson, Director
Delaware, District of Columbia, Maryland, New Jersey, Pennsylvania, Puerto Rico, Virginia, Virgin Islands, West Virginia

Southern Regional Office
456 King Street
Charleston, SC 29403
(803) 722-8552
Susan Kidd, Director
Alabama, Arkansas, Florida, Georgia, Kentucky, Louisiana, Mississippi, North Carolina, South Carolina, Tennessee

Midwest Regional Office
53 West Jackson Boulevard
Suite 1135
Chicago, IL 60604
(312) 939-5547
Tim Turner, Director
Illinois, Indiana, Iowa, Michigan, Minnesota, Missouri, Ohio, Wisconsin

Mountains/Plains Regional Office
511 16th Street
Suite 700
Denver, CO 80202
(303) 623-1504
Barbara Pahl, Director
Colorado, Kansas, Montana, Nebraska, North Dakota, Oklahoma, South Dakota, Wyoming

Texas/New Mexico Field Office
500 Main Street
Suite 606
Fort Worth, TX 76102
(817) 332-4398
Libby Barker Willis, Field Office Coordinator

Texas, New Mexico

Western Regional Office
1 Sutter Street
Suite 707
San Francisco, CA 94104
(415) 956-0610
Kathryn Burns, Director
Alaska, Arizona, California, Guam, Hawaii, Idaho, Micronesia, Nevada, Oregon, Utah, Washington

HISTORIC PROPERTIES

Chesterwood (constructed 1898, 1900–01; architect, Henry Bacon)
P.O. Box 827
Stockbridge, MA 01262-0827
(413) 298-3579
Paul Ivory, Director

Decatur House (constructed 1818–19; architect, Benjamin H. Latrobe)
748 Jackson Place, N.W.
Washington, DC 20006
(202) 842-0920
Vicki Sopher, Director

Drayton Hall (constructed 1738–42)
3380 Ashley River Road
Charleston, SC 29414
(803) 766-0188
George McDaniel, Director

Lyndhurst (constructed 1838, 1864–65; architect, Alexander J. Davis)
635 South Broadway
Tarrytown, NY 10591
(914) 631-0046
Susanne Brendel Pandich, Director

Montpelier (constructed around 1760; additions 1797, 1809–12 [architect, William Thornton], 1902)
P.O. Box 67
Montpelier Station, VA 22957
(703) 672-2728
Christopher Scott, Executive Director

Pope-Leighey House (constructed 1940; architect, Frank Lloyd Wright)
P.O. Box 37
Mount Vernon, VA 22121
(703) 780-4000
Linda Goldstein, Director

The Shadows-on-the-Teche (constructed 1831–34; architect, David Weeks)
P.O. Box 9703
New Iberia, LA 70562-9703
(318) 369-6446
Shereen Minvielle, Director

Woodlawn Plantation (constructed 1800–03; architect, William Thornton)
9000 Richmond Highway
Mount Vernon, VA 22309
(703) 780-4000
Linda Goldstein, Acting Director

NATIONAL REGISTER OF HISTORIC PLACES

Carol Shull

"The Secretary of the Interior is authorized to expand and maintain a National Register of Historic Places composed of districts, sites, buildings, structures, and objects significant in American history, architecture, archeology, engineering, and culture."

Section 101(a)(1)(A), National Historic Preservation Act of 1966, as amended

National Register of Historic Places
Interagency Resources Division
National Park Service
U.S. Department of the Interior
P.O. Box 37127
Washington, D.C. 20013-7127
(202) 343-9536
Carol D. Shull, Chief

STAFF

The National Register of Historic Places is the official list of the nation's historic and cultural resources worthy of preservation. The National Register was authorized by the National Historic Preservation Act of 1966 and is part of the national effort to identify, evaluate, and protect our architectural and archaeological resources. The program is administered by the National Park Service (see separate listing) under the Secretary of the Interior. Properties listed in the National Register include buildings, structures, sites, districts, and objects significant in American history, architecture, archaeology, engineering, and culture.

NATIONAL REGISTER NOMINATION PROCESS

1. Nominations may be prepared by property owners, interested citizens, preservation organizations, or state historic preservation officers (SHPO). Federal agencies submit nominations for their properties through their agency preservation officers. Eligible properties include districts, sites, buildings, structures, and objects of local, state, and national significance. Nominations may also be made for multiple historic resources, a flexible and efficient means of registering a number of significant properties linked by a common historical context or property type.

2. Standard National Register forms (available from the SHPO) must be used and submitted to the SHPO, which reviews the nomination against the National Register evaluation criteria. Certified local governments may also review nominations prior to submission to the SHPO.

3. The SHPO notifies property owners and local officials 30 to 75 days in advance that the state will present a nomination to a state review board. If a property meets the evaluation criteria, the board recommends it for nomination. The nomination form is signed by the SHPO and forwarded to the National Register office in Washington, D.C. The SHPO also may send the nomination for consideration even if it has been rejected by the board.

4. Private property owners, including the owners in a historic district, have the opportunity to concur in or object to National Register nominations. If the owner of an individual property or the majority of owners in a historic district object to listing by means of notarized letters, the property or district will not be listed but may be determined eligible for listing. Properties eligible for the National Register receive the same review by the Advisory Council on Historic Preservation as accorded to listed properties.

5. The staff of the National Register reviews nominations submitted by the SHPO. Provided the documentation is adequate and all procedures have been followed in accordance with federal regulations, a decision must be made on the eligibility of the property within 45 days of receipt. If found acceptable, the nomination form is signed and a notice is sent to members of Congress and the SHPO.

NATIONAL REGISTER EVALUATION CRITERIA

"The quality of significance in American history, architecture, archeology, and culture is present in districts, sites, buildings, structures, and objects that possess integrity of location, design, setting, materials, workmanship, feeling, and association, and:

1. That are associated with events that have made a significant contribution to the broad patterns of our history; or

2. That are associated with the lives of persons significant in our past; or

3. That embody the distinctive characteristics of a type, period, or method of construction, or that represent the work of a master, or that possess high artistic values, or that represent a significant and distinguishable entity whose components may lack individual distinction; or

4. That have yielded, or may be likely to yield, information important in prehistory or history.

"Ordinarily cemeteries, birthplaces, or graves of historical figures, properties owned by religious institutions or used for religious purposes, structures that have

Sunnyside, Washington Irving House, Irvington, New York.
(P. Burkhart, S. Dornbusch, HABS)

been moved from their original locations, reconstructed historic buildings, properties primarily commemorative in nature, and properties that have achieved significance within the past 50 years shall not be considered eligible for the National Register. However, such properties will qualify if they are integral parts of districts that do meet the criteria or if they fall within the following categories:

1. A religious property deriving primary significance from architectural or artistic distinction or historical importance; or

2. A building or structure removed from its original location but which is significant primarily for architectural value, or which is the surviving structure most importantly associated with a historic person or event; or

3. A birthplace or grave of a historical figure of outstanding importance if there is no other appropriate site or building directly associated with his productive life; or

4. A cemetery that derives its primary significance from graves of persons of transcendent importance, from age, from distinctive

design features, or from association with historic events; or

5. A reconstructed building when accurately executed in a suitable environment and presented in a dignified manner as part of a restoration master plan, and when no other building or structure with the same association has survived; or

6. A property primarily commemorative in intent if design, age, tradition, or symbolic value has invested it with its own historical significance; or

7. A property achieving significance within the past 50 years if it is of exceptional importance."

All historic areas in the National Park System are automatically listed in the National Register. Three nationally significant properties were exempted under the 1966 act: the buildings and grounds of the White House, the U.S. Supreme Court, and the U.S. Capitol.

BENEFITS OF NATIONAL REGISTER LISTING

1. Recognition that a property is of significance to the nation, state, or community.

2. Consideration in the planning for federally assisted projects, including review by the Advisory Council on Historic Preservation under Section 106 of the National Historic Preservation Act of 1966.

3. Eligibility for certain federal tax benefits, such as the investment tax credit for rehabilitation of income-producing buildings and the charitable deductions for donation of easements.

4. Qualification for federal preservation grants when funding is available.

5. Consideration in the issuance of surface coal mining permits.

National Historic Landmarks
Certain National Register properties are designated as National Historic Landmarks. They represent the nation's most important historic and cultural resources. National Historic Landmarks include buildings, sites, structures, and objects that possess exceptional value or quality in American history, architecture, archaeology, engineering, and culture.

FURTHER READING

Guidelines for Completing National Register of Historic Places Forms. National Register of Historic Places. Washington D.C.: U.S. Department of the Interior, National Park Service.

National Register of Historic Places: Cumulative List 1966–1988. Nashville: American Association for State and Local History, 1989.

National Register of Historic Places Index on CD-ROM. Mineral, Va.: Buckmaster Publishing, 1989.

National Register of Historic Places Subscription. Nashville: American Association for State and Local History, 1988–90.

INTERNATIONAL AND FOREIGN PROGRAMS

Julie Griffith

"In many respects preservation programs here and abroad are more remarkable for their similarities than for their differences."

Robert E. Stipe, *Historic Preservation in Foreign Countries*, 1983

INTERNATIONAL

Aga Khan Program for Islamic Architecture
School of Architecture
Massachusetts Institute of Technology
77 Massachusetts Avenue
Cambridge, MA 02139

Canadian Parks Service
National Historic Sites Directorate
25 Eddy Street
Ottawa, Ontario K1A 0H3 Hull, Quebec
(819) 994-1808

Council of Europe
Directorate for Environment and
Local Authorities
BP 431 R6
67006 Strasbourg Cedex, France

International Centre for the Study of the Preservation and the Restoration of Cultural Property (ICCROM)
13 Via di San Michele
00153 Rome, Italy
(6)587-901
Telex: 611331141 ICCROM1

U.S. Committee, ICCROM
c/o Advisory Council on Historic
Preservation
1100 Pennsylvania Avenue, N.W.
Suite 809
Washington, DC 20004
(202) 786-0503
Robert D. Bush, Executive Director

International Congress for the Conservation of the Industrial Heritage
Dramstigen 12
S/16138 Bramma, Sweden
Marie Nisser

International Council of Museums (ICOM)
Maison de l'Unesco
1, rue Miollis
75732 Paris Cedex 15, France

U.S. National Committee, ICOM
1225 I Street, N.W.
Suite 200
Washington, DC 20005
(202) 289-1818
Mary Louise Woods, Coordinator

International Council on Monuments and Sites (ICOMOS)
Hotel Saint-Aignan
75, rue du Temple
75003 Paris, France

U.S. Committee, ICOMOS
1600 H Street, N.W.
Washington, DC 20006
(202) 842-1866
Terry B. Morton, President

International Institute for Conservation of Historic and Artistic Works (IIC)
6 Buckingham Street
London WC2N 6BA, England

International Union of Architects
51, rue Raynouard
75016 Paris, France

International Committee
American Institute of Architects
1735 New York Avenue, N.W.
Washington, DC 20006
(202) 626-7315

International Union for Conservation of Nature and Natural Resources
Avenue du Mont-Blanc
CH-1196
Gland, Switzerland

IUCN/US
1400 16th Street, N.W.
Washington, DC 20036
(202)797-5454
Bryan Swift, Executive Director

Casa Carlos Armstrong, Ponce, Puerto Rico. (Summer Team and J. Martinez Canino, HABS)

Organization of American States
Department of Cultural Affairs
Cultural Patrimony Division
1889 F Street, N.W.
Washington, DC 20006
(202) 458-3142
Juan Carlos Torchaa Estrada, Technical
Unit Chief

UNESCO
Physical Property Division and the World
Heritage Center
1, place de Fontenoy
75700 Paris, France

World Monuments Fund
174 East 80th Street
New York, NY 10021
(212) 517-9367
Bonnie Burnham, Executive Director

CHAIRMEN OF ICOMOS NATIONAL COMMITTEES

ALGERIA
Abderrahmanc Khelifa
c/o Agence Nationale d'Archeologie, de la
Promotion des Sites et Monuments
2 avenue Mohamed Taleb
Haute Casbah
Alger

ANGOLA
Eleuterio Freire
Conseil National de la Culture
C.X. Postal 1223
Luanda

ARGENTINA
Maria de las Nieves Arias Incolla
Casilla de Correo 2163
Buenos Aires, 1000

AUSTRALIA
Joan Domicelj
Australia ICOMOS
401/402 Alfred Street
Milson Point 2061 N.S.W.

AUSTRIA
Ernst Bacher
Bundesdenkmalamt
Hofburg Saulenstiege
1010 Wien I

BELGIUM
Andries Van den Abbele
Groot Begijnhof 95
3000 Leuven

BOLIVIA
Teresa Gisbert
Museo de Arte Nacional
Casilla 609
La Paz

BRAZIL
Dalmo Vieira Filho
ICOMOS Brasil
Rua 13 de Maio
CEP 25685 Petropolis
Rio de Janeiro

BULGARIA
Todor Krestev
Comite National Bulgare de l'ICOMOS
44 blvd Dondoukov
1000 Sofia

BURKINA FASO
Boureima Diamitani
08 PZ 11289
Ouagadougou 08

CAMEROON
Mohaman Haman
Secretaire General ICOMOS Cameroun
38 rue Max Dormoy
75018 Paris

CANADA
François LeBlanc
P.O. Box 737
Station B
Ottawa, ONT K1P 5R4

CHILE
Rodriguo Marquez de la Plata
Agustinas 1070
Oficina 420
Santiago

COLOMBIA
A. Corradine Angulo
Apartado Aereo 39610 P.O.
Bogota, D.E. 1

COSTA RICA
Edgar Vargas
Apartado Postal 3866
San Jose 1000

CUBA
Isabel Rigol
CNCRM
Calle Cuba 610 entre Sol y Luz
Habana 1

CYPRUS
A. Papageorghiou
Director of the Department of Antiquities
Box 2024
Nicosia

CZECHOSLOVAKIA
Dobroslav Libal
Palais Palfy
Valdstejnska 14
11801 Praha 1—Mala Strana

DENMARK
Hans Munk Hansen
The Royal Academy of Fine Arts
Kongens Nytorv 1
DK—1050 Copenhagen

DOMINICAN REPUBLIC
Esteban Prieto
Isabel la Catolica n 103
Santo Domingo

ECUADOR
Wilson Herdoiza
Taller de Estudios Sociales
Architectonicos y Urbanisticas
Calle Vargas Machuca 6-25
Cuenca

EGYPT
Egyptian Antiquities Organization
Ministry of Culture
4d. Fakhri Abdel Nour Street
El Cairo

ETHIOPIA
Tadesse Terfa
Ministry of Culture
P.O. Box 1907
Addis Ababa

FINLAND
Panu Kaila
National Board of Antiquities
Department of Historic Monuments
P.O. Box 187
00171 Helsinki

FRANCE
Michel Jantzen
Section Francaise ICOMOS
62 rue Saint Antoine
75004 Paris

GABON
Pierre Ayamine Anguillet
Ministere de la Culture
B.P. 1007
Libreville

GERMANY
Michael Petzet
Bayerisches Landesamt für
Denkmalpflege
Hofgraben 4
8000 München

GHANA
D. S. Kpodo Tay
8 Fish Close
P.O. Box 500
Teshie—Nungua Estate
Accra

GREECE
Nicolas Agriantonis
P.O. Box 17188
100 24 Athenes

GUATEMALA
Blanca Nino Norton
Apartado Postal 625A
Zona 9
01909 Guatemala

HAITI
Daniel Elie
Comite Nationale ICOMOS—Haiti
ISPAN—Rue Cheriez
Port au Prince
Haiti BP 2484

HONDURAS
Gloria de Hasemann
Apartado Postal 1518
Tegucigalpa, D.C.

HUNGARY
Andras Roman
P.O. Box 6
H-1250 Budapest

INDIA
E.F.N. Ribeiro
Chief Planner
Town and Country Planning Organization
I.P. Estate
New Delhi 110002

ISRAEL
Michael Turner
ICOMOS Israel
25 Caspi Street
Jerusalem 93554

ITALY
Gian Franco Borsi
Chiesa Trecentesca di Donnaregina
Vico Donnaregina 26
80128 Napoli

IVORY COAST
Kindo Bouadi
Direction de la Conservation, de la
Protection
et de la Valorisation du Patrimoine
Culturel
B.P.V. 39
Abidjan

JAMAICA
Ann Hodges
P. O. Box 597
Kingston 10
Jamaica—West Indies

JAPAN
Kiyotari Tsuboi
c/o Bunkazai Kougaku Kenkyu-sho
3-9-5-113, Okubo, Shinjuku-ku
Tokyo 169

JORDAN
Adnan Hadidi
The Hashemite Kingdom of Jordan
Department of Antiquities
P.O. Box 88
Amman

KOREA, PEOPLE'S DEMOCRATIC
REPUBLIC OF
Djang Tcheul
Ministere de la Culture et des Arts
Pyong Yang

LEBANON
Joseph Phares
70 rue Saint-Didier
75116 Paris

LUXEMBOURG
Blanche Weicherding
21, route de Diekirch
7220 Walferdange

MEXICO
Carlos Flores-Marini
Mazatlan 190
Col. Condesa
06140 Mexico DF

THE NETHERLANDS
J. M. Hengeveld
Amst. Mij tot Stadsherstel N.V.
Amstelveld 10
1017 JD Amsterdam

NEW ZEALAND
David Reynolds
P.O. Box 37428
Parnell
Auckland 1

NORWAY
Stephan Tschudi-Madsen
Riksantikvaren
Postbox 8196 Dep.
N—0034 OSLO 1

PARAGUAY
Elizabeth Prats Gill
Eligio Ayola 2106
Asuncion

PERU
Jose Correa
Av. Salaverry 2457
Lima 27

POLAND
Olgierd Czerner
The Museum of Architecture
Bernardynska 5
50-153 Wroclaw

PORTUGAL
Sergio Infante
Comissao Nacional Portuguesa do
ICOMOS
Largo da Academia Nacional de Belas
Artes 2-2
1200 Lisbon

ROUMANIA
Radu Popa
Institut d'Archeologie
11 Str. IC Frimu
71119 Bucarest 20

SLOVENIA
Iva Curk
Zavod SR Slovenije za Varstvo
Naravne in Kulturne Dediscine
Plecnikov trg 2
61000 Ljubljana

SPAIN
Alvaro Gomez-Ferrer Bayo
President C.N.E. Cons. Sup. Col.
Arquitectos
Paseo de la Castellana 12
Madrid 1

SRI LANKA
Ashley de Vos
Central Cultural Fund
212 Bauddhaloka Mawatha
Colombo 7

SWEDEN
Margareta Biornstad
Central Board of National Antiquities
Storgatan 41
P.O. Box 5405
S-114 84 Stockholm

SWITZERLAND
Rutis Hauser
c/o Nike
Kaiserhaus
Marktgasse 37
CH—3011 Bern

TANZANIA
A. Mturi
Ministry of Information and Culture
Antiquities Division
P.O. Box 2280
Dar es Salaam

THAILAND
Banjong Choosakulchari
Permanent Secretary for Education
Ministry of Education
Bangkok 10300

TUNISIA
Abdelaziz Daoulatli
33 rue Tourbet El Bey
Tunis

TURKEY
Orhan Semerci
ICOMOS Turkiye Milli Komitesi
Eski Eserler ve Muzeler Genel
Mudurlugo
Ankara

UNIFIED REPUBLICS (former Soviet
ICOMOS)
Sergei Petrov
2 Zatchatievski 2
Korp. 3
G-34 Moscow
Russian Federation

UNITED KINGDOM
Sherban Cantacuzino
ICOMOS UK
10 Barley Mow Passage
Chiswick
London W4 4PH, England

UNITED STATES
John M. Fowler, Chairman
US/ICOMOS
Decatur House
1600 H Street, N.W.
Washington, DC 20006

ZAIRE
Zola Kuandi
Institut des Musees Nationaux du Zaire
B.P. 4249
Kinshasa II

FOREIGN

**L'Association des Vieilles Maisons
Françaises**
93, rue de l'Université
75005 Paris, France

Friends of Vieilles Maisons Françaises
180 Maiden Lane
34th Floor
New York, NY 10038
(212) 734-1651

Association for Industrial Archeology
Church Hill
Ironbridge
Telford, Salop TF8 7RE, England

Australian Council of National Trusts
C.M.L. Building
11th Floor
14 Martin Place
Sydney 20(a), N.S.W., Australia

Civic Trust
17 Carlton House Terrace
London SW1Y 5AW, England

English Heritage
Historic Buildings and Monuments
Commission for England
Fortress House
23 Savile Row
London W1X 2HE, England

American Friends of English Heritage
410 Park Avenue
19th Floor
New York, NY 10022

Environment Canada, Parks
Les Terrasses de la Chaudière
Ottawa, Ontario K1A 1G2, Canada

Heritage Canada Foundation
P.O. Box 1358, Station B
Ottawa, Ontario K1P 5R4, Canada
(613) 237-1066

Historic Houses Association
38 Ebury Street
London SW1, England

**L'Inspection Generale des Monuments
Historiques**
Direction du Patrimonie
Ministère de la Culture
3, rue de Valois
75042 Paris Cedex 01, France

Irish Georgian Society
Castletown House
Celbridge
County Kildare, Ireland

National Monuments Record
Fortress House
23 Savile Row
London W1X 1AB, England

**National Trust for Places of Historic
Interest or Natural Beauty**
36 Queen Anne's Gate
London SW1H 9AS, England

Northern Ireland Region
Rowallane House
Saintfield, Ballynahinch
County Down BT24 7LH, Northern
Ireland

Royal Oak Foundation
285 West Broadway
Suite 400
New York, NY 10013
(212) 966-6565
Damaris Horan

**National Trust for Scotland for Places of
Historic Interest or Natural Beauty**
5 Charlotte Square
Edinburgh EH2 4DU, Scotland

Royal Institute of British Architects
British Architectural Library
66 Portland Place
London W1N 4AD, England

Save Britain's Heritage
68 Battersea High Street
London SW11, England

Scottish Civic Trust
24 George Square
Glasgow G21 1EF, Scotland

Scottish Georgian Society
39 Castle Street
Edinburgh EH2 3BH, Scotland

**Society for the Protection of Ancient
Buildings**
37 Spital Square
London E1 6DY, England

Victorian Society
Priory Gardens
Bedford Park
London W4 1TT, England

WORLD HERITAGE CONVENTION

**"Welcomed throughout the world
as a new departure in international
cooperation, [the World Heritage
Convention] establishes a system
whereby the international community can participate actively in protecting those aspects of the cultural
and natural heritage which have
great universal value."**

Mesa Verde Museum Association, *World
Heritage Sites*, 1984

The Convention Concerning the
Protection of the World Cultural
and Natural Heritage was adopted
in 1972 as the most powerful international legal instrument for protecting cultural and natural
resources. The United States was
the first nation to adopt this treaty,
and in 1988 the 100th nation rat-
ified it. In 1991 the World Heritage
List included 339 significant cultural and natural properties in 69
countries; 17 of the properties are
in the United States.

Participating nations have agreed to
inventory, recognize, and protect
irreplaceable properties of outstanding international significance.
Each country assumes primary
responsibility for protecting and
interpreting its own properties,
while pledging to cooperate with
other nations as required. Through
a World Heritage Fund, made up of
contributions from member
nations, the World Heritage Committee provides financial and technical aid to properties and national
governments for threatened sites as
well as for preparation of new
nominations.

Each participating country may
nominate a property within its
jurisdiction. Cultural properties
must meet criteria such as:

■ unique artistic achievement

■ exceptional examples of vanished civilizations

■ association with events or ideas
of universal significance

Natural areas must demonstrate
significance in areas such as:

■ evolutionary history of the Earth

■ ongoing geological or biological
processes

■ exceptional natural phenomena,
formations, or scenic or scientific features

Proposals are evaluated by the
21-member World Heritage Committee, which is assisted by the
International Council on Monuments and Sites for cultural areas
and by the International Union for
Conservation of Nature for natural
areas.

In the United States the U.S.
Department of the Interior is
responsible for directing and coordinating participation in the World
Heritage Convention, through a
Federal Interagency Panel for World
Heritage and the National Park Service. A U.S. World Heritage Committee also has been established to
obtain support from business, conservation, and cultural interests.

U.S. World Heritage Sites
Cultural Sites

Cahokia Mounds State Historic
Site, Ill.

Chaco Culture National Historical
Park, N.M.

Independence Hall, Pa.

La Fortaleza and San Juan Historic
Site, P.R.

Mesa Verde, Colo.

Monticello and the University of Virginia, Va.

Statue of Liberty, N.Y.

Natural Sites

Everglades National Park, Fla.

Grand Canyon National Park, Ariz.

Great Smoky Mountains National Park, N.C. and Tenn.

Hawaii Volcanoes National Park, Hawaii

Mammoth Cave National Park, Ky.

Olympic National Park, Wash.

Redwood National Park, Calif.

Wrangell-St. Elias National Park, Alaska

Yellowstone National Park, Wyo.

Yosemite National Park, Calif.

CONTACTS

U.S. Department of the Interior
National Park Service
Federal Interagency Panel for World Heritage
P.O. Box 37127
Washington, DC 20013-7127
(202) 343-7063
Richard Cook, International Cooperation Specialist

U.S. World Heritage Committee
c/o World Wildlife Fund/The Conservation Foundation
1250 24th Street, N.W.
Washington, DC 20037
(202) 778-9512
Russell E. Train, Chairman

US/ICOMOS (regarding cultural property)
1600 H Street, N.W.
Washington, DC 20006
(202) 842-1866
Terry B. Morton, President

IUCN/US (regarding natural property)
1400 16th Street, N.W.
Washington, DC 20036

UNESCO
World Heritage Committee
7, Place de Fontenoy
75700 Paris, France

FURTHER READING

Historic Preservation in Other Countries, Vol. 2, Austria, Switzerland, the Federal Republic of Germany. Margaret T. Will; Robert E. Stipe, ed. Washington, D.C.: US/ICOMOS, 1984.

Historic Preservation in Other Countries, Vol. 3, Poland. Paul H. Gleye and Waldemar Szczerba; Robert E. Stipe, ed. Washington, D.C.: US/ICOMOS, 1989.

Historic Preservation in Other Countries, Vol. 4, Turkey. Jo Ramsay Liemenstoll; Robert E. Stipe, ed. Washington, D.C.: US/ICOMOS, 1990.

International Membership Directory and Resource Guide. Association for Preservation Technology. 2nd ed. Ottawa, 1988.

Our World's Heritage. Carol Lutyk, ed. Washington, D.C.: National Geographic Society, 1987.

The Yearbook of International Organizations. Brussels: K. G. Saur (rue Washington 40-1050). Annual.

STATE, LOCAL, AND INTERNATIONAL CONTACTS

Presented on the following pages are some 3,800 organizations and individuals who constitute a key portion of the grass-roots preservation movement.

Listings are drawn from the National Trust's Preservation Forum membership, a program aimed at administrators of historical societies, neighborhood and community preservation groups, statewide preservation organizations, historical sites and museums, as well as city administrators, planners, owners and developers of historic properties, architects, archeologists, and others who must keep up-to-date with events in the field of historic preservation. Benefits include receipt of the regular National Trust publications, including the magazine, as well as a special subscription to the *Preservation Forum* newsletter and journal. Members also participate in financial assistance and insurance programs, are eligible for technical advice, and receive discounts on conferences and publications.

THE STATES AND TERRITORIES

Each state or territorial section is introduced by a list of its eight major preservation contacts. Following is a brief summary of the work carried out by these offices and individuals:

State Historic Preservation Offices
As the key state and territorial government preservation officials, the SHPOs conduct cultural resource surveys, prepare statewide preservation plans, nominate properties to the National Register of Historic Places, review federal undertakings for effects on landmarks, administer grants-in-aid, help certify projects for federal tax incentives, provide public education, cooperate with related state agencies, administer historic properties, and supervise archeological activities.

Statewide Preservation Organizations
Private statewide preservation groups serve as the network centers and representatives of local preservation activities within their states. They work with the SHPOs,

assist local groups, intervene in preservation issues, advocate state legislative support, provide membership and educational programs, issue publications, engage in real estate and revolving fund programs, and serve as preservation clearinghouses.

National Trust Regional Offices
The seven regional and field offices of the National Trust represent Trust programs and services and provide leadership to preservationists within their regions. They work to strengthen local and state preservation organizations, provide on-site technical assistance, advocate preservation positions, respond to inquiries, operate and advise on Trust financial aid programs, hold conferences, and develop special projects to address key regional preservation issues. Preservationists are encouraged to refer inquiries to their National Trust regional office before contacting Trust headquarters.

Lithograph of Chestnut Street, Philadelphia, c. 1879, showing commercial buildings on the north side and south side (opposite) between

National Trust Advisors
A Board of Advisors composed of one or two members in each state and territory assists the National Trust by advising on preservation concerns and serving as a channel for the expression of local opinions and interests. See the National Trust chapter for a listing of officers.

Preservation Action Coordinators
The national lobbying organization Preservation Action has appointed coordinators in the states to provide grass-roots lobbying outreach and support for national preservation issues. They are part of a network of individuals who keep informed about pending legislation and public policies affecting preservation and participate in lobbying efforts as issues arise.

AIA State Preservation Coordinators
As members of the Committee on Historic Resources of the American Institute of Architects, the preservation coordinators serve as state-level contacts dealing with preservation issues of concern to the AIA, working with preservation organizations and SHPOs and speaking out on various issues.

National Park Service Regional Offices
The regional offices of the National Park Service, U.S. Department of the Interior, listed here are specially designated to handle rehabilitation tax credit applications and similar inquiries involving cultural resources not owned or managed by the National Park Service. Among their responsibilities are to

certify the significance of buildings within historic districts, state and local statutes, and rehabilitation applications. Other National Park Service regional offices responsible for Park Service sites and programs are listed in the Federal Programs chapter in Part III.

Advisory Council on Historic Preservation Divisions
Two divisions have been designated by the federal Advisory Council on Historic Preservation for initial review of federal and federally assisted projects that may affect historic properties and sites. Potential cases should be referred to the director of the division noted for each state and territory.

Following the state and territorial entries is a list of Forum members located outside the United States.

7th and 8th Streets. (*Baxter's Panoramic Views*, 1857–82)

ALABAMA

STATE HISTORIC PRESERVATION OFFICE

Alabama Historical Commission
725 Monroe Street
Montgomery, AL 36130
(205) 242-3184
(205) 240-3158 (Fax)
F. Lawerence Oaks, SHPO

STATEWIDE PRESERVATION ORGANIZATION

Alabama Preservation Alliance
P.O. Box 2228
Montgomery, AL 36102
(205) 432-1303
Douglas Purcell, President

NATIONAL TRUST REGIONAL OFFICE

Southern Regional Office
456 King Street
Charleston, SC 29403
(803) 722-8552
(803) 722-8652 (Fax)
Susan A. Kidd, Director

NATIONAL TRUST ADVISORS

Douglas C. Purcell
Historic Chattahoochee Commission
P.O. Box 33
Eufaula, AL 36072-0033
(205) 687-6631

Marjorie L. White
Birmingham Historical Society
1 Sloss Quarters
Birmingham, AL 35222
(205) 251-1880

PRESERVATION ACTION COORDINATORS

Marjorie White
Birmingham Historical Society
1 Sloss Quarters
Birmingham, AL 35222
(205) 251-1880

AIA STATE PRESERVATION COORDINATOR

Nicholas H. Holmes, Jr., FAIA
Holmes and Holmes
257 North Conception Street
Box 864
Mobile, AL 36603
(205) 432-8871
Harvie P. Jones, FAIA, Deputy

NATIONAL PARK SERVICE REGIONAL OFFICE

Southeast Regional Office
Richard B. Russell Federal Building
75 Spring Street, S.W.
Atlanta, GA 30303
(404) 331-2632

ADVISORY COUNCIL ON HISTORIC PRESERVATION DIVISION

Eastern Office of Project Review
1100 Pennsylvania Avenue, N.W.
Suite 809
Washington, DC 20004
(202) 786-0503
Don L. Klima, Director

Alexander City

Robinson Iron Corporation
Robinson Road
P.O. Box 1119
Alexander City, AL 35010
(205) 329-8486
Scott Howell

Andalusia

Covington Historical Society
P.O. Box 1582
Andalusia, AL 36420
John Tisdale

Anniston

City of Anniston
Housing Rehabilitation Division
P.O. Box 670
Anniston, AL 36202
(205) 237-8760
Sam Gaston

Glenda E. Knight
1125 Noble Street
P.O. Box 1850
Anniston, AL 36202
(205) 237-9586

Bessemer

City of Bessemer
1800 3rd Avenue North
Bessemer, AL 35020
(205) 424-4060
Jim Byram

Birmingham

Alabama Theatre
1811 3rd Avenue North
Birmingham, AL 35203
(205) 252-2262
Cecil Whitmire

Birmingham Historical Society
1 Sloss Quarters
Birmingham, AL 35222
(205) 251-1880
Marjorie L. White

Operation New Birmingham
2025 3rd Avenue North
Birmingham, AL 35203
(205) 324-8797
Diane Hairston

Sloss Furnaces Historic Landmark
P.O. Box 11781
Birmingham, AL 35202
(205) 324-1911
Randall Lawrence

Southside Community Development
Corporation
1724 16th Avenue South
Birmingham, AL 35205
(205) 934-3500
Betty Bock

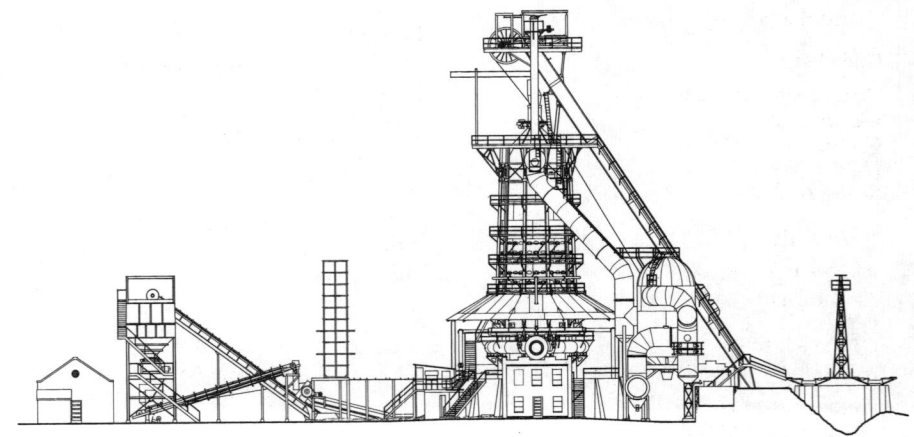

Sloss-Sheffield Steel and Iron Company Furnaces, Birmingham, Alabama.
(J.Y. Hunt, HABS)

Jim H. Waters, Jr.
209 22nd Street North
Birmingham, AL 35203
(205) 254-8037

Clanton

Chilton County Historical Society
P.O. Box 644
Clanton, AL 35045
(205) 755-0736
(205) 755-0750
Mary Richardson

Demopolis

City of Demopolis
P.O. Box 580
Demopolis, AL 36732
Austin Caldwell

Marengo County Historical Society
P.O. Box 159
Demopolis, AL 36732
(205) 289-1666
Gwynn Turner

Dothan

Dothan Landmarks Foundation
P.O. Box 6362
Dothan, AL 36302
(205) 794-3450
Sam Kates

Eufaula

Historic Chattahoochee Commission
211 North Eufaula Avenue
P.O. Box 33
Eufaula, AL 36072-0033
(205) 687-9755
(205) 687-6631
Douglas Purcell

Fairhope

Bates and Associates
P.O. Box 1244
Fairhope, AL 36533
(205) 990-9815
Glenn Bates

Gadsden

Downtown Gadsden
P.O. Box 8501
Gadsden, AL 35901
(205) 549-4324
Susan Andrew

Huntsville

Burritt Museum And Park
3101 Burritt Drive, S.E.
Huntsville, AL 35801
(205) 536-2882
Melinda Herzog

Constitution Hall Village
404 Madison Street
Huntsville, AL 35801
(205) 532-7551
Dana L. Tatum

The Holbec Company
P.O. Box 7162
Huntsville, AL 35807
(205) 883-6093
Mike Holbrook

Karl W. Leo
200 Randolph Avenue
Suite 200
Huntsville, AL 35801
(205) 539-6000

Old Town Historic Preservation
Commission
703 Holmes Avenue
Huntsville, AL 35801
(205) 539-7258
Dean Westerfield

Linden

City of Linden
211 North Main Street
Linden, AL 36748
(205) 298-5051
Dennis Mason

Mobile

Robert H. Allen
P.O. Box 1945
Mobile, AL 36604
(205) 432-1303

Historic Mobile Preservation Society
300 Oakleigh Place
Mobile, AL 36604
(205) 473-6161
Jean G. Wentworth

Mobile Historic Development
Commission
P.O. Box 1827
Mobile, AL 36633
(205) 438-7281
Michael C. McDonald

U.S.S. Alabama Battleship Memorial
Park
P.O. Box 65
Mobile, AL 36601
F. K. Smallwood

Montgomery

Alabama Department of Archives and
History
624 Washington Avenue
Montgomery, AL 36130
(205) 261-4361
Edwin C. Bridges

Alabama Historical Commission
725 Monroe Street
Montgomery, AL 36130
Larry Oaks

Alabama League of Municipalities
P.O. Box 1270
Montgomery, AL 36102
Perry Roquemore

Alabama Main Street Program
Alabama Historical Commission
725 Monroe Street
Montgomery, AL 36104
(205) 261-3184
Alta C. Hodgson

Alabama Preservation Alliance
P.O. Box 2228
Montgomery, AL 36102
(205) 279-9881
Marilyn Sullivan

City Of Montgomery
Planning Controls Division
P.O. Box 1111
Montgomery, AL 36102

Prattville

Historic Prattville
Redevelopment Authority
127 North Wildwood Drive
Prattville, AL 36067
(205) 365-7199
George Stathopoulos

Selma

Olde Towne Association
P.O. Box 728
Selma, AL 36702
Jenny Avritt

Selma and Dallas County Historic
Preservation Society
P.O. Box 586
Selma, AL 36702
(205) 872-1026
Elise Blackwell

Theodore

Bellingrath Gardens and Home
Route 1
P.O. Box 60
Theodore, AL 36582
John M. Brown

Tuscaloosa

Tuscaloosa County Preservation
Society
P.O. Box 1665
Tuscaloosa, AL 35403
(205) 758-2238
Marvin L. Harper

ALASKA

STATE HISTORIC PRESERVATION OFFICE

Alaska Department of Natural Resources
Office of History and Archeology
Division of Parks
P.O. Box 107001
Anchorage, AK 99510-7001
(907) 762-2622
(907) 762-2535 (Fax)
Judith Bittner, SHPO
Robert Shaw, Deputy SHPO

STATEWIDE PRESERVATION ORGANIZATION

Alaska Association for Historic Preservation
Old City Hall
524 West 4th Avenue
Suite 203
Anchorage, AL 99501
(907) 333-4746
Russ Sackett, President
William Coghill, Executive Director

NATIONAL TRUST REGIONAL OFFICE

Western Regional Office
1 Sutter Street
Suite 707
San Francisco, CA 94104
(415) 956-0610
(415) 956-0837 (Fax)
Kathryn Burns, Director

NATIONAL TRUST ADVISORS

Judith Bittner
2323 Hialeah Drive
Anchorage, AK 99517-1366
(907) 762-2622

Renee Blahuta
Box 80353
Fairbanks, AK 99708
(907) 457-6165

PRESERVATION ACTION COORDINATOR

Kerry Hoffman
Anchorage Historic Properties
524 West 4th Avenue
Anchorage, AK 99504
(907) 274-3600

AIA STATE PRESERVATION COORDINATOR

Edwin B. Crittenden, FAIA
801 Lincoln Street
Sitka, AK 99835
(907) 747-2543

NATIONAL PARK SERVICE REGIONAL OFFICE

Alaska Regional Office
2525 Gambell Street
Anchorage, AK 99503
(907) 257-2687

ADVISORY COUNCIL ON HISTORIC PRESERVATION DIVISION

Western Office of Project Review
730 Simms Street
Room 401
Golden, CO 80401
(303) 231-5320
Claudia Nissley, Director

Anchorage

Alaska Association for Historic Preservation
524 West 4th Street
Suite 203
Anchorage, AK 99501
Donna Lane

Anchorage Historic Properties, Inc.
524 West 4th Avenue
Anchorage, AK 99501
Kerry Hoffman

Anchorage Museum of History and Art
121 West 7th Avenue
Anchorage, AK 99501
Patricia B. Wolf

Judith E. Bittner
2323 Hialeah Drive
Anchorage, AK 99517-1366

Wrangell Mountains Center
3018 Alder Circle
Anchorage, AK 99508

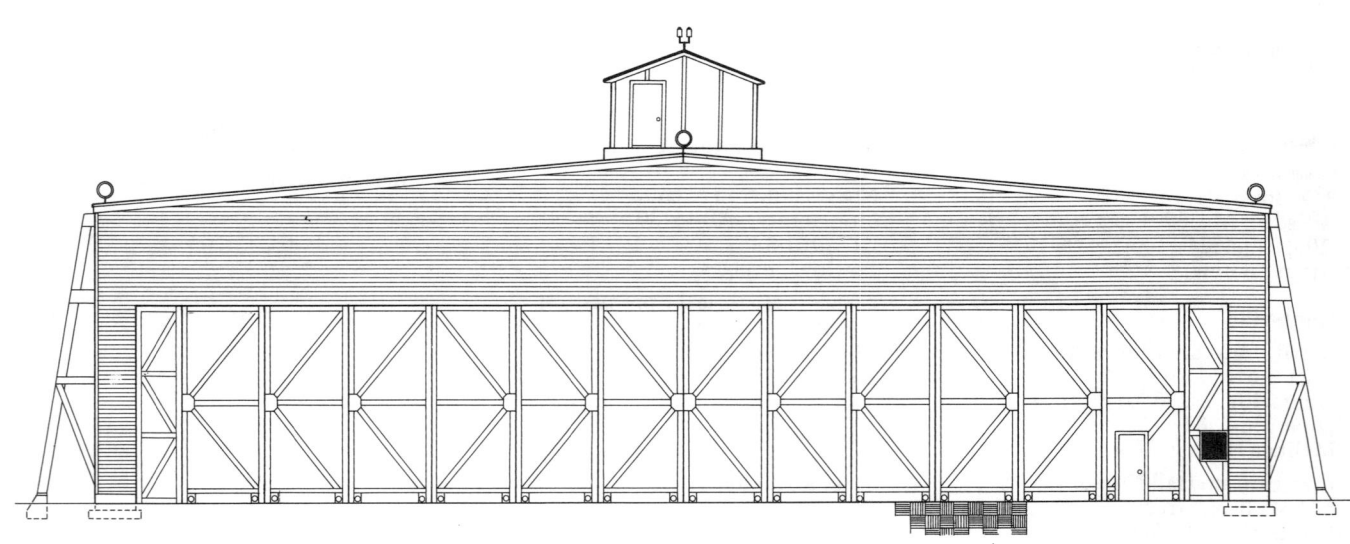

Kodiak T-Hangar, Ladd Field, Fairbanks, Alaska. (I.N. Thomas, HABS)

Bethel

Orutsararmiut Native Council
P.O. Box 927
Bethel, AK 99559
(907) 543-2608
Mary Pavil

Fairbanks

Fairbanks Downtown Association
547 3rd Avenue
Fairbanks, AK 99701
(907) 452-8671
Karen Lavery

Fairbanks Historical Preservation
Foundation, Inc.
P.O. Box 70552
Fairbanks, AK 99707
(907) 452-7295
Jonathan Link

Joint Commission on Historic
Preservation
P.O. Box 71267
Fairbanks, AK 99707
(907) 452-4761
Rose Cade

Tanana Yukon Historical Society
P.O. Box 71336
Fairbanks, AK 99707
(907) 457-7834
Jane Haigh

Juneau

Alaska Division of Tourism
P.O. Box E
Juneau, AK 99811
(907) 465-2012
Peter Carlson

City and Borough of Juneau
155 South Seward Street
Juneau, AK 99801
(907) 586-5238
Gary Gillette

Ketchikan

Historic Ketchikan
P.O. Box 3364
Ketchikan, AK 99901
(907) 225-5515
Karen Stanley

Palmer

Matanuska-Susitna Borough
Historic Preservation Commission
350 East Dahlia Avenue
Palmer, AK 99645-6488
(907) 745-9681
Victoria A. Cole

Sitka

Sitka Historical and Cultural Trust
City and Borough of Sitka
304 Lake Street
Sitka, AK 99835
(907) 747-3294
Stuart Denslow

Skagway

City of Skagway
P.O. Box 415
Skagway, AK 99840

Wrangell

Wrangell Historical Society
126 2nd Street
Box 1050
Wrangell, AK 99929
Pat Ockert

ARIZONA

STATE HISTORIC PRESERVATION OFFICE

Arizona State Parks
800 West Washington
Suite 415
Phoenix, AZ 85007
(602) 542-4009
(602) 542-4180 (Fax)
(Vacant), SHPO
Teresa Hoffman, Deputy SHPO

STATEWIDE PRESERVATION ORGANIZATION

Arizona Preservation Foundation
P.O. Box 13492
Phoenix, AZ 85002
(602) 256-7525
Cindy Myers, President

NATIONAL TRUST REGIONAL OFFICE

Western Regional Office
1 Sutter Street
Suite 707
San Francisco, CA 94104
(415) 956-0610
(415) 956-0837 (Fax)
Kathryn A. Burns, Director

NATIONAL TRUST ADVISORS

Grady Gammage
Gammage & Burnham
2 North Central Avenue
Suite 1800
Phoenix, AZ 85004
(602) 256-0566
(602) 256-4475 (Fax)

Robert C. Giebner
2109 East 5th Street
Tucson, AZ 85719
(602) 621-6735
(602) 621-8700 (Fax)

PRESERVATION ACTION COORDINATORS

Elisabeth Ruffner
1403 Barranca Drive
Prescott, AZ 86301
(602) 445-5644

AIA STATE PRESERVATION COORDINATOR

Donald Ryden, AIA
P.O. Box 1432
Phoenix, AZ 85001
(602) 253-5381

NATIONAL PARK SERVICE REGIONAL OFFICE

Southwest Regional Office
P.O. Box 728
Santa Fe, NM 87504
(505) 988-6100

ADVISORY COUNCIL ON HISTORIC PRESERVATION DIVISION

Western Office of Project Review
730 Simms Street
Room 401
Golden, CO 80401
(303) 231-5320
Claudia Nissley, Director

Bisbee

Bisbee Revitalization
86 Pima Drive
Bisbee, AZ 85603
Sally Holcomb

City of Bisbee
118 Arizona Street
Bisbee, AZ 85603
(602) 432-5446
Bob Jackson

Buckeye

Buckeye Main Street Project
Bravo
P.O. Box 352
Buckeye, AZ 85326
Sammie York

Flagstaff

Center for Excellence in Education
Northern Arizona University
Box 5774
Flagstaff, AZ 86011
(602) 526-9340
Peggy VerVelde

Flagstaff Historic Sites Commission
City of Flagstaff
211 West Aspen Avenue
Flagstaff, AZ 86001
(602) 779-7632
Ursula Montano

Main Street Flagstaff Foundation
P.O. Box 1882
Flagstaff, AZ 86002
(602) 774-1330
Kent Burns

Glendale

City of Glendale
Economic Development Department
5850 West Glendale Avenue
Glendale, AZ 85301
(602) 435-4169
Gregory J. Marek

Globe

Globe Main Street Project
P.O. Box 662
Globe, AZ 85502
(602) 425-9340
Rayna Barela

Holbrook

Holbrook Main Street Program
P.O. Box 237
Holbrook, AZ 86025
(602) 524-3294
Eloise Wycott

Kingman

Main Street Kingman
P.O. Box 6
Kingman, AZ 86402-6711
(602) 753-6644
Eve Clark

Mesa

Mesa Redevelopment Commission
P.O. Box 1466
Mesa, AZ 85201-1466
(602) 834-2691
Harold L. Stewart

Page

City of Page
P.O. Drawer HH
Page, AZ 86040
(602) 645-8861
John Beasley

Phoenix

Arizona Main Street
Department of Commerce
Community Development Program
3800 North Central
Suite 1400
Phoenix, AZ 85012
(602) 280-1350
Rod Keeling

Arizona Preservation Foundation
P.O. Box 13492
Phoenix, AZ 85002
(602) 445-5644
Doug Kupel

Janus Associates
P.O. Box 20132
Phoenix, AZ 85036-0132
James Garrison

Phoenix Planning Department
Historic Preservation Officer
125 East Washington Street
Phoenix, AZ 85004

Phoenix Public Library
12 East McDowell Road
Phoenix, AZ 85004

Mission San Xavier del Bac, Tucson, Arizona. (W.M. Collier, Jr., L. Williams, HABS)

Rosson House
Heritage Square Foundation
113 North 6th Street
Phoenix, AZ 85004
(602) 261-8948
Carolyn Fraker

Prescott

Prescott Preservation Commission
City of Prescott
P.O. Box 2059
Prescott, AZ 86302
(602) 778-9797
E. Birmingham

Yavapai Heritage Foundation
P.O. Box 61
Prescott, AZ 86302
(602) 445-5644

Safford

Safford Main Street Program
620 Central Ave
P.O. Box 535
Safford, AZ 85548
(602) 428-0094
Tonya L. Williams

Scottsdale

Inn Development Company
48 Casa Blanca East
Scottsdale, AZ 85253
(602) 947-2724
L. Kittler

Kevin Norton
7730 East Evans Road
Scottsdale, AZ 85260
(602) 998-3913

Show Low

Show Low Main Street
P.O. Box 3044
Show Low, AZ 85901
(602) 537-8181
Ann Ledget

Tombstone

City of Tombstone
City Hall
315 East Fremont
Tombstone, AZ 85638
(602) 457-2202
(602) 457-3562
George Kruse

Tombstone Historic Districts
Commission
P.O. Box 339
Tombstone, AZ 85638
(602) 457-2202
Phyllis Leonard

Tombstone Restoration Commission
P.O. Box 606
Tombstone, AZ 85638
Catherine Cline

Tucson

Arizona State Museum
The University Of Arizona
Tucson, AZ 85721

Armory Park Neighborhood Association
P.O. Box 2132
Tucson, AZ 85702
(602) 624-4750
Barbara Porter

Stephen K. Brigham
2913 East Hawthorne Street
Tucson, AZ 85716

City of Tucson
Planning Office
P.O. Box 27210
Tucson, AZ 85726
Rudy P. Gallego

Joana D. Diamos
55 Calle Clara Vista
Tucson, AZ 85716

Downtown Development Corporation
177 North Church Street
Suite 1010
Tucson, AZ 85701
(602) 623-5427
Cathy McCoskey

Pima County Historical Commission
Clerk's Office
P.O. Box 27210
Tucson, AZ 85726

Travelers Aid Society of Tucson
40 West Veterans Boulevard
Tucson, AZ 85713
(602) 622-8900

Willcox

Willcox Main Street Program
1500 North Circle I Road
Willcox, AZ 85643
(602) 384-2995
Edward Browning

Williams

Main Street Williams
P.O. Box 235
820 West Bill Williams Avenue
Williams, AZ 86046
Renee Basile-Bearse

Window Rock

Navajo Nation Historic Preservation
Department
P.O. Box 2898
Window Rock, AZ 86515
(602) 871-6437
Alan Downer

Yuma

Yuma Main Street Project
Yuma Downtown Redevelopment
333 South Main Street
Yuma, AZ 85364
(602) 782-4397
Connie Robles

ARKANSAS

STATE HISTORIC PRESERVATION OFFICE

Arkansas Historic Preservation
Program
The Heritage Center
225 East Markham
Suite 200
Little Rock, AR 72201
(501) 324-9346
(501) 324-9345 (Fax)
Cathryn H. Buford, SHPO
Ken Grunewald, Deputy SHPO
(501) 324-9346

STATEWIDE PRESERVATION ORGANIZATION

Historic Preservation Alliance of
Arkansas
P.O. Box 305
Little Rock, AR 72203
(501) 372-4757
Gary Clements, President
Carey Walker, Executive Director

NATIONAL TRUST REGIONAL OFFICE

Southern Regional Office
456 King Street
Charleston, SC 29403
(803) 722-8552
(803) 722-8652 (Fax)
Susan A. Kidd, Director

NATIONAL TRUST ADVISORS

Carl H. Miller, Jr.
P.O. Box 1411
Little Rock, AR 72203
(501) 374-6377

Cyrus Sutherland
Haskell Heights
Fayetteville, AR 72701
(501) 442-9983

PRESERVATION ACTION COORDINATORS

Missy Whitfield
Historic Preservation Alliance
P.O. Box 305
Little Rock, AR 72203
(501) 676-5749

AIA STATE PRESERVATION COORDINATOR

John K. Mott, FAIA
Mott Mobley McGowan and Griffin
302 North 6th Street
Fort Smith, AR 72901
(501) 782-1051

NATIONAL PARK SERVICE REGIONAL OFFICE

Southwest Regional Office
P.O. Box 728
Santa Fe, NM 87504
(505) 988-6100

ADVISORY COUNCIL ON HISTORIC PRESERVATION DIVISION

Western Office of Project Review
730 Simms Street
Room 401
Golden, CO 80401
(303) 231-5320
Claudia Nissley, Director

Blytheville

Main Street Blytheville
116 West Walnut
Blytheville, AR 72315
(501) 763-2525
Lisa Morris

Camden

Main Street Camden
P.O. Box 734
Camden, AR 71701
(501) 836-6105
Teresa Rogers

Conway

Conway Main Street
P.O. Box 1765
Conway, AR 72032
(501) 329-0908
Sue Boze

El Dorado

South Arkansas Historical Foundation
P.O. Box 144
El Dorado, AR 71730
Raymond Higgins

Fayetteville

Cyrus A. Sutherland
Haskell Heights
Fayetteville, AR 72701
(501) 442-9983

Gurdon

Gurdon Historic Preservation
Associates
P.O. Box 198
Gurdon, AR 71743
(501) 353-4435
Martha Harwell

Harrison

Main Street Harrison
P.O. Box 2049
Harrison, AR 72602
(501) 741-4889
Bettie Davidson

Helena

Helena Public Library and Museum
Association
623 Pecan Street
Helena, AR 72342
(501) 338-6474
Deborah King

Hot Springs

Taylor Kempkes Architects
210 Central
Suite 2A
Hot Springs, AR 71901
(501) 624-5679
Renee Marshall

Little Rock

Arkansas Historic Preservation
Program
225 East Markham Street
Suite 200
Little Rock, AR 72201
(501) 371-2763

Arkansas Municipal League
P.O. Box 38
North Little Rock, AR 72115
Don A. Zimmerman

Entrepreneurial Services Center
1123 South University
Suite 608
Little Rock, AR 72204
(501) 686-2509
Linda Robinett

Historic Preservation Alliance of
Arkansas
P.O. Box 305
Little Rock, AR 72203
(501) 372-4757
Frances Whitfield

Main Street Arkansas
225 East Markham Street
Suite 300
Little Rock, AR 72201
(501) 324-9346
Sandra Hanson

Cheryl G. Nichols
1721 South Gaines
Little Rock, AR 72206
(501) 375-2686

Quapaw Quarter Association
1315 South Scott
P.O. Box 165023
Little Rock, AR 72216-5023
(501) 371-0075
Elissa Gross

Magnolia

Main Street Magnolia
P.O. Box 1153
Magnolia, AR 71753
(501) 234-2513
Cathe Nipper

Elkhorn Tavern, Pea Ridge, Arkansas. (D.T. Jones, HABS)

Pine Bluff

Heckatoo Heritage Foundation
P.O. Box 7471
Pine Bluff, AR 71611

Pine Bluff Downtown Development
512 South Pine
Suite 301
Pine Bluff, AR 71601
(501) 536-8742
Montine McNulty

Rogers

Main Street Rogers Program
P.O. Box 935
Rogers, AR 72756
(501) 636-4130
Jenny Harmon

Siloam Springs

Main Street Siloam Springs
P.O. Box 88
205 East Main
Siloam Springs, AR 72761
(501) 624-6140
(501) 524-6148
Marsha Emanuelson

Washington

Pioneer Washington Restoration
Foundation
P.O. Box 127
Washington, AR 71862
(501) 777-8871
J. H. Pilkinton

West Memphis

Main Street West Memphis
P.O. Box 2302
West Memphis, AR 72303
(501) 735-8814
Linda Avery

CALIFORNIA

STATE HISTORIC PRESERVATION OFFICE

Office of Historic Preservation
Department of Parks & Recreation
P.O. Box 942896
Sacramento, CA 94296-9824
(916) 653-6624
(916) 653-9824 (Fax)
Steade R. Craigo, Acting SHPO

STATEWIDE PRESERVATION ORGANIZATION

California Preservation Foundation
1615 Broadway
Suite 705
Oakland, CA 94612
(510) 763-0972
(510) 763-4724 (Fax)
William Delvac, President
John Merritt, Executive Director

NATIONAL TRUST REGIONAL OFFICE

Western Regional Office
1 Sutter Street
Suite 707
San Francisco, CA 94104
(415) 956-0610
(415) 956-0837 (Fax)
Kathryn A. Burns, Director

NATIONAL TRUST ADVISORS

Linda Dishman
1703 Oak Street
Apartment 4
San Francisco, CA 94117
(415) 556-7741

Christy McAvoy
Historic Resources Group
1728 North Whitley Avenue
Hollywood, CA 90028
(213) 469-2349
(213) 469-0491 (Fax)

PRESERVATION ACTION COORDINATORS

John Merritt
California Preservation Foundation
1615 Broadway
Suite 705
Oakland, CA 94612
(415) 763-0972

Alex Stillman
905 H Street
Arcata, CA 95521
(707) 822-2269

Marian Mitchell-Wilson
3851 Chapman Place
Riverside, CA 92506
(714) 782-5676

AIA STATE PRESERVATION COORDINATOR

Dan L. Peterson, AIA
Dan L. Peterson and Associates
146 Bishop Alley
Richmond, CA 94801
(415) 235-2753

NATIONAL PARK SERVICE REGIONAL OFFICE

Western Regional Office
600 Harrison Street
Suite 600
San Francisco, CA 94107
(415) 556-7741

ADVISORY COUNCIL ON HISTORIC PRESERVATION DIVISION

Western Office of Project Review
730 Simms Street
Room 401
Golden, CO 80401
(303) 231-5320
Claudia Nissley, Director

Alameda

Alameda Historical Museum
1023 Morton
Alameda, CA 94501
Greg Beattie

City Of Alameda
P.O. Box 33823
Alameda City Hall
Room 315
Alameda, CA 94501
(415) 748-4554
Arnold B. Jonas

Pacific States Carpentry and Millwork
3249 Briggs Avenue
Alameda, CA 94501
(415) 523-0906
Jim Spaulding

Altadena

Altadena Heritage
P.O. Box 218
Altadena, CA 91001
(818) 798-1268
Tim Gregory

Asti

El Carmelo Corporation
P.O. Box 811
Asti, CA 95425-9998
(707) 528-3288
Carlo Rossi

Auburn

Auburn Main Street Program
933 Lincoln Way
P.O. Box 9171
Auburn, CA 95604
(916) 888-0109
Carol Pepper-Kittredg

Placer County Courthouse Restoration Committee
P.O. Box 9044
Auburn, CA 95604
(916) 885-5564
Barbara Wylie

Placer County Department of Museums
Historical Advisory Board
175 Fulweiler Avenue
Auburn, CA 95603
(916) 885-9570
David Tucker

Bakersfield

Bakersfield Historic Preservation Commission
Redevelopment Agency
1501 Truxton Avenue
Bakersfield, CA 93301
(805) 326-3765
Donna Barnes

Barstow

City of Barstow
City Planning Department
220 East Mountain View Street
Barstow, CA 92311
Paul Warner

Belmont

Mary Ellen Boyling
College of Notre Dame
Belmont, CA 94002
(415) 593-1601

Benicia

Benicia Main Street Project
831 1st Street
Benicia, CA 94510
(707) 745-9791
Patricia Frye

William Jenkins
P.O. Box 843
Benicia, CA 94510
(707) 745-8513

Meridian Architects
131 West D Street
Benicia, CA 94510
(707) 745-8817
Elizabeth A. Pidgeon

Ben Lomond

Karen Nilsen
150 Racoon Lane
Ben Lomond, CA 95005
(408) 336-2971

Berkeley

Berkeley Architectural Heritage
Association
P.O. Box 1137
Main Post Office
Berkeley, CA 94701
(415) 841-2242
Anthony Bruce

Berkeley Landmarks Preservation
Commission
2180 Milvia Street
Berkeley, CA 94704
(415) 644-6570
Mark Paez

Robert Dahlstrom
258 Hillcrest Road
Berkeley, CA 94705
(415) 655-0361

Downtown Berkeley Association
2230 Shattuck Avenue
Suite H
Berkeley,CA 94704
(415) 549-2230
Dan Craig

Manager House
2307 Piemont Avenue
Berkeley, CA 94704

Rainshadow Associates
1131 Hillview Road
Berkeley, CA 94708
(415) 849-1438
Ana B. Koval

University of California
Design Construction Services
2000 Carleton Street
Room 16
Berkeley, CA 94720
(415) 642-2662
Zandras F. LeDuff

Beverly Hills

Harold W. Levitt
221 North Robertson Boulevard A
Beverly Hills, CA 90211
(213) 272-2280

Blue Lake

Dell'Arte
P.O. Box 816
131 H Street
Blue Lake, CA 95525
(707) 668-5664
Jane Hill

Brea

City of Brea
1 Civic Center Circle
Brea, CA 92621
(714) 990-7724
Lisa Donnell

Burbank

Burbank Historical Society
1015 West Olive Avenue
Burbank, CA 91506
(818) 848-4721
Ted Garcia

Burlingame

Victor Glushko
2504 Hale Drive
Burlingame, CA 94010

Calabasas

Leonis Adobe Association
23537 Calabasas Road
Calabasas, CA 91302

California City

East Kern Historical Museum Society
P.O. Box 2305
California City, CA 93504
(619) 373-4811
Patricia V. Gorden

Campbell

Campbell Historical Museum
51 North Central
Campbell, CA 95008
(408) 866-2119
Patricia Leach

City of Campbell
70 North 1st Street
Campbell, CA 95008
(408) 866-2128
Robert Kass

Camp Pendleton

Joint Public Affairs Office
History and Museum Department
Building 1160
Camp Pendleton, CA 92055-5001
(619) 725-5566

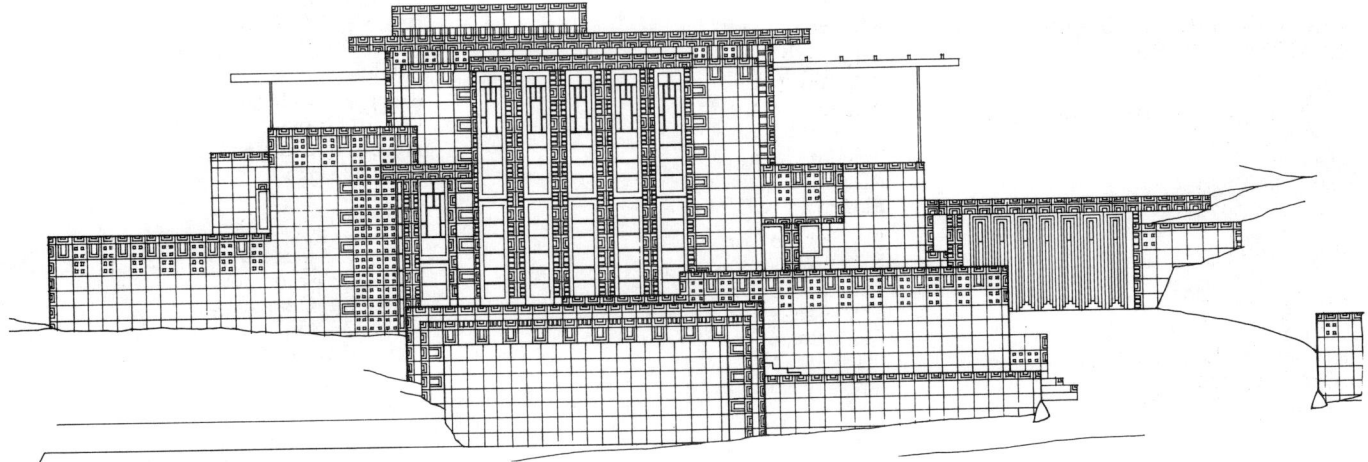

Storer House, Hollywood, California. (J.R. Bateman, HABS)

217

Carlsbad

Housing and Redevelopment
Department
2075 Las Palmas Drive
Carlsbad, CA 92009
(619) 438-5593
Deborah K. Fountain

Carmel

Carmel Heritage
P.O. Box 701
Carmel, CA 93921
(408) 624-4447
B. Schaffer

Marjorie Fisher
27035 Meadow Way
Carmel, CA 93923
(408) 624-4412

Robinson Jeffers Tor House Foundation
P.O. Box 1887
Carmel, CA 93921
(408) 624-1813
Hadley Osborn

Chico

Chico Heritage Association
P.O. Box 2078
Chico, CA 95927
(916) 345-2005
Giovanna Jackson

Claremont

Claremont Heritage
590 West Bonita Avenue
P.O. Box 742
Claremont, CA 91711
(714) 621-0848
Ginger Elliott

Cloverdale

Cloverdale Depot Association
P.O. Box 941
Cloverdale, CA 95425
(707) 894-2442
Li Keiser

Coloma

Marshall Gold Discovery State Historic
Park
P.O. Box 265
310 Back Street
Coloma, CA 95613
(916) 623-3470
M. S. Sugarman

Colusa

City of Colusa
425 Webster Street
P.O. Box 1063
Colusa, CA 95932
(916) 458-4941
Joyce McCullough

Coronado

Coronado Historic Association
P.O. Box 180393
Coronado, CA 92178
(619) 435-5993
Nancy Cobb

Culver City

Culver City Community Development
Program
P.O. Box 507
Culver City, CA 90230
(213) 202-5777
Jody Hall-Esser

Danville

Museum of the San Ramon Valley
P.O. Box 39
Danville, CA 94526
(415) 820-8344
Irma Dotson

Death Valley Junction

Amargosa Opera House
P.O. Box 8
Death Valley Junction, CA 92328
(619) 852-4316
Marta Becket

Downey

Lawrence E. Winans
P.O. Box 4164
Downey, CA 90241
(213) 560-3098

Duarte

Duarte Historical Museum
P.O. Box 263
Duarte, CA 91009
(818) 357-9411
Dorothy L. Montgomery

El Centro

Downtown El Centro Association
107 South 5th Street
Suite 244
P.O. Box 2624
El Centro, CA 92244
(619) 353-6771
Sheryell Tockey

El Cerrito

Rosalie G. Flores
8186 Terrace Drive
El Cerrito, CA 94530

El Monte

City of El Monte
11333 Valley Boulevard
El Monte, CA 91734
(818) 580-2001
Gregory Korduner

Escondido

City of Escondido
Planning Department
201 North Broadway
Escondido, CA 92025
(619) 741-4671
Dawn Suitts

Downtown Escondido Business
Association
P.O. Box 489
Escondido, CA 92033-0489

Escondido Historical Society
P.O. Box 263
Escondido, CA 92033
(619) 743-8207
Norm Syler

Eureka

Eureka Heritage Society
P.O. Box 1354
Eureka, CA 95501

Fort Bragg

City Of Fort Bragg
416 North Franklin Street
Fort Bragg, CA 95437
(707) 961-2825
Gary D. Milliman

Fort Bragg Main Street Program
416 North Franklin Street
P.O. Box 2563
Fort Bragg, CA 95437-2563
(707) 961-0360
Jay Turner

Fremont

City of Fremont
39700 Civic Center Drive
Fremont, CA 94538
(415) 790-6813
Mary Prisco

City of Fremont
Community Development Department
39700 Civic Center Drive
Fremont, CA 94538
(415) 790-6746
Amy Rakley

Fresno

Bruce A. Owdom
718 East Carmen Avenue
Fresno, CA 93728
(209) 435-5250

Fullerton

California State University
Fullerton Arboretum
Fullerton, CA 92634
(714) 773-8357
David Walkington

Georgetown

Georgetown Divide Recreation
c/o Pete Field
P.O. Box 1617
Georgetown, CA 95634

Gilroy

Gilroy Visitors Bureau
7780 Monterey Street
Gilroy, CA 95020
(408) 842-6436
J. C. Dowell

Glendale

The Glendale Historical Society
P.O. Box 4173
Glendale, CA 91202

Charla B. Janecek
1660 Ard Eevin Avenue
Glendale, CA 91202
(213) 931-1321

John W. McKenna
633 East Broadway
Room 104
Glendale, CA 91206
(818) 956-2144

Glendora

Eric A. Magallon
736 North Vista Avenue
Glendora, CA 91740
(818) 335-9720

Grass Valley

Grass Valley Downtown Association
P.O. Box 1986
Grass Valley, CA 95945
(916) 272-8315
Mark Winkler

Half Moon Bay

City of Half Moon Bay
501 Main Street
P.O. Box 338
Half Moon Bay, CA 94019
(415) 726-5566
Allen Harry

Johnston House Foundation
P.O. Box 789
Half Moon Bay, CA 94019

Healdsburg

Dennis E. Harris
1879 Toyon Drive
Healdsburg, CA 95448

Hollister

City of Hollister
375 5th Street
Hollister, CA 95023
(408) 637-8221
Hugh R. Riley

Hollister Downtown Association
446 San Benito Street, Suite 209
Hollister, CA 95023
(408) 636-8406
Dan Craig

Hollywood

Hollywood Heritage
P.O. Box 2586
Hollywood, CA 90078
(213) 874-4005
Kari E. Johnson

Huntington Beach

Ronald Hagan
Community Services Department
2000 Main Street
Huntington Beach, CA 92648

Huntington Beach Historical Society
19820 Beach Boulevard
Huntington Beach, CA 92648
(714) 926-5777
Barbara Milkovich

Huntington Beach Public Library
Periodicals Department
7111 Talbert Avenue
Huntington Beach, CA 92648

Michael Mudd
Cultural Affairs
P.O. Box 190
Huntington Beach, CA 92648
(714) 526-5486

Independence

Eastern California Museum
P.O. Box 206
Independence, CA 93526
(619) 878-2411
William H. Michael

Industry

Willdan Associates
12900 Crossroads Parkway South
Suite 200
Industry, CA 91746-3499
(213) 921-8215
Jeff Bruyn

Workman and Temple Family
Homestead Museum
15415 East Don Julian Road
Industry, CA 91745
(818) 968-8492
Karen G. Wade

Irvine

Irvine Historical Society
5 San Joaquin
Irvine, CA 92715
(714) 786-4112
Judy Liebeck

Lakewood

Angelo M. Iacoboni Library
Room 509
4990 Clark Avenue
Lakewood, CA 90712

La Mesa

City of La Mesa
Planning Department
8130 Allison Avenue
La Mesa, CA 92041
(619) 462-0171
David E. Witt

Dan T. Wheeler
5526 Lake Park Way
La Mesa, CA 91942
(619) 584-7046

La Mirada

La Mirada Historical Committee
Leisure Services Department
12900 Bluefield Avenue
La Mirada, CA 90638
(213) 943-7277

La Verne

City of La Verne
Planning Department
3660 D Street
La Verne, CA 91750
(714) 596-8706
Arlene Banks

Livermore

Livermore Area Recreation and Park
District
71 Trevarno Road
Livermore, CA 94550
(415) 447-7300
Mike Nicholson

Lodi

City of Lodi
Community Development Department
221 West Pine Street
Lodi, CA 95240
(209) 333-6711
Eric Veerkamp

Loma Linda

City of Loma Linda
Community Development Department
25541 Barton Road
Loma Linda, CA 92354
(714) 799-2830
Dan Smith

Long Beach

City of Long Beach
City Hall
333 West Ocean Boulevard
13th Floor
Long Beach, CA 90802
(213) 590-6864
Ruthann Lehrer

Coalition to Preserve Historic Long
Beach
P.O. Box 92521
Long Beach, CA 90809
Nancy Latimer

Leonard M. Kliwinski
3933 East Broadway
Suite 203
Long Beach, CA 90803
(213) 438-2093

Long Beach Redevelopment Agency
333 West Ocean Bouleveard
Long Beach, CA 90802
(213) 590-6168
Tom Anderson

Rancho Los Cerritos
4600 Virginia Road
Long Beach, CA 90807
(213) 424-9423
Ellen Calomiris

Los Altos

Los Altos Historical Commission
1 North San Antonio Road
Los Altos, CA 94022

Los Angeles

A. T. Heinsbergen and Company
7415 Beverly Boulevard
Los Angeles, CA 90036
(213) 934-1134
Dawn Heinsbergen

William C. Baer
Associate Professor
School of Planning
University of Southern California
Los Angeles, CA 90089-0042

Franklyn J. Bergen
Senior Planner
Caltrans District 7
120 South Spring Street
Los Angeles, CA 90012-3606
(213) 620-4978

City of Los Angeles
Department of Recreation and Parks
City Hall East
200 North Main Street
Room 1330
Los Angeles, CA 90012

Rachel S. Cox
352 South Las Palmas
Los Angeles, CA 90020
(213) 937-0970

DRW Associates
120 South Vignes
Suite 105
Los Angeles, CA 90012
(213) 687-8053
Lynn Bryant

Samuel Freeman House
School of Architecture
University of Southern California
1962 Glencoe Way
Los Angeles, CA 90068
(213) 850-6278
Jeffrey M. Chusid

Historical Society Of Centinela Valley
7634 Midfield Avenue
Los Angeles, CA 90045
Gladys Waddingham

Johnson Research Associates
3103 Lindo Street
Los Angeles, CA 90068
(213) 851-8854
Christy McVoy

Peter L. Lassen
1448 North Boylston Street
Los Angeles, CA 90012
(213) 221-0793

Los Angeles Conservancy
727 West 7th Street
Suite 955
Los Angeles, CA 90017
(213) 623-2489
Jay Rounds

National Society of Colonial Dames of
America
Los Angeles Pasadena Committee
425 Cascada Way
Los Angeles, CA 90049
Joseph D. Vaccaro

Norman Neuerburg
4153 Tracy Street
Los Angeles, CA 90027
(213) 663-6052

The Power of Place
8318 Ridpath Drive
Los Angeles, CA 90046
(213) 654-9833
Donna Graves

Preservation Team
Community Redevelopment Agency
354 South Spring Street
Los Angeles, CA 90013
John Kaliski

Ginger Rockwell
12036 Benmore Terrace
Los Angeles, CA 90049

UNOCAL Company
Room M-10
P.O. Box 7600
Los Angeles, CA 90051
(213) 977-6459
Barry Lane

J. A. Zink
930 Palm Avenue
Suite 215
Los Angeles, CA 90069
(213) 652-8041

Lynwood

City of Lynwood
11330 Bullis Road
Lynwood, CA 90262
Michael T. Heriot

Manhattan Beach

Melvyn Green
1145 Artesia Boulevard
Suite 204
Manhattan Beach, CA 90266
(213) 374-6424

McCloud

Edward L. Simons
237 Main Street
McCloud, CA 96057
(916) 964-2747

Mill Valley

MPA Designs
7 Dawn Place
Mill Valley, CA 94941

Milpitas

City of Milpitas
Cultural Resources Preservation Board
455 East Calaveras Boulevard
Milpitas, CA 95035
(408) 942-2379
Steve Burkey

Toni Webb
807 Folsom Circle
Milpitas, CA 95035
(408) 263-0452

Modesto

Wayne Mathes
3812 Pan Am Drive
Modesto, CA 95356
(209) 577-5344

Montebello

City of Montebello
1600 West Beverly Boulevard
Montebello, CA 90640
(213) 725-1200
Paul Deibel

Montebello Library-601
1550 West Beverly Road
Montebello, CA 90640

Monterey

City of Monterey
Planning Department
City Hall
Monterey, CA 93940
(408) 646-3851
Bruce Kibby

Lowel I. Figen
P.O. Box 1627
Monterey, CA 93940

James R. Wright
P.O. Box 805
Monterey, CA 93942
(408) 372-2608

Mountain View

John Miller
196 Castro Street
Mountain View, CA 94041
(415) 967-9584

Napa

City of Napa
Cultural Heritage Commission
P.O. Box 660
Napa, CA 94559
(707) 257-9530
Rebecca A. Yerger

Napa County Landmarks
P.O. Box 702
Napa, CA 94559
(707) 255-1836
Rebecca Yerger

John Whitridge
578 Montecito Boulevard
Napa, CA 94559
(707) 252-1021

Nevada City

Truckee Historical Preservation
Advisory Council
c/o Planning Department
950 Maidu Avenue
P.O. Box 6100
Nevada City, CA 95959-6100

Newark

Rose of Sharon Church
7160 Graham Avenue
Newark, CA 94560
(415) 793-4761
Marcelo Maghirang

Nuevo

R. Denzil Lee, AIA
31242 Orange Avenue
Nuevo, CA 92367
(714) 928-0275

Oakdale

Oakdale Museum
212 West F Street
Oakdale, CA 95361
(209) 847-9229
Glenn E. Burghardt

Oakland

Academy of Chinese Culture and
Health Sciences
420 14th Street
Oakland, CA 94612
(510) 763-7787
Wei Tsuei

Bramalea Pacific
1221 Broadway
Suite 1800
Oakland, CA 94612
(415) 464-8256
Suzanne Hirshen

California College of Arts and Crafts
663 Oakland Avenue
Oakland, CA 94611
Charles K. Fiske

California Preservation Foundation
1615 Broadway
Suite 705
Oakland, CA 94612
(415) 763-0972
John F. Merritt

East Bay Regional Park District
11500 Skyline Boulevard
Oakland, CA 94619
(415) 531-9300
Susan Williams

First Unitarian Church of Oakland
685 14th Street
Oakland, CA 94612
(415) 465-2200
Kerry Parker

Frederick C. Hertz
7045 Chabot Road
Oakland, CA 94618
(415) 428-2252

Kahn Mortimer Associates
478 Santa Clara Avenue
Oakland, CA 94610
(415) 451-5954
Larry J. Mortimer

Lauri Kibby
1084 Trestle Glen
Oakland, CA 94610
(510) 763-3113

Mills College
5000 MacArthur Boulevard
Oakland, CA 94613
(415) 430-2364
Donna Raub

Neighborhood Housing Services of
America
1970 Broadway
Suite 470
Oakland, CA 94612-2216

Oakland Heritage Alliance
P.O. Box 12425
Oakland, CA 94604
(415) 763-9218
Helen Lore

Oakland Landmarks Preservation
Advisory Board
Planning Department
1 City Hall Plaza
6th Floor
Oakland, CA 94612
(415) 273-3941

Pardee Home Museum
672 11th Street
Oakland, CA 94607
(415) 444-2187
David L. Casebolt

Oceanside

City of Oceanside
300 North Hill Street
Oceanside, CA 92054
(619) 439-7272
Rita Baker

J. G. Morrissey
3557 Evening Canyon
Oceanside, CA 92056
(619) 941-3602

Ontario

City of Ontario
Planning Department
303 East B Street
Ontario, CA 91764
Joyce Babicz

Oxnard

Conrad and Okuma, Architects
167 Lambert Street
Oxnard, CA 93030
(805) 983-0053
A. A. Okuma

Steven Kinney
Director of Redevelopment
300 West 3rd Street
Oxnard, CA 93030

Pacific Grove

Heritage Society of Pacific Grove
P.O. Box 1007
Pacific Grove, CA 93950
(408) 372-2898
Adam Weiland

Pacific Palisades

Mark Levinson
632 Swarthmore Avenue
Pacific Palisades, CA 90272

Palm Springs

Historic Site Preservation Board
City of Palm Springs
P.O. Box 1786
Palm Springs, CA 92263
(619) 323-8245
Richard E. Patenaude

Palo Alto

City Of Palo Alto
HRB Planning Department
P.O. Box 10250
Palo Alto, CA 94303
George Zimmerman

William N. Fenerin
1102 Guinda Street
Palo Alto, CA 94301
(415) 3227440
Peter Fenerin

Palo Alto Stanford Heritage
343 Shasta Drive
Palo Alto, CA 94306
(415) 856-2440
Elizabeth Kittas

Cherilyn Widell
738 Guinda
Palo Alto, CA 94301
(415) 326-4016

Pasadena

Pasadena Heritage
80 West Dayton Street
Pasadena, CA 91105
(818) 793-0617
(818) 578-0865
Claire Bogaard

Urban Conservation Pasadena
City Hall
Room 111
Pasadena, CA 91109
(818) 405-4228

Paso Robles

City of Paso Robles
1030 Spring Street
P.O. Box 307
Paso Robles, CA 93447
(805) 238-1529
Robert A. Lata

Paso Robles Main Street
608 1/2 12th Street
Paso Robles, CA 93446
(805) 238-4103
James Mullen

Pismo Beach

City of Pismo Beach
1000 Bello
Pismo Beach, CA 93449
(805) 773-4658
Carolyn Johnson

Placentia

Placentia Historical Committee
401 East Chapman
Placentia, CA 92670
(714) 528-2702
Jack Slota

Point Richard

Dan Peterson, AIA
146 Bishop Alley
Point Richard, CA 94801

Port Costa

Port Costa Conservation Society
1 Plaza El Mambre
P.O. Box 36
Port Costa, CA 94569
(415) 787-2880
Mary Powell

Porterville

Gail Craig, AIA
P.O. Box 593
Porterville, CA 93258
(209) 784-1496

Main Street Porterville
180 North Main
Suite D
Porterville, CA 93257
(209) 781-0880
Denise L. Marchant

Quincy

Plumas County Museum Association
P.O. Box 10776
Quincy, CA 95971
(916) 283-6320
Scott Lawson

Quincy Main Street Project
P.O. Box 155
Quincy, CA 95971
(916) 283-0188
Robyn Grey

Rancho Cucamonga

City of Rancho Cucamonga
Planning Department
P.O. Box 807
Rancho Cucamonga, CA 91730
(714) 989-1858
Anthea Hartig

Redlands

Kimberly Crest House and Gardens
P.O. Box 206
Redlands, CA 92373
(714) 792-2111
Steven T. Spiller

Redlands Area Historical Society
P.O. Box 1024
Redlands, CA 92373

Redlands Town Center Corporation
P.O. Box 3005
Redlands, CA 92373-1505
(714) 798-7548
Marjorie Pettus

San Bernardino County Museum
2024 Orange Tree Lane
Redlands, CA 92373
(714) 825-4825
Allan D. Griesemer

Glenda Saul
454 Cajon Street
Redlands, CA 92373
(714) 792-3780

Redwood City

City of Redwood City
P.O. Box 391
Redwood City, CA 94064
(415) 485-1971
Charles Jany

Redwood City Heritage Association
P.O. Box 1273
Redwood City, CA 94064
(415) 657-6755
Brian Martinson

Richmond

Richmond Museum Association
Box 1267
Richmond, CA 94802
(415) 222-9200
Lois Boyle

Riverside

Robert T. Andersen
Administrative Center
4080 Lemon Street
11th Floor
Riverside, CA 92501-3652

City of Riverside
Development Department
3900 Main Street
Riverside, CA 92522
(714) 782-5584
Marguerett S. Gulati

Ellen L. McPeters
3324 Brockton Avenue
Riverside, CA 92501

Mission Inn Foundation
3639 6th Street
Riverside, CA 92501
(714) 781-8241

Old Riverside Foundation
P.O. Box 601
Riverside, CA, 92502

Riverside Commercial Investors
3516 9th Street
Suite F
Riverside, CA 92501
(714) 788-6100
Michael R. Lawson

Riverside County Historical
Commission
P.O. Box 3507
Riverside, CA 92519
(714) 275-4310
Paul Romero

Sacramento

California Department of
Transportation
650 Howe Avenue
Suite 400
Sacramento, CA 95825
(916) 322-9548
John W. Snyder

California Main Street
801 K Street
Suite 1700
Sacramento, CA 95814
(916) 322-1502
Patricia Noyes

California State Capitol Museum
State Capitol
Room B-27
Sacramento, CA 95814
(916) 324-0312
Robert Wood

California State Park Rangers
Association
P.O. Box 28366
Sacramento, CA 95828-0366
(916) 383-2530
Doug Bryce

Jeffrey W. Groska
8879-E Salmon Faus Drive
Sacramento, CA 95826
(916) 551-3249

Salinas

City of Salinas
Redevelopment
200 Lincoln Avenue
Salinas, CA 93901
(408) 758-7387

E. L. Evans
296 Corral de Tierra
Salinas, CA 93908
(408) 484-1339

Monterey County Agricultural and
Rural Life Museum
Box 367
Salinas, CA 93902
(408) 424-1971
Meg Welden

Monterey County Historical Society
P.O. Box 3576
Salinas, CA 93912
(408) 757-8085
Betty Brusa

San Anselmo

Alan J. Almquist
6 Alice Way
San Anselmo, CA 94960
(415) 459-7292

San Bernardino

City of San Bernardino
City Planning Department
300 North D Street
San Bernardino, CA 92418
(714) 384-5057
Vincent Bautista

San Diego

Patrick Crowley, AIA, AICP
1985 Guy Street
San Diego, CA 92103
(619) 295-5210

Foote Development Company
1751 Hancock Street
San Diego, CA 92110
(619) 291-2100

Galvin-Cristilli, Architects
2150 West Washington Street
Suite 102
San Diego, CA 92110
(619) 298-8344
Joseph Nichols

Gaslamp Quarter Foundation
410 Island Avenue
San Diego, CA 92101
(619) 233-4694

Historical Site Board of San Diego
525 B Street
Suite 2002
Mail Stop 660 C
San Diego, CA 92101
(619) 533-3694
Ron Buckley

Kelley-Markham Consulting
121 Broadway
Suite 252
San Diego, CA 92101-5021
(619) 238-0555
Kathleen Kelley-Markham

Milford, Wayne, Donaldson and
Associates
846 5th Avenue
San Diego, CA 92101
Wayne Donaldson

San Diego Maritime Museum
1306 North Harbor Drive
San Diego, CA 92101
(619) 234-9153
David Brierley

Save Our Heritage Organization
P.O. Box 3571
San Diego, CA 92103
(619) 297-9327

Desiree Scott
4525 42nd Street
San Diego, CA 92116
(619) 284-3481

Ione R. Stiegler
4191 Stephens Street
San Diego, CA 92103
(619) 296-1145

San Francisco

Architectural Resources Group
Pier 9, The Embarcadero
San Francisco, CA 94111
(415) 421-1680
Bruce D. Judd

Carey and Company Architecture
300 Brannen
Suite 402
San Francisco, CA 94107
(415) 957-0100
Alice Ross Carey, AIA

City of San Francisco
Planning Division
P.O. Box 711
San Francisco, CA 94083
Maureen K. Morton

Linda M. Dishman
1703 Oak Street
Suite 4
San Francisco, CA 94117

G. Bland Platt Associates
362 Ewing Terrace
San Francisco, CA 94118
Bland Platt

Ellen R. Gibson
249 Mullen Avenue
San Francisco, CA 94110

Hillary Gitelman
115 Haight Street
Suite 3
San Francisco, CA 94102

GSA-Planning Staff (9PL)
525 Market Street
35th Floor
San Francisco, CA 94105
(415) 744-5258
Joan Byrens

H. J. Degenkolb Associates
350 Sansome Street
San Francisco, CA 94104
(415) 392-6952
David R. Bonneville

Kaplan McLaughlin Diaz
222 Vallejo Street
San Francisco, CA 94111
Herb McLaughlin

McCall Design Group
455 North Point
San Francisco, CA 94133
(415) 474-6100
Michael McCall

National Liberty Ship Memorial
S.S. Jeremiah O'Brian
Fort Mason Center
Building A
San Francisco, CA 94123-1382
(415) 441-3101
Marci Hooper

National Maritime Museum
Association
Presidio of San Francisco
Crissy Field
Building 275
San Francisco, CA 94129
(415) 929-0202
William J. Whalen

Page and Turnbull
364 Bush Street
San Francisco, CA 94104
Charles H. Page

Plant Brothers Corporation
300 Newhall Street
San Francisco, CA 94124-1426
Thomas Plant

Antonio Rossmann
380 Hayes Street
San Francisco, CA 94102
(415) 861-1401

St. Paulus Lutheran Church
950 Gough Street
San Francisco, CA 94102
(415) 493-3997
Mark S. Johnson

Sanchez-Ruschmeyer
2343 North Point
San Francisco, CA 94123
(415) 563-5477

Skidmore, Owings and Merrill
Architects-Engineers
333 Bush Street
San Francisco, CA 94104

Felix M. Warburg
2844 Broderick Street
San Francisco, CA 94123
(415) 243-8489

San Jose

George Barnes
1585 North 4th Street
Suite D
San Jose, CA 95112

Preservation Action Council of San Jose
P.O. Box 2287
San Jose, CA 95109-2287
(408) 971-0940
Karita Hummer

San Jose Historic Landmarks
Commission
801 North 1st Street
Room 400 Annex
San Jose, CA 95110
Nancy Hemmen

San Jose Redevelopment Agency
50 West San Fernando Street
Suite 1500
San Jose, CA 95113
(408) 277-5823
Leon Kimura

San Juan Capistrano

City of San Juan Capistrano
32400 Paseo Adelanto
San Juan Capistrano, CA 92675
(714) 493-1171
Jennifer S. Williams

San Leandro

Alta Mira Club and Docents
561 Lafayette Avenue
San Leandro, CA 94577

Basin Research Associates
Cultural Resource Services
14731 Catalina Street
San Leandro, CA 94577
(415) 357-0566
Colin I. Busby

Lene Cortsen-Diaz
1675 Hays Street
Suite 106
San Leandro, CA 94577
(415) 352-8362

San Leandro Historical Railway Society
P.O. Box 1212
San Leandro, CA 94577
(415) 785-0778
John Carbino

Thoroughbred Systems
1933 Davis Street
Suite 202
San Leandro, CA 94577
(415) 636-0252
K. Johnson

San Luis Obispo

City of San Luis Obispo
Cultural Heritage Commission
P.O. Box 8100
San Luis Obispo, CA 93403
(805) 549-7178
Terry Sanville

San Luis Obispo County Historic
Society
696 Monterey Street
San Luis Obispo, CA 93401
(805) 543-0638
Mark P. Hall-Patton

San Mateo

San Mateo County Historic Association
1700 West Hillsdale Boulevard
San Mateo, CA 94402
(415) 574-6441
Mitch Postel

San Quentin

San Quentin Museum Association
P.O. Box 205
San Quentin, CA 94964
(415) 454-1460
R. A. Nelson

San Ramon

City of San Ramon
Parks and Community Services
Commission
P.O. Box 5148
San Ramon, CA 94583
Paula L. Reilly

Santa Ana

Orange County Historical Commission
P.O. Box 4048
Santa Ana, CA 92702
(714) 834-5560
Robert Selway

Santa Barbara

Michael Neal Arnold, MAI
200 West Victoria Street
Santa Barbara, CA 93101
(805) 966-0869

Janice M. Hubbell, AICP
1311 Salinas Place
Suite 5
Santa Barbara, CA 93103
(805) 564-5470

Lenvik and Minor Architects
315 West Haley Street
Santa Barbara, CA 93101
(805) 963-3357
Edwin A. Lenvik

Monterey Hotel
P.O. Box 91110
Santa Barbara, CA 93190
Carl Johnson

Renovation/Preservation Task Force
980 Canon Road
Santa Barbara, CA 93110
(815) 732-6168
Nancy J. Miller

Santa Clara

Jeff Pedersen
1500 Warburton Avenue
Santa Clara, CA 95050
(408) 984-3114

Santa Cruz

Santa Cruz Historical Trust
1543 Pacific Avenue
Suite 220
Santa Cruz, CA 95060-3928

Victoria H. Scharnikou
2825 Warren Street
Santa Cruz, CA 95062
(408) 479-8479

Santa Fe Springs

City of Santa Fe Springs
11710 Telegraph Road
Santa Fe Springs, CA 90670-3658
(213) 868-0511
Robert G. Orpin

Santa Maria

Jay P. Younger
3107 Bunfill Drive
Santa Maria, CA 93455
(805) 937-0404

Santa Monica

Kenneth A. Breisch
1017 2nd Street
Suite 104
Santa Monica, CA 90403
(213) 458-5984

City Of Santa Monica
Landmarks Commission
1685 Main Street
Suite 212
Santa Monica, CA 90401
(213) 458-8341
Amanda Schachter

Getty Grant Program
401 Wilshire Boulevard
Suite 1000
Santa Monica, CA 90401
(213) 393-4244
Tim Whalen

Santa Rosa

City of Santa Rosa
Department of Community
Development
City Hall
P.O. Box 1678
Santa Rosa, CA 95402
(707) 576-5295
Roy Anderson

Saratoga

City of Saratoga
Saratoga Heritage Preservation
Commission
13777 Fruitvale Avenue
Saratoga, CA 95070
(408) 867-3438
George White

Sausalito

Stewart Brand
27B Gate 5 Road
Sausalito, CA 94965
(415) 332-9741

Sierra Madre

Sierra Madre Main Street
49 South Baldwin Avenue
Unit K
Sierra Madre, CA 91024
(818) 355-5111
Ron Hutson

Simi Valley

Simi Valley Historical Society
P.O. Box 351
Simi Valley, CA 93062
(805) 526-0879
Patricia Havens

South Lake Tahoe

U.S. Forest Service
Lake Tahoe Basin Management Unit
870 Emerald Bay Road
Suite 1
South Lake Tahoe, CA 96150
Penny Rucks

South San Francisco

South San Francisco History Room
306 Walnut Avenue
South San Francisco, CA 94080
(415) 877-8533
Kathleen Kay

Suisun City

City of Suisun City
701 Cedar Street
Suisun City, CA 94585
(707) 429-2900
Thomas E. Bland

Sunnyvale

City of Sunnyvale
456 West Olive Avenue
Sunnyvale, CA 94086

Susanville

Lassen County Heritage Commission
490425 Highway 139
Susanville, CA 96130
(916) 257-9505
Margret Swickard

Thousand Oaks

Sherwood Architects
320 West Stafford Road
Thousand Oaks, CA 91361
(805) 495-6667
Edwin Meals

Tiburon

Belvedere-Tiburon Landmarks Society
1600 Juanita Lane
Box 134
Tiburon, CA 94920
(415) 435-1853
Joan Bergsund

Town of Tiburon
1155 Tiburon Boulevard
Tiburon, CA 94920
(415) 435-0956
R. Kleinert

Louise Wilson
5 Saint Gabrielle Court
Tiburon, CA 94920
(415) 435-9216

Topanga

Charles Lockwood
20030 Valleyview Drive
Topanga, CA 90290

Truckee

Truckee-Donner Historical Society
P.O. Box 893
Truckee, CA 95734
(916) 587-2876
Ruth Everingham

Turlock

City of Turlock
Community Developement
Department
P.O. Box T
Turlock, CA 95381
(209) 668-5565
Steven L. Hallam

Ukiah

Held Poage Research Library
603 West Perkins Street
Ukiah, CA 95482
(707) 462-6969
Lila J. Lee

Judy Pruden
304 South Hortense
Ukiah, CA 95482
(707) 462-4945

Upland

MSAI
123 East 9th Street
Suite 204
Upland, CA 91786
(714) 981-0894

Vacaville

City of Vacaville
Planning and Community
Development Department
650 Merchant Street
Vacaville, CA 95688
Greg Werner

Ventura

County of Ventura Cultural Heritage
Board
800 South Victoria Avenue
Ventura, CA 93009
(805) 654-3968
Katherine E. Garner

Visalia

City of Visalia
707 West Acequia
Visalia, CA 93291
(209) 738-3348
Andrew J. Chamberlain

Vista

Vista Town Center Association
200 East Jefferson
P.O. Box 322
Vista, CA 92085
(619) 724-8822
Vicki Barringer

Walnut Creek

Mara Thiessen Jones
60 Castle Crest
Walnut Creek, CA 94595
(415) 935-1630

Watsonville

Edna E. Kimbro
184 Old Adobe Road
Watsonville, CA 95076
(408) 728-5257

Weaverville

Trinity County Historical Society
P.O. Box 333
Weaverville, CA 96093
Herbert Woods

West Hollywood

City of West Hollywood
Community Development Department
8611 Santa Monica Boulevard
West Hollywood, CA 90069
David Amorena

City of West Hollywood
Community Development Department
8611 Santa Monica Boulevard
West Hollywood, CA 90069
Debbie Potter

Westlake Village

Daniel Keehrer
IB Magazine
875 South Westlake Boulevard
Suite 211
Westlake Village, CA 91361
(805) 496-6156

West Sacramento

California Department of Parks and
Recreation
Interpretive Collections Section
1280 Terminal Street
West Sacramento, CA 95691

Whittier

City of Whittier
13230 Penn Street
Whittier, CA 90602
Elvin H. Porter

Craig Hill
6211 Painter Avenue
Whittier, CA 90601
(213) 696-6438

Woodacre

Pagliacco Turning and Milling
P.O. Box 225
Woodacre, CA 94973
(415) 488-4333
Steve Evans

Woodland Hills

L. R. Evron
22345 De Grasse Drive
Woodland Hills, CA 91364
(818) 887-1166

Yorba Linda

Vicki Solheid
5015 Twilight Canyon Road
Suite G
Yorba Linda, CA 92687
(714) 692-5318

Yuba City

Community Memorial Museum
P.O. Box 1555
Yuba City, CA 95992
(916) 741-7141
Jackie Lowe

COLORADO

STATE HISTORIC PRESERVATION OFFICE

Colorado Historical Society
1300 Broadway
Denver, Co 80203
(303) 866-2136
James E. Hartmann, SHPO
Susan Collins, Deputy SHPO

STATEWIDE PRESERVATION ORGANIZATION

Colorado Preservation
P.O. Box 843
Denver, CO 80201-0843
(303) 920-5090
Roxanne Eflin, President

NATIONAL TRUST REGIONAL OFFICE

Mountains/Plains Regional Office
511 16th Street
Suite 700
Denver, CO 80202
(303) 623-1504
(303) 825-8073 (Fax)
Barbara J. Pahl, Director

NATIONAL TRUST ADVISORS

Betty M. Chronic
4705 Shawnee Place
Boulder, CO 80303
(303) 494-7553

Jon Schler
222 South 6th Street
Room 409
Grand Junction, CO 81501
(303) 248-7310
(303) 248-7317 (Fax)

PRESERVATION ACTION COORDINATORS

Mona Ferrugia
556 Circle Drive
Denver, CO 80206
(303) 355-0834

AIA STATE PRESERVATION COORDINATOR

Daniel J. Havekost, FAIA
Havekost and Lee Architects
1121 Grant Street
Denver, CO 80203
(303) 861-1121

NATIONAL PARK SERVICE REGIONAL OFFICE

Rocky Mountain Regional Office
P.O. Box 25287
Denver, CO 80225
(303) 969-2875

ADVISORY COUNCIL ON HISTORIC PRESERVATION DIVISION

Western Office of Project Review
730 Simms Street
Room 401
Golden, CO 80401
(303) 231-5320
Claudia Nissley, Director

Arvada

City of Arvada
8101 Ralston Road
Arvada, CO 80002
(303) 431-3020
Ed Talbot

Aspen

Aspen Historical Society
620 West Bleeker Street
Aspen, CO 81611
(303) 925-3721
Renee Allard Betts

City of Aspen
Historic Preservation Commission
City Hall
130 South Galena
Aspen, CO 81611
(303) 920-5090
Roxanne Eflin

Aurora

Aurora History Museum
Aurora Preservation Commission
15001 East Alamada Drive
Aurora, CO 80012
(303) 360-8545
Virginia R. Steele

Black Hawk

City of Black Hawk
P.O. Box 327
Black Hawk, CO 80422
(303) 582-5221
Joanne Lah

Boulder

Kathryn H. Barth
2940 20th Street
Boulder, CO 80304
(303) 440-0881

Betty M. Chronic
4705 Shawnee Place
Boulder, CO 80303
(303) 494-7553

Colorado Chautauqua Association
Chautauqua Park
Boulder, CO 80302
(303) 442-3282
Elaine Scanlon

Community Services Collaborative
1315 Broadway Avenue
Boulder, CO 80302
(303) 442-3601
John D. Feinberg

Buena Vista

Melanie I. Milam
Box 1884
Buena Vista, CO 81211
(303) 979-1270

Central City

Gilpin County Historical Society
P.O. Box 244
Central City, CO 80427
(303) 582-5283
Linda Jones

Colorado Springs

Historic Property Alliance
1014 North Weber
Colorado Springs, CO 80903
(719) 632-2800
Mark Mahler

North End Homeowners Association
1238 Wood Avenue
Colorado Springs, CO 80903
(719) 578-9289
Mary J. Lain

Craig

City of Craig
300 West 4th Street
Craig, CO 81625
(303) 824-8151
Greg Potter

Cripple Creek

City of Cripple Creek
337 East Bennett
P.O. Box 430
Cripple Creek, CO 80813
(303) 689-2778
Brian Levine

Denver

Agape Christian Church
2501 California Street
Denver, CO 80205
(303) 296-2454
Robert E. Woolfolk

Capitol Hill Community Center
1290 Williams Street
Denver, CO 80218
(303) 399-1324
Nancy Schoyer

Central City Opera House Association
621 17th Street
Suite 1601
Denver, CO 80202
(303) 292-6500
Daniel Rule

Colfax on the Hill Business
Improvement District
P.O. Box 18853
Denver, CO 80218
(303) 832-2086
Marty Amble

Colorado Historical Society
Colorado Heritage Center
1300 Broadway
Denver, CO 80203
(303) 866-4684
Julie Morgan

Colorado Preservation
P.O. Box 843
Denver, CO 80201-0843
(303) 866-4678
Sally Pearce

Council of Public Television
Channel 6
1261 Glenarm Place
Denver, CO 80204
(303) 892-6666
A. Zimmerman

Del Norte Neighborhood Development
Corporation
3500 Navajo Street
Denver, CO 80211-3039
(303) 433-0924
Marvin Kelly

Hammer, Siler, George Associates
1638 Pennsylvania Street
Denver, CO 80203
(303) 860-9996
Daniel R. Guimond

Historic Denver
1330 17th Street
Denver, CO 80202
(303) 534-1858
Jennifer T. Moulton

J. R. Harris and Company
1580 Lincoln
Suite 770
Denver, CO 80203
(303) 860-9021
James R. Harris

National Conference of State
Legislatures
1560 Broadway
Suite 700
Denver, CO 80202-5140
(303) 830-2200
Laura Loyacono

Temple Events Center Uptown
1595 Pearl Street
Denver, CO 80203
(303) 860-9400
Marcia Johnson

Mike Teskey
2088 Jasmine Street
Denver, CO 80207
(303) 393-7623

John B. Woodward III
3436 East Kentucky Avenue
Denver, CO 80209

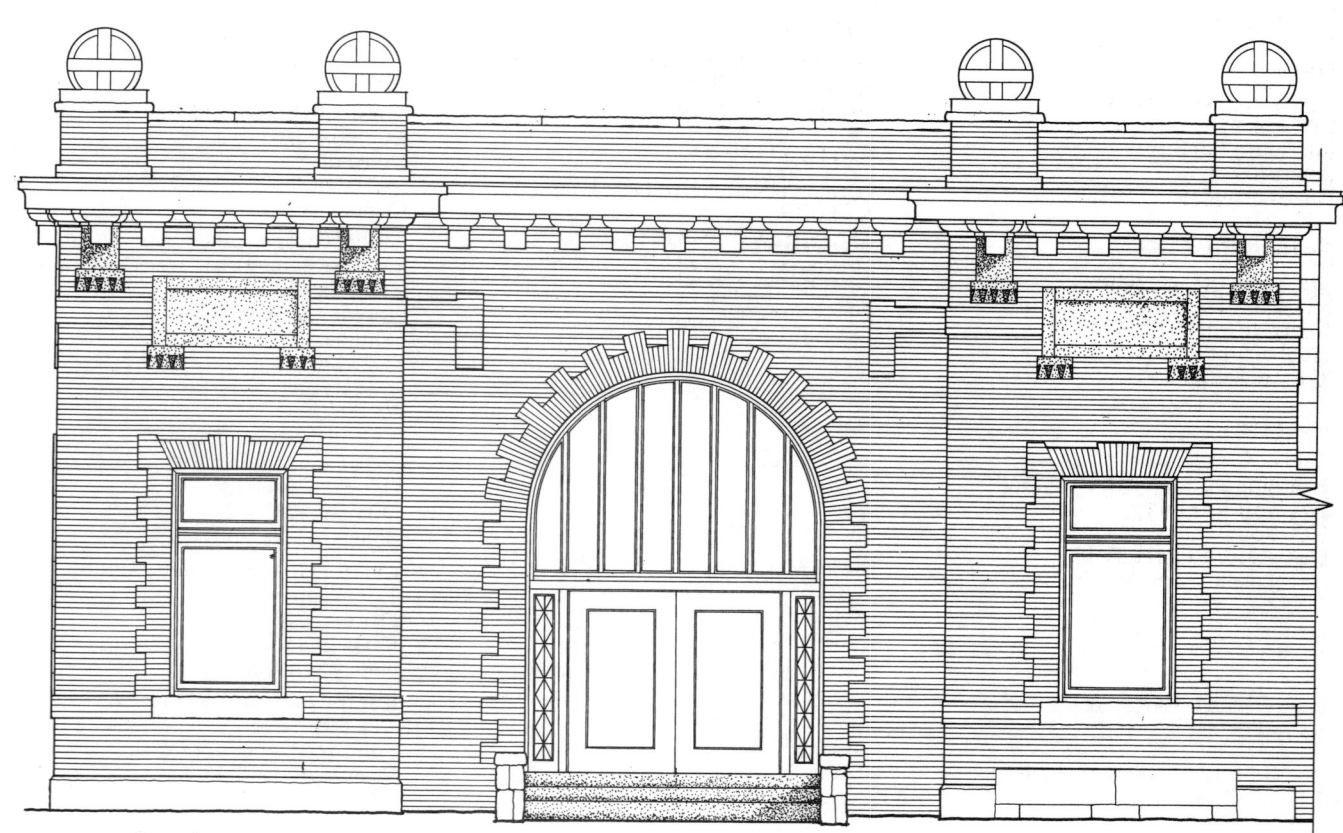

Moffat Station, Denver, Colorado. (D.R. Mighell, HABS)

Durango

City of Durango
Community Development
949 East 2nd Avenue
Durango, CO 81301
(303) 247-5622
Gregory Hoch

Fort Collins

City of Fort Collins
Historic Preservation Office
P.O. Box 580
Fort Collins, CO 80522
(303) 221-6756
Carol Tunner

Colorado State University
Historic Preservation Department
Fort Collins, CO 80521
Liston Leyendecker

Fort Collins Downtown Development
Authority
102 Remington
Fort Collins, CO 80524
Garry Estrada

Fort Lupton

South Platte Valley Historical Society
P.O. Box 633
147 South Denver Avenue
Fort Lupton, CO 80621
(303) 857-2471
John Martin

Frisco

Frisco Historic Society
P.O. Box 820
Frisco, CO 80443
(303) 668-5276
Odette Baehler

Georgetown

The Town of Georgetown
P.O. Box 426
Georgetown, CO 80444
(303) 623-6882
Tammy Sanford

Gunnison

City of Gunnison
Planning Department
201 West Virginia
P.O. Box 239
Gunnison, CO 81230

Idaho Springs

Idaho Springs Historical Society
Box 1318
Idaho Springs, CO 80452
(303) 657-4100
Larry Cox

Lakewood

U.S. Department of Agriculture
Forest Service
Recreation
P.O. Box 25127
Lakewood, CO 80225

Lamar

City of Lamar
P.O. Box 270
Lamar, CO 81052

Las Animas

Pioneer Historical Society of Bent
County
P.O. Box 68
Las Animas, CO 81054
(719) 456-0220
Galen E. Moss

Littleton

Judy Lee Kinney
5200 Yellowstone
Littleton, CO 80123
(303) 794-8657

Littleton Historic Museum
6028 South Gallup
Littleton, CO 80120

Loveland

Loveland Downtown Development
Authority
427 North Railroad Avenue
Loveland, CO 80537
(303) 667-6130
Felicia Harmon

Manitou Springs

City of Manitou Springs
Historic Preservation Commission
606 Manitou Avenue
Manitou Springs, CO 80829
(303) 685-4398

Pueblo

City of Pueblo, REDO
P.O. Box 1427
Pueblo, CO 81002
(719) 543-6006
William J. Zwick

Southern Colorado Heritage
Conservancy
P.O. Box 4407
Pueblo, CO 81003
(719) 546-6100
Vicki Burkland

Silver Plume

Town of Silver Plume
P.O. Box 457
Silver Plume, CO 80476
(303) 569-2363
Colleen Skinner

Silverton

San Juan County Historical Society
P.O. Box 154
Silverton, CO 81433
(303) 387-5838
Beverly Rich

Telluride

Sheridan Arts Foundation
P.O. Box 2680
Telluride, CO 81435
(303) 728-6363

Trinidad

Trinidad Downtown Area Development
Corporation
309 Nevada Avenue
Trinidad, CO 81082-2596
(303) 846-4550
Janie Kurtz

CONNECTICUT

STATE HISTORIC PRESERVATION OFFICE

Connecticut Historical Commission
59 South Prospect Street
Hartford, CT 06106
(203) 566-3005
John W. Shannahan, SHPO
Dawn Maddox, Deputy SHPO

STATEWIDE PRESERVATION ORGANIZATION

Connecticut Trust for Historic
Preservation
940 Whitney Avenue
Hamden, CT 06517-4002
(203) 562-6312
C. Roderick O'Neil, Chairman
Laura Clarke, Executive Director

NATIONAL TRUST REGIONAL OFFICE

Northeast Regional Office
7 Faneuil Hall Marketplace
5th Floor
Boston, MA 02109
(617) 523-0885
(617) 523-1199 (Fax)
Vicki Sandstead, Director

NATIONAL TRUST ADVISORS

Terry J. Tondro
103 North Beacon Street
Hartford, CT 06105
(203) 232-4660
(203) 241-7666 (Fax)

PRESERVATION ACTION COORDINATOR

Frederick Biebesheimer
Interdesign
Shore Road
Old Lyme, CT 06371
(203) 434-8083

AIA STATE PRESERVATION COORDINATOR

Frederick C. Biebesheimer, FAIA
Interdesign
101 Shore Road
P.O. Box 250
Old Lyme, CT 06371
(203) 434-8083

NATIONAL PARK SERVICE REGIONAL OFFICE

North Atlantic Regional Office
15 State Street
Boston, MA 02109-3572
(617) 223-5001

ADVISORY COUNCIL ON HISTORIC PRESERVATION DIVISION

Eastern Office of Project Review
1100 Pennsylvania Avenue, N.W.
Suite 809
Washington, DC 20004
(202) 786-0503
Don L. Klima, Director

Berlin

John W. Wilcox
121 Sunset Lane
Berlin, CT 06037

Bristol

Bristol Development Authority
111 North Main Street
Bristol, CT 06010
(203) 584-7971
Walter J. Murphy

Centerbrook

Robert L. Harper
38 Oak Drive
Centerbrook, CT 06409
(203) 767-0175

Cos Cob

Historical Society of the Town of
Greenwich
Bush-Holley House
39 Strickland Road
Cos Cob, CT 06807
(203) 869-6899
Peter H. Conze

Danbury

Danbury Scott-Funton Museum and
Historical Society
43 Main Street
Danbury, CT 06810
(203) 743-5200
Lucye Boland

Danielson

Brooklyn Housing Authority
P.O. Box 156
Danielson, CT 06239
(203) 774-7020
Sal Zito

Town of Killingly
172 Main Street
Danielson, CT 06239
(203) 774-8601
Denise M. Rose

Farmington

C.D.C. Financial Corporation
17 Talcott Notch Road
Farmington, CT 06032
(203) 236-6234
Margo Kelleher

Hill-Stead Museum
35 Mountain Road
Farmington, CT 06032

Town of Farmington
1 Monteith Drive
Farmington, CT 06032
(203) 673-8221
Jeffrey Ollendorf

Greenwich

William E. Hegarty
Mead's Point
Greenwich, CT 06830
(203) 869-5563

Lorraine C. Waechter
190 Lake Avenue
Greenwich, CT 06830
(203) 661-5616

Guilford

C. A. Hollingsworth
64 Fair Street
Guilford, CT 06437
(203) 453-6956

Hamden

Connecticut Trust
940 Whitney Avenue
Hamden, CT 06517
(203) 562-6312
Charles L. Granquist

Hamden Historical Society
P.O. Box 5512
Hamden, CT 06518
(203) 288-7678
Brian Poivier

Hartford

Jerry Arnstein and Associates
7 Sequassen Street
Hartford, CT 06106

Greater Hartford Architecture
Conservancy
278 Farmington Avenue
Hartford, CT 06105
(203) 525-0279
Michael J. Kerski

Hartford Public Library
500 Main Street
Hartford, CT 06103

Mark Twain Memorial
351 Farmington Avenue
Hartford, CT 06105
(203) 247-0998
John Boyer

Middletown Preservation Trust
518 Prospect Avenue
Hartford, CT 06105
(203) 346-1646
Ann C. Street

First Church of Christ, Congregational, New Haven, Connecticut. (L. Robinson, W. Morton, HABS)

Smith Edwards Architects
331 Wethersfield Avenue
Hartford, CT 06114
Jared T. Edwards

State of Connecticut
Office of the Comptroller
55 Elm Street
Hartford, CT 06106
J. Caldwell

Stowe-Day Foundation
77 Forest Street
Hartford, CT 06105
(203) 522-9258

Harwinton

Harwinton Historic District
Commission
Town of Harwinton
100 Bentley Drive
Harwinton, CT 06791
(203) 485-1381
Frank Chiarmonte

Litchfield

Litchfield Historical Society
P.O. Box 385
Litchfield, CT 06759
(203) 567-4501
Catherine Keene

Manchester

The Little Theater of Manchester
9 Laurel Street
Cheney Hall
Manchester, CT 06040
(203) 646-0657
Mary E. Blish

Middletown

Municipal Development Office
P.O. Box 1300
Middletown, CT 06457
(203) 344-3419
William M. Kuehn

New Canaan

Richard Bergmann
63 Park Street
New Canaan, CT 06840-4598
(203) 966-9505

New Canaan Historical Society
13 Oenoke Ridge
New Canaan, CT 06840
(203) 966-1776
Janet Lindstrom

New Haven

Paul B. Bailey, AIA
110 Audubon Street
New Haven, CT 06510
(203) 776-8888

Cesar Pelli and Associates
1056 Chapel Street
New Haven, CT 06510
(203) 777-2515
Jack A. Gold

F.J. Dahill Company
P.O. Box 9578
New Haven, CT 06535
(203) 469-6454
James D. McAdam

Neighborhood Housing Services of New
Haven
333 Sherman Avenue
New Haven, CT 06511
(203) 562-0598
James A. Paley

New Haven Downtown Council
P.O. Box 1456
195 Church Street
New Haven, CT 06506
Matthew Nemerson

New Haven Preservation Trust
P.O. Box 1671
New Haven, CT 06507
(203) 562-5919
Preston Maynard

New London

Garde Arts Center
325 Captain's Walk
New London, CT 06320
(203) 444-6766
Steve Sigel

Newtown

Heritage Preservation Trust
39 Deep Brook Road
Newtown, CT 06470
(203) 748-6517
Donald Studley

North Haven

Donald Baerman
42 Wayland Street
North Haven, CT 06473
(203) 288-8911

Norwalk

Lockwood-Mathews Mansion Museum
295 West Avenue
Norwalk, CT 06850
(203) 838-1434
David Byrnes

Old Lyme

Interdesign
P.O. Box 250
Old Lyme, CT 06371
(203) 434-8083
Frederick Biebesheimer

Portland

Town of Portland
P.O. Box 71
265 Main Street
Portland, CT 06480
(203) 342-2880
Raymond Carpentino

Salisbury

Henry W. Burgess
Box 218
Salisbury, CT 06068

Simsbury

East Weatogue Historic District
57 East Weatogue Street
Simsbury, CT 06070
(203) 658-1190
Anita L. Mielert

Galliher and Baier, Architects
The Courtyard 10
Simsbury, CT 06070
(203) 658-7617
Roger Galliher

South Windsor

Charles S. Rotenberg
P.O. Box 174
809 Main Street
South Windsor, CT 06074-0174
(203) 289-1759

Southbury

Henry Bassett
Gilbert Road
Southbury, CT 06488

Southington

The Greater Southington Chamber of
Commerce
51 North Main Street
Southington, CT 06489
(203) 628-8036
Pauline B. Levesque

Stamford

Auram Freedberg
992 High Ridge Road
Stamford, CT 06905
(203) 322-3992

Thomas Yanicky
53 Westover Avenue
Stamford, CT 06902
(203) 323-2721

Stonington

Betty Richards
18 School Street
Stonington, CT 06378
(203) 535-0432

Storrs

Public Archaeology Survey Team
P.O. Box 209
Storrs, CT 06268
(203) 456-4264
Mary Soulsby

Stratford

Richard J. Hershner, Jr.
P.O. Box 973
Stratford, CT 06490
(203) 375-5155

Torrington

Torrington Historical Society
192 Main Street
Torrington, CT 06790
(203) 482-8260
Mark McEachern

Waterbury

Waterbury Development Agency
101 South Main Street
Waterbury, CT 06706
(203) 757-9621
Howard Plomann

West Hartford

Town of West Hartford
Town Hall Common
West Hartford, CT 06107-2428
(203) 236-3231
Donald Foster

Westport

Landscapes
Box 2425
Saugatuck Station
Westport, CT 06880
(203) 227-3310
Patricia M. O'Donnell

Westport Historical Society
25 Avery Place
Westport, CT 06880
(203) 227-0528
Barbara Shriver

Woodbury

Straus-Edwards Associates
331 Main Street, South
Woodbury, CT 06798
(203) 263-0018

DELAWARE

STATE HISTORIC PRESERVATION OFFICE

Division of Historical and Cultural Affairs
Hall of Records
P.O. Box 1401
Dover, DE 19901
(302) 739-5313
Daniel Griffith, SHPO
Joan Larrivee, Deputy SHPO

STATEWIDE PRESERVATION ORGANIZATION

Historical Society of Delaware
505 Market Street Mall
Wilmington, DE 19081
(302) 655-7161
Barbara Benson, Director

NATIONAL TRUST REGIONAL OFFICE

Mid-Atlantic Regional Office
Cliveden
6401 Germantown Avenue
Philadelphia, PA 19144
(215) 438-2886
(215) 438-2892 (Fax)
Patricia D. Wilson, Director

NATIONAL TRUST ADVISORS

Kim Rogers Burdick
Carvel State Office Building
820 North French Street
11th Floor
Wilmington, DE 19801
(302) 577-6517

Eldon duP. Homsey
2003 North Scott Street
Wilmington, DE 19806
(302) 656-4491
(302) 656-5956 (Fax)

PRESERVATION ACTION COORDINATOR

Joan Robinson-Medland
Jonathan Jenkins House
11 North Main Street
Camden, DE 19934
(302) 734-2509

AIA STATE PRESERVATION COORDINATOR

Vacant

NATIONAL PARK SERVICE REGIONAL OFFICE

Mid-Atlantic Regional Office
143 South 3rd Street
Philadelphia, PA 19106
(215) 597-7018

ADVISORY COUNCIL ON HISTORIC PRESERVATION DIVISION

Eastern Office of Project Review
1100 Pennsylvania Avenue, N.W.
Suite 809
Washington, DC 20004
(202) 786-0503
Don L. Klima, Director

Dover

Central Dover Business Association
c/o Delaware Made
214 South State Street
Dover, DE 19901
(302) 736-1419
Tom Smith

Division of Historical and Cultural Affairs
Hall of Records
Dover, DE 19901
Daniel Griffith

Kent County Planning Office
414 Federal Street
Room 320
Dover, DE 19901

Old Town Hall, New Castle, Delaware.
(S. Barnette, HABS)

Hockessin

Paul R. Stambaugh, Sr.
514 Hemingway Drive
Hockessin, DE 19707
(302) 998-1131

Lewes

New Devon
P.O. Box 516
Lewes, DE 19971
(30) 26-456466
Dale Jenkins

Newark

City of Newark
200 Elkton Road
P.O. Box 390
Newark, DE 19715-0390
(302) 366-7030
Marguerite Ashley

New Castle County Department of Planning
2701 Capitol Trail
Newark, DE 19711
(302) 366-7780
Wayne W. Grafton

Wilmington

City of Wilmington
Mayor's Office
Planning Division
800 North French Street
7th Floor
Wilmington, DE 19801
(302) 571-4030
Herb M. Inden

Friends of Rockwood
610 Shipley Road
Wilmington, DE 19809
(302) 571-7776
Frederick T. Haase

Hagley Museum and Library
Acquisitions Department
Buck Road
P.O. Box 3630
Wilmington, DE 19807

Eldon Homsey
2003 North Scott Street
Wilmington, DE 19806

International Group for Historic Aircraft Recovery
1121 Arundel Drive
Wilmington, DE 19808
(302) 994-4410
Patricia Thrasher

Mohawk Corporation
901 Washington Street
Wilmington, DE 19801
(302) 658-8760
Charles M. Weymouth

DISTRICT OF COLUMBIA

STATE HISTORIC PRESERVATION OFFICE

Department of Consumer and
Regulatory Affairs
District Building
1350 Pennsylvania Avenue, N.W.
Washington, DC 20004
Robert Mallett, SHPO
(202) 727-6365

Stephen J. Raiche
Division Chief
Historic Preservation Division
Deputy SHPO
614 H Street, N.W.
Suite 305
Washington, DC 20004
(202) 727-7360
(202) 727-8040 (Fax)

STATEWIDE PRESERVATION ORGANIZATION

D.C. Preservation League
918 F Street, N.W.
Suite 301
Washington, DC 20004
(202) 737-1519
James Rogers, President
Steve Callcott, Acting Executive
Director

NATIONAL TRUST REGIONAL OFFICE

Mid-Atlantic Regional Office
Cliveden
6401 Germantown Avenue
Philadelphia, PA 19144
(215) 438-2886
(215) 438-2892 (Fax)
Patricia D. Wilson, Director

NATIONAL TRUST ADVISOR

Charles Cassell
3003 Van Ness Street, N.W.
Washington, DC 20006
(202) 362-0858

Carol Thompson-Cole
2032 Belmont Street
Suite 207
Washington, DC 20009
(202) 626-7260

PRESERVATION ACTION COORDINATOR

Vacant

AIA STATE PRESERVATION COORDINATOR

M. Hamilton Morton, AIA
4813 Falstone Avenue
Chevy Chase, MD 20815-5541
(202) 872-0082

NATIONAL PARK SERVICE REGIONAL OFFICE

National Capital Region
1100 Ohio Drive, S.W.
Washington, DC 20242
(202) 619-7005

ADVISORY COUNCIL ON HISTORIC PRESERVATION DIVISION

Eastern Office of Project Review
1100 Pennsylvania Avenue, N.W.
Suite 809
Washington, DC 20004
(202) 786-0503
Don L. Klima, Director

William Adair
P.O. Box 50156
Washington, DC 20004
(202) 638-4660

Anacostia Historical Society
1395 Morris Road, S.E.
Washington, DC 20020

Archetype
1841 Columbia Road, N.W.
Washington, DC 20009
(202) 265-7565
Belinda Reeder

Association Of Collegiate Schools of
Architecture
1735 New York Avenue, N.W.
Washington, DC 20006

James Bayley Associates, PC
514 10th Street, N.W.
Suite 1000
Washington, DC 20004
(202) 737-4477
James Bayley

The Brass Knob
2311 18th Street, N.W.
Washington, DC 20009
(202) 332-3370
Donetta George

Capitol Hill Restoration Society
1002 Pennsylvania Avenue, S.E.
Washington, DC 20003
Pat Schauer

Cleveland Park Historical Society
3030 Newark Street, N.W.
Washington, DC 20008
(202) 966-3107
Susan Hornbostel

T. Allan Comp
1847 Vernon Street, N.W.
Washington, DC 20009

Morgan A. Conolly
2100 Connecticut Avenue, N.W.
Washington, DC 20036
(202) 785-4591

D.C. Preservation League
918 F Street, N.W.
Suite 310
Washington, DC 20004
(202) 737-1519

David Scott Dederick
1616 18th Street, N.W.
Suite 1009
Washington, DC 20009
(202) 682-7169

W. Kirk Denton
2001 16th Street, N.W.
Suite 4
Washington, DC 20009
(202) 667-3564

Educational Organization for United
Latin Americans
1842 Calvert Street, N.W.
Washington, DC 20009
(202) 483-5800

Federal Reserve System
Board of Governors
Research Library
Washington, DC 20551
(202) 452-3390

Marc Fetterman
3318 Ordway Street, N.W.
Washington, DC 20008

Foundation for the Preservation of
Historic Georgetown
Georgetown Station
P.O. Box 3603
Washington, DC 20007
(202) 822-8888
C. D. Langhorne

Fried, Frank, Harris, Shriver and
Jacobson
1001 Pennsylvania Avenue, N.W.
Suite 800
Washington, DC 20004
(202) 639-7010
Leonard A. Zax

Friends of The Eastern Market
14 7th Street, S.E.
Washington, DC 20003
(202) 547-3320
Shelley R. Larson

John Gearon
4801 Connecticut Avenue, N.W.
Suite 707
Washington, DC 20008
(202) 363-4983

Geohistory Associates
1323 Wallach Place, N.W.
Washington, DC 20009
F. R. Doleman

Georgetown University
Design and Engineering
New South Building
3700 O Street, N.W.
Washington, DC 20057
(202) 637-1317
Andrew E. Harris

Grace Church, Georgetown, Washington, D.C. (E. Villatoro, G. Rueblinger, HABS)

Gomez Foundation for Mill House
1001 Connecticut Avenue, N.W.
Suite 1004
Washington, DC 20036
(202) 872-1401
Michael H. Cardozo

Roy Eugene Graham, AIA
3133 Connecticut Avenue, N.W.
Suite 1122
Washington, DC 20008
(202) 462-2011

Hartman-Cox Architects
1025 Thomas Jefferson St., N.W.
Suite 170
Washington, DC 20007
(202) 333-6446
Warren J. Cox

Historic Mount Pleasant
P.O. Box 3176
Washington, DC 20010
(202) 462-1880
Ellen Jeurling

Kim Hoagland
423 4th Street, S.E.
Washington, DC 20003
(202) 343-9607

William West Hopper
5455 Broad Branch Road
Washington, DC 20015
(202) 363-4189

Harry Weese and Associates
955 L'Enfant Plaza North, S.W.
Washington, DC 20024
(202) 863-1090

H Street Community Development
Corporation
611 H Street, N.E.
Washington, DC 20002
(202) 544-8353
William J. Barrow

International Masonry Institute
823 15th Street N.W.
Washington, DC 20005
Ray Lackey

George S. Jenkins
1769 Church Street, N.W.
Washington, DC 20036
(202) 429-6946

Jewish Historical Society
701 3rd Street, N.W.
Washington, DC 20001
(202) 789-0900
Julian Feldman

Kaempfer Company Investment
Builders
1250 24th Street, N.W.
Suite 300
Washington, DC 20037
(202) 331-4300
William N. Herman

Kalorama Design Associates
1745 Lanier Place, N.W.
Washington, DC 20009
(202) 265-8577
Patricia D. Bonbrest

Karr Associates
1916 Biltmore Street, N.W.
Washington, DC 20009
(202) 387-3413
Lawrence Karr

The L'Enfant Trust
2013 O Street, N.W.
Washington, DC 20036
(202) 347-1814
Margaret S. Dean

Le Droit Park Preservation Society
317 U Street, N.W.
Washington, DC 20001
(202) 232-8239

League of Historic American Theatres
1511 K Street, N.W.
Suite 923
Washington, DC 20005
(202) 783-6966
Mary M. Schoenfeld

Lutheran Resources Commission
5 Thomas Circle, N.W.
Washington, DC 20005
Arlene Rosheim

Manna
614 S Street, N.W., Rear
Washington, DC 20001
(202) 232-2844
Kay Schultz

Moorland-Spingarn Research Center
Howard University
Washington, DC 20059
(202) 636-7241
Elinor D. Sinnette

Mount Vernon College Library
2100 Foxhall Road, N.W.
Washington, DC 20007

Mount Zion United Methodist Church
1334 29th Street, N.W.
Washington, DC 20007-3351

Muse-Wiedemann Architects
5630 Connecticut Avenue, N.W.
Washington, DC 20015
(202) 966-1266
Stephen Muse

National Center For Preservation Law
1333 Connecticut Avenue, N.W.
Suite 300
Washington, DC 20036
Stephen N. Dennis

National Conference of State Historic
Preservation Officers
444 North Capitol Street, N.W.
Washington, DC 20001
Eric Hertfelder

National Geographic Society Traveler
Magazine
17th and M Streets, N.W.
Washington, DC 20036-3208
(202) 857-7000

National Institute for the Conservation
of Cultural Property
3299 K Street, N.W. Suite 403
Washington, DC 20007
(202) 625-1495
L. L. Reger

National Parks and Conservation
Association
1015 31st Street, N.W.
Washington, DC 20007
(202) 944-8572
Bruce Craig

National Preservation Institute
401 F Street
Room 301
Judiciary Square, N.W.
Washington, DC 20001
(202) 393-0038
Carol S. Gould

Neighborhood Housing Services
2401 Virginia Avenue, N.W.
Suite 302
Washington, DC 20037
(202) 429-2994
David B. Wiley

North American Society for Oceanic
History
P.O. Box 18108
Washington, DC 20036

Notter Finegold and Alexander
1350 Connecticut Avenue, N.W.
Suite 400
Washington, DC 20036
(202) 296-2700
George M. Notter

Preservation Action
1350 Connecticut Avenue, N.W.
Suite 401
Washington, DC 20036
(202) 659-0915
Nellie L. Longsworth

Darrel Rippeteau, AIA
1530 14th Street, N.W.
Washington, DC 20005
(202) 265-0777

Thomas D. Saunders
1843 Mintwood Place N.W.
Suite 202
Washington, DC 20009
(202) 483-3461

Save The Tivoli
P.O. Box 44930
Washington, DC 20026-4930
(202) 462-2792
Patricia A. Meyer

Shaw Heritage Trust
700 7th Street, S.W.
Washington, DC 20024
Steve Newman

Sheridan-Kalorama Historic
Association
2144 California Street, N.W.
Suite 812
Washington, DC 20008
(202) 483-4866
Howard S. Berger

Sumner School Museum and Archives
1201 17th Street, N.W.
Washington, DC 20036
(202) 727-3419
Richard Hurlbut

Taylor Real Estate
5506 Connecticut Avenue, N.W.
Washington, DC 20015
(202) 362-0300
Keene Taylor

Traceries
1606 20th Street, N.W.
Washington, DC 20009
Emily H. Eig

Tudor Place Foundation
1605 32nd Street, N.W.
Washington, DC 20007
(202) 965-0400
Osborne P. Mackie

United States Coast Guard
Commandant (G-NSR-1)
2100 2nd Street, S.W.
Washington, DC 20593-0001
Lt. J.F. Brooks

United States Department of
Agriculture
Farmers Home Administration
14th and Independence Avenue, S.W.
South Building, Room 5302
Washington, DC 20250
Warren Clayman

United States Senate Commission on
Art and Antiquities
U.S. Capitol Building
Room S-411
Washington, DC 20510
(202) 244-2955
James R. Ketchum

U.S. Committee/International Council
on Monuments and Sites
1600 H Street, N.W.
Washington, DC 20006
(202) 842-1866
Terry B. Morton, Hon. AIA

Anne Vytlacil
1511 33rd Street, N.W.
Washington, DC 20007
(202) 338-7291

West End Corporation
1817 M Street, N.W.
Washington, DC 20036
(202) 659-5660
Robert I. Kling

Jack C. White
P.O. Box 18997
Washington, DC 20036
(202) 265-1427

Daniel A. Witt
1810 35th Street, N.W.
Washington, DC 20007

Kathleen S. Wood
3101 Highland Place, N.W.
Washington, DC 20008

World Wildlife Fund
The Conservation Foundation
Library
1250 24th Street, N.W.
Washington, DC 20037

FLORIDA

STATE HISTORIC PRESERVATION OFFICE

Division of Historical Resources
Department of State
R.A. Gray Building
500 South Bronough Street
Tallahassee, FL 32399-0250
(904) 488-1480
(904) 488-3353 (Fax)
George W. Percy, SHPO
Suzanne Walker, Deputy SHPO

STATEWIDE PRESERVATION ORGANIZATION

Florida Trust for Historic Preservation
P.O. Box 11206
Tallahassee, FL 32302
(904) 224-8128
(904) 224-1359
Stephanie Ferrell, President

NATIONAL TRUST REGIONAL OFFICE

Southern Regional Office
456 King Street
Charleston, SC 29403
(803) 722-8552
(803) 722-8652 (Fax)
Susan A. Kidd, Director

NATIONAL TRUST ADVISOR

Janet Snyder Matthews
P.O. Box 5343
Sarasota, FL 34277
(813) 922-6960

Arva Moore Parks
1601 South Miami Avenue
Miami, FL 33129
(305) 854-3501
(305) 854-8087

PRESERVATION ACTION COORDINATORS

Katherine Dickenson
1240 Cocoanut Road
Boca Raton, FL 33432
(407) 391-4372

AIA STATE PRESERVATION COORDINATOR

Susan Tate, AIA
University of Florida
College of Architecture
331 Architecture
Gainsville, FL 32611
(904) 392-7003
Herschel E. Shepard, FAIA, Deputy

NATIONAL PARK SERVICE REGIONAL OFFICE

Southeast Regional Office
Richard B. Russell Federal Building
75 Spring Street, S.W.
Atlanta, GA 30303
(404) 331-2632

ADVISORY COUNCIL ON HISTORIC PRESERVATION DIVISION

Eastern Office of Project Review
1100 Pennsylvania Avenue, N.W.
Suite 809
Washington, DC 20004
(202) 786-0503
Don L. Klima, Director

Boca Raton

City of Boca Raton
Community Development Department
201 West Palmetto Park Road
Boca Raton, FL 33432
(305) 393-7781
Jesse W. Moore

Boynton Beach

City of Boynton Beach
100 East Boynton Beach Boulevard
P.O. Box 310
Boynton Beach, FL 33425-0310
(407) 738-7493
Johnetta Broomfield

Bradenton

Manatee Company Historical
Commission
604 15th Street, East
Bradenton, FL 34208
(813) 749-7165
Cathy Slusser

Chokoloskee

Ted Smallwood's Store
P.O. Box 367
Chokoloskee, FL 33925
(813) 695-2989
Lyn S. McMillin

Clearwater

James J. Thornton
1124 Drew Street
Clearwater, FL 34615
(813) 442-7917

Coral Gables

City of Coral Gables
Historic Preservation Division
Coral Gables City Hall
405 Biltmore Way
Coral Gables, FL 33134
(305) 460-5216
Ellen Uguccioni

Organized Fisherman of Florida
Cortez Chapter
P.O. Box 33
Cortez, FL 34213
(813) 748-0280
Allen Garner

Dania

City of Dania
100 West Dania Beach Boulevard
P.O. Box 1708
Dania, FL 33004
(305) 921-8700
Michael W. Smith

Daytona Beach

Community Development Office
P.O. Box 551
Daytona Beach, FL 32015
James Huger

DeLand

Main Street DeLand
P.O. Box 3194
DeLand, FL 32721
(904) 738-0649
Kerry Hassen

Delray Beach

Community Redevelopment Agency
1 S.E. 4th Avenue
Suite 204
Delray Beach, FL 33483
(407) 276-8640
Chris Brown

Historic Preservation Board
Planning Office
100 N.W. 1st Avenue
Delray Beach, FL 33444
(407) 278-2921
Pat Healy

Dunnellon

Greater Dunnellon Historical Society
P.O. Box 336
Dunnellon, FL 32630

Fort Lauderdale

Bonnet House
900 North Birch Road
Fort Lauderdale, FL 33304
(305) 563-5393

Fort Lauderdale Historical Society
219 S.W. 2nd Avenue
Fort Lauderdale, FL 33301
(305) 463-4431
Daniel T. Hobby

Historic Broward County Preservation
Board
P.O. Box 1181
Fort Lauderdale, FL 33302
Paul George

Ralph Monroe House, Miami, Florida. (S. Feller, HABS)

Fort Myers

Edison Winter Home
2350 McGregor Boulevard
Fort Myers, FL 33901
(813) 334-7419
Mary C. Fitzpatrick

Lee County Planning Division
P.O. Box 398
Fort Myers, FL 33902
(813) 335-2479
G. Cook

Fort Pierce

St. Lucie County Historical
Commission
609 South 9th Street
Fort Pierce, FL 34950
(407) 464-6682
Lucille Rights

Gainesville

Friends of the Matheson House
810 East University Avenue
Gainesville, FL 32601
(904) 376-0600
Mark Barrow

University Of Florida
College of Architecture
Gainesville, FL 32611
(904) 392-4836
R. H. Schneider

Haines City

City of Haines City
P.O. Box 1507
Haines City, FL 33845
(813) 422-4986
Donald Hansen

Hollywood

Hollywood Community
Redevelopment Agency
2600 Hollywood Boulevard
Hollywood, FL 33020
(305) 921-3017

Jacksonville

Kathleen Bissa
1604 Wood Hill Place
Jacksonville, FL 32256

City of Jacksonville
Public Buildings Division
930 Liberty Street
Jacksonville, FL 32208
(904) 630-3525
Jose Papa

Historic Springfield Community
Council of Jacksonville
1823 North Pearl Street
Jacksonville, FL 32206
(904) 355-5012
Steve Runnels

National Society of Colonial Dames in
America
Ortega Station
P.O. Box 34
Jacksonville, FL 32210
Joseph Weed

Riverside-Avondale Preservation
904 King Street
Jacksonville, FL 32205
(904) 389-2449
Christopher Jones

Springfield Ecumenical Ministries
157 East 8th Street
Suite E
Jacksonville, FL 32206
(904) 355-2645
Bill Robertson

Springfield Neighborhood Housing
Services
P.O. Box 40681
Jacksonville, FL 32203-0681
(904) 355-1248
Richard Harrill

Springfield Preservation and
Restoration
P.O. Box 3192
Jacksonville, FL 32206-3192

Jacksonville Beach

Beaches Area Historical Society
P.O. Box 50646
Jacksonville Beach, FL 32250
(904) 249-8261
Dennie Carter

Key Largo

Florida Park Service
Historic Preservation Office
P.O. Box 2660
Key Largo, FL 33037
(305) 451-3005
Rudd M. Long

Key West

Audubon House and Gardens
Whitehead and Greene Streets
Key West, FL 33040
Mitchell Wolfson

Historic Florida Keys Preservation
Board
510 Greene Street
Key West, FL 33040
(305) 292-6718
Wright Langley

Key West Maritime Historical Society
P.O. Box 695
Key West, FL 33041
(305) 292-7903

Key West Planning Department
P.O. Box 1409
Key West, FL 33041
(305) 292-8178
Eugene E. Burr

Lake City

Main Street Lake City
P.O. Box 255
Lake City, FL 32056
(904) 755-9023
Marci S. Monchek

Lakeland

Florida Southern College
111 Lake Hollinsworth Drive
Lakeland, FL 33801
(813) 680-4116
Ray Fischer

Historic Lakeland
P.O. Box 3347
Lakeland, FL 33802
(813) 688-9482
John C. White

Historic Lakeland
P.O. Box 3347
Lakeland, FL 33802
(813) 683-5229
C. E. Vilushis

Lakeland Public Library
Membership Magazines
100 Lake Norton Drive
Lakeland, FL 33801

Polk Theatre
127 South Florida Avenue
Lakeland, FL 33801
(813) 682-8227
DeNeece Ham

Lake Wales

Bok Tower Gardens
P.O. Box 3810
Lake Wales, FL 33859-3810
(813) 676-1408
Karen Driscoll

Lake Wales Downtown
130 East Stewart Avenue
P.O. Box 1320
Lake Wales, FL 33859-1320
(813) 676-2533
Ronali A. Wood

Largo

Heritage Park-Pinellas County
Historical Museum
11909 125th Street, North
Largo, FL 34644
(813) 462-3474
Kendrick T. Ford

Leesburg

City of Leesburg
Planning Department
P.O. Box 490630
Leesburg, FL 34749
Carl H. Hammons

Marathon

Florida Keys Land and Sea Trust
P.O. Box 536
Marathon, FL 33050
(305) 743-3900
Chuck Olson

Miami

Dade Heritage Trust
190 S.E. 12th Terrace
Miami, FL 33131
(305) 358-9572
Louise Yarbrough

Nancy Kalthleen Enzler
4648 S.W. 12th Street
Miami, FL 33134
(305) 445-0494

Suzanne P. Johnson
PSC 04, Box 354
APO AA
Miami, FL 34004-5000

Metro-Dade Historic Preservation
Division
Office of Community and Economic
Development
Warner Place
Suite 101
111 S.W. 5th Avenue
Miami, FL 33130
(305) 545-4228

Miami Artisans
4712 S.W. 74th Avenue
Miami, FL 33155
(305) 261-4127
Wade Foy

Miami Heritage Conservation Board
City of Miami Planning Department
P.O. Box 330708
Miami, FL 33133
(305) 579-6086
Sarah Eaton

New Birth Corporation
18330 N.W. 86th Avenue
Miami, FL 33015
(305) 823-4123
Jefferson P. Rogers

University of Miami
8230 S.W. 99th Avenue
Miami, FL 33173
A. Millas

Miami Beach

Jud Kurlancheek
Historic Preservation and Urban Design
Director
1700 Convention Center Drive
Miami Beach, FL 33139

Miami Beach Development
Corporation
1205 Drexel Avenue, 2nd Floor
Miami Beach, FL 33139
(305) 538-0090
Denis Russ

Miami Design Preservation League
P.O. Bin L
Miami Beach, FL 33119
(305) 672-2014
Nancy Liebman

Micanopy

Micanopy Historical Society
P.O. Box 462
Micanopy, FL 32667
(904) 466-3848
Diana Cohen

Monticello

Jefferson Historical Association
P.O. Box 496
Monticello, FL 32344

New Smyrna Beach

Community Redevelopment Agency
210 Sams Avenue
New Smyrna Beach, FL 32069
(904) 427-4166
Pamela Graham-Mathis

Cornerstone Restoration
509 North Peninsula Avenue
New Smyrna Beach, FL 32169
(904) 427-2150
Doug Davich

New Smyrna Beach Main Street
115 Canal Street
New Smyrna Beach, FL 32168
(904) 428-2449
Kim Dryfoos

Ocala

Historic Ocala Preservation Society
P.O. Box 3123
Ocala, FL 32671
(904) 351-7200
Bill Mansfield

Oldsmar

Cline Contec South
P.O. Box 1689
Oldsmar, FL 34677
(813) 855-6645
Rocco F. Capabianco

Orlando

Orlando Historic Preservation Board
400 South Orange Avenue
6th Floor
Orlando, FL 32801
(407) 849-2300
Jodi Rubin

Sons of Confederate Veterans
Florida Division
7213 Grace Road
Orlando, FL 32819

Yeilding and Provost
112 East Concord Street
Orlando, FL 32801
(407) 423-4919
Chalmers Yeilding

Palatka

Palatka Main Street
Putnam County Chamber of
Commerce
P.O. Box 550
Palatka, FL 32178
(904) 328-1503
John Mayfield

Palm Beach

Henry Morrison Flagler Museum
Whitehall Way
P.O. Box 969
Palm Beach, FL 33480
(407) 655-2833
Charles B. Simmons

Preservation Foundation of Palm Beach
356 South County Road
Palm Beach, FL 33480-4442
(407) 832-0731
Polly Earl

Pensacola

Downtown Improvement Board
P.O. Box 653
Pensacola, FL 32593
(904) 434-5371
L. A. Maygarden

Historic Pensacola Preservation Board
120 East Church Street
Pensacola, FL 32501
(904) 444-8905
John P. Daniels

Pensacola Heritage Foundation
P.O. Box 12424
Pensacola, FL 32582
(904) 438-6505
Gwen Roland

Plant City

Plant City Main Street Program
P.O. Box 2538
104 North Evers Street
Plant City, FL 34289-2538
(813) 752-9161
Sherie Brezina

Ponce Inlet

Ponce De Leon Inlet Lighthouse
Preservation Association
4931 South Peninsula Drive
Ponce Inlet, FL 32019
Ann Caneer

Punta Gorda

Lynn Harrell Jones
512 Edmund Street
Punta Gorda, FL 33950
(813) 639-4953

Quincy

Quincy Main Street
P.O. Box 1106
Quincy, FL 32351
(904) 627-2346
Jennifer E. Maglievaz

St. Augustine

Historic St. Augustine Preservation
Board
48 King Street
Box 1987
St. Augustine, FL 32084

Lightner Museum
75 King Street
P.O. Box 334
St. Augustine, FL 32085-0334
(904) 824-2874
Robert W. Harper

St. Petersburg

City of St. Petersburg
Planning Department
P.O. Box 2842
St. Petersburg, FL 33731
Ralph Stone

Vinoy Development Corporation
111 2nd Avenue, N.E., Suite 1401
St. Petersburg, FL 33701
(813) 894-6445
Craig McLaughlin

Sanford

Community Improvement Association
of Sanford
612 West 9th Street
Sanford, FL 32771
(407) 660-8666

Sanibel

Island Historic Museum
800 Dunlap
Sanibel, FL 33957
(813) 472-3373
Richard Noon

Sarasota

Jon B. Brannen
2667 Briar Oak Circle
Sarasota, FL 34232
(913) 377-9121

City Of Sarasota
P.O. Box 1058
Sarasota, FL 33578
(813) 954-4195
Jane N. Robinson

City of Sarasota
Planning and Development
Department
1565 1st Street
P.O. Box 1058
Sarasota, FL 34230
(813) 954-4195
Michael Taylor

Janet S. Matthews
P.O. Box 5343
Sarasota, FL 34277

Sarasota Alliance for Historic
Preservation
P.O. Box 1754
653 South Orange Avenue
Sarasota, FL 34236
(813) 951-1547
Betty Cookson

Tallahassee

Lacy F. Bullard
Route 3, Box 649
Tallahassee, FL 32308
(904) 893-1517

David E. Ferro
1951 North Meridian Road
Suite 12
Tallahassee, FL 32303
(904) 386-3709

Florida Main Street Program
Bureau of Historic Preservation
500 South Bronough Street
R.A. Gray Bldg
Room 411
Tallahassee, FL 32399-0250
(904) 487-2333
Robert Trescott

Florida Trust for Historic Preservation
P.O. Box 11206
Tallahassee, FL 32302
(904) 224-8128
Charles Olson

Historic Tallahassee Preservation Board
329 North Meridian Street
Tallahassee, FL 32301
(904) 488-3901

Trust for Public Land
Southeast Region
1310 Thomasville Road
Tallahassee, FL 32303-5608

Tampa

Historic Tampa/Hillsborough County
Preservation Board
2009 North 18th Street
Tampa, FL 33605
(813) 272-3843
Stephanie E. Ferrell

International Academy of Merchants
DES/200 South Hoover Boulevard
211 Mariner Square
Tampa, FL 33609

The Italian Club
P.O. Box 5054
Tampa, FL 33675
(813) 248-3316
Jan Pardo

Joseph Pedalino, Jr.
P.O. Box 2969
Tampa, FL 33601
(813) 978-8043

Rowe Holmes Hammer Russell
Architects
100 Madison Street
Tampa, FL 33602
(813) 221-8771

Tampa Union Station Preservation
Redevelopment Corporation
1328 Autumn Drive
Tampa, FL 33613
(813) 978-1378
Jim Shephard

Tarpon Springs

City of Tarpon Springs
P.O. Box 5004
Tarpon Springs, FL 34688-5004
(813) 938-3711
Kathleen Monahan

Tarpon Springs Cultural Center
101 South Pinellas Avenue
Tarpon Springs, FL 34689
(813) 942-5605
K. Monahan

Titusville

River Centre Association
P.O. Box 6023
Titusville, FL 32782
(407) 269-7363
Barbara Fine

Venice

City of Venice
401 West Venice Avenue
Venice, FL 34285
(813) 485-3311
Betty Arnall

Venice Main Street
P.O. Box 602
Venice, FL 34284-0602
(813) 484-6722
Doug Mann

White Springs

Florida Folklife Program
U.S. 41 and S.R. 136
White Springs, FL 32096
Iris Greene

GEORGIA

STATE HISTORIC PRESERVATION OFFICE

Office of Historic Preservation
205 Butler Street, S.E.
1456 Floyd Towers East
Atlanta, GA 30334
(404) 656-2840
(404) 656-2285 (Fax)
Elizabeth A. Lyon, SHPO
Richard Cloues, Deputy SHPO
Carole Griffith, Deputy SHPO

STATEWIDE PRESERVATION ORGANIZATION

Georgia Trust for Historic Preservation
1516 Peachtree Street, N.W.
Atlanta, GA 30309
(404) 881-9980
Janice Biggers, Chairman
Gregory B. Paxton, President

NATIONAL TRUST REGIONAL OFFICE

Southern Regional Office
456 King Street
Charleston, SC 29403
(803) 722-8552
(803) 722-8652 (Fax)
Susan A. Kidd, Director

NATIONAL TRUST ADVISORS

H. Ben Grace
111 East Washington Street
Thomasville, GA 31972
(912) 226-5156
Roy Mann
P.O. Box 1595
Rome, GA 30162
(404) 295-7099
(404) 295-0688 (Fax)

PRESERVATION ACTION COORDINATORS

Stephanie Churchill
Historic Savannah Foundation
41 West Broad Street
P.O. Box 1733
Savannah, GA 31402
(912) 233-7787

AIA STATE PRESERVATION COORDINATOR

Eugene L. Surber, AIA
Surber Barber Moony Architects
1389 Peachtree Street, N.W.
Suite 350
Atlanta, GA 30309
(404) 874-2400
Linda Ramsey, AIA, Deputy

NATIONAL PARK SERVICE REGIONAL OFFICE

Southeast Regional Office
Richard B. Russell Federal Building
75 Spring Street, S.W.
Atlanta, GA 30303
(404) 331-2632

ADVISORY COUNCIL ON HISTORIC PRESERVATION DIVISION

Eastern Office of Project Review
1100 Pennsylvania Avenue, N.W.
Suite 809
Washington, DC 20004
(202) 786-0503
Don L. Klima, Director

Americus

Americus Main Street Program
P.O. Box M
Americus, GA 31709
(912) 924-4421
Jo Childers

Mr. Frank Brown, III
6594 Mistletoe Plantation
Americus, GA 31709-6594

Athens

Athens Clarke County Preservation Commission
P.O. Box 329
155 East Washington Street
Athens, GA 30605
(404) 354-2950
Jennifer Clower

Athens-Clarke Heritage Foundation
489 Prince Avenue
Fire Hall Suite 2
Athens, GA 30601
(404) 353-1801
Sheila Hackney

Bertis Downs IV
738 Cobb Street
Athens, GA 30606
(404) 543-5033

Georgia Alliance of Preservation Commissions
609 Caldwell Hall
University of Georgia
Athens, GA 30602
(404) 542-4731
Pratt Cassity

Larry V. McLeod
P.O. Box 8108
Athens, GA 30603
(404) 549-9400
Betty J. Perry

Northeast Georgia Area Planning and Development Commission
305 Research Drive
Athens, GA 30610
(404) 548-3141
William G. Moffat

University of Georgia
School of Environmental Design
Caldwell Hall
Athens, GA 30602
(404) 542-3030
Claris Ingersoll

Atlanta

Adair Park Today
836 Oakhill Avenue, S.W.
Atlanta, GA 30310
(404) 753-8820
Richard Mueller

American College for the Applied Arts
3330 Peachtree Road, N.E.
Atlanta, GA 30326-1001
(404) 231-9000
Lisa Beers

Atlanta Historical Society
3101 Andrews Drive, N.W.
Atlanta, GA 30305
(404) 261-1837
John H. Ott

Atlanta Preservation Center
401 The Flatiron Building
84 Peachtree Street, N.W.
Atlanta, GA 30303
(404) 522-4345
Anne H. Farrisee

Atlanta Urban Design Commission
55 Trinity Avenue
Suite 3400
Atlanta, GA 30335-0331
(404) 330-6200
Karen Huebner

Designx
100 Colony Square
Suite 200
Atlanta, GA 30361
(404) 870-9028
Lloyd J. Lewis

Georgia Department of Community Affairs
Office of Rural Development
1200 Equitable Building
100 Peachtree Street
Atlanta, GA 30303
(404) 656-9790

Georgia Department of Natural Resources
Historic Preservation Division
Floyd Tower East
Suite 1462
205 Butler Street, S.E.
Atlanta, GA 30334
(404) 656-2840
Elizabeth A. Lyons

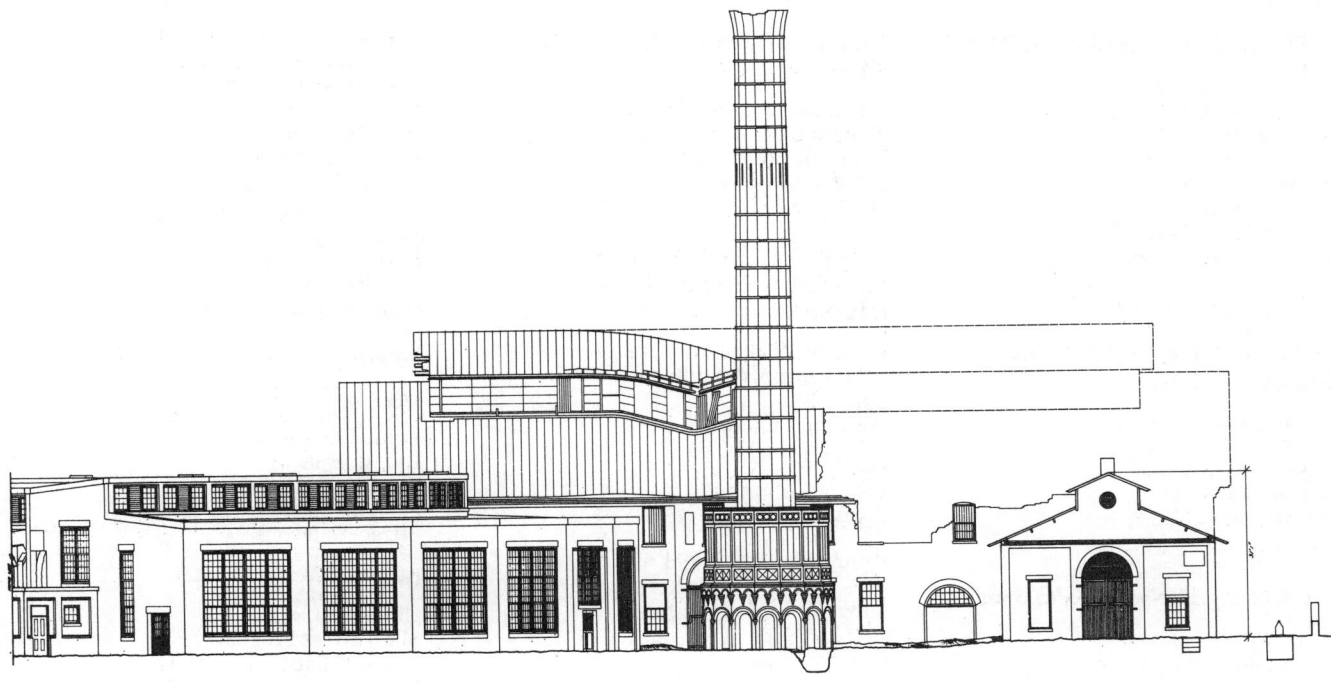

Central of Georgia Railway Repair Shops, Savannah, Georgia. (P. Dubin, HAER)

Georgia Main Street Program
100 Peachtree Street
Suite 1200
Atlanta, GA 30303
(404) 656-9790
Frank McIntosh

Georgia Municipal Association
201 Pryor Street, S.W.
Atlanta, GA 30303
(404) 688-0472
Mary Draut

Georgia Tech
College of Architecture
Atlanta, GA 30332
John H. Myers

Georgia Trust for Historic Preservation
1516 Peachtree Street, N.W.
Atlanta, GA 30309
(404) 881-9980
Gregory B. Paxton

Georgia Trust for Historic Preservation
1516 Peachtree Street
Atlanta, GA 30309
(404) 881-9980
Debra J. Kraybill-Bevin

Historic District Development
Corporation
449 Auburn Avenue, N.E.
Atlanta, GA 30312
(404) 524-1956
(404) 526-8931
Ella M. Brayboy

Lord Aeck and Sargent
127 Peachtree Street
Suite 1717
Atlanta, GA 30303

Marketek
2596 Forrest Way, N.E.
Atlanta, GA 30305
(404) 233-1885
Angela H. Cassidy

Phillips, Hinchey, and Reid
233 Peachtree Street
Suite 1200
Atlanta, GA 30303
(404) 659-6000
Robert Crewdson

Richard Rauh and Associates
3300 Piedmont Road
Atlanta, GA 30305
(404) 233-9447
Richard Rauh

Eugene L. Surber
1389 Peachtree Street, N.E.
Suite 350
Atlanta, GA 30309
(404) 874-2400

Augusta

Helen Bryngelson
1511 John's Road
Augusta, GA 30904

Downtown Revitalization
Historic Augusta
P.O. Box 32
Augusta, GA 30903
(404) 724-0454
Samm Fusselle

Haltermann Partners
Real Estate Development
405 Lamar Building
753 Broad Street
Augusta, GA 30901
(404) 722-3961
Bryan M. Haltermann

Historic Augusta
P.O. Box 37
Augusta, GA 30903
(404) 733-6768
Erick D. Montgomery

Summerville Neighborhood
Association
P.O. Box 12212
Augusta, GA 30904
(404) 738-7527
Walter Alexanderson

Avondale Estates

City of Avondale Estates
21 North Avondale Plaza
Avondale Estates, GA 30002
(404) 294-5400
Phyllis Flowers

Bainbridge

Bainbridge Main Street Association
107 South Broad Street
Bainbridge, GA 31717
(912) 248-1941
Kimberly Mills

Baxley

Altamaha Georgia Southern Regional
Development Center
P.O. Drawer 459
Baxley, GA 31513
(912) 367-3648
Robin Nail

Bostwick

City of Bostwick
P.O. Box 129
Bostwick, GA 30623
(404) 342-1251
June Whittaker

Brunswick

Coastal Georgia Regional Development
Center
P.O. Box 1917
Brunswick, GA 31521
(912) 264-7363
William D. Waters

Mainstreet Brunswick
P.O. Box 684
Brunswick, GA 31521
(912) 265-4032
Cuffy Hise

Old Town Brunswick Preservation
Association
1229 Newcastle Street
Old City Hall
Brunswick, GA 31520

Calhoun

City of Calhoun
P.O. Box 248
226 South Wall Street
Calhoun, GA 30701
(404) 629-0151
Eddie Peterson

Canton

Cherokee County Historical Society
P.O. Box 1287
Canton, GA 30114
(404) 345-6663
Elaine Hubbard

Carnesville

Franklin County Historical Society
P.O. Box 531
Carnesville, GA 30521
(404) 549-0825
David Kidd

Cartersville

Cartersville Downtown Development
Authority
P.O. Box 1015
Cartersville, GA 30120
(404) 386-6458
Ann E. Arnold

Noble Hill-Wheeler Memorial
Heritage Center
105 Fite Street
Cartersville, GA 30120

Columbus

Elena D. Amos
Columbus, GA 31999

James W. Biggers, Jr.
2316 19th Street
Columbus, GA 31906

City of Columbus
Planning Division
P.O. Box 1340
Columbus, GA 31993
(404) 571-4767
James W. Jones

Confederate Naval Museum
P.O. Box 1022
Columbus, GA 31902
(404) 327-9798
Kevin J. Foster

Historic Columbus Foundation
P.O. Box 5312
Columbus, GA 31906
(404) 323-7979
Stephen M. Howard

Commerce

Commerce Main Street
P.O. Box 717
Commerce, GA 30529
(404) 335-2954
Jennifer Sanders

Covington

City of Covington
Main Street Program
2111 Conyers Street
P.O. Box 1527
Covington, GA 30209
(404) 786-5324
Debbie Rushton

Dalton

Downtown Dalton Development
Authority
P.O. Box 707
Dalton, GA 30722-0707
(404) 278-3332
Gaile R. Jennings

North Georgia Regional Development
Center
503 West Waugh Street
Dalton, GA 30720
(404) 272-2300
Mary McLeod

Whitfield Murray Historical Society
Crown Gardens and Archives
715 Chattanooga Avenue
Dalton, GA 30720
(404) 278-0217
Polly Boggess

Danielsville

Madison County Chamber of
Commerce
Highway 29 North
P.O. Box 381
Danielsville, GA 30633
(404) 795-3473
Barbariann G. Russell

Decatur

Druid Hills Civic Association
P.O. Box 363
Decatur, GA 30031
(404) 373-3007
Elizabeth Jacobs

Douglas

City of Douglas
Main Street Program
P.O. Box 248
Douglas, GA 31533
(912) 384-3302
Patti Vick

Dublin

Main Street Dublin
P.O. Box 969
Dublin, GA 31040
(912) 272-6421
Joan Kilian

Duluth

Lynn Heriford
1752 Mitzi Court
Duluth, GA 30136

Catherine Wilson Martin
3503 Wood Acres Boulevard
Duluth, GA 30136
(404) 621-0138

Eatonton

Eatonton-Putnam County Historical
Society
104 Church Street
Eatonton, GA 31024
(404) 485-4532
James P. Marshall

Elberton

Main Street Elberton
P.O. Box 6447
Elberton, GA 30635
(404) 213-0626
Kay P. Shiver

Ellaville

Schley County Commissioners
P.O. Box 352
Ellaville, GA 31806
(912) 937-2609
W. C. Holt

Franklin

Chattahoochee-Flint Area Planning and
Development Commission
P.O. Box 110
Franklin, GA 30217-0110
Julie Turner

Chattahoochee-Flint Regional
Development Center
P.O. Box 1600
Franklin, GA 30217
(404) 675-6721
Julie Turner

Gainesville

Jaeger/Pyburn
119 Washington Street
Gainseville, GA 30501
(404) 534-7024
Dale Jaeger

Mountains Regional Development
Center
P.O. Box 1720
Gainesville, GA 30503
(404) 536-3431
Cathleen Turner

Homer

Banks County Historical Society
P.O. Box 473
Homer, GA 30547
(404) 778-1900
Phillip Brooks

Jackson

Butts County Historical Society
P.O. Box 215
Jackson, GA 30233
(404) 775-6734
Deryle Lamb

Jasper

Marble Valley Historical Society
P.O. Box 567
Jasper, GA 30143
Joy Lankford

Jekyll Island

Jekyll Island Museum
375 Riverview Drive
Jekyll Island, GA 31520
(912) 935-2236
Thomas Rhodes

La Grange

Troup County Archives
136 Main Street
La Grange, GA 30241
(404) 884-1828
Kaye L. Minchew

Troup County Planning Commission
900 Dallis Street
La Grange, GA 30240
(404) 883-1655
Linda Straub

Lexington

Oglethorpe County
Oglethorpe County Courthouse
Lexington, GA 30648
Judy Paul

Lumpkin

Charles Junction Historic Preservation
Route 2B
Box 15
Lumpkin, GA 31815
(904) 942-0958
Julia Ridley

Macon

Macon-Bibb Company
Planning and Zoning
682 Cherry Street
Suite 1000
Macon, GA 31201

Macon Heritage Foundation
P.O. Box 6092
Macon, GA 31208
(912) 742-5084
Maryel Battin

Middle Georgia Historical Society
935 High Street
Macon, GA 31201
(912) 743-3851
Kitty Oliver

D. T. Walton, Jr.
591 Cotton Avenue
Macon, GA 31201
(912) 742-8478

Madison

Madison Downtown Business Council
P.O. Box 826
Madison, GA 30650
(404) 342-4454

Madison-Morgan County Chamber of
Commerce
P.O. Box 826
Madison, GA 30650

Marietta

Cobb County Historic Preservation
Commission
100 Cherokee Street
Suite 500
Marietta, GA 30090
(404) 421-2010
Joan Cole

Cobb Landmarks and Historical Society
156 Church Street
Marietta, GA 30060
(404) 426-4982
Marcelle B. David

Nancy R. Edwards
81 Whitlock Avenue
Marietta, GA 30064
(404) 424-1277

Metter

Candler County Historical Society
P.O. Box 325
Metter, GA 30439
(912) 685-6375
Laurine Witmer

Milledgeville

Milledgeville Main Street
P.O. Box 1708
Milledgeville, GA 31061
(912) 453-1928
Jacquelyn Decell

Oglethorpe

Macon County Commissioners
P.O. Box 297
Oglethorpe, GA 31068
(912) 472-7040
John Luckie

Plains

City of Plains
P.O. Box 190
Plains, GA 31780-0190
(912) 824-5445
Karen Oliver

Rome

Coosa Valley Regional Development
Center
P.O. Box 1793
Rome, GA 30163-1001
(404) 295-6485
Kitty Houston

Rome Downtown Development
Authority
P.O. Box 346
428 Broad Street
Rome, GA 30162
(404) 236-4520
Paul B. Pitts

Roswell

Bulloch Hall Reference Library
Roswell Historical Society
227 South Atlanta Street
Roswell, GA 30075
Martha W. Stovall

City of Roswell
Historic District
38 Hill Street
Roswell, GA 30075
(404) 641-3727
Jane Pruett

Roswell Historic Preservation
Commission
P.O. Box 1309
Roswell, GA 30077
(404) 992-1731
Bill W. Gray

St. Simons Island

Fort Frederica Association
Fort Frederica National Monument
Route 9, Box 286-C
St. Simons Island, GA 31522

Savannah

City of Savannah
P.O. Box 1027
Savannah, GA 31402
Arthur A. Mendonsa

Coastal Heritage Society
601 West Harris Street
Savannah, GA 31401
Scott Smith

Economic Development Authority
Planning and Community
Development
P.O. Box 1027
Savannah, GA 31402
Robert A. Haywood

National Society of the Colonial Dames
of America
329 Abercorn Street
Savannah, GA 31401

Summerville

Chattooga County Historical Society
200 South Commerce Street
Summerville, GA 30747
(404) 857-2553
Jerry Sprayberry

Thomasville

Main Street Thomasville
P.O. Box 1140
Thomasville, GA 31799
(912) 228-7673
Sharlene Celaya

Thomasville Landmarks
P.O. Box 1285
Thomasville, GA 31792
(912) 226-6016
Nancy Tinker

Thomson

Wrightsboro Foundation
P.O. Box 1816
633 Hemlock Drive
Thomson, GA 30824
(404) 595-5584
Dorothy M. Jones

Toccoa

Main Street Toccoa
P.O. Box 579
Toccoa, GA 30577
(404) 886-8451
Connie Tabor

Tybee Island

Tybee Island Historical Society
P.O. Box 366
Tybee Island, GA 31328
(912) 786-5801
Jody Owens

Valdosta

South Georgia Regional Development
Center
P.O. Box 1223
Valdosta, GA 31603
(912) 333-5277
Donna De Weese

Valdosta-Lowndes County Main Street
P.O. Box 1125
216 East Central Avenue
Valdosta, GA 31601
(912) 242-3987
Phylis Lowther

Valdosta Town Committee of Colonial
Dames
821 Millpond Road
Valdosta, GA 31602

Vidalia

Downtown Vidalia Association
Main Street Program
P.O. Box 1605
Vidalia, GA 30474
(912) 537-8033
Brad Morris

Waycross

Southeast Georgia Regional
Development Center
3395 Harris Road
Waycross, GA 31501
(912) 285-6097
Randy Randolph

Waycross Downtown Development
Authority
P.O. Box 158
Waycross, GA 31502
(912) 283-7787
Nancy L. Campbell

Waynesboro

Burke County Historical Society
536 Liberty Street
Waynesboro, GA 30836
(404) 554-4889
Ella L. Bargeron

HAWAII

STATE HISTORIC PRESERVATION OFFICE

Department of Land and Natural Resources
P.O. Box 621
Honolulu, HI 96809
(808) 548-6550
William W. Paty, SHPO
Ralston Nagata, Deputy SHPO
Keith Anne, Deputy SHPO
Don Hibbard, Administrator
State Historic Preservation Division
33 South King Street
6th Floor
Honolulu, HI 96809
(808) 587-0047

STATEWIDE PRESERVATION ORGANIZATION

Historic Hawai'i Foundation
P.O. Box 1658
Honolulu, HI 96806
(808) 537-9564
(808) 526-3989 (Fax)
Charles Pietsch III, Chairman
Phyllis G. Fox, President

NATIONAL TRUST REGIONAL OFFICE

Western Regional Office
1 Sutter Street
Suite 707
San Francisco, CA 94104
(415) 956-0610
(415) 956-0837 (Fax)
Kathryn A. Burns, Director

NATIONAL TRUST ADVISORS

Barnes Riznik
4767 Hoomana Road
Lihue, Kauai, HI 96766
(808) 245-2681

Gerald Takano
60 North Beretania
Suite 1810
Honolulu, HI 96817

PRESERVATION ACTION COORDINATOR

Phyllis G. Fox
Historic Hawai'i Foundation
P.O. Box 1658
Honolulu, HI 96806
(808) 537-9564

AIA STATE PRESERVATION COORDINATOR

Spencer Leineweber, AIA
Spencer Mason Architects
1050 Smith Street
Honolulu, HI 96817
(808) 536-3636

NATIONAL PARK SERVICE REGIONAL OFFICE

Western Regional Office
600 Harrison Street
Suite 600
San Francisco, CA 94107
(415) 744-3988

ADVISORY COUNCIL ON HISTORIC PRESERVATION DIVISION

Western Office of Project Review
730 Simms Street
Room 401
Golden, CO 80401
(303) 231-5320
Claudia Nissley, Director

Ewa

Friends for Ewa
P.O. Box 1356
Ewa, HI 96706
(808) 524-4977
Penny Pagliaro

Haleiwa

Haleiwa Main Street Business Association
P.O. Box 878
Haleiwa, HI 96712
(808) 637-4558
Beth Wotkyns

Hanalei

Hanalei Community Association
P.O. Box 369
Hanalei, HI 96714
(808) 826-9370
Barbara Robeson

Ho'Opulapula Haraguchi Rice
P.O. Box 427
Hanalei, HI 96714
(808) 826-6202
Karol Haraguchi

Hilo

Hilo Main Street Program
252 Kamehameha Avenue
Hilo, HI 96720
(808) 935-8850
Tom Hudson

Honolulu

Nancy Bannick
871 Kapiolani Boulevard
Room 3
Honolulu, HI 96813
(808) 533-7168

Bishop Museum
1525 Bernice Street
P.O. Box 19000-A
Honolulu, HI 96817
(808) 847-3511

Min-yung Chiu
1837 Kalakaua Avenue
Suite A31
Honolulu, HI 96815

Daughters Of Hawaii
2913 Pali Highway
Honolulu, HI 96817
(808) 595-6291
Mildred Nolan

Friends Of 'Iolani Palace
P.O. Box 2259
Honolulu, HI 96804
(808) 522-0824
Alice F. Guild

David Hajino
State Capitol
Room 425
Honolulu, HI 96813

Hawaii Army Museum Society
P.O. Box 8064
Honolulu, HI 96830-0064

Hawaii Maritime Center
Pier 7
Honolulu, HI 96813
(808) 525-6151
Evarts Fox

Historic Hawai'i Foundation
P.O. Box 1658
Honolulu, HI 96806
(808) 537-9564
Phyllis G. Fox

Spencer Leineweber, AIA
1050 Smith Street
Honolulu, HI 96817

Mission Houses Museum
553 South King Street
Honolulu, HI 96813
(808) 531-0481
Deborah Pope

Mullahey and Mullahey
P.O. Box 1348
Honolulu, HI 96807
(808) 533-0777
Ramona K. Mullahey

University of Hawaii at Manoa
American Studies Department
1890 East-West Road
Moore 324
Honolulu, HI 96822
(808) 948-8570
Paul F. Hooper

Rod Wilson
700 Richards Street
Suite 2408
Honolulu, HI 96813

Windom Constuction Company
1200 College Walk
Suite 202
Honolulu, HI 96817
(808) 523-5579
Robert Windom

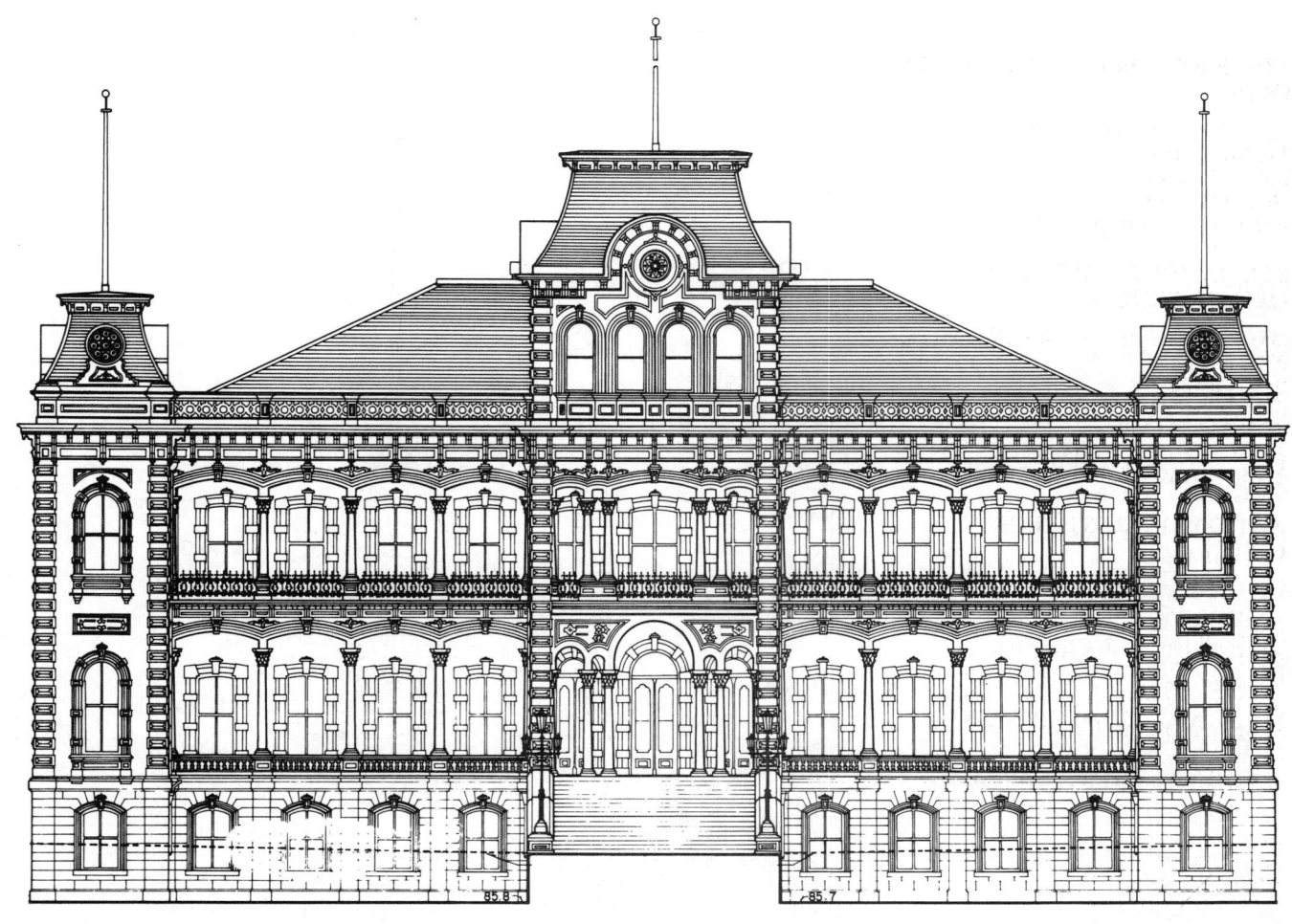

'Iolani Palace, Honolulu, Hawai'i. (E.D. Chauviere, HABS)

Kapaau

Iole Mission Homestead Foundation
P.O. Box 1000
Kapaau, HI 96755
(808) 531-4501
Norene Alexander

Lahaina

The Lahaina Restoration Foundation
P.O. Box 338
Lahaina, HI 86761
James C. Luckey

Lawai

National Tropical Botanic Garden
Allenton Gardens
P.O. Box 340
Lawai, HI 96765
(808) 332-7324
William L. Theobaid

Wailuku

County of Maui
200 South High Street
Wailuku, HI 96793
(808) 244-7735
Christopher L. Hart

R. Keoni Fairbanks
P.O. Box 1291
Wailuku, HI 96793
(808) 242-8970

Maui Historical Society
2375 A Main Street
Wailuku, HI 96793
(808) 244-3326
John Cooper

Wailuku Main Street Association
P.O. Box 981
Wailuku, HI 96793
(808) 244-3888
Jocelyn Perreira

IDAHO

STATE HISTORIC PRESERVATION OFFICE

Idaho State Historical Society
210 Main Street
Boise, ID 83702
(208) 334-2682
Kenneth J. Swanson, SHPO

STATEWIDE PRESERVATION ORGANIZATION

Idaho Historic Preservation Council
P. O. Box 1495
Boise, ID 83701
(208) 336-9880
(208) 342-2263 (Fax)
Robert Aldridge, President
Lorraine Gross, Executive Director

NATIONAL TRUST REGIONAL OFFICE

Western Regional Office
1 Sutter Street
Suite 707
San Francisco, CA 94104
(415) 956-0610
(415) 956-0837 (Fax)
Kathryn A. Burns, Director

NATIONAL TRUST ADVISORS

Robert L. Aldridge
1209 North 8th Street
Boise, ID 83702-4297
(208) 336-9880
(208) 342-2263 (Fax)

John L. Bertram
417 South 13th Street
Boise, ID 83702
(208) 336-1438

PRESERVATION ACTION COORDINATOR

Sheri Freemuth
1119 North 21st Street
Boise, ID 83702
(208) 336-4984

AIA STATE PRESERVATION COORDINATOR

Ernest J. Lombard, AIA
1221 Shoreline Lane
Boise, ID 83702
(208) 345-6677

NATIONAL PARK SERVICE REGIONAL OFFICE

Pacific Northwest Regional Office
83 South King Street
Suite 212
Seattle, WA 98104
(206) 553-5565

Pioneer Lodge No. 1, International Order of Odd Fellows Hall, Idaho City, Idaho. (M. Wellen, J. Schafer, HABS)

ADVISORY COUNCIL ON HISTORIC PRESERVATION DIVISION

Western Office of Project Review
730 Simms Street
Room 401
Golden, CO 80401
(303) 231-5320
Claudia Nissley, Director

Boise

Ada County Development Services
650 Main Street
Boise, ID 83702
(208) 383-4424
Sheri Freemuth

Boise City Historic Preservation
Commission
Boise City Planning Department
P.O. Box 500
Boise, ID 83701
(208) 384-4366
William Von Tagen

Fund Raisers
P.O. Box 8836
Boise, ID 83706
(208) 345-6447
Karen Lockner

Idaho Heritage Trust
1310 Ranch Road
Boise, ID 83702
(208) 384-0176
John Barnes

Idaho Power Company
P.O. Box 70
Boise, ID 83707
(208) 383-2729
Mark Druss

Caldwell

Canyon County Parks and Recreation
P.O. Box 44
1115 Albany St.
Caldwell, ID 83605
(208) 454-7435
Tom Bicak

Coeur D'Alene

Coeur D'Alene Downtown Association
P.O. Box 64
Coeur D'Alene, ID 83814
(208) 667-4040
Shauna Nelson

Idaho City

Idaho City Historical Foundation
Box 358
Idaho City, ID 83631
(208) 392-4263
Byron Johnson

Idaho Falls

Idaho Falls Historic Preservation
Commission
P.O. Box 50220
Idaho Falls, ID 83405
(208) 529-1000
Mark Platt

J. Stephen Julian Associates
353 12th Street
Idaho Falls, ID 83403
(208) 529-2277
John S. Julian

Lewiston

LCSC Center for the Arts
5th and Main
Lewiston, ID 83501
(208) 799-2243
Leslie Esselburn

Port City Action Corporation
P.O. Box 1702
Lewiston, ID 83501
(208) 746-4845
Ronda Laybourn

Moscow

Latah County
P.O. Box 8068
Moscow, ID 83843
P. Peterson

Pocatello

City of Pocatello
Community Development and
Research
P.O. Box 4169
Pocatello, ID 83205
(208) 234-6184
George Ramjoue

Pocatello Historic Preservation
Commission
P.O. Box 4169
Pocatello, ID 83205
(208) 234-6184
Lee A. Dutton

Priest River

Priest River Restoration and
Revitalization
P.O. Box 1284
Priest River, ID 83856
(208) 448-1221
Donna Jones

Wallace

Silver Valley Development Corporation
P.O. Box 469
Wallace, ID 83873
Jim Hays

Weiser

ICC&M Architectural Preservation
Committee
P.O. Box 185
Weiser, ID 83672
(208) 549-0205

ILLINOIS

STATE HISTORIC PRESERVATION OFFICE

Illinois Historic Preservation Agency
1 Old State Capitol Plaza
Springfield, IL 62701-1512
(217) 785-9045
(217) 524-7525 (Fax)
William L. Wheeler, SHPO
Theodore Hild, Deputy SHPO

STATEWIDE PRESERVATION ORGANIZATION

Landmarks Preservation Council of Illinois
53 West Jackson Boulevard
Suite 752
Chicago, IL 60604
(312) 922-1742
Howard Decker, President
Carol Wyant, Executive Director

NATIONAL TRUST REGIONAL OFFICE

Midwest Regional Office
52 West Jackson Boulevard
Suite 1135
Chicago, IL 60604
(312) 939-5447
(312) 939-5651 (Fax)
Tim Turner, Director

NATIONAL TRUST ADVISOR

Amy R. Hecker
3300 Lake Shore Drive
Suite 8E
Chicago, IL 60657
(312) 281-1771

PRESERVATION ACTION COORDINATORS

Susan Baldwin
2800 North Lake Shore Drive
Suite 2604
Chicago, IL 60657
(312) 321-0707

Roberta B. Deering
Chicago Office of Illinois Preservation Agency
100 West Randolph Street
Suite 4-900
Chicago, IL 60601
(312) 814-1409

AIA STATE PRESERVATION COORDINATOR

Walker C. Johnson, AIA
Holabird and Root
300 West Adams Street
Chicago, IL 60606
(312) 726-5960
Harry J. Hunderman, AIA, Deputy

NATIONAL PARK SERVICE REGIONAL OFFICE

Midwest Regional Office
1709 Jackson Street
Omaha, NE 68102
(402) 221-3431

ADVISORY COUNCIL ON HISTORIC PRESERVATION DIVISION

Eastern Office of Project Review
1100 Pennsylvania Avenue, N.W.
Suite 809
Washington, DC 20004
(202) 786-0503
Don L. Klima, Director

Alton

Alton Area Landmarks Association
P.O. Box 232
Alton, IL 62002
(618) 474-0148
Kerry Miller

Antioch

Antioch Redevelopment Commission
874 Main Street
Antioch, IL 60002
(312) 395-1000
Peg Ramsom

Arlington Heights

Building Blocks
1845 East Rand Road
Suite 106
Arlington Heights, IL 60004
(312) 506-0500
Michael T. Miske

Historic Society and Museum of Arlington Heights
500 North Vail Avenue
Arlington Heights, IL 60004

Aurora

Aurora Historical Museum
P.O. Box 905
Aurora, IL 60506

Aurora Preservation Commission
44 East Downer Place
Aurora, IL 60507
(312) 844-3625
Dana Jenkins

Barrington

Barrington Area Historical Society
212 West Main Street
Barrington, IL 60010
(708) 381-1730
Barbara Benson

National Association of Dealers in Antiques
P.O. Box 421
Barrington, IL 60010
(815) 877-4282
Shirley Kowing

Village of Barrington
206 South Hough Street
Barrington, IL 60010
Janice B. Hill

Belleville

Belleville Public Library
121 East Washington Street
Belleville, IL 62220
R. B. Kirchgraber

City of Belleville
Planning and Development Department
111 South 6th Street
Belleville, IL 62220
(618) 233-6817
Mike Pierceal

Belvidere

Belvidere Historic Preservation Commission
Belvidere/Boone Planning Department
601 North Main
Belvidere, IL 61008
(815) 544-4746
Roseann Coots

Bloomington

Bloomington Unlimited
Illinois House
207 West Jefferson
Suite 300
Bloomington, IL 61701
Sharon McGinnis

McLean County Historical Society
201 East Grove Street
Bloomington, IL 61701
(309) 827-0428
Greg Koos

Broadview

Clark Roofing Company
2700 West Cermak Road
Broadview, IL 60153
(708) 681-2200
Jeffrey L. Suess

Brookfield

Brookfield Historical Society
P.O. Box 342
Brookfield, IL 60513
(312) 485-3420
Mary Kircher

Carbondale

Carbondale Preservation Commission
City of Carbondale
P.O. Box 2047
Carbondale, IL 62902-2047
(618) 549-5302
Donald D. Monty

City of Carbondale
P.O. Box 2047
Carbondale, IL 62902
(618) 549-5302
Tom Redmond

Cary

Village of Cary
255 Stonegate Road
Cary, IL 60013
George E. Kraus

Champaign

Illinois 4-H Foundation
302 East John Street
Suite 1908
Champaign, IL 61820
John Geissal

Markstahler
General Contractors
321 Fremont
Champaign, IL 61820
(217) 352-3498
Michael Markstahler

Preservation and Conservation
Association
Station A
P.O. Box 2555
Champaign, IL 61824
(217) 328-7222
Karen L. Kummer

Charleston

Charleston Downtown Development
2230 Seneca Drive
Charleston, IL 61920
Fred Preston

Chicago

American Planning Association
1313 East 60th Street
Chicago, IL 60637
(312) 955-9100
Carolyn Torma

Auditorium Theatre
50 East Congress
Chicago, IL 60605
(312) 431-2395
Barbara Corrigan

Baldwin Development
401 South La Salle
Suite 1701
Chicago, IL 60605
(312) 786-4600
Bill Taki

Republic Building, Chicago, Illinois. (P. Gardner, HABS)

Andrew P. Bocek
6090 North Caldwell Avenue
Chicago, IL 606464

Booth/Hansen and Associates
555 South Dearborn Street
Chicago, IL 60605
(312) 427-0300
Paul Hansen

Canal Corridor Association
220 South State Street
Suite 1880
Chicago, IL 60604
(312) 427-3688
Lauren Scott

Chicago Park District
425 East McFetridge
Suite 4 West
Chicago, IL 60604
(312) 294-2226
Julia Sniderman

Chicago Tribune Company
435 North Michigan Avenue
Chicago, IL 60611
Colleen Dishon

Commission on Chicago Landmarks
320 North Clark Street
Room 516
Chicago, IL 60610
(312) 744-3200
W. McLenahan

Richard Daspit, Jr.
929 West Margate Terrace
Suite 3W
Chicago, IL 60640-3826
(312) 929-5028

Charles E. Gregersen, AIA
11433 South St. Lawrence Avenue
Chicago, IL 60628
(312) 468-5231

Hasbrouck Peterson Associates
711 South Dearborn
2nd Floor
Chicago, IL 60605
(312) 922-7211
Wilbert Hasbrouck

Illinois Historic Preservation Agency
100 West Randolph Street
Suite 4-900
Chicago, IL 60601
(312) 814-1409
(312) 814-1422 (Fax)
Roberta Deering

Landmarks Preservation Council of
Illinois
53 West Jackson
Suite 752
Chicago, IL 60604
(312) 922-1742
Carol S. Wyant

Magazine Elks
425 West Diversey
Chicago, IL 60614
(312) 528-4500
Rudy R. Wight

Norwood Park Historical Society
5831 North Nickerson
Chicago, IL 60631
(312) 774-7440
Susan J. Kroll

KPMG Peat Marwick
303 East Wacker Drive
Chicago, IL 60601
(312) 938-1000
Michael J. Lins

Roosevelt University
Physical Plant Department
430 South Michigan Avenue
Chicago, IL 60605
(312) 341-3600
L. T. Ryan

Roosevelt University
430 South Michigan Avenue
Room 815
Chicago, IL 60605
(312) 341-3808
David Lansing

James Shorts
WLVP Suite 3210
875 North Michigan
Chicago, IL 60611
(312) 591-9587
Kevin Matthews

Judith Wood Spock
1925 North Fremont Street
Chicago, IL 60614
(612) 525-7623

Three Arts Club of Chicago
1300 North Dearborn Street
Chicago, IL 60610
(312) 944-6250
Nancy W. Moore

Uptown Chicago Commission
4753 North Broadway
Suite 610
Chicago, IL 60640
(312) 561-3978
Patricia Reskey

Vintage Realty
2329 West Chicago Avenue
Chicago, IL 60622
(312) 235-1200
Lilia Kulas-Zaparaniu

Visibility
The Travel Industry Resource
8 South Michigan Avenue
Suite 2500
Chicago, IL 60603
(312) 419-0303
Sandy M. Guettler

Harry Weese
10 West Hubbard Street
Chicago, IL 60610
(312) 467-7030

Decatur

Historic and Architectural Sites
Commission
1 Civic Center Plaza
Decatur, IL 62523
(217) 424-7777
David A. Clark

Dekalb

Ellwood House Museum
509 North 1st Street
Dekalb, IL 60115
(815) 756-4609
Gerald J. Brauer

Des Plaines

Prime Restorations
P.O. Box 2041
Des Plaines, IL 60017
(708) 437-5219
Michael Oiszowka

Dolton

Helen J. Berg, Chairman
Historical Commission
14014 Park Avenue
Dolton, IL 60419

Dundee

Laura Anderson
36W429 Oak Hill Drive
Dundee, IL 60118
(708) 428-1844

East Moline

Nixalite of America
1025 16th Avenue
East Moline, IL 61244
(309) 755-8771
Marie Gellerstedt

Edwardsville

Goshen Preservation Alliance
P.O. Box 303
Edwardsville, IL 62025
(618) 656-0292
(618) 656-4984
Carol A. Keene

Elgin

Elgin Area Historical Society
360 Park Street
Elgin, IL 60120
(312) 742-4248

Elgin Heritage Commission
150 Dexter Court
Elgin, IL 60120-5555
(708) 695-6500
Mary Buchheid

Elmhurst

Elmhurst Historical Museum
120 East Park Avenue
Elmhurst, IL 60126
(312) 833-1457
Brian F. Bergheger

Elmhurst Historical Society
120 East Park
Elmhurst, IL 60126
(708) 530-0776
Pat Zubak

Elsah

Historic Elsah Foundation
P.O. Box 117
Elsah, IL 62028
(618) 374-1059

Evanston

Evanston Preservation Commission
2100 Ridge Avenue
Evanston, IL 60204

Junior League of Evanston
Historic Preservation Chairman
225 Greenwood
Evanston, IL 60201

Teska Associates
627 Grove Street
Evanston, IL 60201
(708) 869-2015
Lee M. Brown

Flora

City of Flora
City Hall
122 North Main Street
Flora, IL 62839
(618) 662-2155
Jack Thatcher

Galesburg

City of Galesburg
Community Development Department
161 South Cherry Street
P.O. Box 1387
Galesburg, IL 61401
(309) 343-4181
Roy A. Parkin

Geneva

Kane County Development
Department
Planning Division
719 Batavia Avenue
Geneva, IL 60134
(708) 232-3451
Christine M. Poll

Glencoe

Historic Preservation Commission
Village of Glencoe
675 Village Court
Glencoe, IL 60022
(708) 835-4111
Mark C. Meyers

Glen Ellyn

Debora Helledy
455 Taylor Street
Glen Ellyn, IL 60137

Jeffrey Mansfield
290 South Milton Avenue
Glen Ellyn, IL 60137

Leland Marks
475 Hawthorne Street
Glen Ellyn, IL 60137

Bud Martin
583 North Main Street
Glen Ellyn, IL 60137

Sarah L. Poeppel
67 Stephanie Lane
Glen Ellyn, IL 60137

Patrick Schaur
754 North Main Street
Glen Ellyn, IL 60137

Alice Vanest
505 Greenbrier Road
Glen Ellyn, IL 60137

Village Of Glen Ellyn
Historical Sites Commission
535 Duane Street
Glen Ellyn, IL 60137
Robert J. Pilipiszyn

Hanover Park

The Hanover Park-Ontarioville
Historical Commission
2121 West Lake Street
Hanover Park, IL 60103
(312) 837-3800

Highland Park

City of Highland Park
1707 St. Johns Avenue
Highland Park, IL 60035
(312) 432-0800
Mary Herr

Highland Park Economic Development
Department
1707 St. Johns Avenue
Highland Park, IL 60035
(312) 432-0800
Peggy Blanchard

Hinsdale

Hinsdale Public Library
20 East Maple Street
Hinsdale, IL 60521

Old Metal Shop
427 South Lincoln Street
Hinsdale, IL 60521
Barbara T. Maxwell

Salt Creek Greenway Association
1 South Monroe
Hinsdale, IL 60521
(312) 865-8736
Valerie Spale

Jacksonville

Jacksonville Historic Preservation
Commission
Municipal Building
Jacksonville, IL 62650
(217) 245-4557
Steve Hardin

Joliet

City of Joliet
150 West Jefferson Street
Joliet, IL 60431
(815) 740-2433
Barbara Newberg

Joliet Public Library
150 North Ottawa Street
Joliet, IL 60431
(815) 740-2668
James R. Johnston

Universalist Unitarian Church
156 North Chicago Street
Joliet, IL 60431
(815) 722-0836
Janet Emmons

Kankakee

Riverview Historic District
834 South Greenwood Avenue
Kankakee, IL 60901
(815) 935-8534
June Boisrert

Kenilworth

Teresa Conway
532 Sterling Road
Kenilworth, IL 60043
(312) 251-6416

Kenilworth Historical Society
240 Cumberland Avenue
Kenilworth, IL 60043
(708) 251-2565
Jacob R. Suker

LaFox

Garfield Farm Museum
Box 403
3N016 Garfield Road
LaFox, IL 60147
(312) 584-8485
Jerome M. Johnson

LaGrange

James Leatherberry
211 South LaGrange Road
LaGrange, IL 60525
(312) 354-6335

Main Street LaGrange
26 South LaGrange Road
LaGrange, IL 60525
(312) 354-7261

Village of LaGrange
53 South LaGrange Road
LaGrange, IL 60525
(312) 579-2318
Scott Randall

Lake Forest

Lake Forest Foundation For Historic
Preservation
P.O. Box 813
Lake Forest, IL 60045

Lemont

Downtown Lemont
Lemont Chamber of Commerce
101 Main Street
Lemont, IL 60439
(708) 257-2522
Mary Hunt

Lemont Area Chamber of Commerce
101 Main Street
Lemont, IL 60439
(708) 257-5997
Thomas Sniegoski

Lemont Area Historical Society
Box 126
Lemont, IL 60439
(312) 257-5924
Nancy Thornton

Libertyville

David Adler Cultural Center
1700 North Milwaukee Avenue
Libertyville, IL 60048
(708) 367-0707
Douglas Miller

Lazzaretto Construction Company
P.O. Box 465
Libertyville, IL 60048
(708) 688-7356
Nick Lazzaretto

Main Street Libertyville
507 North Milwaukee Avenue
Libertyville, IL 60048
(708) 680-0336
Ellen Brin

Village of Libertyville
Planning Department
200 East Cook
Libertyville, IL 60048
(708) 362-2430
Joseph Duffy

Lincoln

Downtown Lincoln Association
c/o Bylines
527 Pulaski Street
Lincoln, IL 62656
(217) 732-1619
Valecia Crisafulli

Lockport

Lockport Area Development
Commission
222 East 9th Street
Lockport, IL 60441
(815) 744-2328
Ann G. Hintze

Main Street Lockport
132 East 9th Street
Lockport, IL 60441
Sara Rickert

Lombard

Village of Lombard
255 East Wilson Street
Lombard, IL 60555
(708) 620-5743
Jo Charlton

Macomb

Rural Economic Technical Assistance
Center
318 Stipes Hall
Macomb, IL 61455
(800) 526-9943
Dan Walker

Maple Park

Ronald G. Chandler
45 W 521
Route 38
Maple Park, IL 60151
(312) 365-5310

Moline

Rock Island County Historical Society
Box 632
Moline, IL 61265
(130) 976-4859
(130) 978-8710
James R. Sampson

Naperville

Darlene Ebel
215 South Columbia
Naperville, IL 60540
(312) 413-3617

New Berlin

Nancy E. M. Schneider
Walnut Hill
R.R. 2
Box 67
New Berlin, IL 62670
(217) 488-7482

Northbrook

Northbrook Public Library
1201 Cedar Lane
Northbrook, IL 60062
Nancy Unzicker

Oak Brook

Oak Brook Historical Society
Box 3821
Oak Brook, IL 60521-3821

Oak Park

Historical Society of Oak Park and
River Forest
P.O. Box 771
Oak Park, IL 60303
(708) 848-6755

Carol R. Kelm
Carl A. Lind
930 Columbian
Oak Park, IL 60302
(708) 386-8927

Oak Park Public Library
Technical Services Department
834 Lake Street
Oak Park, IL 60301

Pleasant Home
217 Home Avenue
Oak Park, IL 60302
(708) 383-2612
Darinka Dimitrijevic

Unity Temple Restoration Foundation
Box 2273
Oak Park, IL 60303
(312) 848-6225
Frank Pond

Ottawa

Main Street Ottawa
617 LaSalle Street
Ottawa, IL 61350
(815) 433-6246
Nick Calogeresis

Ottawa Historic Preservation League
P.O. Box 295
Ottawa, IL 61350
(815) 433-2775
Keith Johnson

Pekin

Pekin Public Library
Adult Department
301 South 4th Street
Pekin, IL 61554

Peoria

Diana C. Johnson
131 West Columbia Terrace
Peoria, IL 61603

Peotone

Historical Society of Greater Peotone
213 West North Street
Peotone, IL 60468
(312) 747-7500
Michael R. Morrison

Pontiac

Pontiac Redeveloping Our United
Downtown
P.O. Box 622
Pontiac, IL 61764-0622
(815) 844-6692
Karen Grimm

Princeton

Owen Lovejoy Homestead
906 South Main Street
Princeton, IL 61356
(815) 875-8491
Carolyn Shipp

Quincy

City of Quincy
507 Vermont
Quincy, IL 62301
(217) 228-4545
Verne Hagstrom

Gardner Museum of Architecture and
Design
4th and Maine Streets
Quincy, IL 62301
(217) 224-6873
John W. Gardner

Quincy Preservation Commission
200 North 8th
Suite 102
Quincy, IL 62301
(217) 228-4515
Ann Sloniger

Riverside

Riverside Historic Commission
99 Lawton Road
Riverside, IL 60546
(708) 447-4120
Dorothy Unger

Rockford

Virginia Gregory
3050 Jacqueline Drive
Suite 6
Rockford, IL 61109
(815) 987-5600

Restoration Education
P.O. Box 17121
Rockford, IL 61107
(815) 399-5828
Richard J. Marsh

Rockford Historic Preservation
Commission
425 East State Street
Rockford, IL 61104
(815) 987-5600
Virginia Gregory

Julie Snively
1618 National Avenue
Rockford, IL 61103

Tinker Swiss Cottage
411 Kent Street
Rockford, IL 61102
(815) 964-2424
Scott G. Prine

Rock Island

City of Rock Island
Preservation Commission
Community and Economic
Development Department
1528 3rd Avenue
Rock Island, IL 61201
(309) 793-3442
Alan M. Carmen

Col. Davenport Historic Foundation
3403 35th Avenue
Rock Island, IL 61201
(309) 788-7983
Rita Lyons

Rolling Meadows

Architects Group
3887 Industrial Avenue
Rolling Meadows, IL 60067
(708) 392-0603
Leslie Rice

St. Charles

Restoration of Kane County
P.O. Box 903
St. Charles, IL 60174
(708) 377-6424
Elizabeth Safanda

Salem

Keep Salem Beautiful
401 North Jefferson
Salem, IL 62881
(618) 548-0893
Patricia Bauer

Skokie

Skokie Public Library
5215 Oakton Street
Skokie, IL 60077

Springfield

City of Springfield Historic Sites
Commission
1331 South Dial Court
Springfield, IL 62704
(217) 785-4263
Enrique J. Unanue

Dana Thomas House Foundation
P.O. Box 7123
Springfield, IL 62791
(217) 753-3262
Shirley Stevens

Illinois Historic Preservation Agency
Old State Capitol
Springfield, IL 62701
(217) 785-9045
William Wheeler

Legislative Space Needs Commission
W.G. Stratton Building
Room 602
Springfield, IL 62706
(217) 782-7963
Mal Hildebrand

Tinley Park

Village of Tinley Park
16250 South Oak Park Avenue
Tinley Park, IL 60477
(708) 532-7700
Brad L. Bettenhausen

Urbana

URBANA Group
110 South Race Street
P.O. Box 1028
Urbana, IL 61801
(217) 344-7526
Lachlan F. Blair

Villa Park

Villa Park Historical Preservation
Commission
Village of Villa Park
20 South Ardmore Avenue
Villa Park, IL 60181
(312) 834-8500
Frank Carlson

Waukegan

Waukegan Downtown Association
Box 191
Waukegan, IL 60079
(312) 662-7081
Debbie Rakestraw

Wayne

Jane M. Schroeder
Box 542
Wayne, IL 60184
(708) 377-0858

West Chicago

City of West Chicago
Historical Museum
132 Main Street
West Chicago, IL 60185
(312) 231-3376
LuAnn Bombard

Wheaton

City of Wheaton
303 West Wesley Street
Wheaton, IL 60189
(708) 260-2019
Anne Wollensak

Conservation Foundation of Dupage
County
703 Warrenville Road
Wheaton, IL 60187
(708) 682-3505
Jean Connell

Wilmette

Preservation Board of Wilmette
Village Hall
1200 Wilmette Avenue
Wilmette, IL 60091
(708) 256-8314
Belinda Blanchard

Village of Wilmette
1200 Wilmette Avenue
Wilmette, IL 60091
(708) 251-2700
Tracy Einspanjer

Wood Dale

Wood Dale Historical Society
Yesterday's Farm Museum
7N040 Wood Dale Road
P.O. Box 13
Wood Dale, IL 60191
(708) 595-8777
Debra Morgan

Woodstock

City of Woodstock
P.O. Box 190
Woodstock, IL 60098
(815) 338-4300
Dennis L. Anderson

INDIANA

STATE HISTORIC PRESERVATION OFFICE

Department of Natural Resources
402 West Washington Street
Indiana Government Center South
Room C-256
Indianapolis, IN 46204
(317) 232-4020
(317) 232-8036 (Fax)
Patrick R. Ralston, SHPO

James A. Glass, Deputy SHPO
Director
Division of Historic Preservation
Room 274
(317) 232-1646

STATEWIDE PRESERVATION ORGANIZATION

Historic Landmarks Foundation of
Indiana
340 West Michigan Street
Indianapolis, IN 46202-3204
(317) 639-4534
(317) 639-6734 (Fax)
Randall T. Shepard, Chairman
J. Reid Williamson, Jr., President/CEO

NATIONAL TRUST REGIONAL OFFICE

Midwest Regional Office
52 West Jackson Boulevard
Suite 1135
Chicago, IL 60604
(312) 939-5447
(312) 939-5651 (Fax)
Tim Turner, Director

NATIONAL TRUST ADVISORS

Kent Schuette
604 Wyandotte Avenue
Lafayette, IN 47905
(317) 494-1324
(317) 494-0391 (Fax)

J. Reid Williamson, Jr.
Historic Landmarks Foundation of
Indiana
340 West Michigan
Indianapolis, IN 46202
(317) 639-4534
(317) 639-6734 (Fax)

PRESERVATION ACTION COORDINATOR

David B. Frederick
1028 North Delaware Street
Indianapolis, IN 46202
(317) 638-5264

AIA STATE PRESERVATION COORDINATOR

H. Roll McLaughlin, FAIA
950 West 116 Street
Carmel, IN 46032
(317) 846-8020

NATIONAL PARK SERVICE REGIONAL OFFICE

Midwest Regional Office
1709 Jackson Street
Omaha, NE 68102
(402) 221-3431

ADVISORY COUNCIL ON HISTORIC PRESERVATION DIVISION

Eastern Office of Project Review
1100 Pennsylvania Avenue, N.W.
Suite 809
Washington, DC 20004
(202) 786-0503
Don L. Klima, Director

Anderson

City Of Anderson
Planning Department
P.O. Box 2100
120 East 8th Street
Anderson, IN 46018
(317) 646-9690
Joyceann Yelton

Auburn

Auburn-Cord Duesenberg Museum
P.O. Box 271
Auburn, IN 46706
(219) 925-1444
Skip Marketti

Bloomington

Bloomington Restorations
P.O. Box 1522
Bloomington, IN 47402
Michael Flory

City of Bloomington
P.O. Box 100
Bloomington, IN 47402
Glenda E. Morrison

Cook Finance Company
405 North Rogers Street
Bloomington, IN 47401
(812) 332-0053
Chris Taylor

Wylie House
317 East 2nd Street
Bloomington, IN 47405

Cambridge City

Historic Landmarks Foundation of
Indiana
Eastern Regional Office
P.O. Box 284
Cambridge City, IN 47327

Corydon

Main Street Corydon
P.O. Box 69
Corydon, IN 47112
(812) 738-1917
Jim Urban

Elkhart

Elkhart Centre
616 South Main Street
Elkhart, IN 46515
(219) 295-8701
Sandy Geiser

Elkhart Historic and Cultural
Preservation Commission
229 South 2nd Street
Elkhart, IN 46516
(219) 294-5471
Pam Gundlach

Ruthmere Museum
302 East Beardsley Avenue
Elkhart, IN 46514
(219) 264-0330
Robert B. Beardsley

Evansville

Bill Gaisser Designs/Architecture
219 Oak Street
Evansville, IN 47713
(812) 423-3777
H. Gaisser

Center City Corporation
329 Main Street
Suite 612
Evansville, IN 47708
(812) 424-2986
Theo Boots

Historic Preservation Services
Department of Metropolitan
Development
216 Washington Avenue
306 Civic Center Complex
Evansville, IN 47708
(812) 428-0737
Joan C Marchand

The Willard Library of Evansville
21 1st Avenue
Evansville, IN 47710
(812) 425-4309
William A. Goodrich

Farmersburg

The Westerly Group
R.R. 1
Box 141
Farmersburg, IN 47850
(812) 696-2415
C. B. Fife

Fort Wayne

ARCH
P.O. Box 11383
Fort Wayne, IN 46857
Janet V. Nahrwold

Nob Sales Corporation
4936 Nob Road
Fort Wayne, IN 46825
(219) 483-2126
Bob Benhower

Frankfort

Frankfort Main Street
16 North Main Street
P.O. Box 311
Frankfort, IN 46041
(317) 654-4081
Rich Young

Franklin

Franklin College Library
Franklin, IN 46131

Gary

City of Gary
401 Broadway
Gary, IN 46402

Main Street Gary
656 Broadway
Gary, IN 46402
(219) 885-2937
Jihad T. Muhammad

Greencastle

Historic Landmarks Foundation
2 South Jackson
Greencastle, IN 46135
(317) 653-4927

Hammond

Hammond Historic Preservation
Commission
649 Conkey Street
Hammond, IN 46324
(219) 853-6398
City Planner

Hanover

The Duggan Library
P.O. Box 287
Hanover, IN 47243
(812) 866-7160
W. D. Morrill

Indianapolis

William B. Clark
7530 North Penn
Indianapolis, IN 46240
(317) 255-3728

Department of Natural Resources
608 State Office Building
Indianapolis, IN 46204
James Glass

Historic Landmarks Foundation of
Indiana
340 West Michigan Street
Indianapolis, IN 46202-3204
(317) 639-4534
Reid Williamson

Historic Landmarks Foundation of
Indiana
340 West Michigan Street
Indianapolis, IN 46202-4534
(317) 639-4534
David B. Frederick

Indiana Main Street
Indiana Department of Commerce
1 North Capitol
Suite 700
Indianapolis, IN 46204
(317) 232-8908
William A. Dory

Indianapolis Historic Preservation
Commission
1821 City-County Building
Indianapolis, IN 46204
(317) 236-4406
David L. Baker

Indianapolis Museum of Art
1200 West 38th Street
Indianapolis, IN 46208
(317) 923-1331
Charles Gleaves

Jeffersonville

Historic Landmarks Foundation
Southern Regional Office
111 West Chestnut Street
Jeffersonville, IN 47130
(812) 284-4534
Mark A. Dollase

Howard Steamboat Museum
P.O. Box 606
Jeffersonvlle, IN 47130
(812) 283-3728
Louise M. Schildroth

Jeffersonville Main Street
113 West Chestnut Street
Jeffersonville, IN 47130
(812) 283-0301
Holly Johnson

Kokomo

Kokomo Public Library
Adult Services
220 North Union Street
Kokomo, IN 46901

Lafayette

Downtown Business Center
P.O. Box 1005
Lafayette, IN 47902
(317) 742-2313
Carolann Signorella

Lafayette Community Development
Department
324 Ferry Street
Room 204
Lafayette, IN 47901
(931) 774-21145
Lillian Cote

Wabash Valley Trust for Historic
Preservation
P.O. Box 1354
Lafayette, IN 47902
(317) 474-4505
Leonard Boyll

Madison

Cornerstone Society
P.O. Box 92
Madison, IN 47250
(812) 265-6507
Maryanne Ines

Historic Madison
500 West Street
Madison, IN 47250
(812) 265-2967
John E. Galvin

Marion

Main Street Marion
301 South Branson
Marion, IN 46952
(317) 668-4458
Therese M. Groff

Mishawaka

The Troyer Group
415 Lincolnway East
Mishawaka, IN 46544
(219) 259-9976
Phil Cartwright

Muncie

City of Muncie
Community Development
220 East Jackson Street
Muncie, IN 47305
(317) 747-4825
Gretchen B. Cheesemen

Andrea Urbas
216 Pasture Lane
Muncie, IN 47304
(317) 285-1900

Nappanee

Amish Acres
1600 West Market
Nappanee, IN 46550
(219) 773-4188
Richard Pletcher

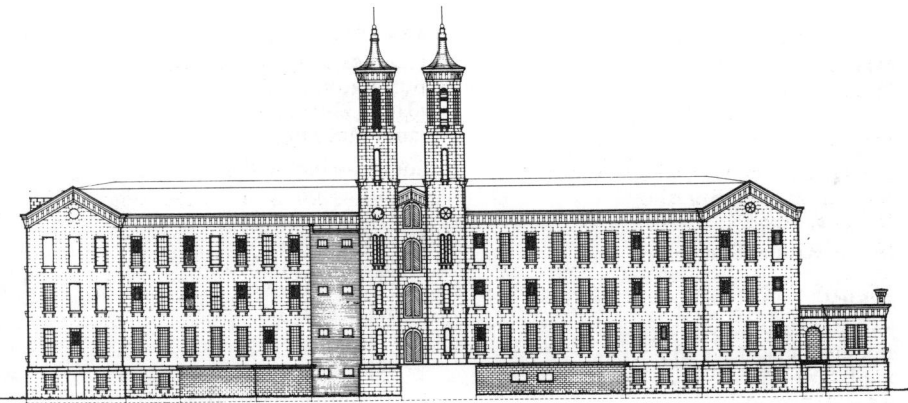

Cannelton Cotton Mills, Cannelton, Indiana. (M. Boles, L.D. Schaaf, HAER)

New Albany

Develop New Albany,
230 Pearl Street
New Albany, IN 47150
(812) 941-0018
Marsha Conner

Scott B. Wood
175 Woodbourne Drive
New Albany, IN 47150
(812) 949-8335

Newburgh

Historic Newburgh
P.O. Box 543
Newburgh, IN 47630
(812) 853-2815
Rose Polster

New Harmony

Historic New Harmony
344 Church Street
Box 579
New Harmony, IN 47631
(812) 464-9595
James A. Sanders

Noblesville

Noblesville Main Street Program
16 South 10th Street
Noblesville, IN 46060
(317) 773-4616
Jane Howey

Pendleton

Bob Post
106 1/2 West State Street
Pendleton, IN 46064
(317) 778-7778

Plainfield

Guilford Township Historical
Collection
Plainfield Public Library
1120 Stafford Road
Plainfield, IN 46168
(317) 839-6602
Susan M. Carter

Rensselaer

Zacher Afix Corporation
R.R. 3
Box 128
Rensselaer, IN 47978
William C. Zacher

Richmond

Indiana Historical Museum
Wayne County
1150 North A Street
Richmond, IN 47374
(317) 962-5756
Michele Bottorff

Rushville

Rush County Heritage
Rural Route 4
Box 121
Rushville, IN 46173
(317) 629-2386
Larry L. Stout

South Bend

Center City Associates
122 South Michigan Street
South Bend, IN 46601
(219) 282-1110
Carter Wolf

Historic Landmarks Foundation
Northern Regional Office
914 Lincolnway West
South Bend, IN 46616
(219) 232-4534
Karen L. Kiemnec

Historic Preservation Commission of
South Bend and St. Joseph City
County/City Building
Room 1123
South Bend, IN 46601
(219) 284-9578
Lori Dixon

South Bend Heritage Foundation
914 Lincolnway West
South Bend, IN 46616
(219) 289-1066

Southhold Restorations
322 West Washington Street
South Bend, IN 46601-1524
(219) 234-3441
Barbara Steele

Terre Haute

Immanuel Lutheran Church
645 Poplar Street
Terre Haute, IN 47807
(812) 232-4972
Philip Meyer

Vigo Preservation Alliance
P.O. Box 2020
Terre Haute, IN 47802
Mike Moreo

Valparaiso

Porter County Public Library Systems
103 Jefferson Street
Valparaiso, IN 46383
(219) 462-0524
Don Johnson

Vincennes

Knox County Public Library
502 North 7th Street
Vincennes, IN 47591

Warsaw

Warsaw Community Development
Corporation
P.O. Box 1223
Warsaw, IN 46581-1223
(219) 267-6419

IOWA

STATE HISTORIC PRESERVATION OFFICE

State Historical Society of Iowa
Capitol Complex
East 6th and Locust Streets
Des Moines, IA 50319
(515) 281-8837
(515) 282-0502 (Fax)
David Crosson, SHPO

James E. Jacobsen, Deputy SHPO
Chief, Historic Preservation Bureau
(515) 281-4358

STATEWIDE PRESERVATION ORGANIZATION

Iowa Historic Preservation Alliance
P. O. Box 532
West Branch, IA 52358
(319) 643-5327
Jan Nash, Director
906 South Lucan Street
Iowa City, IA 52240

NATIONAL TRUST REGIONAL OFFICE

Midwest Regional Office
52 West Jackson Boulevard
Suite 1135
Chicago, IL 60604
(312) 939-5447
(312) 939-5651 (Fax)
Tim Turner, Director

NATIONAL TRUST ADVISORS

Harry S. Budd
907 Burnett
Ames, IA 50010
(515) 239-1391

PRESERVATION ACTION COORDINATOR

Rebecca Conard
Box 717
Lake View, IA 51450

AIA STATE PRESERVATION COORDINATOR

William Wagner, FAIA
2272 240th Street
Dallas Center, IA 50063
(515) 992-3023

NATIONAL PARK SERVICE REGIONAL OFFICE

Midwest Regional Office
1709 Jackson Street
Omaha, NE 68102
(402) 221-3431

ADVISORY COUNCIL ON HISTORIC PRESERVATION DIVISION

Eastern Office of Project Review
1100 Pennsylvania Avenue, N.W.
Suite 809
Washington, DC 20004
(202) 786-0503
Don L. Klima, Director

Ackley

Ackley Main Street Program
208 State Street
Ackley, IA 50601
(515) 847-3332
Terri L. Schroeder

Amana

Amana Heritage Society
P.O. Box 81
Amana, IA 52203
(319) 622-3567
Lanny Haldy

Ames

City of Ames
Department of Planning and Housing
P.O. Box 811
Ames, IA 50010
Ray D. Anderson

Susan Minks
216 North Riverside
Ames, IA 50010
(515) 232-5761

Anamosa

Main Street Anamosa
124 East Main Street
Anamosa, IA 52205
(319) 462-4267
Maureen Williams

Bettendorf

Scott Community College
Library-500
Belmont Road
Bettendorf, IA 52722

Bonaparte

Main Street Bonaparte
103 Main Street
P.O. Box 69
Bonaparte, IA 52620-0069
(319) 592-3580
Gianna Barrow

Burlington

Art Guild of Burlington
P.O. Box 5
Burlington, IA 52601
Lois Rigdon

Main Street of Burlington
P.O. Box 901
Burlington, IA 52601
(319) 752-0015
Janet McCannon

Southwest Iowa Regional Planning
Commission
610 Main
Box 397
Burlington, IA 52601
(319) 753-5107
Beth Eveleth

Cedar Falls

Community Main Street
128 Main
Cedar Falls, IA 50613
(319) 277-0213

Centerville

Centerville Main Street
100 West Maple
P.O. Box 211
Centerville, IA 52544
(515) 856-8088
Roxeene Jones

Clear Lake

Clear Lake Area Chamber of Commerce
205 Main Avenue
P.O. Box 188
Clear Lake, IA 50428
(515) 357-2159
Cyndi Chizek

Clinton

Main Street Clinton
River City Chamber of Commerce
334 4th Avenue South
Clinton, IA 52732
(319) 242-5702
Cindy Steinhauser

Conrad

Conrad Main Street
P.O. Box 414
Conrad, IA 50621
(515) 366-2108
Claudia Jones

Corning

Corning Main Street Program
601 6th Street
Corning, IA 50841
(515) 322-3243
Alec Turner

Council Bluffs

Historical Society of Pottawattamie
County
Box 2
Council Bluffs, IA 51502
Laurence G. Roberts

Historic General Dodge House
621 3rd Street
Council Bluffs, IA 51503-6614
(712) 322-2406
Catherine White

Cresco

Malcolm Pirnie
315 North Elm Street
Cresco, IA 52136
Barbara B. Long

Davenport

City of Davenport
Planning Office
226 West 4th Street
Davenport, IA 52801
Charles Heston

Midwest Wood Products
1051 South Rolff Street
Davenport, IA 52802
(319) 323-4755
Jim Englander

Palmer College of Chiropractic
1000 Brady Street
Davenport, IA 52803
(319) 326-9891
Dennis R. Peterson

Svendsen Tyler
215 Main Street
Suite 901
Davenport, IA 52801
(319) 324-1910
Marlys Svendsen

Van Allen and Son Store, Clinton, Iowa. (HABS)

Decorah

Norwegian-American National
Museum
502 West Water Street
Decorah, IA 52101
Marion J. Nelson

Des Moines

American Institute of Architects, Iowa
Chapter
512 Walnut
Des Moines, IA 50309
(515) 244-7502
Suzanne Schwengels

City of Des Moines
Planning Department
Armory Building, East
1st and Des Moines Streets
Des Moines, IA 50309
(515) 283-4182
Patricia Zingsheim

Main Street Iowa
Iowa Department of Economic
Development
200 East Grand Avenue
Des Moines, IA 50309
(515) 242-4733
Thomas D. Guzman

Judith Ann McClure
654 19th Street
Suite 1
Des Moines, IA 50314
(515) 280-3015

RDG Bussard/Dikis
303 Locust Street
Des Moines, IA 50309
(515) 288-3141
Ron Siggelkow

Sherman Hill Association
756 16th Street
Des Moines, IA 50314
(515) 284-5717
Jack C. Porter

State Historical Society of Iowa
Capitol Complex
600 East Locust Street
Des Moines, IA 50319
(515) 281-8837
David Crosson

Wallace House Foundation
756 16th Street
P.O. Box 7063
Des Moines, IA 50309
(515) 243-7063
Kent J. Newman

Dubuque

Department of Community
Development
City Hall
13th and Central
Dubuque, IA 52001
(319) 589-4210
Pamela Myhre-Gonyier

Dubuque County Historical Society
P.O. Box 305
Dubuque, IA 52004-0305
(319) 557-9545

Dunlap

Dunlap Community Development
Corporation
P.O. Box 33
Dunlap, IA 51529
(712) 643-2161
Carol Mulligan

Elkader

Main Street Elkader
207 North Main
Elkader, IA 52043

Fort Dodge

Fort Dodge Main Street
Department of Community
Development
819 1st Avenue, South
Fort Dodge, IA 50501
(515) 573-5097
Ellen Diehl

Fort Madison

Chamber of Commerce
P.O. Box 277
955 Avenue H
Fort Madison, IA 52627
(319) 372-5471
Mitch Zynowski

Fort Madison Historic Preservation
Commission
City Hall
811 Avenue E
Fort Madison, IA 52627
(319) 372-7700

Grinnell

Main Street Grinnell
P.O. Box 338
Grinnell, IA 50112
(515) 236-6555
Karen B. Baxter

Hampton

Hampton Main Street
P.O. Box 21
Hampton, IA 50441
(515) 456-668

Harlan

Harlan Main Street
712 Court Street
P.O. Box 424
Harlan, IA 51537
(712) 755-2692
Maureen Williams

Independence

Buchanan County Historical Society
516 6th Avenue, N.W.
Independence, IA 50644
(319) 334-4998
Jasper Riskedahl

Iowa City

City of Iowa City
410 East Washington Street
Iowa City, IA 52240
(319) 356-5243
Robert Miklo

Friends of Historic Preservation
P.O. Box 2001
Iowa City, IA 52244

Jan Nash
906 South Lucas Street
Iowa City, IA 52240
(319) 351-5156

Steve Vander Woude
730 North Van Buren
Iowa City, IA 52245
(319) 354-0953

Iowa Falls

Main Street Iowa Falls
412½ Washington Avenue
Iowa Falls, IA 50126
(515) 648-3654
Mary Murphy

Keokuk

Keokuk Main Street
401 Main Street
Keokuk, IA 52632
(319) 524-5055
Joyce Glasscock

Knoxville

Eunice M. Kuyper
Rural Route 3
Box 180
Knoxville, IA 50138
(515) 842-6626

Main Street Knoxville
P.O. Box 7
105 East Main Street
Knoxville, IA 50138
(515) 842-7243
Melanie L. Waldon

Le Claire

Buffalo Bill Museum of Le Claire
P.O. Box 284
Le Claire, IA 52753
(319) 289-5580
Ella Ross

Maquoketa

Maquoketa Chamber of Commerce
Marquoketa Main Street
213 South Main Street
Maquoketa, IA 52060
(319) 652-4602

Muscatine

Muscatine Main Street Program
Central Business District Commission
319 East 2nd Street
P.O. Box 297
Muscatine, IA 52761
(319) 263-8897

Ogden

Ogden Rural Main Street Program
P.O. Box 3
Ogden, IA 50212
(515) 275-2902
Susan Wallace

Oskaloosa

Main Street Oskaloosa
104-A Avenue E
Oskaloosa, IA 52577
(515) 673-7401

Sibley

Sibley Main Street Program
P.O. Box 67
Sibley, IA 51249
(712) 754-4646
Lila Hatting

Sigourney

Sigourney Main Street
112 East Washington
Sigourney, IA 52591
(515) 622-2288
Amy Butz

Sioux City

Castle on the Hill Association
P.O. Box 1776
Sioux City, IA 51102
Douglas Batcheller

City of Sioux City
Community Development Department
405 6th Street
P.O. Box 447
Sioux City, IA 51102-0447
(712) 279-6159
Gretchen Schalge

Main Street Sioux City
417 4th Street
Suite 290
Sioux City, IA 51101
(712) 252-0014
Carl Kaden

Siouxland Historical Railroad
Association
P.O. Box 1355
Sioux City, IA 51102
(712) 276-6432
Larry Obermeyer

Spencer

Spencer Main Street Company
c/o S.A.A.B.I.
420 2nd Avenue West
P.O. Box 3047
Spencer, IA 51301
(712) 262-7246
Doug Pals

Storm Lake

Storm Lake Chamber of Commerce
P.O. Box 584
Storm Lake, IA 50588
(712) 732-3780
Sandy Knoke

Waterloo

City of Waterloo
Planning Department
715 Mulberry Street
Waterloo, IA 50703
(319) 291-4366
Bob Stevenson

Waverly

Waverly Main Street
P.O. Box 813
Waverly, IA 50677
(319) 352-5939
Michael Hahn

Webster City

Hamilton County SEED
Box 474
Webster City, IA 50595
(515) 832-6708
Ann Wing-Vogelbache

Main Street Webster City
P.O. Box 310
Webster City, IA 50595
(515) 832-5130
Christine Mollenkopf

West Branch

Iowa Historic Preservation Alliance
Box 532
West Branch, IA 52358
(319) 351-5156

Wilton

Wilton Historical Society
Box 258
Wilton, IA 52778

KANSAS

STATE HISTORIC PRESERVATION OFFICE

Kansas State Historical Society
120 West 10th
Topeka, KS 66612
(913) 296-3251
(913) 296-1005 (Fax)
Ramon S. Powers, SHPO

Richard D. Pankratz, Deputy SHPO
Director, Historic Preservation
Department
(913) 296-4788

STATEWIDE PRESERVATION ORGANIZATION

Kansas Preservation Alliance
1927 Vermont
Manhattan, KS 66502
(913) 539-1811
Ray Weisenburger, President

NATIONAL TRUST REGIONAL OFFICE

Mountains/Plains Regional Office
511 16th Street
Suite 700
Denver, CO 80202
(303) 623-1504
(303) 825-8073 (Fax)
Barbara J. Pahl, Director

NATIONAL TRUST ADVISORS

J. Eric Engstrom
901 Wiley Street
Wichita, KS 67203
(316) 267-7361
(316) 267-1754 (Fax)

Dean Graves, FAIA
5328 West 67th Street
Prairie Village, KS 66208
(913) 341-0007
(913) 383-1454 (Fax)

PRESERVATION ACTION COORDINATOR

Ray Weisenburger
1927 Vermont Street
Manhattan, KS 66502
(913) 532-5958

AIA STATE PRESERVATION COORDINATOR

Bernd Foerster, FAIA
920 Ratone Street
Manhattan, KS 66502
(913) 532-5953
Ray Weisenburger, AIA, Deputy

NATIONAL PARK SERVICE REGIONAL OFFICE

Midwest Regional Office
1709 Jackson Street
Omaha, NE 68102
(402) 221-3431

ADVISORY COUNCIL ON HISTORIC PRESERVATION DIVISION

Western Office of Project Review
730 Simms Street
Room 401
Golden, CO 80401
(303) 231-5320
Claudia Nissley, Director

Atchison

Amelia Earhart Birthplace
P.O. Box 118
Atchison, KS 66002
(913) 367-1480
Marilyn D. Boehler

Baldwin City

The Baker University
8th and Grove
Baldwin City, KS 66006
David Pittman

Chanute

Southeast Kansas Regional Planning
Commission
P.O. Box 664
Chanute, KS 66720-0664
(316) 431-0080
Steve Robb

Cottonwood Falls

Chase County Twin Cities
P.O. Box 362
Cottonwood Falls, KS 66845
(316) 273-8469
Tim Vaughn

Dodge City

Downtown Dodge
P.O. Box 963
Dodge City, KS 67801
(316) 225-2284
Kirk Schraeder

Downtown Fort Scott Program
23 South Main
P.O. Box 226
Fort Scott, KS 66701
(316) 223-6088
Mary Asher

Great Bend

Great Bend Downtown Development
1920 16th Street
Suite 105
Great Bend, KS 67530
(316) 793-7700
Gail E. Johnson

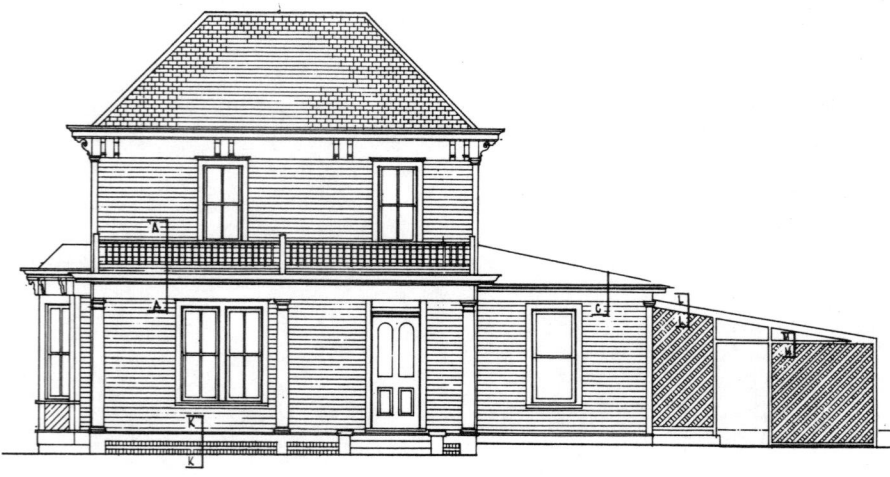

Eisenhower Boyhood Home, Abilene, Kansas. (T. Simmons, HABS)

Hays

Greater Downtown Hays Business
Improvement District
1301 Pine
P.O. Box 220
Hays, KS 67601
(913) 628-8201
Wayne Billinger

Herrington

Herrington Main Street
17 North Broadway
Herrington, KS 67449
(913) 258-2271
Jean Warta

Hutchinson

Hutchinson Downtown Development
Department
P.O. Box 1567
Hutchinson, KS 67504-1567
(316) 665-3727
Howard Woodward

Independence

Independence Main Street
106 East Myrtle
Independence, KS 67301
(316) 331-2300
Judy Holroyd

Kansas City

Prosoco
P.O. Box 1578
Kansas City, KS 66117
(913) 281-2700
Gerald E. Boyer

Lawrence

Lawrence Preservation Alliance
P.O. Box 1073
Lawrence, KS 66044
(913) 842-5241
Oliver Finney

University of Kansas
Division of Continuing Education
Continuing Education Building
Lawrence, KS 66045-2607
(913) 964-3284
Paul Forio

University of Kansas
Historic Mount Oread Fund
Lawrence, KS 66044

Manhattan

City of Manhattan
Redevelopment Office
Box 748
Manhattan, KS 66502
(913) 537-9683
Karen Davis

Kansas Preservation Alliance
1927 Vermont
Manhattan, KS 66502
Ray Weisenburger

Kansas State University
Department of Clothing and Interior
Design
Justin Hall
Suite 225
Manhattan, KS 66506

Manhattan Main Street
413 Poyntz Avenue
Manhattan, KS 66502
(913) 537-9683
Jeanne Stinson

Ness City

Ness County Bank Building Foundation
P.O. Box 6
Ness City, KS 67560
(913) 848-2351
Basil C. Marhofer

Oberlin

Oberlin Main Street
132 South Penn Street
Oberlin, KS 67749
(913) 475-3475
Reva Marshall

Olathe

Mahaffie Farmstead and
Stagecoach Stop
Box 768
Olathe, KS 66061
(913) 782-6972
Michael E. Duncan

Peabody

Peabody Main Street
P.O. Box 21
Peabody, KS 66866
(316) 983-2174
Julie Eberhard

Prairie Village

Kansas City Architectural Foundation
5328 West 67th Street
Prairie Village, KS 66208
(913) 262-0691
Ginny Graves

Salina

City of Salina
City Planning Department
P.O. Box 736
Salina, KS 67401
(913) 827-8781
Roy Dudark

Salina Downtown
P.O. Box 1065
Salina, KS 67402-1065
(913) 825-0535
Judy Ewalt

Smoky Hill Museum
P.O. Box 101
Salina, KS 67402
(913) 827-3958
Tom Pfannenstiel

Shawnee Mission

Johnson County Museum System
6305 Lackman Road
Shawnee Mission, KS 66217
(913) 631-6709
Janet B. Campbell

Topeka

Historic Topeka
Box 903
Topeka, KS 66601
(913) 354-8982
Thomas C. Flett

Kansas Department of Commerce
Kansas Main Street Program
400 S.W. 8th Street
5th Floor
Topeka, KS 66603-3957
(913) 296-3485
Brenda Spencer

Kansas State Historical Society
Historic Preservation Department
120 West 10th
Topeka, KS 66612
(913) 296-4788
Richard Pankrotz

Prebound Periodicals
914 Jefferson Street
Topeka, KS 66607
Juanita Wallen

Slemmons Associates
1 Townsite Plaza
Suite 1030
Topeka, KS 66603
(913) 235-9244
Robert S. Slemmons

Wamego

Wamego Main Street
P.O. Box 34
Wamego, KS 66547
(913) 456-7849
Rosemary Crilly

Wichita

Wichita-Sedgwick County
Metropolitan Area Planning
Department
455 North Main Street
10th Floor
Wichita, KS 67202
(316) 268-4421
Bob Beardsley

KENTUCKY

STATE HISTORIC PRESERVATION OFFICE

Kentucky Heritage Council
300 Washington Street
Frankfort, KY 40601
(502) 564-7005
(502) 564-6578
David Morgan, SHPO

STATEWIDE PRESERVATION ORGANIZATION

Commonwealth Preservation
Advocates
254 South Ashland Avenue
Lexington, KY 40502
(606) 268-0900
Mary Breeding, President

NATIONAL TRUST REGIONAL OFFICE

Southern Regional Office
456 King Street
Charleston, SC 29403
(803) 722-8552
(803) 722-8652 (Fax)
Susan A. Kidd, Director

NATIONAL TRUST ADVISORS

Richard L. Holland
P.O. Box 2632
Paducah, KY 42002-2632
(502) 443-9284
(502) 442-9152 (Fax)

Ann Early Sutherland
207 East Flaget Avenue
Bardstown, KY 40004-1519
(502) 348-3827

PRESERVATION ACTION COORDINATOR

Richard T. Jett
119 North Galt Avenue
Louisville, KY 40206
(502) 564-7005

AIA STATE PRESERVATION COORDINATOR

Robert L. Lape, AIA
31 Pike Street
Covington, KY 41011
(606) 261-7805

NATIONAL PARK SERVICE REGIONAL OFFICE

Southeast Regional Office
Richard B. Russell Federal Building
75 Spring Street, S.W.
Atlanta, GA 30303
(404) 331-2632

ADVISORY COUNCIL ON HISTORIC PRESERVATION DIVISION

Eastern Office of Project Review
1100 Pennsylvania Avenue, N.W.
Suite 809
Washington, DC 20004
(202) 786-0503
Don L. Klima, Director

Ashland

Department of Planning and
Community Development
P.O. Box 1839
Ashland, KY 41105-1839
(606) 329-9650
J. M. Gurnee

Bellevue

City of Bellevue
616 Poplar Street
Bellevue, KY 41073
(606) 431-8866
Karen Keown

David Royalty
Ann Royalty
256 Ward Avenue
Bellevue, KY 41073
(606) 291-9980

Bowling Green

Riverview House Musuem
1100 West Main Avenue
P.O. Box 10059
Bowling Green, KY 42102-4859
(502) 843-5565
W. S. Terry

Burlington

Boone County Historic Preservation
P.O. Box 960
Burlington, KY 41005

Dinsmore Homestead Foundation
P.O. Box 453
Burlington, KY 41005
(606) 586-6127
William Kreuger

Cadiz

Cadiz, The Place To Be
P.O. Box 735
Cadiz, KY 42211
(502) 522-6343
Linda Stover

Covington

City of Covington
Economic Development Department
Old Town Plaza
18 West Pike Street
Covington, KY 41011
(606) 292-2111
Leah Konicki

Danville

Mack Jackson
P.O. Box 85
412 South 4th Street
Danville, KY 40422

Dawson Springs

Dawson Springs Main Street
P.O. Box 107
Dawson Springs, KY 42408
(502) 797-2235
Susan Mestan

Elkton

Esther B. Dickinson
1969 Pond River Road
Elkton, KY 42220

Frankfort

Downtown Frankfort
234 West Main
Frankfort, KY 40601
(502) 223-2261
Cathy Noel

Kentucky Historical Society
Museum Division
Old State House Broadway
Frankfort, KY 40601
(502) 564-3016

Kentucky Heritage Council
667 Comanche Trail
Frankfort, KY 40601-1967
(502) 564-7005
Bill Steiner

Kentucky Housing Corporation
1231 Louisville Road
Frankfort, KY 40601
(502) 564-7630
Sheila Etchen

Hopkinville

Hopkinville Historic Preservation
Commission
P.O. Box 1161
Hopkinsville, KY 42241
(502) 885-0007
John C. Mahre

Lexington

Bluegrass Trust for Historic
Preservation
253 Market Street
Lexington, KY 40507
(606) 253-0362

Henry Clay Memorial Foundation
120 Sycamore Road
Lexington, KY 40502
(606) 266-8581
Bettie Kerr

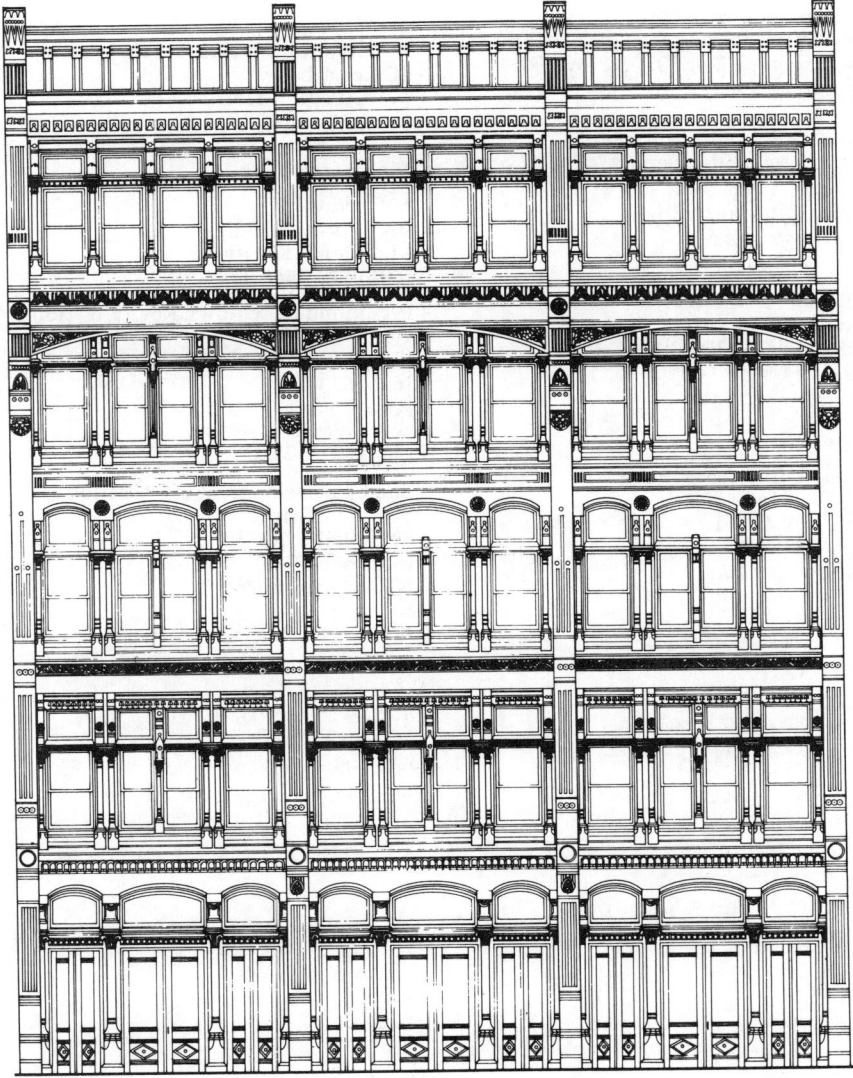

Hart Block, Louisville. (C. Alexander, HABS)

Kentucky League of Cities
2201 Regency Road
Suite 100
Lexington, KY 40503
Edwin L. Griffin

Kentucky Mansions Preservation
Foundation
578 West Main Street
P.O. Box 132
Lexington, KY 40501
(606) 233-9999
Lou Holden

Lexington-Fayette County Historic
Preservation Office
200 East Main Street
Lexington, KY 40507
(606) 258-3265

Ward Hall
2090 North Limestone
Lexington, KY 40505
(502) 863-1619
(606) 299-1318
Frances Susong Jenkins

Waveland Historic Site
225 Higbee Mill Road
Lexington, KY 40503-4778
(606) 272-3611
Sara Farley

Louisville

Belle of Louisville
4th and River Road
Louisville, KY 40202

Broadway/Brown Partnership
21 Theatre Square
Louisville, KY 40202
(502) 584-6633
Richard H. Dinsmore

City of Louisville Landmarks
Commission
City Hall Annex
609 West Jefferson Street
2nd Floor
Louisville, KY 40202
(502) 625-3501
Ann S. Hassett

Colonial Dames of America
Historicl Activities Committee
2 Woodhill Road
Louisville, KY 40207
W. B. Davis

Filson Club
1310 South 3rd Street
Louisville, KY 40208
(502) 635-5083
James R. Bentley

Historic Homes Foundation
414 Baxter Avenue
Louisville, KY 40204-1180
(502) 568-6397
Anne Bertrand

Jefferson County Office of Historic
Preservation and Archives
810 Barrett Avenue
Louisville, KY 40204
(502) 625-5761
Leslee Keys

Preservation Alliance
716 West Main Street
Louisville, KY 40202
(502) 583-8622
Monie Mitchell

Professional Bank Services
6200 Dutchman's Lane
Suite 305
Louisville, KY 40205
(502) 451-6633
George Freibert

River Fields
643 West Main Street
Suite 200
Louisville, KY 40202-2921
Meme S. Runyon

Madisonville

Discover Downtown
140 South Main Street
Madisonville, KY 42431
(502) 821-3435
Charlotte Sellers

Mt. Sterling

Chamber of Commerce
Main Street Program
51 North Maysville
Mt. Sterling, KY 40353
(606) 498-5343
Del White

Newport

Newport Historic Preservation
Commission
330 York Street
Newport, KY 41071
(606) 292-3630
Teresa Brum

Owensboro

Commonwealth Preservation
Advocates
P.O. Box 604
Owensboro, KY 42302
(502) 633-5029
Nancy Alexander

Downtown Owensboro
205 West 4th Street
Owensboro, KY 42301
Glenda Thacker

Paducah

Paducah McCracken County Growth
P.O. Box 2632
Paducah, KY 42002-2632
(502) 443-9284
Richard Holland

Paducah Main Street
316 Broadway
Paducah, KY 42001
(502) 444-8649
Denise P. McNutt

Paris

Paris Main Street Program
P.O. Box 164
Paris, KY 40361
(606) 987-6818
Jennifer Carlson

Pikeville

Preservation Council of Pike County
10 Cedar Lane
Pikeville, KY 41501
Betsy Venters

Richmond

Betty Powers
102 Burrier Building
Richmond, KY 40475
(606) 622-3445

Richmond Downtown Revitalization
P.O. Box 250
Richmond, KY 40475
(606) 623-1000
Nina J. Poage

Shelby

Shelby Development Corporation
500 Main Street
Shelbyville, KY 40065
(502) 633-5029
Nancy Alexander

LOUISIANA

STATE HISTORIC PRESERVATION OFFICE

Office of Cultural Development
Department of Culture, Recreation and Tourism
P.O. Box 44247
Baton Rouge, LA 70804
(504) 342-8200
(504) 342-3207 (Fax)
Gerri Hobdy, SHPO
W. Edwin Martin, Jr., Deputy SHPO

Jonathan Fricker, Deputy SHPO
Director, Division of Historic Preservation
(504) 342-8160

STATEWIDE PRESERVATION ORGANIZATION

Louisiana Preservation Advocates
P.O. Box 1587
Baton Rouge, LA 70821
(504) 383-4476 or 928-9304
Paul St. Martin, President
Mary Wolbrette, Executive Director

NATIONAL TRUST REGIONAL OFFICE

Southern Regional Office
456 King Street
Charleston, SC 29403
(803) 722-8552
(803) 722-8652 (Fax)
Susan A. Kidd, Director

NATIONAL TRUST ADVISORS

Winifred Byrd
3737 Essen Lane
17 Jamestowne Court
Baton Rouge, LA 70809
(504) 926-3782

Joseph F. Newell
933 Burgundy Street
New Orleans, LA 70116
(504) 525-4057

PRESERVATION ACTION COORDINATOR

Patty Gay
Preservation Resource Center
604 Julia Street
New Orleans, LA 70130
(504) 581-7032

AIA STATE PRESERVATION COORDINATOR

George M. Leake, AIA
218 Freidrichs Avenue
Metairie, LA 70005
(504) 835-5318

NATIONAL PARK SERVICE REGIONAL OFFICE

Southwest Regional Office
P. O. Box 728
Santa Fe, NM 87504
(505) 988-6100

ADVISORY COUNCIL ON HISTORIC PRESERVATION DIVISION

Western Office of Project Review
730 Simms Street
Room 401
Golden, CO 80401
(303) 231-5320
Claudia Nissley, Director

Baton Rouge

Downtown Development District
251 Florida Street
Suite 314
Baton Rouge, LA 70801
Davis Rhorer

East Baton Rouge Parish Library
Main Library
7711 Goodwood Boulevard
Baton Rouge, LA 70806

Marion E. Fairchild
243 Cloud Drive
Baton Rouge, LA 70806
(504) 927-2012

Foundation For Historical Louisiana
900 North Boulevard
Baton Rouge, LA 70802
(504) 387-2464
Carolyn G. Bennett

Louisiana Black History Hall of Fame
P.O. Box 66384
Baton Rouge, LA 70896
(504) 356-9673
B. L. Williams

Louisiana Municipal Association
P.O. Box 4327
Baton Rouge, LA 70821
Charles J. Pasqua

Louisiana Preservation Alliance
P.O. Box 1587
Baton Rouge, LA 70821
(504) 383-4476
Mary Boston

Louisiana State University
Rural Life Museum
6200 Burden Lane
Baton Rouge, LA 70808
(504) 765-2437
John E. Dutton

Louisiana State University
School of Architecture
Baton Rouge, LA 70803-5710
(504) 388-4259
(504) 388-6885
Barrett Kennedy

Office of Cultural Development
Division of Historic Preservation
P.O. Box 44247
Baton Rouge, LA 70804
Jonathan Fricker

Janice W. Rutherford
2865 Zeeland Avenue
Baton Rouge, LA 70808
(504) 343-4341

State of Louisiana
Department of Culture, Recreation and Tourism
P.O. Box 94361
Baton Rouge, LA 70804-9361

Cheneyville

Loyd Hall Plantation
292 Loyd Bridge Road
Cheneyville, LA 71325
(318) 776-5641
Anne Y. Fitzgerald

Denham Springs

Capital Resource Conservation and Development
P.O. Box 1296
Denham Springs, LA 70727
(504) 389-0730
Danny Clement

Destrehan

Destrehan Manor
P.O. Box 5
999 River Road
Destrehan, LA 70047
(504) 764-9315
Irene Tastet

Franklin

Franklin Main Street Project
619 2nd Street
P.O. Box 567
Franklin, LA 70538
(318) 828-3631
Michael Dominque

Grand Coteau

Academy of the Sacred Heart
P.O. Box 310
1821 Academy Road
Grand Coteau, LA 70541
(318) 662-5275
Liz DeJean

Hammond

CBDA of Hammond
Box 2307
Hammond, LA 70404
(504) 542-6331
Jeanne G. Gewalt

Louis Lanoix House, New Orleans. (G. Embry, HABS)

Hammond Downtown Development
District
P.O. Box 2788
Hammond, LA 70404
(504) 542-3471
Marguerite Watter

Houma

Terrebonne Historical and Cultural
Society
P.O. Box 2095
Houma, LA 70361
(504) 851-0154
C. L. Olivier

Kenner

Philip L. Batson
3700 Arizona Avenue
Kenner, LA 70063
(504) 466-7442

Lafayette

City of Lafayette
P.O. Box 4017-C
Lafayette, LA 70502
(318) 261-8402
Phillip A. Lank

Downtown Development Authority
556 Jefferson Street
Drawer 15
Lafayette, LA 70501
(318) 261-8402
Catherine C. Webre

Lake Charles

Calcasieu Preservation Society
P.O. Box 938
Lake Charles, LA 70602

Curtis Enterprises
626 Broad Street
Lake Charles, LA 70601-4337
(318) 439-3859

Marksville

City Of Marksville
503 North Main Street
Marksville, LA 71351

Metairie

John K. Wildgen
915 Zinnia Street
Metairie, LA 70001
(504) 888-0477

Minden

Minden Main Street Project
511 Main
Room 207
Minden, LA 71055
(318) 377-6437
Madelyn Harden

New Orleans

David L. Campbell
755 Magazine
New Orleans, LA 70130
(504) 581-5141

Eskew Filson Architects
1008 North Peters Street
New Orleans, LA 70116
(504) 561-8686
Ronald C. Filson

John Geiser, III
1620 8th Street
New Orleans, LA 70115
(504) 897-6510

Historic District Landmarks
Commission
830 Julia Street
New Orleans, LA 70113
(504) 523-7501
Saundra K. Levy

Louisiana Historical Society
10 Trianon Plaza
New Orleans, LA 70125
(504) 866-1134
Bruns D. Redmond

Louisiana Landmarks Society
1440 Moss Street
New Orleans, LA 70119

Louisiana State Museum
751 Chartres Street
New Orleans, LA 70116
(504) 568-6968
James F. Sefcik

McGlinchey, Stafford, Cellini
Law Library
643 Magazine Street
New Orleans, LA 70130
Pat Musso

Matilda Geddings Gray Foundation
1315 Royal Street
New Orleans, LA 70116
(504) 581-5396
Wayne Foret

Jerry K. Nicholson
431 Broadway
New Orleans, LA 70118

Preservation Resource Center of New
Orleans
604 Julia Street
New Orleans, LA 70130
(504) 581-7032
Patricia H. Gay

H. P. St. Martin
1514 Prytania
New Orleans, LA 70130

Save Our Cemeteries
P.O. Box 15770
New Orleans, LA 70175
(504) 561-8850
Bret Clesi

Tulane University Library
Serials Section
New Orleans, LA 70118

University of New Orleans
College of Urban and Public Affairs
New Orleans, LA 70148
(504) 286-6277
Jane S. Brooks

Vieux Carré Commission
516 Chartres Street
New Orleans, LA 70130
(504) 528-3950
Stephen B. Hand

Plaquemine

W. Beverly Middleton, III
317 Plaquemine Street
Plaquemine, LA 70764
(504) 687-2226

Ruston

Lincoln Parish Museum
3 Honeysuckle Lane
Ruston, LA 71270
(318) 255-7934
William D. Green

Shreveport

Downtown Development Authority
400 Edwards Street
Shreveport, LA 71101
(318) 222-7403
Bobby G. Masters

Shreveport Landmark Rehabilitation
P.O. Box 84
Shreveport, LA 71101
(318) 222-4839
Annie Johnson

Vacherie

Oak Alley Foundation
3645 LA 18
Vacherie, LA 70090
(504) 265-2151
Zeb Mayhew

MAINE

STATE HISTORIC PRESERVATION OFFICE

Maine Historic Preservation
Commission
55 Capitol Street
Station 65
Augusta, ME 04333
(207) 289-2132
(207) 289-2861 (Fax)
Earle G. Shettleworth, Jr., SHPO
Robert L. Bradley, Deputy SHPO

STATEWIDE PRESERVATION ORGANIZATION

Maine Citizens for Historic
Preservation
P.O. Box 1198
Portland, ME 04101
(207) 775-3652
Nancy Barba, President
Thomas Johnson, Executive Director

NATIONAL TRUST REGIONAL OFFICE

Northeast Regional Office
7 Faneuil Hall Marketplace
5th Floor
Boston, MA 02109
(617) 523-0885
(617) 523-1199 (Fax)
Vicki Sandstead, Director

NATIONAL TRUST ADVISOR

Pamela Plumb
104 Park Street
Portland, ME 04101
(207) 772-2680

PRESERVATION ACTION COORDINATOR

Martha Deprez
Greater Portland Landmarks
165 State Street
Portland, ME 04101
(207) 774-5561

AIA STATE PRESERVATION COORDINATOR

Malcolm L. Collins, AIA
6 Fox Farm Road
P.O. Box 152
South Freeport, ME 04078
(207) 865-9269

NATIONAL PARK SERVICE REGIONAL OFFICE

North Atlantic Regional Office
15 State Street
Boston, MA 02109-3572
(617) 223-5001

ADVISORY COUNCIL ON HISTORIC PRESERVATION DIVISION

Eastern Office of Project Review
1100 Pennsylvania Avenue, N.W.
Suite 809
Washington, DC 20004
(202) 786-0503
Don L. Klima, Director

Augusta

State of Maine
Treasury Department
State House Station 39
Augusta, ME 04333
(207) 289-2771

Bangor

Bangor Center Management
Corporation
6 State Street
Bangor, ME 04401-5112
Carol LaBella

Bath

Maine Maritime Museum
Washington Street
Bath, ME 04530
(207) 443-1316
Robert L. Webb

Sagadahoc Preservation
Box 322
Bath, ME 04530
(207) 442-8544
R. A. Gaul

Bethel

Bethel Historical Society
P.O. Box 12
Bethel, ME 04217
(207) 824-2908
Stanley R. Howe

Blue Hill

Jonathan Fisher Memorial
P.O. Box 537
Blue Hill, ME 04614
(207) 374-2459
(207) 374-2757
Margaret Beardsley

Boothbay

Boothbay Railway Museum
P.O. Box 123
Boothbay, ME 04537
(207) 633-4727
Robert Ryan

Brunswick

Pejepscot Historical Society
159 Park Row
Brunswick, ME 04011
(207) 729-6606
F. Browne

Greenville

Moosehead Marine Museum
P.O. Box 1151
Greenville, ME 04441
Alice Elliott

Houlton

Town of Houlton
21 Water Street
Houlton, ME 04730
Charles J. Upton

Kennebunk

Brick Store Museum
P.O. Box 177
Kennebunk, ME 04043
(207) 985-4802
Susan Edwards

Kennebunkport

New England Electric Railway
Historical Society
Seashore Trolley Museum
Drawer A
Kennebunkport, ME 04046

Kingfield

Stanley Museum
P.O. Box 280
Kingfield, ME 04947
(207) 265-2729
Susan S. Davis

Lewiston

Development Department
City Building
Lewiston, ME 04240
Robert Thompson

New Gloucester

New Gloucester Public Library
P.O. Box 105
New Gloucester, ME 04260
Susan Hawkins

Poland Spring

United Society of Shakers
Sabbathday Lake
Poland Spring, ME 04274
(207) 926-4597
Leonard L. Brooks

Portland

City of Portland
Historic Preservation Committee
389 Congress Street
Room 211
Portland, ME 04101
(207) 874-8300
Philip L. Meyer

Greater Portland Landmarks
165 State Street
Portland, ME 04101
(207) 774-5561
Martha B. Deprez

Maine Preservation
P.O. Box 1198
Portland, ME 04104
(207) 775-3652
Thomas B. Johnson

Philip L. Meyer
77 Montrose Avenue
Portland, ME 04103
(207) 775-7760

Neal Dow Memorial
714 Congress Street
Portland, ME 04102
(207) 773-7773
Christene E. Barton

Victorian Society of Maine
109 Danforth Street
Portland, ME 04101
(207) 772-4841
Diane Boas

Rockport

Camden-Rockport Historical Society
P.O. Box 747
Rockport, ME 04856
Marlene A. Hall

Saco

Saco Historic Preservation Commission
City Hall
300 Main Street
Saco, ME 04072
(207) 282-3487
Walter Smart

South Portland

James T. Bratton
73 Clemons Street
South Portland, ME 04106-3702
(214) 528-6651

Topsham

Town of Topsham
Planning Office
22 Elm Street
Topsham, ME 04086
(207) 725-2454
Frank Fiori

Wells

Laudholm Trust
P.O. Box 1007
Wells, ME 04090
(207) 646-4521
Mort Mather

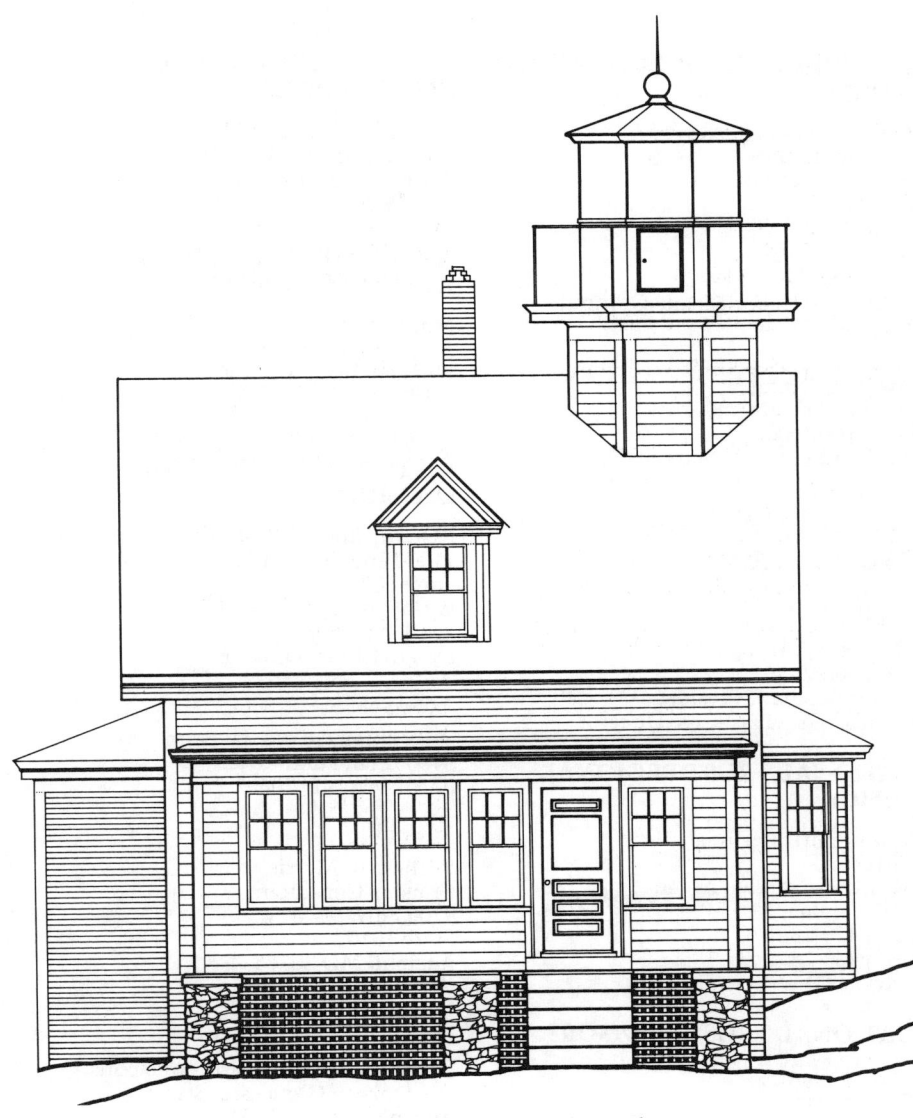

St. Croix River Lighthouse, Dochet Island, Maine. (HABS)

Yarmouth

Yarmouth Historical Society
P.O. Box 107
Yarmouth, ME 04096
(207) 846-6259
Marilyn Hinkley

York

York Historic District Commission
York Town Hall
P.O. Box 9
York, ME 03909
(207) 363-3802
Thomas L. Hinkle

MARYLAND

STATE HISTORIC PRESERVATION OFFICE

Department of Housing and
Community Development
100 Community Place
3rd Floor
Crownsville, MD 21401
(410) 514-7662
J. Rodney Little, SHPO
Director of Historical and Cultural
Programs

Mark Edwards, Deputy SHPO
Deputy Director
Maryland Historical Trust
(410) 514-7600
William Pencek, Deputy SHPO
Chief, Office of Preservation Services

STATEWIDE PRESERVATION ORGANIZATION

Preservation Maryland
24 West Saratoga Street
Baltimore, MD 21201
(410) 685-2886
Robert Kershaw, President
David Chase, Executive Director

NATIONAL TRUST REGIONAL OFFICE

Mid-Atlantic Regional Office
Cliveden
6401 Germantown Avenue
Philadelphia, PA 19144
(215) 438-2886
(215) 438-2892 (Fax)
Patricia D. Wilson, Director

NATIONAL TRUST ADVISORS

Grant Dehart
Program Open Space
138 Lafayette Avenue
Annapolis, MD 21401
(410) 974-3581
(410) 974-3158 (Fax)

Michael F. Trostel, FAIA
1307 Bolton Street
Baltimore, MD 21217
(410) 669-3964

PRESERVATION ACTION COORDINATORS

Grant Dehart
Program Open Space
138 Lafayette Avenue
Annapolis, MD 21401
(410) 974-3581

Douglas A. Dunn
Preservation Maryland
24 West Saratoga Street
Baltimore, MD 21201
(410) 685-2886

AIA STATE PRESERVATION COORDINATOR

Michael F. Trostel, FAIA
1307 Bolton Street
Baltimore, MD 21217
(410) 669-3964

NATIONAL PARK SERVICE REGIONAL OFFICE

Mid-Atlantic Regional Office
143 South 3rd Street
Philadelphia, PA 19106
(215) 597-7018

ADVISORY COUNCIL ON HISTORIC PRESERVATION DIVISION

Eastern Office of Project Review
1100 Pennsylvania Avenue, N.W.
Suite 809
Washington, DC 20004
(202) 786-0503
Don L. Klima, Director

Accokeek

The National Colonial Farm
3400 Brian Point Road
Accokeek, MD 20607

Adelphi

William R. Morris
Housing Consultant
9200 Edwards Way
Suite 1010
Adelphi, MD 20783
(301) 445-5221

Annapolis

Charles Carroll House of Annapolis
109 Duke of Gloucester Street
Annapolis, MD 21401
(410) 263-2969
Robert L. Worden

City of Annapolis
Historic District Commission
160 Duke of Gloucester Street
Annapolis, MD 21401
(410) 263-7961
Jacquelyn Rouse

Department of Housing and
Community Development
45 Calvert Street
Annapolis, MD 21401
(410) 269-2945
Michael Siegel

Department of Housing and
Community Development
DHCD/Maryland Historic Trust
21 State Circle
Annapolis, MD 21401
Mary L. De Sarran

Hammond-Harwood House Association
19 Maryland Avenue
Annapolis, MD 21401
(410) 269-1714
Barbara Brand

H. Grant Dehart Associates
138 Lafayette Avenue
Annapolis, MD 21401
(410) 280-6272
H. G. Dehart

Historic Annapolis Foundation
194 Prince George Street
Annapolis, MD 21401
(410) 267-7619
Sarah Filkins

Historic Naval Ships Association of
North America
U.S. Naval Academy Museum
Annapolis, MD 21402
(410) 267-2109
James W. Cheevers

Maryland Environmental Trust
275 West Street
Suite 322
Annapolis, MD 21401
(410) 974-5350
H. G. Dehart

Baltimore

Alexander and Alexander
111 Market Place
Baltimore, MD 21202
Doris S. Small

Baltimore City Life Museums
225 Holliday Street
Baltimore, MD 21202
(410) 396-3523

Baltimore Heritage
11 1/2 West Chase Street
Baltimore, MD 21201
(410) 625-2585
Fred Shoken

Baltimore Museum of Industry
1415 Key Highway
Baltimore, MD 21230
(410) 727-4808
Dennis Zembala

Catonsville Community College
Library
800 South Rolling Road
Baltimore, MD 21228
(410) 455-4399

Cho, Wilks and Benn
218 West Saratoga Street
Baltimore, MD 21201
(410) 576-0440
David Benn

Chubb Group of Insurance Companies
200 St. Paul Plaza
Suite 2500
Baltimore, MD 21202
(410) 659-6500
Matthew Mueller

Church Services Restorations
Contractors
7312 Old Harford Road
Baltimore, MD 21234
(410) 444-7616
Stephan Fernardi

Commission for Historical and
Architectural Preservation
417 East Fayette Street
Suite 1037
Baltimore, MD 21202-3431
(410) 396-4866
Kathleen Kotarba

E. I. Kane Office Movers
4546 Annapolis Road
Baltimore, MD 21227
(410) 636-4000
James H. Durfee

Engineering Society of Baltimore
11 West Mount Vernon Place
Baltimore, MD 21201
(410) 539-6914
Donald J. Blum

Grieves, Worrall, Wright and
O'Hatnick
5 East Read Street
Baltimore, MD 21202
(410) 332-1009
Phillip Worrall

Higartner Natural Stone Company
101 West Cross Street
Baltimore, MD 21230

Hord Coplan Macht
2526 St. Paul Street
Baltimore, MD 21218
(410) 467-7011
Edward M. Hord

Melissa C. Marsh
500 Woodlawn Road
Baltimore, MD 21210-2313

Maryland Society for the Prevention of
Cruelty to Animals of Baltimore City
3300 Falls Road
Baltimore, MD 21211
(410) 235-8826
John R. Holland

Deborah A. Maufer
2708 Superior Avenue
Baltimore, MD 21234

Matthew John Mosca
2513 Queen Anne Road
Baltimore, MD 21216
(410) 466-5325

9 Front Street Foundation
Baltimore, MD 21202
(410) 837-5424
Emogene F. Williams

Preservation Maryland
24 West Sarasota Street
Baltimore, MD 21201
(410) 685-2886
Douglas A. Dunn

Hattie F. Russell
2111 Liberty Heights Avenue
Baltimore, MD 21217
(410) 523-9180

Baltimore and Ohio Railway Station, Point of Rocks, Maryland. (T. Wolosz, HAER)

Shirish Shah
5605 Purlington Way
Baltimore, MD 21212
(410) 323-0803

Society for the Preservation of Federal
Hill and Fells Point
812 South Ann Street
Baltimore, MD 21231
(410) 675-6750
Carolyn M. Donkervoet

Star-Spangled Banner Flag House and
1812 Museum
844 East Pratt Street
Baltimore, MD 21202
(410) 837-1793

Martavius R. Tyler
1510 McCulloh Street
Baltimore, MD 21217
(410) 669-7031

U.S.S. Constellation Foundation
Pier 1
Pratt Street
Baltimore, MD 21202

Bel Air

Town of Bel Air
Historic District Commission
705 Churchville Road
Bel Air, MD 21014
(301) 879-9500
Carol L. Deibel

Bladensburg

Maryland National Capital Park and
Planning Commission
History Division
4302 Baltimore Avenue
Bladensburg, MD 20710
(301) 779-2011
John Walton

Prince George's Heritage
4703 Annapolis Road
Bladensburg, MD 20710
(301) 927-7150
Dick Charlton

Bowie

Elizabeth Watson
5103 Crain Highway
Bowie, MD 20715
(301) 267-9150

Camp Springs

AMR Group
6104 Tyburn Street
Camp Springs, MD 20748
(301) 899-7422
Joseph F. Jennings, Jr.

Catonsville

Friends of the President
Street Station
P.O. Box 9382
Catonsville, MD 21228
(301) 356-3434
John W. Foster

Chesapeake City

North Chesapeake City Neighborhood
Association
P.O. Box 257
Chesapeake City, MD 21915
(410) 885-5421
Diane Casho

Chestertown

Historical Society of Kent County
P.O. Box 665
Chestertown, MD 21620
(410) 778-3499
Nancy Nunn

Janes United Methodist Church
Cannon and Cross Streets
Chestertown, MD 21620
(410) 778-3542
Victor Harrison

Preservation Incorporated
Route 3
Box 209
Chestertown, MD 21620

Town of Chestertown
P.O. Box 38
Chestertown, MD 21620
William S. Ingersoll

Cheverly

Bob Scarfo
3121 Belleview Avenue
Cheverly, MD 20785
(301) 322-2750

Chevy Chase

Stanley and Mildred Rosenbaum House
3107 Rolling Road
Chevy Chase, MD 20815
(301) 654-1988
Alvin Rosenbaum

Clinton

Surratt Society
9110 Brandywine Road
Clinton, MD 20735
(301) 868-1121
Laurie Verge

College Park

College Park Planning
45 Knox Road
College Park, MD 20740
(301) 277-8898
Terry Chum

Cumberland

Allegany Community College
Library Periodicals Department
Willowbrook Road
Cumberland, MD 21502-2596

Allegany County Historical Society
218 Washington Street
Cumberland, MD 21502
(301) 777-8678
Helen L. Baldwin

Preservation Society of Allegany
County
P.O. Box 1648
Cumberland, MD 21502
(301) 722-3717
Mel Collins

Edgewater

London Town Publik House
839 Londontown Road
Edgewater, MD 21037
(410) 956-4900
Michael J. Menard

Ellicott City

Ellicott City Historic Preservation
3430 Court House Drive
Ellicott City, MD 21043
(410) 313-2357
Alice A. Wetzel

Ellicott City Restoration Foundation
P.O. Box 92
Ellicott City, MD 21043
(410) 465-0980
Jane Trolinger

Historic Ellicott City
P.O. Box 244
Ellicott City, MD 21043
(410) 465-0980
Gerald Talbert

Richard B. Nettler
3811 Grosvenor Drive
Ellicott City, MD 21043
(410) 461-9227

Oella Company
732 Oella Avenue
Ellicott City, MD 21043
(410) 465-1700
Charles Wagandt

Office of Planning and Zoning
3430 Court House Drive
Ellicott City, MD 21043
(410) 313-2357
Alice Wetzel

Fairplay

Larry Ardinger
Route 3
Box 153
Fairplay, MD 21733

Forestville

Heritage International
3939 Penn Bell Place
Forestville, MD 20747
(301) 967-1081
John Telling

Frederick

Frederick City Historic District
Commission
217 West Patrick Street
Frederick, MD 21701
(301) 663-6149
G. B. Callan

Greater Frederick Development
Corporation
P.O. Box 4198
Frederick, MD 21701
Steve Goff

Skyline Engineers of Maryland
5405 Beall Drive
Frederick, MD 21701
(301) 831-8800
Daniel T. Quinn

Gaithersburg

Gaithersburg Commerical
Development
31 South Summit Avenue
Gaithersburg, MD 20877
(301) 258-6310
Linda Michael

Germantown

Smith and Faass Consulting Engineers
12800 Middlebrook Road
Suite 206
Germantown, MD 20874-5201
(301) 770-2111
Kenneth J. Anderer

Glen Echo

Glen Echo Park Foundation
7300 MacArthur Boulevard
Glen Echo, MD 20818
(301) 320-2330
Diane Leatherman

Glyndon

Historic Glyndon
312 Central Avenue
Glyndon, MD 21071
(301) 833-7466
Arlen Herb

Grantsville

Spruce Forest Artisan Village
P.O. Box 203
Grantsville, MD 21536
(301) 895-3315
Jack Dueck

Greenbelt

City of Greenbelt
25 Crescent Road
Greenbelt, MD 20770
(301) 474-8000
Celia Wilson

Coakley and Williams
7500 Greenway Center Drive
Suite 1600
Greenbelt, MD 20770
(301) 345-9730
Charles R. Perell

Greenhorne and O'Mara
9001 Edmonston Road
Greenbelt, MD 20770
(301) 982-2800
Julieanne Mueller

Hagerstown

City of Hagerstown
City Hall
1 East Franklin Street
Hagerstown, MD 21740-4817
(301) 791-3080
David Barnhart

City of Hagerstown
City Hall
1 East Franklin Street
Hagerstown, MD 21740
Stephen L. Wolfenberger

City of Hagerstown
Department of Planning
City Hall
1 East Franklin Street
Hagerstown, MD 21740-4855
(301) 790-3200
Richard L. Kautz

Carl W. Disque
1108 Fry Avenue
Hagerstown, MD 21740
(301) 790-3920

Havre de Grace

Ruth Hendricksen
222 Robin Hood Road
Havre de Grace, MD 21078

Hollywood

Sotterley Museum Foundation
P.O. Box 67
Hollywood, MD 20636
(410) 373-2280
Donna Ely

Hyattsville

Prince Georges County Memorial
Library System
Serials System
6532 Adelphi Road
Hyattsville, MD 20782

Kensington

Kensington Historical Society
Box 453
Kensington, MD 20895

Laurel

City of Laurel
Historic District Commission
Department of Planning and Zoning
350 Municipal Square
Laurel, MD 20707
(301) 725-5300
Jacquelyn K. Bateman

Laytonville

Oak Grove Designs
5815 Riggs Road
Laytonville, MD 20882
(301) 948-6412
Ann Handler

Mt. Airy

Duracon
6825 Runkles Road
Mt. Airy, MD 21771
(301) 829-1982
Cliff Durant

Oxon Hill

Oxon Hill Manor Foundation
6901 Oxon Hill Road
Oxon Hill, MD 20745
(301) 839-7782
Ingrid Britt

Pasadena

Great Inns of America
P.O. Box 1535
Pasadena, MD 21122
W. E. Gilbert

Potomac

Old Home Action League
P.O. Box 59404
Potomac, MD 20859
(301) 299-8100
Greg Hubbard

Prince Frederick

Department of Planning and Zoning
Calvert County Courthouse
Prince Frederick, MD 20678
(301) 535-2348
Frank Jaklitsch

Rockville

City of Rockville
Maryland Avenue at Vinson Street
Rockville, MD 20850
(301) 424-8000
Douglas F. Horne

HUD User
P.O. Box 6091
Rockville, MD 20850
Jane Bennett

Peerless Rockville Historic
Preservation
P.O. Box 4262
Rockville, MD 20850
(301) 762-0096

St. Mary's City

St. Mary's City Commission
P.O. Box 39
St. Mary's City, MD 20686
(410) 862-0960
Burt Kummerow

St. Mary's Parish
P.O. Box 145
St. Mary's City, MD 20686

St. Michaels

Chesapeake Bay Maritime Museum
P.O. Box 636
St. Michaels, MD 21663
(410) 745-2916
John Valliant

St. Michaels Historic Area Commission
P.O. Box 206
St. Michaels, MD 21663
Tracy Smith

Seabrook

Robert T. Faass
Consulting Engineer
6812 96th Place
Seabrook, MD 20706
(301) 731-5772

Silver Spring

Friends of Walter Reed at Forest Glen
Save Our Seminary Committee
P.O. Box 8274
Silver Spring, MD 20907
(301) 587-7356
Richard Schaffer

Hammer, Siler, George Associates
1111 Bonifest Street
Silver Spring, MD 20910
(301) 565-5200
Susan Bishop

Sudlersville

Trustee of Dudley's Chapel
R.R. 1
Box 29B
Sudlersville, MD 21668
(410) 556-6792
Robert Smith

Takoma Park

Historic Takoma
P.O. Box 5781
Takoma Park, MD 20912
Karen Fishman

Thurmont

Catoctin Furnace Historical Society
12320 Auburn Road
Thurmont, MD 21788
(301) 271-2306
Clement E. Gardiner

Thurmont Historical Society
P.O. Box 251
Thurmont, MD 21788
(301) 271-7147
Sterling Kelbaugh

Timonium

Whitney Bailey Cox and Magnani
1850 York Road
Suite C
Timonium, MD 21093
(410) 252-6060
Richard W. Magnani

Towson

Baltimore County Historical Trust
P.O. Box 10067
Towson, MD 21285-0067
(410) 574-9061
Lauri Fitzgerald

Historic Towson
P.O. Box 10072
Towson, MD 21285-0072
(410) 321-6725
Judith S. Kremen

Uniontown

Historic Uniontown
Uniontown, MD 21157
Diane Wiebe

Upper Marlboro

Prince George's County Planning
Department
Historic Preservation Commission
14741 Governor Oden Bowie Drive,
4th Floor
Upper Marlboro, MD 20772
(301) 952-4392

Walkersville

Town of Walkersville
P.O. Box 249
30 West Frederick Street
Walkersville, MD 21793
(301) 845-4500
Jan Coffey

Westminster

Historical Society of Carroll County
210 East Main Street
Westminster, MD 21157
(301) 848-6494
Joseph Getty

Union Mills Homestead Foundation
3311 Littlestown Pike
Westminster, MD 21157
(301) 848-2288
Esther L. Shriver

Western Maryland College
Hoover Library
Westminster, MD 21157

MASSACHUSETTS

STATE HISTORIC PRESERVATION OFFICE

Massachusetts Historical Commission
80 Boylston Street
Suite 310
Boston, MA 02116
(617) 727-8470
(617) 727-5128 (Fax)
Judith McDonough, SHPO
Brona Simon, Deputy SHPO
Director of Technical Services
(617) 727-8470

STATEWIDE PRESERVATION ORGANIZATIONS

Architectural Conservation Trust for
Massachusetts
45 School Street
Boston, MA 02108
(617) 523-8678
Raynor Warner, President
Alan Schwartz, Executive Director

Historic Massachusetts
45 School Street
Boston, MA 02108
(617) 723-3383
Robert H. Kuehn, Jr., President
Alan Schwartz, Executive Director

NATIONAL TRUST REGIONAL OFFICE

Northeast Regional Office
7 Faneuil Hall Marketplace
5th Floor
Boston, MA 02109
(617) 523-0885
(617) 523-1199 (Fax)
Vicki Sandstead, Director

NATIONAL TRUST ADVISORS

Philip Herr
Philip Herr and Asssociates
447 Centre Street
Newton, MA 02158
(617) 969-1805

Thomas M. Menino
Boston City Council
City Hall
Boston, MA 02201
(617) 725-3510
(617) 723-1258 (Fax)

PRESERVATION ACTION COORDINATOR

Patricia L. Weslowski
115 Randlett Park
West Newton, MA 02165
(617) 527-4273

AIA STATE PRESERVATION COORDINATOR

Robert G. Neiley, AIA
Robert G. Neiley Architects
286 Congress Street
Boston, MA 02210
(617) 426-9720

NATIONAL PARK SERVICE REGIONAL OFFICE

North Atlantic Regional Office
15 State Street
Boston, MA 02109-3572
(617) 223-5001

ADVISORY COUNCIL ON HISTORIC PRESERVATION DIVISION

Eastern Office of Project Review
1100 Pennsylvania Avenue, N.W.
Suite 809
Washington, DC 20004
(202) 786-0503
Don L. Klima, Director

Acton

R. Rheome
30 Wethersbee
Acton, MA 01720-5526

Andover

Andover Historical Society
97 Main Street
Andover, MA 01810
(508) 475-2236
Barbara Thibault

Arlington

Schwamb Mill Preservation Trust
Old Schwamb Mill
17 Mill Lane
Arlington, MA 02174
(617) 643-0554
Patricia C. Fitzmaurice

Ayer

Ayer Historical Commission
Town of Ayer
P.O. Box 547
59 Main Street
Ayer, MA 01432

Barnstable

Henry E. Blair
Box 648
Barnstable, MA 02630

Berlin

Chestnut Street Meeting House and
Cemetery Association
264 Linden Street
P. O. Box 178
Berlin, MA 01503
(508) 838-2471
C. R. Wickstrom

Billerica

Town of Billerica
365 Boston Road
Billerica, MA 01821
(508) 671-0942
Paul F. Talbot

Boston

American Meteorological Society
45 Beacon Street
Boston, MA 02108

Appalachian Mountain
5 Joy Street
Boston, MA 02108
Gwen Wilcox

Architectural Heritage Foundation
Old City Hall
45 School Street
Boston, MA 02108
(617) 523-8678
Roger Webb

Boston Aging Concerns-Young and Old
United
67 Newbury Street
Boston, MA 02116
(617) 266-2257
Joanne R. Potter

Boston Center for the Arts
539 Tremont Street
Boston, MA 02116
(617) 426-5000
Janet Langsam

Boston City Council
Boston City Hall
Boston, MA 02201
(617) 725-3510
Thomas M. Menino
Peggie Cronin

Bostonian Society
Old State House
206 Washington Street
Boston, MA 02109
(617) 720-1713
Joan C. Hull

Boston Redevelopment Authority
1 City Hall Square
9th Floor
Boston, MA 02201
(617) 722-4300
Maxine Strickland

Boston University
American and New England Studies
226 Bay State Road
Boston, MA 02215
(617) 353-2948
Richard Candee

Boston University Library
Serials Department
771 Commonwealth Avenue
Boston, MA 02215

Department of Enviromental
Management
100 Cambridge Street
Suite 1404
Boston, MA 02202-0001
Mark Cullinan

First Baptist Church of Boston
110 Commonwealth Avenue
Boston, MA 02116
(617) 267-3148
Milton P. Ryder

First Church of Christ Scientist
Archives and Library
175 Huntington Avenue
Boston, MA 02115
(617) 450-3505
Lynn Mercer

Friends of the Boston Harbor Islands
P.O. Box 9025
Boston, MA 02114
(617) 523-8386
Suzanne G. Marsh

Hamlen, Collier and Company
10 Liberty Square
Boston, MA 02109
(617) 482-0110

Historic Boston
3 School Street
Boston, MA 02108
(617) 227-4679
Stanley M. Smith

Historic Massachusetts
Old City Hall
45 School Street
Boston, MA 02108
(617) 723-3383
Alan G. Schwartz

Kenny Development Company
120 Fulton Street
Boston, MA 02109

Massachusetts Historical Commission
80 Boylston Street
Suite 310
Boston, MA 02116
(617) 727-8470
Sandra A. Curro

Massachusetts Main Street Center
Executive Office of Community
Development
100 Cambridge Street
Room 1803
Boston, MA 02202
(617) 727-7180

Museum of Afro-American History
46 Joy Street
Boston, MA 02114
(617) 742-1854
Monica Fairbairn

Nature Conservancy
Eastern Region
201 Devonshire Street
5th Floor
Boston, MA 02110
Dennis Wolkoff

North Bennet Street School
39 North Bennet Street
Boston, MA 02113
(617) 227-0156
Walter H. McDonald

David O'Leary
28 Seaverns Avenue
Suite 3
Boston, MA 02130

Paul Revere Memorial Association
19 North Square
Boston, MA 02113
(617) 523-2338
Nina Zannieri

Joel F. Pierce
19 Garden Street
Boston, MA 02114

Shepley Bulfinch Richardson
40 Broad Street
Boston, MA 02109
(617) 423-1700
Katherine Meyer

Sierra Club
3 Joy Street
Boston, MA 02108

Society for the Preservation of New
England Antiquities
141 Cambridge Street
Boston, MA 02114
(617) 227-3956
Nancy R. Coolidge

Society for the Preservation of New
England Antiquities
Museums Department
141 Cambridge Street
Boston, MA 02114
(617) 227-3956
Rob Kret

South End Historical Society
532 Massachusetts Avenue
Boston, MA 02118
(617) 536-4445
Arthur Howe

Trinity Church
City of Boston
Copley Square
Boston, MA 02116
Archivist

Trust for Public Land
67 Batterymarch Street
Boston, MA 02110-3306

U.S.S. Constitution Museum
P.O. Box 1812
Boston, MA 02129
(617) 426-1812
Ellen J. Kraft

Women's City Club of Boston
40 Beacon Street
Boston, MA 02108
(617) 451-3480
D. G. Pulos

Boxford

Barbara Latty
166 Main Street
Boxford, MA 01921
(508) 887-8553

Braintree

Braintree Historical Society
786 Washington Street
Braintree, MA 02184
(617) 848-1640

Brighton

Trombadore Contracting
414 Market Street
Brighton, MA 02135
(617) 783-9427

Brockton

Gardner R. Gibson
55 Lenox Street
Brockton, MA 02401
(508) 583-1461

Brookline

Steve Grindl
9 Bartlett Crescent
Brookline, MA 02146
(617) 277-0563

Longyear Historical Society
120 Seaver Street
Brookline, MA 02146
(617) 277-8943
Loretta L. Lewis

Burlington

Burlington Historical Commission
Town Hall
Center Street
Burlington, MA 01803
(617) 272-6700
Pauline R. Keans

Cambridge

John D. Bruce
6 Craigie Circle
Suite 31
Cambridge, MA 02138
(617) 876-2680

Cambridge Historical Commission
City Hall Annex
57 Inman Street
Cambridge, MA 02139
(617) 498-9040
Charles Sullivan

Cambridge Seven Associates
Library
1050 Massachusetts Avenue
Cambridge, MA 02138

Thomas Martin
174 Appleton Street
Cambridge, MA 02138
(617) 491-3459

Mount Auburn Cemetery
580 Mount Auburn Street
Cambridge, MA 02138
(617) 547-7105
William Clendaniel

Townscape Institute
2 Hubbard Park
Cambridge, MA 02138
(617) 491-8952
Ronald L. Fleming

Albert B. Wolfe
28 Bradbury Street
Cambridge, MA 02138
(617) 547-6824

Carver

Melvin E. Fryer
56-5 South Meadow Village
Carver, MA 02330
(508) 866-5014

Charlestown

Anne Booth
28 Cordis Street
Charlestown, MA 02129
(617) 241-9629

Charlestown Preservation Society
P.O. Box 201
Charlestown, MA 02129
(617) 572-9121
Jane Philippi

William A. Woodhead
465-D Medford Street
Charlestown, MA 02129
(617) 242-4480

Chelmsford

Stephen Stowell
22 Pleasant Street
Chelmsford, MA 01824
(508) 251-3464

Chelsea

Governor Bellingham Cary House
Association
34 Parker Street
Chelsea, MA 02150
(617) 889-3523
Ronald M. Hansen

Chester

First Congregational Church of Chester
Skyline Trail
Chester, MA 01011
(413) 354-9697
Lynn Fisk

Chicopee

Betsy J. Doyon, Advisor
City Hall
Market Square
Chicopee, MA 01013
(413) 594-4711

Clinton

Town of Clinton
Community and Economic
Development
Town Hall
242 Church Street
Clinton, MA 01510
(508) 365-4317
Philip R. Boyce

Concord

Fannin/Lehner Preservation
Consultants
271 Lexington Road
Concord, MA 01742
(508) 369-6703
Minxie Fannin

Louisa May Alcott Memorial
Association
P.O. Box 343
Concord, MA 01742
(617) 369-4118
Brenda K. Bhaunani

Thoreau County Conservation Alliance
100 Barretts Mill Road
Concord, MA 01742
Thomas Blanding

South Family Dwelling and Washshed, Harvard Shakers, Worcester, Massachusetts. (R.T. Newman, HABS)

Danvers

Danvers Historical Society
P.O. Box 381
Danvers, MA 01923
(508) 777-1666
Joan Reedy

James R. Dumke
79 Water Street
Danvers, MA 01923

Deerfield

Historic Deerfield
P.O. Box 321
Deerfield, MA 01342
(413) 774-5581
Donald Friary

Historic Deerfield Library
P.O. Box 53
Deerfield, MA 01342
Sharman Prouty

Duxbury

Duxbury Rural and Historical Society
Box 2865
Duxbury, MA 02331
(617) 934-6106
Alexandra Earle

Easthampton

Easthampton Historical Society
21 Lovefield Street
Easthampton, MA 01027
(413) 527-2211
Edward Dwyer

Essex

Essex Historical Society
28 Main Street
Essex, MA 01929
(508) 768-7541
Diana Stockton

Fall River

Save Architecturally Valued Edifices
P.O. Box 3526
Fall River, MA 02722
(508) 672-3332
Kristie Gardiner

Franklin

Downtown Franklin
13-25 Main Street
Franklin, MA 02038
(508) 528-8753
Ginny Bissig

Gloucester

Roberta Arena
19 Harbor Loop
Gloucester, MA 01930
(508) 281-9781

Cape Ann Historical Association
27 Pleasant Street
Gloucester, MA 01930
(508) 283-0455
Judith McCulloch

Gloucester Adventure
P.O. Box 1306
Gloucester, MA 01930
(50) 828-18079
Cayte Ward

Greenfield

Greenfield Community College
Library
Greenfield, MA 01301
(413) 774-3131
Margaret Howland

Hadley

Porter Phelps Huntington Foundation
130 River Drive
Hadley, MA 01035
(413) 584-4699
Susan J. Lisk

Harwich

Town of Harwich
P.O. Box 888
Harwich, MA 02645
(508) 432-8012
Michael J. Pessolano

Hingham

Hingham Historical Commission
7 East Street
Hingham, MA 02043
(617) 749-0612
Alexander Macmillan

Hull Main Street Association
c/o Feingold Associates
63 Pleasant Street
Hingham, MA 02043
(617) 925-5665
John Feingold

Holyoke

Greater Holyoke
187 High Street
Suite 203
Holyoke, MA 01040
(413) 536-4611
Ann M. Burke

Hyannis

Barnstable Historical Commission
Town Hall
Hyannis, MA 02601
(508) 775-1120
Patricia Jones Anderson

Main Street Hyannis
PO Box 1327
Hyannis, MA 02601
(508) 790-2711
Richard A. Bishop

Jefferson

Susan McDaniel Ceccacci
360 Causeway Street
Jefferson, MA 01522
(508) 829-6640

Kingston

Kingston Historical Commission
6 Maple Avenue
Kingston, MA 02364
Marion Lanagan

Lenox

Edith Wharton Restoration
P.O. Box 974
Lenox, MA 01240
(413) 637-1899
Thomas S. Hayes

Lexington

De Bonte Restoration/Construction
271 Marrett Road
Lexington, MA 02173-7008
(617) 861-8523
Patricia De Bonte

Lincoln

Andervett Associates
11 Juniper Ridge Road
Lincoln, MA 01773-1502
(617) 259-8807
Richard H. Anderson

Friends of the Battle Road
P.O. Box 95
Lincoln, MA 01773
(601) 259-9744
Paul E. Marsh

John Williams
42 Old Sudbury Road
Lincoln, MA 01773
(617) 259-8724

Lowell

Muckle and Associates
354 Andover Street
Lowell, MA 01852
(508) 937-2747
Richard Muckle

Lowell Historic Preservation
Commission
U.S. Department of the Interior
Old City Hall
222 Merrimack Street
Suite 310
Lowell, MA 01852
(508) 458-7653
Charles Parrott

Unitas
48 Lawrence Street
Lowell, MA 01852
(508) 458-8793
Dalia Calvo

Lynn

Department of Community
Development
City Hall
Room 315
Lynn, MA 01901
(617) 598-4000
Jansi D. Chandler

Marblehead

David L. Bittermann
31 Sewall Street
Marblehead, MA 01945
(508) 458-7653

Medfield

Medfield Historical Commission
Town House
Main Street
Medfield, MA 02052
(617) 359-8505
Donald MacDonald

Medford

Royall House Association
15 George Street
Medford, MA 02155
(617) 396-9032
William A. Slagle

Mendon

Preservation Services
59 North Avenue
P.O. Box 18
Mendon, MA 01756
(508) 473-4884
R. C. Noonan

Nantucket

First Congregational Church of
Nantucket
Box 866
Nantucket, MA 02554
Howard Lewis

Michael Kepenash
P.O. Box 2351
Nantucket, MA 02584
(508) 325-4657

Nantucket Historical Association
P.O. Box 1016
Nantucket, MA 02554
(617) 228-1894
Wynn Lee

Nantucket Historic District
Commission
4 North Water Street
Nantucket, MA 02554
(508) 228-7231
Patricia Butler

Preservation Institute of Nantucket
P.O. Box 477
11 Centre Street
Nantucket, MA 02554
(508) 228-2429
Susan Tate

Needham

C. Daniel McLaughlin
188 Charles River Street
Needham, MA 02192
(61) 744-94374

New Bedford

Carolton J. Dasent
109 8th Street
New Bedford, MA 02742
(508) 997-5578

Downtown New Bedford
700 Pleasant Street
5th Floor
New Bedford, MA 02740
(508) 990-2600
Emmy Hahn

Massachusetts Schooner Ernestina
Commission
30 Union Street
New Bedford, MA 02740
(508) 990-1493
Joseph C. Cardozo

Carol Ann Nelson, AIA
24 North Waters Street
New Bedford, MA 02740
(508) 997-5977

New Bedford Historical Commission
City Hall
133 Williams Street
New Bedford, MA 02740
(508) 999-2931
Antone G. Souza

Waterfront Historic Area League
(WHALE)
13 Centre Street
New Bedford, MA 02740
(617) 996-6912
Therese S. Kelly

Newburyport

Historical Society of Old Newbury
98 High Street
Newburyport, MA 01950

Newburyport Historical Commission
City Hall
Newburyport, MA 01950
(508) 465-4400
Anne C. Tuthill

Newton

Alex L. Hillman
40 Cross Street
Newton, MA 02165-2102

Robert Imperato
145 Washington Street
Newton, MA 02158
(617) 969-6475

North Adams

Cummings Studios
P.O. Box 427
North Adams, MA 01247
Bill Cummings

Northampton

Historic Northampton
46 Bridge Street
Northampton, MA 01060
(413) 584-6011
Pamela Toma

Northampton Historical Commission
City Hall
210 Main Street
Northampton, MA 01060
(413) 586-6950
Marcia Berkley

North Andover

North Andover Historical Society
153 Academy Road
North Andover, MA 01845

Robert D. Stevens
Box 116
North Andover, MA 01845

North Attleborough

North Attleborough Historical
Commission
P.O. Box 951
North Attleborough, MA 02761
(617) 695-4466
(617) 695-1835
Ann J. Chapdelaine

North Falmouth

Janice Stanley
P.O. Box 1457
North Falmouth, MA 02556
(508) 563-7349

Peabody

City of Peabody
24 Lowell Street
Peabody, MA 01960
(617) 532-3000
Dennis DiZoglio

Pepperell

Daniel Tocci
14 Independence Road
Pepperell, MA 01463
(508) 433-9750

Pittsfield

Pittsfield Central
P.O. Box 725
141 North Street
Pittsfield, MA 01202
(413) 443-6501

Plymouth

Plimoth Plantation
Public Relations Department
P.O. Box 1620
Plymouth, MA 02360
(508) 746-1622

Plymouth Development and Industrial
Commission
P.O. Box 3321
Plymouth, MA 02361
(508) 746-6680
Michael A. Gallerani

Plymouth Downtown Harbor
Corporation
P. O. Box 3196
Plymouth, MA 02361
(508) 747-7606
Robert Dawley

Provincetown

Cape Cod Pilgrim Memorial
Association
P.O. Box 1125
Provincetown, MA 02657
(508) 487-1310
Clive E. Driver

Quincy

Quincy Planning Department
1305 Hancock Street
Quincy, MA 02169
Roberta M. Fitzgerald

Karen Yourell
175 Centre Street
Suite 1403
Quincy, MA 02169

Roslindale

Kathleen McCabe
592 Beech Street
Roslindale, MA 02131-4942
(617) 469-2108

Salem

Essex Institute
132 Essex Street
Salem, MA 01970
(617) 744-3390
Anne Farnam

First Universalist Society in Salem
211 Bridge Street
Salem, MA 01970
Dale E. Yale

Historic Salem
P.O. Box 865
Salem, MA 01970

Northfields Preservation Associates
10 Barr Street
Salem, MA 01970
(617) 741-4355
Kim W. Brengle

Salem Historical Commission
City of Salem Planning Department
1 Salem Green
Salem, MA 01970
(617) 744-4580
Jane Guy

Salem Partnership
13 Washington Square, West
Salem, MA 01970
(508) 741-8100
Maura Smith

Somerville

Historic Preservation Commission
93 Highland Avenue
City Hall
3rd Floor
Somerville, MA 02143
(617) 625-6600
Margen Kelsey

South Boston

St. Augustine's Save the Chapel
Committee
9 F Street
South Boston, MA 02127
(617) 268-1230
Thomas J. McDonnell

Southbridge

Southbridge Historical Commission
41 Elm Street
Southbridge, MA 01550
(617) 764-8121
Helen E. Walkowiak

South Hadley

Mount Holyoke College Library
Serials Division
South Hadley, MA 01075
Lois J. McCabe

Southwick

Southwick Historical Society
P.O. Box 323
Southwick, MA 01077
(413) 569-0221
Lee D. Hamburg

South Yarmouth

Priscilla D. Gregory
115 Old Main Street
South Yarmouth, MA 02664
(508) 398-8399

Spencer

Spencer Historical Commission
246 Main Street
Spencer, MA 01562
(508) 885-3675
A. M. Hughes

Springfield

McKnight Homeowners Association
25 St. James Avenue
Room 102
Springfield, MA 01109
(413) 736-8583
Hazel H. Adams

Sargent Design Associates
969 Main Street
Springfield, MA 01103
(413) 732-3030
David K. Sargent

Springfield Preservation Trust
979 Main Street
Springfield, MA 01103

Sturbridge

Old Sturbridge Village
1 Old Sturbridge Village Road
Sturbridge, MA 01566
(508) 347-3362
Crawford Lincoln

Sudbury

Richard F. Cook, Jr.
76 Pokonoket Avenue
Sudbury, MA 01776

Taunton

Taunton Historic District Commission
City Hall
15 Summer Street
Taunton, MA 02780
(508) 823-1826
Philip La France

Townsend

Town of Townsend
Memorial Hall Restoration Committee
P.O. Box 669
Townsend, MA 01469
(508) 597-2837
David Vanderwerf

Townsend Historical Society
P.O. Box 95
Townsend, MA 01469
(617) 597-2106

Truro

Truro Historical Society
P.O. Box 486
Truro, MA 02666
(508) 349-2809
E. Allen

Waltham

Society for the Preservation of New
England Antiquities
Conservation Center
185 Lyman Street
Waltham, MA 02154
(617) 891-1985
Patricia L. Weslowski

Wellesley

William C. S. Remsen
31 College Road
Wellesley, MA 02181

Westborough

New England Power
Technical Information Center
25 Research Drive
Westborough, MA 01581

Westfield

Westfield 2000
P.O. Box 1867
Westfield, MA 01085
(413) 562-7221
John Ryan

Westminster

Westminster Historical Society
Box 177
Westminster, MA 01473
(508) 630-1497
(508) 874-5569
Betsy Hannula

Weston

Golden Ball Tavern Trust
Box 223
Weston, MA 02193
(617) 894-1751
Joan Bines

West Roxbury

Judith Stoessel
54 Brook Farm Road
West Roxbury, MA 02132

West Springfield

Pioneer Valley Planning Commission
26 Central Street
West Springfield, MA 01089
(413) 781-6045
Gregory Farmer

West Stockbridge

David R. Bixby
Swamp Road
West Stockbridge, MA 01266
(212) 864-4030

Whitinsville

Lanee Hamer
100 Main Street
Whitinsville, MA 01588
(617) 234-6324

Williamstown

Williams College
Hopkins Forest Farm Museum
Center for Environmental Studies
Library
Box 632
Williamstown, MA 01267
(413) 597-2500
Marcella Rauscher

Wilmington

Wilmington Historical Commission
Town Hall
Wilmington, MA 01887

Winchester

Allen C. Hill
25 Englewood Road
Winchester, MA 01890
(617) 729-0748

Stirling/Brown Architects
1 Mount Vernon Street
Winchester, MA 01890
(617) 721-1310
David W. Stirling

Woburn

First Congregational Church
322 Main Street
Woburn, MA 01801
(617) 667-3523
W. A. Goodwin

Worcester

College of the Holy Cross
Department of History
P.O. Box 134A
Worcester, MA 01610
(508) 793-3448
Ross W. Beales

Terrence S. O'Connell
15 Pineland Avenue
Worcester, MA 01604
(508) 798-0932

Preservation Worcester
71 Pleasant Street
Worcester, MA 01609
(508) 754-8760
Janet K. McCorison

MICHIGAN

STATE HISTORIC PRESERVATION OFFICE

Bureau of History
Department of State
717 West Allegan Street
Lansing, Michigan 49654
(517) 373-0511
(517) 373-0851 (Fax)
Kathryn Eckert, SHPO

STATEWIDE PRESERVATION ORGANIZATION

Michigan Historic Preservation
Network
P.O. Box 398
Clarkston, MI 48347
(313) 625-8181
Jennifer L. Radcliff, President

NATIONAL TRUST REGIONAL OFFICE

Midwest Regional Office
52 West Jackson Boulevard
Suite 1135
Chicago, IL 60604
(312) 939-5447
(312) 939-5651 (Fax)
Tim Turner, Director

NATIONAL TRUST ADVISORS

William J. Kimball
Natural Resources Department
323 Natural Resources Building
Michigan State University
East Lansing, MI 48824-1222
(517) 353-1906
(517) 353-8994 (Fax)

Kathleen H. Wendler
Southwest Detroit Business
Association
1601 Clark Street
Detroit, MI 48209
(313) 842-0986
(313) 842-6350 (Fax)

PRESERVATION ACTION COORDINATOR

Jennifer Radcliff
33 North Main
Clarkston, MI 48106
(313) 625-8181

AIA STATE PRESERVATION COORDINATOR

Betty-Lee Seydler-Sweatt, AIA
Preservation Planning
32870 Brandingham
Franklin, MI 48025
(313) 851-7125
Louis A. Goldstein, AIA, Deputy

NATIONAL PARK SERVICE REGIONAL OFFICE

Midwest Regional Office
1709 Jackson Street
Omaha, NE 68102
(402) 221-3431

ADVISORY COUNCIL ON HISTORIC PRESERVATION DIVISION

Eastern Office of Project Review
1100 Pennsylvania Avenue, N.W.
Suite 809
Washington, DC 20004
(202) 786-0503
Don L. Klima, Director

Adrian

Croswell Opera House and Fine Arts
Association
129 East Maumee Street
Adrian, MI 49221
(517) 263-6868
(517) 263-6415
Carol A. Jodis

Downtown Development Authority
100 East Church Street
Adrian, MI 49221
(517) 263-2161
Cindy Helinski

Albion

City of Albion
112 West Cass Street
Albion, MI 49224
(517) 629-5535
William C. Rieske

Allegan

City of Allegan
112 Locust Street
Allegan, MI 49010
(616) 673-5511
Robert Hillard

Allen Park

Great Lakes Lighthouse Keeper's
Association
P.O. Box 580
Allen Park, MI 48101
(313) 426-4150
Richard L. Moehl

Ann Arbor

City of Ann Arbor
Planning Department
100 North 5th Street
P.O. Box 8647
Ann Arbor, MI 48107
(313) 994-2800
Martin Overhiser

Gregory S. Eggleston
406 Westwood
Ann Arbor, MI 48103
(313) 668-7142

Michigan Historic Preservation
Network
312 South Division Street
Ann Arbor, MI 48104
(313) 996-3008
Louisa Pieper

Baraga

Superior Restorations
Route 1
Box 193
Baraga, MI 49908
Ruth M. Lonhgern

Battle Creek

City of Battle Creek
P.O. Box 1717
Battle Creek, MI 49017
(616) 966-3320
Martha Levy

Historial Society of Battle Creek
196 Capital Avenue, N.E.
Battle Creek, MI 49017
(616) 965-2613
Diane Beckley

Birmingham

Baldwin Public Library
300 West Merrill
Birmingham, MI 48012

City of Birmingham
P.O. Box 3001
Birmingham, MI 48012
(313) 644-3865
Patricia McCullough

Cadillac

Cadillac Downtown Development
Authority
200 Lake Street
Cadillac, MI 49601
(616) 775-0181
Mark W. Sochocki

Canton

Charter Township of Canton
1150 South Canton Center Road
Canton, MI 48188
(313) 397-1000
Kim Scherschligt

Robert A. Young
7164 Camelot
Canton, MI 48187
(313) 459-9217

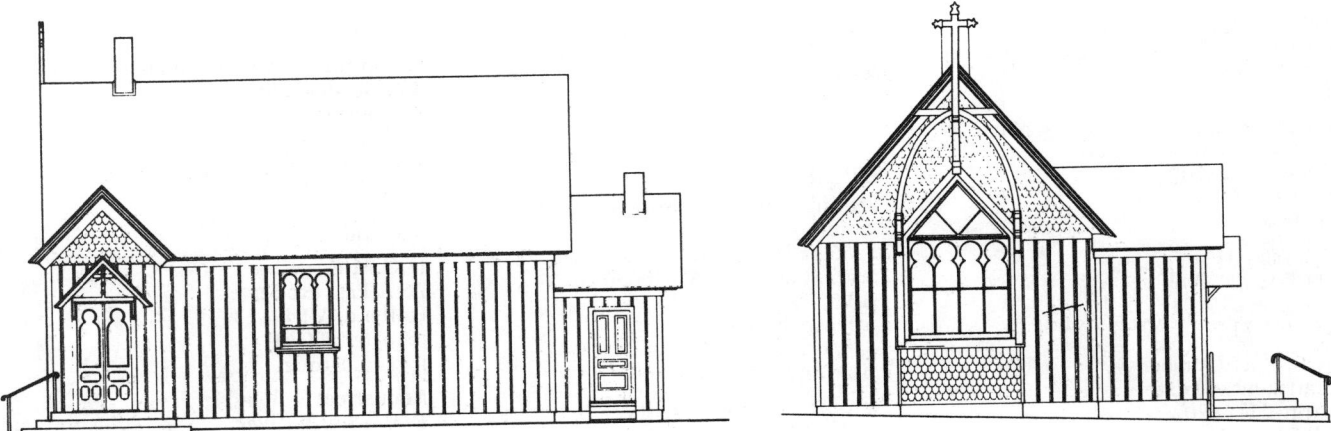

St. Katherine's Episcopal Church, Williamstown, Michigan. (M. Trumbo, HABS)

Centreville

St. Joseph County Cooperative
Extension Service
Centreville, MI 49032
(616) 467-6361
Sally J. Carpenter

Charlotte

Eaton County Historical Commission
100 West Lawrence Avenue
Charlotte, MI 48813
Chris Kent

Chelsea

John Frank
Jacquelyn Frank
138 East Middle Street
Chelsea, MI 48118
(313) 475-7396

Coldwater

Coldwater Downtown Development
Authority
28 West Chicago Street
Coldwater, MI 49036
(517) 278-8324
Lisa Dean

Comstock Park

Alpine Historic Commission
6584 Wahlfield, N.W.
Comstock Park, MI 49321
(616) 784-3262
Mark Stein

Coopersville

Coopersville Downtown Development
Authority
289 Danforth
Coopersville, MI 49404
(616) 837-9731
Tom O'Malley

Dafter

Ted Klevay
Route 1
Box 140A
Dafter, MI 49724

Dearborn

University of Michigan
4901 Evergreen
Dearborn, MI 48128-1491
(313) 593-5590
Donn R. Werling

Detroit

Chene Area Business Association
5851 Chene Street
Detroit, MI 48211-2752
(313) 831-0250
(313) 922-4363
Richard Hodas

Cityscape Detroit
1150 Griswold
Suite 3201
Detroit, MI 48226
(313) 962-7900
(313) 963-0616
Kim Stroud

Detroit Landmarks
4231 Saint Aubin Avenue
Detroit, MI 48207
(313) 224-3487
William M. Worden

Dossin Great Lakes Museum
Detroit Historical Department
100 Strand Belle Isle
Detroit, MI 48207
(313) 267-6440
John Polacsek

Manufacturers Bank, N.A.
Community Relations Division
100 Renaissance Center
Level 2
Detroit, MI 48243
(313) 222-7276
Kathryn A. Reid

Christine Michaels
1771 Balmoral Drive
Detroit, MI 48203-1440
(313) 369-9774

Motor City Theatre Organ Society
17360 Lahser Road
Detroit, MI 48219
(313) 537-1133
(313) 534-1954
Robert D. Duerr

Carl A. Nielbock
2503 Helen Street
Detroit, MI 48207
(313) 922-3755

Old Redford Main Street Project
17221 Lahser Road
Detroit, MI 48219
(313) 534-8100

Period Details
3516 Cadieux Road
Detroit, MI 48224
(313) 885-9237
Cynthia Ruffner

Preservation Detroit
P.O. Box 3061
Detroit, MI 48231
Maryann Maher

Preservation Wayne
The David MacKenzie House
4735 Cass Avenue
Detroit, MI 48202
(313) 577-3559
William Colburn

Rocky Investment Company
7201 West Fort
Detroit, MI 48209
Robert Heide

University Cultural Center Association
4735 Cass Avenue
Detroit, MI 48202
(313) 577-5088
Sue Mosey

Kelly Van Heest
8700 Tireman Avenue
Detroit, MI 48204
(313) 933-7161

Wayne State University
Department of History
Detroit, MI 48202
(313) 577-2525
Charles K. Hyde

Joseph B. Young
Job Consulant Corporation
P.O. Box 10689
Detroit, MI 48210

Zachary and Associates
422 West Congress
Suite 32
Detroit, MI 48226
(313) 963-1410
Ernest Zachary

Dowagiac

Dowagiac Downtown Development
Authority
107 Beeson
Dowagiac, MI 49047
(616) 782-8148
Stephen Guile

Eagle Harbor

Keweenaw County Historical Society
Star Route 1
Eagle Harbor, MI 49951
(906) 296-2561
Clarence J. Monette

East Lansing

Susan D. Elder
6213 Cobblers Drive
East Lansing, MI 48823

Elk Rapids

Village of Elk Rapids
Downtown Development Authority
131 River Street
P.O. Box 398
Elk Rapids, MI 49629
(616) 264-9274
Joseph W. Yuchasz

Farmington

Farmington Downtown Development
Authority
33411 Grand River
Farmington, MI 48335
(313) 473-7276
Wendy Strip-Sittsamer

Marble Institute of America
33505 State Street
Farmington, MI 48024
(313) 476-1630
R. Hund

Farmington Hills

Farmington Hills Historic District
Commission
31555 11 Mile Road
Farmington Hills, MI 48336
(313) 473-9541
Ruth Moehlman

Franklin

Franklin Historical Society
Box 7
Franklin, MI 48025

Grand Rapids

City of Grand Rapids
345 State Street, S.E.
Suite 11
Grand Rapids, MI 49503
(616) 645-63451
Michael Page

Design Plus
48 Fountain, N.W.
Grand Rapids, MI 49503
(616) 458-0875
Wayne Norlin

Heritage Hill Foundation
126 College Avenue, S.E.
Grand Rapids, MI 49503
(616) 459-8950
Jan Earl

Neighborhood Business Specialists
Program
17 Fountain, N.W.
Grand Rapids, MI 49503
(616) 771-0314
Craig Kinnear

Warner, Norcross and Judd
900 Old Kent Building
1 Vandenberg Center
Grand Rapids, MI 49503
(616) 459-6121
John H. Logie

Grosse Pointe

Grosse Pointe Historical Society
18530 Mack Avenue
Grosse Pointe, MI 48236
(313) 884-7010

Hancock

Quincy Mine Hoist Association
Box 265
Hancock, MI 49930
(906) 482-1001
Louis G. Koepel

Harrison

Community Growth Alliance
402 North 1st Street
P.O. Box 408
Harrison, MI 48625
(517) 539-2173
R. A. Volker

Holland

H.O.M.E.
76$^{1}/_{2}$ East 8th Street
Holland, MI 49423
(616) 393-0070
Joyce Kortman

Mainstreet Holland
194$^{1}/_{2}$ South River
Suite 201
Holland, MI 49423
(616) 392-1036
Andrea Wilkins

Horton

William E. Rutter
6707 Cross Road
Horton, MI 49246
(517) 782-5298

Houghton

Upper Peninsula Engineering and
Architectural Associates
322 Shelden Avenue
Houghton, MI 49931
(906) 482-4810
Jerry Flum

Ventures Group
P.O. Box 364
Houghton, MI 49931-0364
Edward J. Koepel

Hudson

City of Hudson
121 North Church Street
Hudson, MI 49247
(517) 448-8983
Frank L. Crosby

Jackson

Ella Sharp Museum
3225 4th Street
Jackson, MI 49203
Robert A. Kret

Jackson Historic District Commission
120 West Michigan Avenue
16th Floor
Jackson, MI 49201
(517) 788-7273
John R. Schaub

Jenison

Jenison Historical Association
P.O. Box 664
Jenison, MI 49429
Kenyon Williams

Kalamazoo

Kalamazoo Neighborhood Housing
Services
1301 East Main
Kalamazoo, MI 49001
(616) 385-2916
Chuck Vliek

Rodger Parzyck
121 Bulkley
Kalamazoo, MI 49007
(616) 385-4701

Tower Pinkster Titus Associates
P.O. Box 3508
Kalamazoo, MI 49003-3508
(616) 343-6133
Robert Tower

Vine Neighborhood Association
913 South Westnedge Avenue
Kalamazoo, MI 49008

Lansing

Lansing Planning Division
Washington Square Annex
119 North Washington Square
2nd Floor
Lansing, MI 48933
(517) 483-4066
James Foulds

Michigan Department of State
Bureau of History
717 West Allegan Street
Lansing, MI 49654
(517) 3730-6362
Kathryn Eckert

Ron Staley
408 Kalamazoo Plaza
Lansing, MI 48933
(517) 482-1488

Turner-Dodge House
100 East North Street
Lansing, MI 48906
(517) 483-4220
Susan Cantlon

Marquette

Marquette City Planning Department
300 West Baraga Avenue
Marquette, MI 49855
(906) 228-8200
Norman L. Gruber

Marshall

Marshall Historical Society
107 North Kalamazoo
P.O. Box 68
Marshall, MI 49068
(616) 781-8544

Mason

Mason Historic District Commission
201 West Ash Street
Mason, MI 48854
Patrick M. Price

Monroe

City of Monroe
Economic Development Department
120 East 1st Street
Monroe, MI 48161
(313) 243-0700
Ann Witkowski

Mount Clemens

Downtown Development Authority
58 North Avenue
Mount Clemens, MI 48043
(313) 469-2510
Sherrie B. Schuster

Brian Korth
58 Lincoln Avenue
Mount Clemens, MI 48043
(313) 465-2306

Richard Spier
20870 Moxon Drive
Mount Clemens, MI 48043

Muskegon

Muskegon County Museum
430 West Clay
Muskegon, MI 49440
John H. McGarry III

Muskegon Heritage Association
561 West Western Avenue
Muskegon, MI 49440
(616) 722-1363
Bonnie Bauman

Negaunee

Economic Development Corporation of
the County of Marquette
184 West U.S. Highway 41
Negaunee, MI 49866
(906) 475-4121
Charles L. Manto

Niles

City of Niles
Planning Department
508 East Main Street
P.O. Box 487
Niles, MI 49120
(616) 683-4700
Debra L. Davino

Fort St. Joseph Museum
City of Niles History Department
508 East Main Street
Niles, MI 49120

Northville

Mary Grace York
221 St. Lawrence Boulevard
Northville, MI 48167
(313) 347-6688

Novi

Novi Historical Commission
45175 West 10 Mile Road
Novi, MI 48375
(313) 349-6774
Cathy Mutch

Port Haven

Jeffrey Bitzinger
Lisa Bitzinger
924 Court Street
Port Haven, MI 48060
(313) 985-5687

Port Huron

Downtown Port Huron
102 Huron Avenue
Port Huron, MI 48060
(313) 985-8843
William R. Gilmer

Portage

City of Portage
7900 South Westnedge Avenue
Portage, MI 49002-5160
(616) 329-4403
Clifford Mulder

Rochester

Michigan Society of Planning Officials
Main Street Plaza
Suite 202
Rochester, MI 48307
(313) 651-3339

Rochester Main Street Project
Downtown Development Authority
308 1/2 Main Street
Rochester, MI 48307
(313) 656-0060
Susan Affleck-Childs

St. Joseph

Elowyn Ann Keech
375 Ridgeway
St. Joseph, MI 49085
(616) 983-1175

Landmark Center
708 Market Street at Main Street
St. Joseph, MI 49085
(616) 983-5062
Montgomery Shepard

Saline

Richard C. Frank, FAIA
302 East Henry Street
Saline, MI 48176
(313) 429-9594

Saline Historic District Commission
P.O. Box 40
Saline, MI 48176
(313) 429-4907
Jerry Austin

Uptown Saline Associates
107½ East Michigan
Saline, MI 48176
(313) 429-2624
M. E. Martindale

Sault Ste. Marie

Eastern Upper Peninsula Planning and
Development Commission
416 Ashmun Street
Sault Ste. Marie, MI 49783

Historical Sites
501 Water Street
Sault Ste. Marie, MI 49783
(906) 632-3658

Southfield

City of Southfield
Parks and Recreation
26000 Evergreen
Southfield, MI 48076
(313) 354-9518
Merrie Carlock

Vilican-Leman and Associates
28316 Franklin Road
Southfield, MI 48034
(313) 356-8181
Bob Donohue

South Haven

City of South Haven
539 Phoenix Street
South Haven, MI 49090
(616) 637-5211
John Marple

Lake Michigan Maritime Museum
P.O. Box 534
South Haven, MI 49090

South Lyon

South Lyon Area Historical Society
P.O. Box 263
South Lyon, MI 48178
(313) 437-9279
Eleanor Donley

Spring Lake

Village of Spring Lake
Village Hall
Spring Lake, MI 49456
(616) 842-1393
Eric R. DeLong

Sturgis

Sturgis Downtown Development
Association
P.O. Box 280
Sturgis, MI 49091
(616) 651-1907
Kim Musolff

Taylor

Wade, Trim and Associates
25185 Goodard
Taylor, MI 48180
(313) 291-5400
Larry L. Bauman

Traverse City

City Opera House Heritage Association
112½ East Front
Traverse City, MI 49684

City of Traverse City
P.O. Box 592
Traverse City, MI 49685
(616) 922-4465
Russell A. Soyring

Troy

Troy Public Library
510 West Big Beaver Road
Troy, MI 48084

Vassar

Cork Pine Preservation Association
205 North West
Vassar, MI 48768
(517) 823-3442

Washington

Friends of the Octagon House
57500 Van Dyke
P.O. Box 94118
Washington, MI 48094
(313) 795-3587
Matthew Schulte

Wyandotte

Wyandotte Museum
2610 Biddle Avenue
Wyandotte, MI 48192
(313) 246-4520
Chet Hunt

Ypsilanti

Eastern Michigan University
Department of Geography and Geology
Historic Preservation Program
Ypsilanti, MI 48197
(313) 487-0232
Marshall McLennan

Ypsilanti Main Street
34 North Washington
Yspilanti, MI 48197
(313) 483-6680
Linda J. Young

Zeeland

Zeeland Historical Society
37 East Main Avenue
Zeeland, MI 49464
(616) 772-4079
Mary Vanderweide

MINNESOTA

STATE HISTORIC PRESERVATION OFFICE

Minnesota Historical Society
690 Cedar Street
St. Paul, MN 55101
(612) 296-2747
Nina Archabal, Director, SHPO
Ian Stewart, Deputy Director
Deputy SHPO
Donn Coddington, Assistant Director
Deputy SHPO
Britta L. Bloomberg, Acting Deputy
SHPO
Acting Head, Historic Preservation
Field Services and Grants Department
345 Kellogg Boulevard West
St. Paul, MN 55102
(612) 296-5434

STATEWIDE PRESERVATION ORGANIZATION

Preservation Alliance of Minnesota
P. O. Box 582804
Minneapolis, MN 55458-2804
(612) 885-1229
Rolf Anderson, President

NATIONAL TRUST REGIONAL OFFICE

Midwest Regional Office
52 West Jackson Boulevard
Suite 1135
Chicago, IL 60604
(312) 939-5447
(312) 939-5651 (Fax)
Tim Turner, Director

NATIONAL TRUST ADVISORS

John L. Kuester
Minnesota Project
1885 University Avenue, West
Suite 315
St. Paul, MN 55104
(612) 645-6159

Charlene Roise
Hess, Roise and Company
710 Grain Exchange Building
Minneapolis, MN 55415
(612) 338-1987

PRESERVATION ACTION COORDINATORS

Dale Helmich
319 West 7th Street
Winona, MN 55987
(507) 896-3234

AIA STATE PRESERVATION COORDINATOR

Robert C. Mack, FAIA
MacDonald and Mack Partnership
712 Grain Exchange Building
Minneapolis, MN 55415
(612) 341-4051

NATIONAL PARK SERVICE REGIONAL OFFICE

Midwest Regional Office
1709 Jackson Street
Omaha, NE 68102
(402) 221-3431

ADVISORY COUNCIL ON HISTORIC PRESERVATION DIVISION

Eastern Office of Project Review
1100 Pennsylvania Avenue, N.W.
Suite 809
Washington, DC 20004
(202) 786-0503
Don L. Klima, Director

Albert Lea

Department of Community
Development
221 East Clark Street
Albert Lea, MN 56007
William Schmitt

Becker

Sherburne County Historical Society
13122 1st Street
Becker, MN 55308
(612) 261-4433
Kurt K. Kragness

Burnsville

New Morning Windows
11921 Portland Avenue, South
Burnsville, MN 55337
(612) 895-6180
Alexis Richman

Chisholm

IRRRB-IRRC
Highway 169 West
P.O. Box 392
Chisholm, MN 55719
(218) 254-3325
Edward P. Nelson

Dakota

Architectural Environments
P.O. Box 177
625 Main Street
Dakota, MN 55925
(507) 643-6765
Robert J. Hurt

Duluth

Glensheen
University of Minnesota
3300 London Road
Duluth, MN 55804
(218) 724-8864
Michael J. Lane

Eden Prairie

Display Sales
6365 Carlson Drive
Eden Prairie, MN 55346
(612) 937-5313
Cheryl L. Legan

Edina

Edina Heritage Preservation Board
4801 West 50th Street
Edina, MN 55424
(612) 927-8861
Joyce Repya

Embarrass

Town of Embarrass
Route 1, Box 58A
Embarrass, MN 55732
(218) 984-2672
Margaret Kinnunen

Fergus Falls

City of Fergus Falls
P.O. Box 868
112 West Washington
Fergus Falls, MN 56537
(218) 739-2251
Gordon A. Hydukovich

Fergus Falls Preservation Commission
1110 West Lincoln Avenue
Fergus Falls, MN 56537

Hastings

City of Hastings
Housing and Redevelopment Authority
100 Sibley Street
Hastings, MN 55033
(612) 437-4127
John Grossman

Hastings Historical Preservation
Commission
Hastings City Hall
100 Sibley Street
Hastings, MN 55033
(612) 437-4127
John Grossman

Pillsbury "A" Mill, Minneapolis. (D. Jacobson, HABS)

Houston

City of Houston
Municipal Center
P. O. Box 667
105 Maple Street
Houston, MN 55943
(507) 896-3234
Dale Helmich

Hutchinson

Mainstreet Hutchinson
37 Washington Avenue, West
Hutchinson, MN 55350
(612) 587-5394
Kay Johnson

Little Falls

Little Falls Main Street
P.O. Box 642
Little Falls, MN 56345
(612) 632-5695
Ann M. Gerbi

Luverne

City of Luverne
203 East Main
Luverne, MN 56156
(507) 283-9422
Steven L. Perkins

Minneapolis

Basilica of St. Mary
88 North 17th Street
Minneapolis, MN 55403
(612) 333-1381
Tom Green

Gary Gustafson
4953 Upton Avenue
Minneapolis, MN 55410
(612) 920-1563

Hess, Roise and Company
710 Grain Exchange Building
Minneapolis, MN 55415
(612) 338-1987
Charlene Roise

Institute for Minnesota Archeology
3300 University Avenue
Suite 202
Minneapolis, MN 55414
(612) 623-0299
Elden Johnson

MacDonald and Mack Partnership
712 Grain Exchange Building
Minneapolis, MN 55415
(612) 341-4051
Robert C. Mack

Minneapolis Community Development
Agency
1215 Marshall Street, N.E.
Minneapolis, MN 55413
(612) 673-5168
William A. Tetzlaff

Minneapolis Heritage Preservation
Commission
210 Minneapolis City Hall
350 South 5th Street
Minneapolis, MN 55415
(612) 348-6538
Martha H. Frey

Preservation Alliance of Minnesota
P.O. Box 582804
Minneapolis, MN 55428-2804
(612) 224-1815
Rolf Anderson

University of Minnesota
Department of History
267 19th Avenue South
Minneapolis, MN 55455
(612) 624-2800
Sue Haskins

Mounds View

Erny Mattila
2335 Hillview Road
Mounds View, MN 55112-1226

Prior Lake

Alex M. Wilson
20586 Panama Avenue
Prior Lake, MN 55372
(612) 447-2832

Red Wing

Goodhue County Preservation Project
1176 Oak Street
Red Wing, MN 55066
(612) 388-4919
Carrie J. Conklin

Red Wing Heritage Preservation
Commission
City Hall
P.O. Box 34
Red Wing, MN 55066
(612) 388-6734
Brian Peterson

Rochester

Heritage Associates
P.O. Box 512
Rochester, MN 55903

Olmsted County Historical Center
P.O. Box 6411
Rochester, MN 55903
(507) 282-9447
William D. Gernes

Roseville

Building Restoration Corporation
1920 Oakcrest Avenue
Suite 1
Roseville, MN 55113
(612) 332-2800
Dale Zoerb

St. Paul

American Civil War Commission
P.O. Box 14327
St. Paul, MN 55114
Lawrence C. Trost

Briggs and Morgan
2200 1st National Bank Building
St. Paul, MN 55101
(612) 291-1215
D. E. Muntean

Robert M. Frame III
P.O. Box 65158
St. Paul, MN 55165
(612) 227-9531
(612) 291-7882

Historic Dayton's Bluff Association
770 East 6th Street
St. Paul, MN 55106
(612) 776-3414
Susan Omoto

Lowertown Redevelopment
Corporation
175 5th Street, East
Box 104
St. Paul, MN 55101-2901
(612) 227-9131
Weiming Lu

Minnesota Department of Trade and
Economic Development
150 East Kellogg Boulevard
St. Paul, MN 55101
(612) 297-1755
Jane Leonard

Minnesota Historical Society
690 Cedar Street
St. Paul, MN 55101
(612) 296-2747
Carolyn A. Kompelien

Minnesota Historical Society
Fort Snelling History Center
St. Paul, MN 55111
(612) 726-1171
Britta Bloomberg

Minnesota Mainstreet Program
150 East Kellogg Boulevard
900 American Center
St. Paul, MN 55101
(612) 297-1755
Peggy Papenfuss

Minnesota Project
1885 University Avenue
Suite 315
St. Paul, MN 55104
John Kuester
(612) 645-6159

St. Paul Heritage Preservation
Commission
1100 City Hall Annex
25 West 4th Street
St. Paul, MN 55102
(612) 228-3399
Beth Bartz

Sanders and Associates
365 East Kellogg Boulevard
St. Paul, MN 55101
(612) 212-0401
William Sanders

Carole Zellie
1466 Hythe Street
St. Paul, MN 55108
(602) 641-1230

St. Peter

City of St. Peter
227 South Front Street
St. Peter, MN 56082
(507) 931-4840
Scott Holmgren

Stillwater

Washington County
14900 North 61st Street
Stillwater, MN 55082
(612) 779-5194
Lyle C. Doerr

Warroad

Marvin Windows
P.O. Box 100
Warroad, MN 56763
(218) 386-1430
Melvin Ortmann

White Earth

White Earth Reservation Tribal Council
P.O. Box 443
White Earth, MN 56591
(218) 983-3285
Bonnie Stewart

MISSISSIPPI

STATE HISTORIC PRESERVATION OFFICE

Mississippi Department of Archives and History
P.O. Box 571
Jackson, MS 39205-0571
(601) 359-6850
Elbert Hilliard, Director, SHPO
Kenneth H. P'Pool, Deputy SHPO
Director, Division of Historic Preservation
(601) 359-6940
(601) 359-6905 (Fax)

STATEWIDE PRESERVATION ORGANIZATION

Mississippi Heritage Trust
P.O. Box 254
Vicksburg, MS 39181-0254
(601) 636-5010
Sam Kaye, Chairperson

NATIONAL TRUST REGIONAL OFFICE

Southern Regional Office
456 King Street
Charleston, SC 29403
(803) 722-8552
(803) 722-8652 (Fax)
Susan A. Kidd, Director

NATIONAL TRUST ADVISORS

Alferdteen Brown Harrison
Margaret Walker Alexander Research Center
P. O. Box 17008
Jackson State University
Jackson, MS 39217
(601) 968-2055

Samuel H. Kaye
114 5th Street South
P.O. Box 48
Columbus, MS 39703-0048
(601) 327-6241

PRESERVATION ACTION COORDINATOR

Nancy Bell
1107 Washington Street
Vicksburg, MS 39180
(601) 638-6514

AIA STATE PRESERVATION COORDINATOR

Samuel H. Kaye, AIA
114 5th Street South
P. O. Box 48
Columbus, MS 39703-0048
(601) 327-6241
Belinda J. Stewart, AIA, Deputy

NATIONAL PARK SERVICE REGIONAL OFFICE

Southeast Regional Office
Richard B. Russell Federal Building
75 Spring Street, S.W.
Atlanta, GA 30303
(404) 331-2632

ADVISORY COUNCIL ON HISTORIC PRESERVATION DIVISION

Eastern Office of Project Review
1100 Pennsylvania Avenue, N.W.
Suite 809
Washington, DC 20004
(202) 786-0503
Don L. Klima, Director

Biloxi

City of Biloxi
P.O. Box 1907
Biloxi, MS 39533
(601) 435-6320
Valerie Lyons

Canton

Canton Mainstreet
202 East Fulton
P.O. Box 74
Canton, MS 39046
(601) 859-1665
Margaret Sholar

Gulfport

Gulf Coast Main Street Program
P.O. Drawer FF
Gulfport, MS 39502
(601) 863-2933
Donna DeWeese

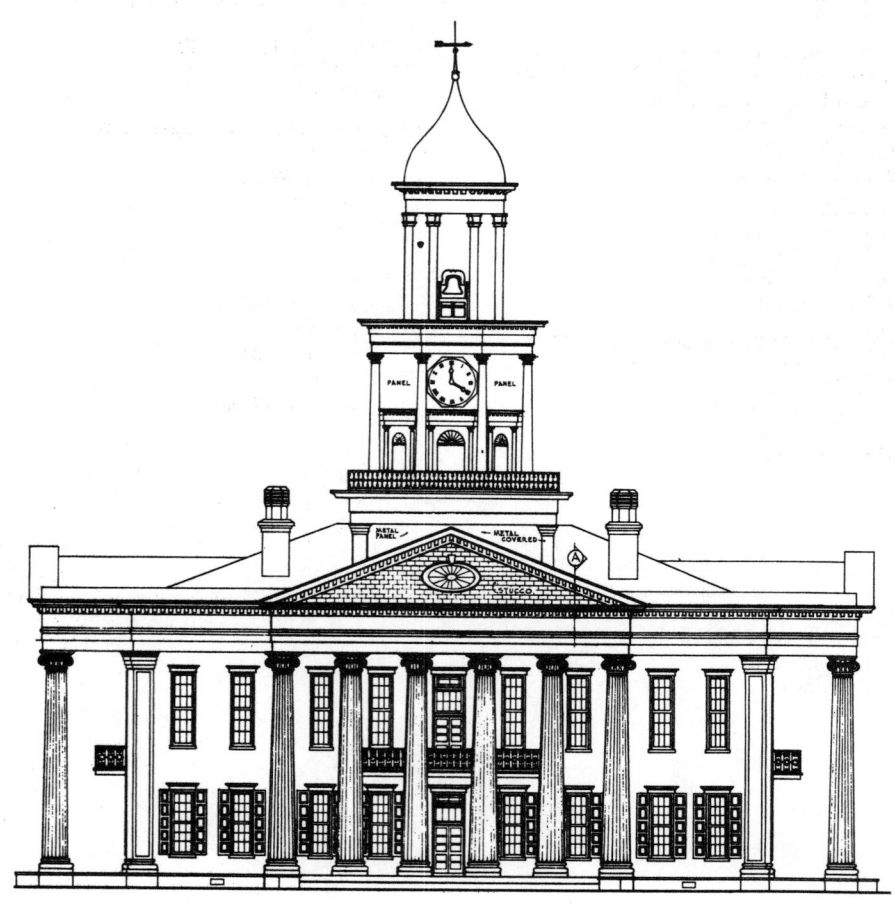

Warren County Courthouse, Vicksburg, Mississippi. (F. Deboe, S. Tuminello, E. Murphy, HABS)

Hattiesburg

City of Hattiesburg
P.O. Box 1898
Hattiesburg, MS 39403-1898
(601) 545-4595
John King

Hattiesburg Area Historical Society
P.O. Box 1573
Hattiesburg, MS 39401

Jackson

City of Jackson
Division of Planning
200 South President Street
P.O. Box 22568
Jackson, MS 39225-2568
(601) 960-1845
Helene Rotwein

Marcie A. Cohen
6295 Old Canton Road
Apartment 16B
Jackson, MS 39211
(601) 362-6357

Jackson Civil War Round Table
905 Pinehurst Place
Jackson, MS 39203
(601) 984-1668
Carroll R. Ball

Jackson State University
Institute for the Study of Black People
1400 John R. Lynch Street
Jackson, MS 39217
(601) 960-2323
James A. Hefner

Manship House
420 East Fortification Street
Jackson, MS 39202
(601) 961-4724
V. A. Patterson

Mississippi Department of Archives
and History
P.O. Box 571
Jackson, MS 39205
(601) 359-1424
Elbert R. Hilliard

Mississippi Department of Archives
and History
Department of Historic Preservation
P.O. Box 571
Jackson, MS 39205
(601) 359-6940
Georgia Jordan

Mississippi Downtown Development
Association
Mississippi Main Street Program
P.O. Box 2719
Jackson, MS 39207
(601) 359-3744
Scott Barksdale

Mississippi Municipal Association
600 East Amite Street
Jackson, MS 39202
(601) 353-5854
A. L. Dunne

Meridian

City of Meridian
P.O. Box 1430
Meridian, MS 39302
(601) 485-1910
Don Farrar

City of Meridian
Community Development Department
P.O. Box 1430
Meridian, MS 39302-1430
(601) 485-1996
John McClure

Grand Opera House of Mississippi
P.O. Box 5792
Meridian, MS 39302
(601) 693-5239
Elliott Street

Natchez

Archaeologists Unlimited
P.O. Box 1756
Natchez, MS 39121
(601) 445-8468
Elizabeth M. Boggess

City of Natchez
Architectural Review Board
City Hall
124 South Pearl Street
Natchez, MS 39120
Tod Williams

Historic Natchez Foundation
P.O. Box 1761
Natchez, MS 39120
(601) 442-2500
Ronald W. Miller

Natchez Chamber of Commerce
P.O. Box 1403
Natchez, MS 39120
(601) 445-4611
William B. Chison

Pilgrimage Garden Club
Box 347
Natchez, MS 39121
Jean S. Porter

Preservation Society of Ellicott Hill
P.O. Box 537
Natchez, MS 39121
(601) 442-6672

Philadelphia

Mississippi Band of Choctaw Indians
P.O. Box 6010
Philadelphia, MS 39350
(601) 656-5251
Kenneth H. Carleton

Vicksburg

Vicksburg Foundation For Historic
Preservation
1107 Washington Street
P.O. Box 254
Vicksburg, MS 39181
(601) 636-5010
Nancy H. Bell

MISSOURI

STATE HISTORIC PRESERVATION OFFICE

State Department of Natural Resources
205 Jefferson
P. O. Box 176
Jefferson City, MO 65102
(314) 751-4422
(314) 751-8656 (Fax)
G. Tracy Mehan III, Director, SHPO
Ron Kucera, Deputy Director, Deputy SHPO
Claire F. Blackwell, Deputy SHPO
Director, Historic Preservation Program
Division of Parks, Recreation and Historic Preservation
(314) 751-7858

STATEWIDE PRESERVATION ORGANIZATION

Missouri Heritage Alliance
P. O. Box 895
Jefferson City, MO 65102
Milton Perry, President
Susan Hoefner, Executive Secretary

NATIONAL TRUST REGIONAL OFFICE

Midwest Regional Office
52 West Jackson Boulevard
Suite 1135
Chicago, IL 60604
(312) 939-5447
(312) 939-5651 (Fax)
Tim Turner, Director

NATIONAL TRUST ADVISORS

Gregory Allen
Allen Financial Corporation
P.O. Box 410737
Kansas City, MO 64141
(816) 756-0034
(816) 561-0642 (Fax)

William Scott Meyer, II
River City Restorations
P.O. Box 1065
Hannibal, MO 64401
(314) 248-0733

PRESERVATION ACTION COORDINATOR

Claire F. Blackwell
Division of Parks and Historic Preservation
1915 Southridge Drive
P.O. Box 176
Jefferson City, MO 65102
(314) 751-2479

AIA STATE PRESERVATION COORDINATOR

W. Phillip Cotton, Jr., AIA
1110 Washington 7th Floor
St. Louis, MO 63101
(314) 421-1667

NATIONAL PARK SERVICE REGIONAL OFFICE

Midwest Regional Office
1709 Jackson Street
Omaha, NE 68102
(402) 221-3431

ADVISORY COUNCIL ON HISTORIC PRESERVATION DIVISION

Western Office of Project Review
730 Simms Street
Room 401
Golden, CO 80401
(303) 231-5320
Claudia Nissley, Director

Blue Springs

City of Blue Springs
903 Main Street
Blue Springs, MO 64015
(816) 228-0124
Karen Gable

Bonne Terre

Bonne Terre Services
P.O. Box 404
Bonne Terre, MO 63628
(314) 358-3515
Paul Williams

Boonville

Friends of Historic Boonville
P.O. Box 1776
Boonville, MO 65233
(816) 882-7977
Judy Shields

Main Street Boonville
417 1/2 Main Street
Boonville, MO 65233
(816) 882-2598
Brian Kolde

Cabool

Cabool Revitalization Group
519 Pine
P.O. Box 285
Cabool, MO 65689
(417) 962-3002
Betty Tate

California

California Progress
P.O. Box 42
California, MO 65018
(314) 796-3050
Linda Martin

Cape Girardeau

Historic Preservation Commission
401 Independence Street
Cape Girardeau, MO 63701
(314) 334-4466
Kent Bratton

Carthage

Carthage Historic Preservation
P.O. Box 375
Carthage, MO 64836
(417) 358-2259
Caryl MacMorran

Main Street Carthage
P.O. Box 545
Carthage, MO 64836
(417) 358-4974
Gail E. Johnson

Clarksville

Historic Clarksville
P.O. Box 155
Clarksville, MO 63336
(314) 242-3376
Ralph Huesing

Clayton

St. Louis County Department of Parks and Recreation
41 South Central Avenue
Clayton, MO 63105
(314) 889-3655
Virginia Stith

Clinton

Clinton Main Street
140 South Main
P.O. Box 173
Clinton, MO 64735
(816) 885-8977
Harry L. Whitaker

Columbia

Central Columbia Association
1015 East Broadway
Suite 210
Columbia, MO 65201
(314) 442-6816
Renee L. Verhoff

Governmental Affairs Program
206 Professional Building
Columbia, MO 65211
(314) 882-6401

Summerhouse, Tower Grove Park, St. Louis, Missouri. (S. Bauer, HABS)

Peckham and Wright Architects
18 North 18th Street
Columbia, MO 65201
(314) 449-2683
Nicholas Peckman

University of Missouri
Missouri Cultural Heritage Center
Conley House
Columbia, MO 65211
(314) 882-6296

Excelsior Springs

Richard Brown
Rural Route 2
Box 512
Excelsior Springs, MO 64024
(816) 637-6700

Ferguson

Ferguson Special Business District
2 South Florissant Road
Ferguson, MO 63135
(314) 521-6666
Jean Montgomery

Hannibal

Hannibal Main Street
Hannibal Chamber of Commerce
623 Broadway
P.O. Box 230
Hannibal, MO 63401
(314) 221-2677
Linda Rex-Dobson

Hermann

Historic Hermann
P.O. Box 88
Hermann, MO 65041
(314) 486-2781
Gennie Tesson

Independence

City of Independence
Planning Department
P.O. Box 1019
Independence, MO 64051
Frank C. Davis

Harry S. Truman Library
24 Highway and Delaware
Independence, MO 64050
(816) 833-1400

Uptown Independence
129 West Lexington
Suite 307
Independence, MO 64050
(816) 833-0505
Cilia Sherman

Jefferson City

Historic Preservation Program
P.O. Box 176
Jefferson City, MO 65102
(314) 751-7858
Claire Blackwell

Jefferson Community Betterment
208-E East High Street
Jefferson City, MO 65101
(314) 635-1171
Eric J. Chism

Jefferson Landing State Historic Site
State Capitol
Room B-2
Jefferson City, MO 65101
(314) 751-2854
M. Shay

Missouri Alliance for Historic
Preservation
P.O. Box 895
Jefferson City, MO 65102

Missouri Department of Economic
Development
Missouri Main Street Program
P.O. Box 118
Jefferson City, MO 65102
(314) 751-7939
Randy Gray

Missouri Mansion Preservation
100 Madison Street
P.O. Box 1133
Jefferson City, MO 65102
(816) 882-6815
Mary P. Abele

Joplin

Joplin Heritage Trust
P.O. Box 1654
Joplin, MO 64802
(417) 781-1508
Blanche McKee

Main Street Joplin
402 Main
Suite 505
Joplin, MO 64801
(417) 624-1060
Wendy Swartz

Kansas City

William L. Bruning
4818 Jarboe
Kansas City, MO 64112
(816) 556-0214

Historic Kansas City Foundation
1003 Broadway
Kansas City, MO 64105
(816) 471-3391
Nancy Ellermann

Kansas City Masonic Temple
Historic Preservation Society
903 Harrison Street
Kansas City, MO 64106-3070
(816) 942-8300
Kirk McDaniel

Landmarks Commission of Kansas City
City Hall
Kansas City, MO 64105
(816) 274-2555
Lisa Briscoe

Spencer Art Reference Library
Nelson-Atkins Museum of Art
4525 Oak Street
Kansas City, MO 64111

Alisha M. Stockton
115 West 73rd Street
Kansas City, MO 64114

Eric C. Youngberg
3809 Campbell Street
Kansas City, MO 64109-2611

Kirkwood

Kirkwood Landmarks Commission
139 South Kirkwood Road
Kirkwood, MO 63122
(314) 822-5808
Rosalind Williams

Kirkwood Main Street
139 South Kirkwood Road
Kirkwood, MO 63122
(314) 822-5807
Charlotte Sparks

Lee's Summit

Downtown Lee's Summit Main Street
123 Southeast 3rd Street
P.O. Box 1688
Lee's Summit, MO 64063
(816) 246-6598
Terry Harris

Lexington

Battle of Lexington
P.O. Box 6
Lexington, MO 64067
(816) 259-2112
Jane Fuller

Liberty

City of Liberty
101 East Kansas
Liberty, MO 64068
(816) 792-6109
Maura Johnson

Moberly

Moberly Downtown Association
P.O. Box 602
Moberly, MO 65270
(816) 263-5251
Deborah H. Dean

Nevada

Main Street Nevada
209 West Walnut
Nevada, MO 64772
(417) 667-8111
John Joslin

Poplar Bluff

Downtown Development Corporation
213 South Main Street
Poplar Bluff, MO 63901
(314) 686-6866
Doris-Ann Shelton

St. Charles

City of St. Charles
Department of City Development
200 North 2nd Street
St. Charles, MO 63301
(314) 925-2000
Patrick Switz

City of St. Charles
Special Business District
200 North 2nd Street
St. Charles, MO 63301
(314) 925-2000
Sue Schneider

St. Joseph

Bill Collier
3517 Mitchell
St. Joseph, MO 64507
(816) 279-6587

Junior League of St. Joseph
304 North 8th Street
St. Joseph, MO 64501
Heidi Hornaday

Mainstreet St. Joseph
P.O. Box 1065
St. Joseph, MO 64502
(816) 233-7171
Kathy Morgan

St. Joseph Preservation
P.O. Box 121
St. Joseph, MO 64506
(816) 364-2327
David D. Denman

St. Louis

Affton Historical Society
Box 28855
St. Louis, MO 63123

Chatillon-DeMenil Mansion
3352 DeMenil Place
St. Louis, MO 63118
(314) 771-5828
Anthony Bommarito

Community Women Against Hardship
1532 Marbella
Apartment 10
St. Louis, MO 63138
(314) 382-8748
Gloria Taylor

Eugene Field House Foundation
P.O. Box 9012
St. Louis, MO 63102
(314) 241-5845
William R. Piper

Landmarks Association of St. Louis
917 Locust Street
Room 1113
St. Louis, MO 63101-1401
(314) 421-6474
Carolyn Toft

Missouri Historical Society
Jefferson Memorial Building
Forest Park
St. Louis, MO 63112
(314) 361-1424
Robert Archibald

National Association of Independent
Fee Appraisers
7501 Murdoch
St. Louis, MO 63119
(314) 781-6688
William H. Steele

Pride Organization
P.O. Box 4998
St. Louis, MO 63108-0998
(314) 776-2400
Stephen E. Bayer

St. Louis County Library
1640 South Lindbergh Boulevard
St. Louis, MO 63131
L. K. Silence

St. Louis Design Alliance
7511 Delmar
St. Louis, MO 63130
(314) 863-1313
David Mastin

Soulard Restoration Group
Soulard Station
P.O. Box 13312
St. Louis, MO 63157
(314) 771-5563

Tower Grove House
Missouri Botanical Garden
2315 Tower Grove Avenue
St. Louis, MO 63110
(314) 577-5150
Victor A. Silber

Washington University Technology
Associates
8220 Brentwood Industial
St. Louis, MO 63144
(314) 645-5230
Patrick F. Rice

Savannah

Main Street Savannah
411 Court Street
Savannah, MO 64485
(816) 324-3976
James C. Boosworth

Springfield

City of Springfield
Community Development
830 Boonville
Springfield, MO 65801
(417) 864-1031
Mary Lilly

Laura J. Derrick
1027 South Weller Avenue
Springfield, MO 65804
(417) 865-8388

James Keet
3534 East Catalpa
Springfield, MO 65804

Museum of the Ozarks
603 East Calhoun Street
Springfield, MO 65802
(417) 869-1976
Julie March

Unionville

Putnam County Foundation
P.O. Box 2000
Unionville, MO 63565
(816) 947-2080
Linda Fettig

Washington

Downtown Washington
323 West Main Street
Washington, MO 63090
(314) 239-1743
Linda Kuenzie

Washington Historic Preservation
Commission
405 Jefferson Street
Washington, MO 63090
(314) 239-6710
James Briggs

Webster Groves

City of Webster Groves
4 East Lockwood
Webster Groves, MO 63119
(314) 961-4100
Joseph D. Parente

MONTANA

STATE HISTORIC PRESERVATION OFFICE

State Historic Preservation Office
Montana Historical Society
225 North Roberts
Helena, MT 59620-9990
(406) 444-7715
(406) 444-2696 (Fax)
Marcella Sherfy, SHPO
Mark F. Baumler, Archeologist, Deputy
SHPO
Pat Bik, Survey and Registration,
Deputy SHPO
Herbert E. Dawson, Historical
Architect, Deputy SHPO

STATEWIDE PRESERVATION ORGANIZATION

Montana Preservation Alliance
P. O. Box 1872
Bozeman, MT 59771
(406) 585-9551
Ellen Sievert, President

NATIONAL TRUST REGIONAL OFFICE

Mountains/Plains Regional Office
511 16th Street
Suite 700
Denver, CO 80202
(303) 623-1504
(303) 825-8073 (Fax)
Barbara J. Pahl, Director

NATIONAL TRUST ADVISORS

Ellen Sievert
1602 3rd Westhill Drive
Great Falls, MT 59404
(406) 761-6955

Lawrence J. Sommer
231 Anderson Boulevard
Helena, MT 59601
(406) 444-2694
(406) 444-2696 (Fax)

PRESERVATION ACTION COORDINATOR

Janet Cornish
Montana Preservation Alliance
305 West Mercury
Butte, MT 59701
(406) 723-4061

AIA STATE PRESERVATION COORDINATOR

John N. DeHaas Jr., FAIA
1021 South Tracey Avenue
Bozeman, MT 59715
(406) 586-2276
Jim McDonald, AIA, Deputy

NATIONAL PARK SERVICE REGIONAL OFFICE

Rocky Mountain Regional Office
12795 West Alameda Parkway
P.O. Box 25287
Denver, CO 80225-0287
(303) 969-2875

ADVISORY COUNCIL ON HISTORIC PRESERVATION DIVISION

Western Office of Project Review
730 Simms Street
Room 401
Golden, CO 80401
(303) 231-5320
Claudia Nissley, Director

Billings

Billings Preservation Society
914 Division Street
Billings, MT 59101
(406) 256-5100
Ruth Towe

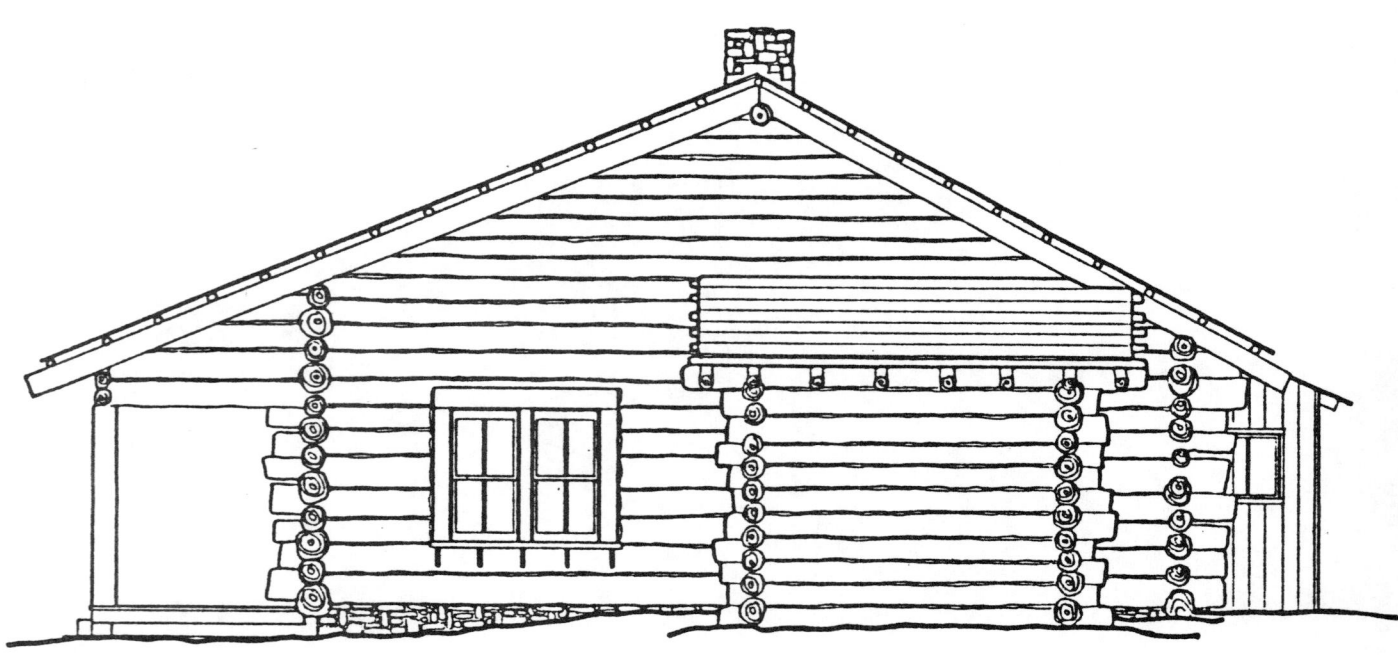

Lubec Ranger Station, East Glacier, Montana. (K. Speece, HABS)

Bozeman

Montana State University
School of Architecture
Bozeman, MT 59717
(406) 224-4255
Paul Gleye

Keith C. Swenson, AICP
603 South Willson Avenue
Bozeman, MT 59715
(406) 587-4434

Butte

Butte Historical Society
Box 3913
Butte, MT 59701

Great Falls

Cascade County Historical Society
1400 1st Avenue, North
Great Falls, MT 59401
(406) 452-3462
Cindy Adams

Great Falls City-County Planning
Board
Box 5021
Great Falls, MT 59403
(406) 727-5881
John Mooney

Ellen Cornwall Sievert
1602 3rd West Hill Drive
Great Falls, MT 59404
(406) 538-7029

Hamilton

Daly Mansion Preservation Trust
P.O. Box 223
Hamilton, MT 59840
(406) 363-6004
Jeanette McKee

Helena

City of Helena
Planning Department
316 North Park Avenue
Helena, MT 59623
Kathy Macefield

State Historical Preservation Office
225 North Roberts
Helena, MT 59620
(406) 444-7715
Marcella Sherfy

Livingston

City of Livingston
414 East Callender Street
Livingston, MT 59047

Livingston Depot Foundation
Box 1319
Livingston, MT 59047
(406) 222-2300
C. Smith

Miles City

City of Miles City
P.O. Drawer 910
Miles City, MT 59301
(406) 232-3462
Mark Richardson

Shelby

City of Shelby
P.O. Box 743
Shelby, MT 59474
(406) 434-5222
Larry J. Bonderud

Walkerville

Town of Walkerville
P.O. Box 257
Walkerville, MT 59701
Bernard Harrington

NEBRASKA

STATE HISTORIC PRESERVATION OFFICE

Nebraska State Historical Society
P.O. Box 82554
Lincoln, NE 68501
(402) 471-4787
Director and SHPO (Vacant)
L. Robert Puschendorf, Interim
Director, Deputy SHPO
(402) 471-4769

STATEWIDE PRESERVATION ORGANIZATION

Nebraska Preservation Council
11711 Arbor
Omaha, NE 68144
(402) 345-3060
George Haecker, President

NATIONAL TRUST REGIONAL OFFICE

Mountains/Plains Regional Office
511 16th Street
Suite 700
Denver, CO 80202
(303) 623-1504
(303) 825-8073 (Fax)
Barbara J. Pahl, Director

NATIONAL TRUST ADVISORS

Nancy Hoch
River Country Industrial Commission
806 1st Avenue
Nebraska City, NE 68410
(402) 873-4293

James C. Seacrest
Points West Box 808
North Platte, NE 69103-1666
(308) 532-6783
(308) 532-3239 (Fax)

PRESERVATION ACTION COORDINATOR

Edward F. Zimmer
Planning Department
555 South 10th Street
Lincoln, NE 68508
(402) 471-7491

AIA STATE PRESERVATION COORDINATOR

Ted A. Ertl, AIA
University of Nebraska
236 Architecture Hall
Lincoln, NE 68588-0107
(402) 472-3592
Jerry L. Berggren, AIA, Deputy

NATIONAL PARK SERVICE REGIONAL OFFICE

Midwest Regional Office
1709 Jackson Street
Omaha, NE 68102
(402) 221-3431

ADVISORY COUNCIL ON HISTORIC PRESERVATION DIVISION

Western Office of Project Review
730 Simms Street
Room 401
Golden, CO 80401
(303) 231-5320
Claudia Nissley, Director

Beatrice

Gage County Historical Society
P.O. Box 793
Beatrice, NE 68310
(402) 228-1679
Kent Wilson

Fremont

Mainstreet Fremont
P.O. Box 471
Fremont, NE 68025
(402) 727-1531
Robert Missel

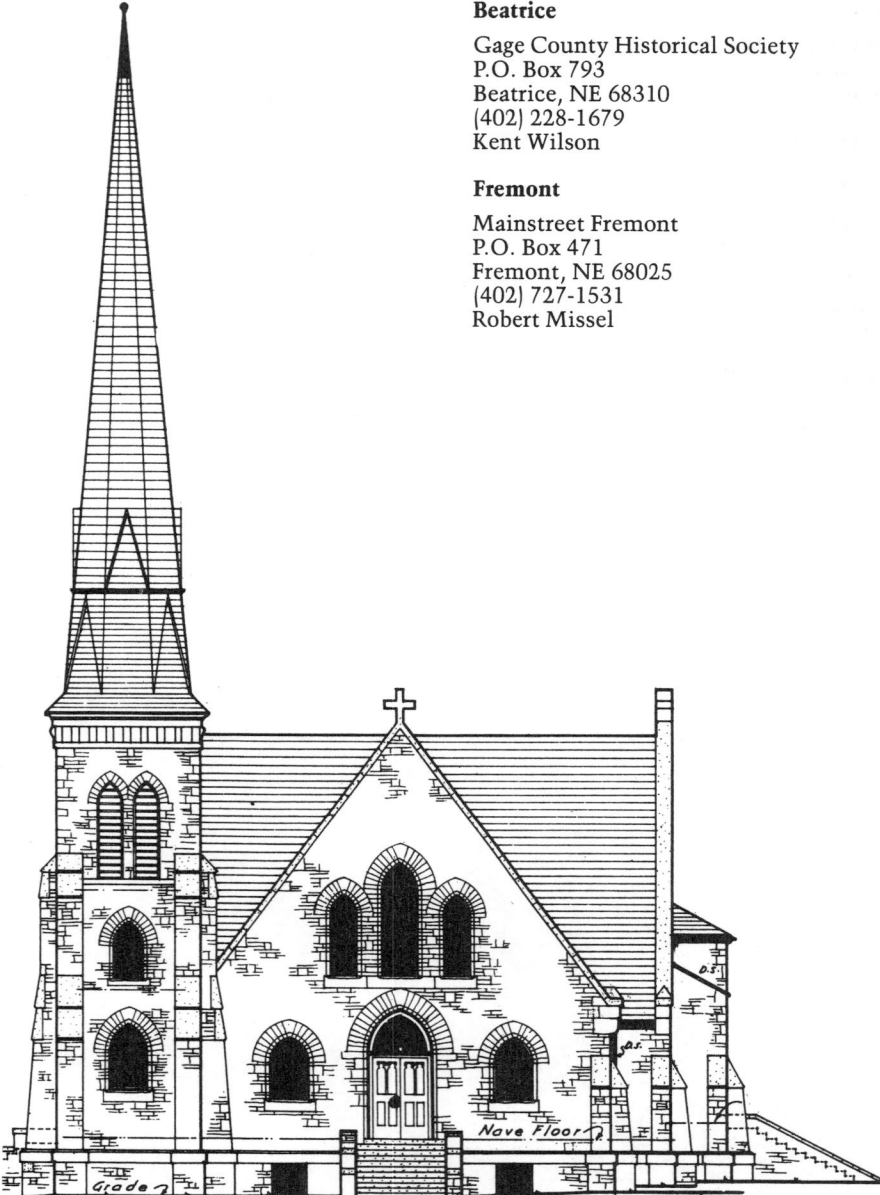

Holy Trinity, Lincoln, Nebraska. (F. Mooberry, HABS)

Gering

Gering Clothing
1455 10th Street
Gering, NE 69341
(308) 436-5096
Steve Taylor

Gordon

Cherished Moments
220 Pine
Gordon, NE 69343
(308) 282-0674
Bea-Lou Annett

Grand Island

Grand Island Downtown Improvement
Office
P.O. Box 1306
Grand Island, NE 68802
(308) 381-5455
Dianne Kelley

Hastings

Lakeway Land and Development
Company
P.O. Box 2032
Hastings, NE 68901
(402) 461-3310
Greg Burke

Lincoln

Department of Administrative Services
State Building Division
State Capitol
10th Floor
Lincoln, NE 68509
(402) 471-3191
Thomas Kasper

Ted A. Ertl
2435 Sewell
Lincoln, NE 68502
(402) 472-3592

Harris Preservation Center
1630 K Street
Lincoln, NE 68508

Lincoln Haymarket Development
Corporation
335 North 8th
Suite B
Lincoln, NE 68508
(402) 435-7496
Lou Shields

Lincoln/Lancaster City Planning
Department
555 South 10th Street
Lincoln, NE 68508
(402) 471-7491
Ed Zimmer

Nebraska Municipal Power Pool
P.O. Box 95124
Lincoln, NE 68509
(402) 474-4759
Corrinne Pedersen

University Place Community
Organization
2723 North 50
Lincoln, NE 68504
(402) 466-1906
Coleen Seng

Ogallala

National Arbor Day Foundation
P.O. Box 117
Ogallala, NE 69153
Robert F. Lute

Omaha

Bahr Vermeer and Haecker Architects
1209 Harney
Omaha, NE 68102
(402) 345-3060
George Haecker

Landmarks
Old Market Passageway
1058 Howard Street
Omaha, NE 68102
(402) 346-1055
Debbi Aliano

Landmarks Heritage Preservation
Commission
Omaha/Douglas Civic Center
1819 Farnam Street
Omaha, NE 68183
(402) 444-5208
Lynn Meyer

Nebraska Preservation Council
11711 Arbor
Omaha, NE 68144
Leonard Sommer

Western Heritage Society
801 South 10th Street
Omaha, NE 68108
(402) 444-5072
Phillip C. Kwiatkowski

NEVADA

STATE HISTORIC PRESERVATION OFFICE

Division of Historic Preservation and Archeology
123 West Nye Lane
Room 208
Carson City, NV 89710
(702) 687-5138
Ronald James, Supervisor, SHPO
Alice Baldrica, Deputy SHPO

STATEWIDE PRESERVATION ORGANIZATION

Nevada Heritage
1405 Joshua Drive
Reno, NV 89509-2202
(702) 887-2180
(702) 887-2202 (Fax)
Rob Joiner, President

NATIONAL TRUST REGIONAL OFFICE

Western Regional Office
1 Sutter Street
Suite 707
San Francisco, CA 94104
(415) 956-0610
(415) 956-0837 (Fax)
Kathryn A. Burns, Director

NATIONAL TRUST ADVISORS

Richard F. Moreno
2355 Pinebrook Drive
Carson City, NV 89701
(702) 687-4322
(702) 882-7998

Elizabeth Warren
1384C Santa Anita
Las Vegas, NV 89119-6228
(702) 382-7198

PRESERVATION ACTION COORDINATOR

Elizabeth Beckham
1647 Noreen Drive
Sparks, NV 89431
(702) 355-7418

AIA STATE PRESERVATION COORDINATOR

Vacant

NATIONAL PARK SERVICE REGIONAL OFFICE

Western Regional Office
600 Harrison Street
Suite 600
San Francisco, CA 94107
(415) 744-3988

ADVISORY COUNCIL ON HISTORIC PRESERVATION DIVISION

Western Office of Project Review
730 Simms Street
Room 401
Golden, CO 80401
(303) 231-5320
Claudia Nissley, Director

Carson City

Carson City Main Street
400 South Carson Street
P.O. Box 3008
Carson City, NV 89703
(702) 885-0411
Gail Thomssen

Nevada Heritage
P.O. Box 352
Carson City, NV 89702
(702) 847-0449
Anna B. Koval

Mimi Rodden
880-A Round House Lane
Carson City, NV 89701
(702) 882-6459

Ely

Mark Henderson
1001 Canyon Street
Ely, NV 89301-2104

Las Vegas

City of Las Vegas
Downtown Redevelopment Agency
400 East Stewart
5th Floor
Las Vegas, NV 89101

Preservation Association of Clark County
P.O. Box 7437
Las Vegas, NV 89125
(702) 383-8897
Garry Hayes

Reno

Don D. Fowler
1010 Foothill Road
Reno, NV 89511
(702) 784-6951

Nevada Historical Society
1650 North Virginia
Reno, NV 89503

Virginia City

Carmen Kuffner
Box 701
Virginia City, NV 89440
(702) 847-0615

Storey County
Drawer D
Virginia City, NV 89440

Winnemucca

Friends of Nixon Opera House
P.O. Box 819
Winnemucca, NV 89445
(702) 623-2712
Gene Wamholt

James D. Roberts House, Carson City, Nevada. (J.T. McCreery, HABS)

NEW HAMPSHIRE

STATE HISTORIC PRESERVATION OFFICE

Division of Historical Resources and State Historic Preservation Office
P.O. Box 2043
Concord, NH 03301
(603) 271-3483
Nancy Muller, Director, SHPO
Linda Ray Wilson, Deputy SHPO
(603) 271-3558

STATEWIDE PRESERVATION ORGANIZATION

Inherit New Hampshire
P.O. Box 268
118 North Main Street
Concord, NH 03302
(603) 224-2281
Linda G. Willett, Chairman

NATIONAL TRUST REGIONAL OFFICE

Northeast Regional Office
7 Faneuil Hall Marketplace
5th Floor
Boston, MA 02109
(617) 523-0885
(617) 523-1199 (Fax)
Vicki Sandstead, Director

NATIONAL TRUST ADVISORS

Barbara B. Pressly
80 Concord Street
Nashua, NH 03060
(603) 880-7752

John Scott
35 Pine Street
P.O. Box 3127
Peterborough, NH 03458-3127
(603) 924-9213

PRESERVATION ACTION COORDINATOR

Martha Fuller Clark
152 Middle Street
Portsmouth, NH 03801
(603) 433-8798

AIA STATE PRESERVATION COORDINATOR

Richard H. Monahon, Jr., AIA
The Granite Block
44 Maine Street
Peterborough, NH 03458
(603) 924-7279

NATIONAL PARK SERVICE REGIONAL OFFICE

North Atlantic Regional Office
15 State Street
Boston, MA 02109-3572
(617) 223-5001

ADVISORY COUNCIL ON HISTORIC PRESERVATION DIVISION

Eastern Office of Project Review
1100 Pennsylvania Avenue, N.W.
Suite 809
Washington, DC 20004
(202) 786-0503
Don L. Klima, Director

Acworth

Acworth Silsby Library
Acworth, NH 03601
(603) 835-2295
Fay Knicely

Alstead

Historic Window and Door Corporation
Junction 123A and 123
P.O. Box 138
Alstead, NH 03602
(603) 835-2918
Richard Pelletier

Boscawen

New Hampshire Art Association
P.O. Box 3096
150 King Street
Boscawen, NH 03303
(603) 796-6414
Elise Kohl

Candia

Christopher W. Closs
255 New Boston Road
P.O. Box 293
Candia, NH 03034

Stephen A. Kull
193 High Street
Candia, NH 03034
(603) 483-8232

Concord

Division of Historical Resources
15 South Fruit Street
P.O. Box 2043
Concord, NH 03302-2043
(603) 271-3558
R.S. Wallace

Elizabeth Durfee Hengen
25 Ridge Road
Concord, NH 03301
(603) 225-7977

Inherit New Hampshire
P.O. Box 268
Concord, NH 03301
(603) 224-2281
John F. Page

New Hampshire Historical Society
30 Park Street
Concord, NH 03301
(603) 225-3381
John Frisbee

Deerfield

Town of Deerfield
P.O. Box 159
8 Old Center Road
Deerfield, NH 03037
(603) 463-8811

Durham

New Hampshire Association of Historic District Commissions
Box 216
Durham, NH 03824
L. F. Heald

Enfield

Museum at Lower Shaker Village
Route YA
Box 25
Enfield, NH 03748
(603) 632-5533
Carolyn Smith

Exeter

American Independence Museum
1 Governor's Lane
Exeter, NH 03833
(603) 772-2622
Richard D. Tobin

Harrisville

Historic Harrisville
Box 79
Harrisville, NH 03450
(603) 827-3333
(603) 827-3722
J.J. Colony

Jaffrey Center

Robert Stephenson
Box 435
Jaffrey Center, NH 03452
(617) 262-0060

Manchester

Currier Gallery of Art
192 Orange Street
Manchester, NH 03104
(603) 669-6144
Michael Komanecky

Shaker Church Family Cow Barn, Enfield, New Hampshire. (P.M. Burkhart, HABS)

Nashua

Nashua Public Library
2 Court Street
Nashua, NH 03060

Andrew T. Singelakis
66 Saint Camille Street
Nashua, NH 03060

Peterborough

Peterborough Historical Society
P.O. Box 58
Peterborough, NH 03458
(603) 924-3235
Ellen S. Derby

Portsmouth

City of Portsmouth
Planning Department
P.O. Box 628
Portsmouth, NH 03802-0628
(603) 431-2000
Regina M. Lammers

Friends of the Music Hall
P.O. Box 1112
Portsmouth, NH 03801
(603) 433-3100
Mary Kelley

MacPheadris Warner House
P.O. Box 895
150 Daniel Street
Portsmouth, NH 03801

Portsmouth Advocates
P.O. Box 4066
Portsmouth, NH 03801

Portsmouth Athenaeum
P.O. Box 848
9 Market Square
Portsmouth, NH 03801

Strawberry Banke
P.O. Box 300
Portsmouth, NH 03801
Vincent Lombardi

Sugar Hill

Sugar Hill Historical Museum
Village Green
Sugar Hill, NH 03585
Jane L. Vincent

NEW JERSEY

STATE HISTORIC PRESERVATION OFFICE

Department of Environmental
Protection
501 East State Street
CN 402
Trenton, NJ 08625
(609) 292-2885
Scott A. Weiner, Commissioner, SHPO

James Hall, Acting Assistant
Commissioner for Natural and Historic
Resources, Deputy SHPO
501 East State Street
CN 404
Trenton, NJ 08625
(609) 292-3541

Gregory A. Marshall, Director
Division of Parks and Forestry
Deputy SHPO
(609) 292-2733

Nancy L. Zerbe, Administrator
Office of New Jersey Heritage
Deputy SHPO
(609) 292-2023

STATEWIDE PRESERVATION ORGANIZATION

Preservation New Jersey
170 Township Line Road
Belle Mead, NJ 08502
(908) 359-4557
Claire Degnan, President

New Jersey Historic Trust
CN 404
Trenton, New Jersey 08625-0404
Harriette Hawkins, Executive Director

NATIONAL TRUST REGIONAL OFFICE

Mid-Atlantic Regional Office
Cliveden
6401 Germantown Avenue
Philadelphia, PA 19144
(215) 438-2886
(215) 438-2892 (Fax)
Patricia A. Wilson, Director

NATIONAL TRUST ADVISORS

Flavia Alaya
520 East 28th Street
Patterson, NJ 07514
(201) 278-6372

Katharine Earnshaw Shuler
8 Copper Hill Road
Bridgewater, NJ 08807
(908) 636-5680

PRESERVATION ACTION COORDINATORS

Katharine Shuler
Preservation New Jersey
180 Township Line Road
Belle Mead, NJ 08502
(908) 636-5680

AIA STATE PRESERVATION COORDINATOR

Michael J. Mills, AIA
Short and Ford and Partners Architects
864 Mapleton Road
Princeton, NJ 08540-9539
(609) 452-1777
Ken Underwood, Deputy

NATIONAL PARK SERVICE REGIONAL OFFICE

North Atlantic Regional Office
15 State Street
Boston, MA 02109-3572
(617) 223-5001

ADVISORY COUNCIL ON HISTORIC PRESERVATION DIVISION

Eastern Office of Project Review
1100 Pennsylvania Avenue, N.W.
Suite 809
Washington, DC 20004
(202) 786-0503
Don L. Klima, Director

Allaire

Allaire Village
P.O. Box 220
Allaire Road
Allaire, NJ 07727
(908) 930-2253
Patrick Clarke

Annandale

Cultivators
76A Sandhill Road
Annandale, NJ 08801
Robert C. Congdon

Atlantic City

Office of Cultural Affairs
Regional Planning and Development
1333 Atlantic Avenue
Atlantic City, NJ 08401
(609) 343-2243
Karen DeRosa

Belle Mead

Preservation New Jersey
180 Township Line Road
Belle Mead, NJ 08502
(908) 359-4557
Katharine Shuler

Boonton

Morris County Trust for Historic
Preservation
P.O. Box 734
Boonton, NJ 07005
(201) 263-1026
Doreen De Roberts

Bridgeton

City of Bridgeton
Community Development Office
City Hall Annex
181 East Commerce Street
Bridgeton, NJ 08302
(609) 455-3230
Ray Maier

Main Street Bridgeton
50 East Broad Street
Bridgeton, NJ 08302
(609) 451-4802
Dennis Campbell

Watson and Henry Associates
12 North Pearl Street
Bridgeton, NJ 08302-1902
(609) 451-1779
Michael C. Henry

Burlington

Burlington County Historical Society
457 High Street
Burlington, NJ 08016
(609) 386-4773
M. M. Pernot

Historic Burlington Preservation
Foundation
P.O. Box 1344
Burlington, NJ 08016
(609) 387-4332
Charles Cheeseman

Camden

Camden Department of Housing and
Community Development
City Hall
10th Floor
Camden, NJ 08101
Barbara S. Olejnik

Department of Policy Planning
City Hall
Room 409
Camden, NJ 08101
(609) 757-7488
John E. Doyle

Vintage Living
600 Benson Street
Camden, NJ 08103
(609) 964-0547
Earl Geertgens

Cape May

Mid-Atlantic Center for the Arts
P.O. Box 340
Cape May, NJ 08204
(609) 884-5404
B. M. Zuckerman

Cape May Court House

Cape May County Cultural and
Heritage Commission
Library Office Building
Cape May Court House, NJ 08210
(609) 465-1005
Jennie A. Snyder

Clifton

City of Clifton
Recreation Department
1232 Main Avenue
Clifton, NJ 07011
(201) 470-2246
Raymond M. Mastroberte

Clinton

Douglas D. Martin
P.O. Box 5156
Clinton, NJ 08809
(908) 735-7180

Cranbury

Cranbury Historical and Preservation
Society
P.O. Box 77
Cranbury, NJ 08512
(609) 655-3736
Betty Wagner

Cranford

Township of Cranford
8 Springfield Avenue
Cranford, NJ 07016-2199
(908) 709-7206
Edward J. Murphy

Watertrol
P.O. Box 163
Cranford, NJ 07016
(908) 245-6622
James Papandrea

Dorchester

Delaware Bay Schooner Project
a.k.a. Schooner Clyde A. Phillips
P.O. Box 57
Dorchester, NJ 08316
(609) 785-2060
Meghan E. Wren

East Brunswick

East Brunswick Museum
P.O. Box 904
East Brunswick, NJ 08816

East Orange

East Orange Public Library
Periodical Department
21 South Arlington Avenue
East Orange, NJ 07018

Elizabeth

City of Elizabeth
50 Winfield Scott Plaza
Elizabeth, NJ 07201
(201) 820-4037
Phyllis Brociner

Union County Office of Cultural and
Heritage Affairs
633 Pearl Street
Elizabeth, NJ 07202
Linda B. McTeague

Phoenix Mill, Paterson, New Jersey. (W. Gavzy, HAER)

Englewood

Englewood Economic Development
Corporation
Main Street Project
2-10 North Van Brunt Street
Englewood, NJ 07631
(201) 871-6665
Peter M. Beronio

Fairlawn

The Ives Group
Architects/Planners
14-25 Plaza Road
Fairlawn, NJ 07410
(201) 791-7444
C. J. Smith

Far Hills

Thomas S. Waller
Box 694
Far Hills, NJ 07931
(201) 781-6027

Flemington

Hunterdon County Cultural and
Heritage Commission
Administration Building
Flemington, NJ 08822
(908) 788-1256
Shirley Watson

Madden/Kummer
260 Route 202/31
Suite 500
Flemington, NJ 08822
(908) 788-8200
Tony Soriano

Freehold

City of Monmouth
Board of Chosen Freeholders of the
County of Monmouth
Freehold, NJ 07728
Owen Redmond

County of Monmouth
Planning Board
Hall of Records Annex
P.O. Box 1255
Freehold, NJ 07728
(201) 431-7460
Robert W. Clark

Monmouth County Historical
Association
70 Court Street
Freehold, NJ 07728
(201) 462-1466
Carol B. O'Donnell

Greenwich

Cumberland County Historical Society
Box 16
Greenwich, NJ 08323
(698) 455-4055
Sara C. Watson

Hackensack

Design Group Architectura
155 Polifly Road
Suite 310
Hackensack, NJ 07601
(201) 343-5522

Division of Cultural and Historic
Affairs
Court Plaza South
21 Main Street
Room 203W
Hackensack, NJ 07601-7000
(201) 646-2780
Ruth Van Wagoner

Haddonfield

Borough of Haddonfield
242 Kings Highway East
Haddonfield, NJ 08033
(609) 428-8383
John J. Tarditi

Haddonfield Preservation Society
120 Warwick Road
Haddonfield, NJ 08033
(609) 429-5486
Joan L. Aiken

High Bridge

Harlan I. Ettinger, Attorney
P.O. Box 209
High Bridge, NJ 08829
(201) 638-5696

Hope

Mark A. Hewitt
Walnut Street
P.O. Box 121
Hope, NJ 07844
(201) 387-8913

Hopewell

Heritage Studies
20 Seminary Avenue
Hopewell, NJ 08525
(609) 466-9606
Constance M. Greiff

I.S.E. Associates
P.O. Box 58
Hopewell, NJ 08525
(609) 466-1515
Ilona S. English

Jersey City

Barrow Mansion Development
Corporation
83 Wayne Street
Jersey City, NJ 07302
(201) 432-6979
Maurice McClelland

Columbus Line
Plaza 2
Harborside Financial Center
Jersey City, NJ 07302
(201) 432-0900
Horst Ullman

County of Hudson
Office of the County Executive
595 Newark Avenue
Administration Building
Jersey City, NJ 07306

Cynthia Mannetta
59 Coles Street
Suite 2
Jersey City, NJ 07302
(201) 653-4372

Lawnside

Lawnside Historical Society
P.O. Box 608
Lawnside, NJ 08045-0608
(609) 520-5929
Linda Waller

Layton

John Bruce Dodd, Architect
P.O. Box 306
Layton, NJ 07851
(201) 948-3268

Little Falls

Main Street Little Falls
5 Paterson Avenue
Little Falls, NJ 07424
(201) 256-3665
Helene Baumann

Long Branch

Office of Community and Economic
Development
279 Broadway
5th Floor
Long Branch, NJ 07740
(201) 222-2019

Long Valley

Washington Township Historic
Preservation Commission
P.O. Box 216
43 Schooley's Mountain Road
Long Valley, NJ 07853
(201) 876-3315
Margaret Nordstrom

Margate

Save Lucy Committee
P.O. Box 3000
Margate, NJ 08402
(609) 822-6519
Josephine L. Harron

Marlton

Township of Evesham
Historic Preservation Commission
125 East Main Street
Marlton, NJ 08053
(609) 983-2900
Bryan Havir

Montclair

Poor Richards
101 Walnut Street
Montclair, NJ 07042
(207) 835-333
J. R. Mordwin

Moorestown

Alice Paul Centennial Foundation
P.O. Box 472
Moorestown, NJ 08057
(609) 829-8969
Barbara Irvine

Morris Plains

Craftsman Farms Foundation
2352 Route 10 West
Box 5
Morris Plains, NJ 07950
Robert P. Guter

Morristown

Fosterfields Living Historical Farm
Morris County Park Commission
P.O. Box 1295
Morristown, NJ 07962-1295
(201) 326-7645
Helen Hesselgrave

Morris County Historical Society
68 Morris Avenue
P.O. Box 170M
Morristown, NJ 07960
(201) 267-3465
Jeanne H. Watson

Women's Club of Morristown
51 South Street
Morristown, NJ 07960
(201) 539-0467
Ann Morris

Mount Holly

National Society of Colonial Dames Of
America in New Jersey
Rural Route 1
Box 1145, Burr Road
Mount Holly, NJ 08060
Judith Perinchief

Newark

Newark Museum
49 Washington Street
P.O. Box 540
Newark, NJ 07101
(201) 596-6550
Samuel C. Miller

Newark Preservation and Landmarks
Committee
P.O. Box 1066
Newark, NJ 07101
(201) 622-4910
Elizabeth Del Tufo

Newark Public Library
Aquisitions Department
P.O. Box 630
Newark, NJ 07101
(201) 733-7800

New Jersey Historical Society
230 Broadway
Newark, NJ 07104
(201) 483-3939

New Jersey Transit Corporation
1 Penn Plaza, East
Newark, NJ 07106
(201) 491-7199
David Koeig

John M. Payne
15 Washington Street
Newark, NJ 07102
(201) 648-5378

New Brunswick

Douglass College Library
Rutgers State University
New Brunswick, NJ 08903

North Caldwell

Andrew J. Flannery
29 Stepping Ridge Road
North Caldwell, NJ 07006
(201) 785-8673

Ocean Grove

Ocean Grove Camp Meeting
Association
54 Pitman Avenue
P.O. Box 126
Ocean Grove, NJ 07756
(201) 775-0035
James D. Lindemuth

Orange

Patrick Morrissy
National Housing Institute
439 Main Street
Orange, NJ 07050
(201) 678-3110

Paterson

City of Paterson
Historic Preservation Commission
65 McBride Avenue
Paterson, NJ 07501
(201) 279-5980
Rae Fronzaglia

Great Falls Development Corporation
2 Market Street
Paterson, NJ 07501
(201) 279-1270
Peggy L. Robertson

Passaic County Department of Parks
Lambert Castle
Valley Road
Paterson, NJ 07503
(201) 881-4832
Norman M. Robertson
R. J. Wright, Jr.

Passaic County Historical Society
Lambert Castle
Valley Road
Paterson, NJ 07503
(201) 881-2761
Susan Pumilia

Perth Amboy

Emil M. Kiyak
200 High Street
Perth Amboy, NJ 08861
(908) 826-8448

Piscataway

Piscataway Township
Municipal Complex
455 Hoes Lane
Piscataway, NJ 08854
(201) 562-6570
Lenore Slothower

Plainfield

Theodore J. Horger
P.O. Box 8
Plainfield, NJ 07061
(201) 953-5117

Alan R. Meltzer
Dianne L. Rosen
963 Hillside Avenue
Plainfield, NJ 07060

Plainsboro

Plainsboro Historical Society
P.O. Box 278
Plainsboro, NJ 08536-0278
(609) 799-0222
Clifford Sohl

Princeton

New Jersey Barn Company
P.O. Box 702
Princeton, NJ 08542
(609) 924-8480
Elric Endersby

Princeton Construction Group
P.O. Box 87
Princeton, NJ 08540
(609) 921-0617
Andrew Ward

Township of Princeton
Historic Preservation Officer
369 Witherspoon Street
Princeton, NJ 08540
(609) 921-1359

Rahway

Jacquelyn Zelinka
1593 Oliver Street
Rahway, NJ 07065
(908) 932-3640

Ramsey

Weber, Muth and Weber
1 Cherry Lane
P.O. Box L
Ramsey, NJ 07446
George J. Cotz

Ridgefield

Brisk Waterproofing Company
720 Grand Avenue
Ridgefield, NJ 07657
(212) 532-4430
Steve Brisk

Ridgewood

Maureen Soter
556 Grove Street
Ridgewood, NJ 07450
(201) 445-0763

Rockaway

Albert C. Levenelm, Jr.
27 Hillsborough Court
Rockaway, NJ 07866
(201) 328-6734

Salem

Preservation Salem
P.O. Box 693
Salem, NJ 08079

Stand Up For Salem
104 Market Street
Salem, NJ 08079
(609) 935-1145
Tom Purchase

Seabrook

Upper Deerfield Township Historical
Commission
P.O. Box 98
Seabrook, NJ 08302

Somers Point

Risley Homestead Committee of the
Atlantic County Historical Society
P.O. Box 301
Somers Point, NJ 08244
(609) 927-5218
Fred Ehrhardt

Somerville

Ken Bosted
101 Harlan School Road
Somerville, NJ 08876-3453
(908) 231-8874

Shive/Spinelli/Perantoni and
Associates/Architects, Planners
148 West End Avenue
P.O. Box 758
Somerville, NJ 08876
(201) 725-7800
Richard B. Shive

Somerset County Cultural and Heritage
Commission
302 Historic Courthouse
P.O. Box 3000
Somerville, NJ 08876
(201) 231-7021
Thomas R. D'Amico

Summit

Planners Diversified
382 Springfield Avenue
P.O. Box 280
Summit, NJ 07901
(201) 273-2600
Fred C. Michaeli

Tenafly

Tenafly Historic Preservation
Commission
401 Tenafly Road
Tenafly, NJ 07670
(201) 568-6100
Virginia T. Mosley

Trenton

Archaeological Society of New Jersey
1626 Riverside Drive
Trenton, NJ 08618-5837
Kurt Kalb

Capital City Redevelopment
Corporation
4 North Broad Street
4th Floor, CN 203
Trenton, NJ 08625-0203
(609) 984-5664

Edwin P. MacNicoll
1095 South Broad Street
Trenton, NJ 08611
(609) 695-7201

New Jersey Historic Trust
506-508 East State Street
2nd Floor, CN 404
Trenton, NJ 08625
(609) 984-0473
Harriette Hawkins

New Jersey Main Street Program
Office of New Jersey Heritage
506-508 East State Street
2nd Floor, CN 404
Trenton, NJ 08625
(609) 292-2023
Barbara Swanda

Office of New Jersey Heritage
506-508 East State Street
2nd Floor, CN 404
Trenton, NJ 08625
(609) 984-0543

Old Barracks Museum
Barrack Street
Trenton, NJ 08608
(609) 396-1776
Cynthia Koch

Union City

City of Union City
Office of the Mayor
3715 Palisades Avenue
Union City, NJ 07087
(201) 348-5755

Watchung

Valley Furniture Shop
Pratt Building
20 Stirling Road
Watchung, NJ 07060
(201) 756-7623
H. S. Kipe

Wayne

Museums Council of New Jersey
Ben Shahn Center
William Patterson College
Wayne, NJ 07470
Don Pettifer

Weehawken

Eric Holtermann
1 Carroll Place
Weehawken, NJ 07087
(201) 392-9389

Weehawken Public Library
Historical Room
49 Hauxhurst Avenue
Weehawken, NJ 07087
Lillie B. Stokes

West Orange

Township of West Orange
Main Street Development Corporation
66 Main Street
West Orange, NJ 07052
(201) 325-4109
Nancy Long

West Trenton

Sarah A. Hantman
481 Grand Avenue
West Trenton, NJ 08628
(609) 883-1827

Whippany

Morris County Free Library
Periodicals/Reference Department
30 East Hanover Avenue
Whippany, NJ 07981

Wildwood

City of Wildwood
4400 New Jersey Avenue
Wildwood, NJ 08260
(609) 729-1231
Peter R. Holcombe

Willingboro

Heidi Lippert
65 Woodhaven Lane
Willingboro, NJ 08046
(609) 835-1366

Woodbury

Gloucester County
Historical Society
P.O. Box 409
Woodbury, NJ 08096
(609) 845-7881
Joanne Bradley

Woodcliff Lake

Kenneth F. Cerullo
105 Kenwood Drive
Woodcliff Lake, NJ 07675
(201) 391-1407

NEW MEXICO

STATE HISTORIC PRESERVATION OFFICE

Historic Preservation Division
Office of Cultural Affairs
Villa Rivera
228 East Palace Avenue
Santa Fe, NM 87503
(505) 827-6320
(505) 827-7308 (Fax)
Thomas W. Merlan, SHPO
Lynne Sebastian, Deputy SHPO
(505) 827-6320

STATEWIDE PRESERVATION ORGANIZATION

New Mexico Preservation Coalition
801 Loma Boreal
Santa Fe, NM 87501
(505) 982-5637
David Snow, President

NATIONAL TRUST REGIONAL OFFICE

Texas/New Mexico Field Office
500 Main Street
Suite 606
Fort Worth, TX 76102
(817) 332-4398
(817) 332-4390 (Fax)
Elizabeth B. Willis, Field Office
Coordinator

NATIONAL TRUST ADVISORS

Victor W. Johnson
459 Camino de las Animas
Santa Fe, NM 87501
(505) 983-5490

Rina Swentzell
Route 10
Box 91B
Santa Fe, NM 87501
(505) 471-8021

PRESERVATION ACTION COORDINATOR

Sam Baca
New Mexico Community Foundation
227 Otero Street
Santa Fe, NM 87501
(505) 982-9521

AIA STATE PRESERVATION COORDINATOR

Steven E. Kells, AIA
Kells and Craig Architects Inc.
912 Roma, N.W.
Albuquerque, NM 87102
(505) 243-2724
Barbara L. Daniels, Deputy

NATIONAL PARK SERVICE REGIONAL OFFICE

Southwest Regional Office
P. O. Box 728
Santa Fe, NM 87504
(505) 988-6100

ADVISORY COUNCIL ON HISTORIC PRESERVATION DIVISION

Western Office of Project Review
730 Simms Street
Room 401
Golden, CO 80401
(303) 231-5320
Claudia Nissley, Director

Albuquerque

Albuquerque Community Foundation
6501 Americas Parkway, N.E.
Room 665
Albuquerque, NM 87110
(505) 883-6240
Laura H. Bass

Albuquerque Conservation Association
P.O. Box 946
Albuquerque, NM 87103
Marquerite Taylor

City of Albuquerque
Redevelopment Division
P.O. Box 1293
Albuquerque, NM 87103
(505) 768-3270
Mary Davis

Friends of the Albuquerque Petroglyphs
(FOTAP)
2920 Carlisle, N.E.
Albuquerque, NM 87110-2855
(505) 889-3779
Ike Eastvold

Paul G. McHenry, Jr., AIA
834 Griegos Road, N.W.
Albuquerque, NM 87107

Roman Catholic Archdiocese of
Santa Fe
4000 St. Joseph Place, N.W.
Albuquerque, NM 87120
(505) 831-8100
Jerome Martinez

Traditions Southwest Magazine
P.O. Box 7725
Albuquerque, NM 87194
(505) 243-7801
Michael Moquin

Chris Wilson
219 Cornell, S.E.
Albuquerque, NM 87106

Alcalde

American Studies Foundation
P.O. Box 489
Alcalde, NM 87511
(505) 852-4717
T. G. Futch

Bernalillo

Bernalillo Main Street Association
P.O. Box 638
Bernalillo, NM 87004
(505) 867-3311
Maria Rinaldi

Corrales

New Mexico Preservation Coalition
P.O. Box 2748
Corrales, NM 87048
David Snow

Deming

Wes Cook
1405 Carlsbad Street
Deming, NM 88030

Deming Luna Mimbres Museum
Luna County Historical Society
301 South Silver
P.O. Box 1617
Deming, NM 88030
(505) 546-8016
T. B. Southerland

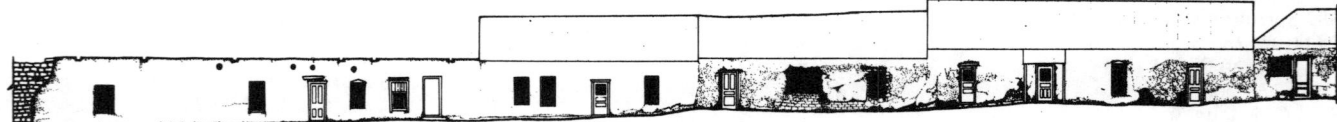

El Cerrito, Upper Pecos River Valley, New Mexico. (M. Lewis, HABS)

Deming Main Street Association
P.O. Box 706
Deming, NM 88031-0706
(505) 546-8848
Dona Irwin

Historic Landmark Commission
1405 Carlsbad Street
Deming, NM 88030
(505) 546-3044
Wesley D. Cook

Gallup

Gallup Downtown Development Group
City Hall
P.O. Box 1270
Gallup, NM 87301
(505) 863-6871
Elmo Baca

High Rolls

Peter L. Eidenbach
Tipi Alley
Box 174
High Rolls, NM 88325
(505) 585-2858

Las Vegas

Citizens Committee for Historic
Preservation
P.O. Box 707
Las Vegas, NM 87701
(505) 454-1401
Adelita M. Medina

City of Las Vegas
P.O. Box 179
Las Vegas, NM 87701
(505) 454-1401
Gwen Spatzier

Kak Slick
P.O. Box 2184
Las Vegas, NM 87701
(505) 425-7581

Lincoln

Lincoln County Heritage Trust
P.O. Box 98
Lincoln, NM 88338
(505) 653-4025
Gary Miller

Roswell

Chaves County Historical Society
200 North Lea Avenue
Roswell, NM 88201
(505) 622-8333
Peggy L. Stokes

Mainstreet Roswell
P.O. Box 1328
Roswell, NM 88202
(505) 622-6706
Vern Popp

Ruidoso

Ruidoso Main Street Program
2537 Sudderth Drive
Ruidoso, NM 88345
(505) 257-6346
Mary Maul

Santa Fe

City of Santa Fe Planning Development
P.O. Box 909
Santa Fe, NM 87901
Linda Tigges

John P. Conron
P.O. Box 935
Santa Fe, NM 87504
(505) 983-6948

Historic Preservation Division
228 East Palace Avenue
Room 101
Santa Fe, NM 87503
(505) 827-8320
Thomas W. Merlan

Historic Santa Fe Foundation
P.O. Box 2535
Santa Fe, NM 87504-2535
(505) 983-2567

New Mexico Community Foundation
227 Otero Street
Santa Fe, NM 87501
(505) 982-9521
Bruce Rolstad

Office of Economic Development and
Tourism
Main Street New Mexico
1100 St. Francis Drive
Santa Fe, NM 87503
(505) 827-0300
(800) 432-4406
Ursula M. Boatright

Silver City

Silver City Main Street Project
P.O. Box 1188
City Hall
Silver City, NM 88062
(505) 538-3731
Paul Cianon

Silver City Museum
312 West Broadway
Silver City, NM 88061
(505) 538-5921
Susan Berry

Taos

Historic Taos Inn
125 Paseo Del Pueblo Norte
Taos, NM 87571
(505) 758-2233
Paula Derevensky

Kit Carson Foundation
P.O. Drawer CCC
Taos, NM 87571
(505) 758-0505
Neil Poese

NEW YORK

STATE HISTORIC PRESERVATION OFFICE

Parks, Recreation and Historic
Preservation
Agency Building 1
Empire State Plaza
Albany, NY 12238
(518) 474-0443
Orin Lehman, Commissioner, SHPO

Julia S. Stokes, Deputy Commissioner
for Historic Preservation, Deputy
SHPO
(518) 474-0468
David Gillespie, Director
Historic Preservation Field Services
(518) 474-0479

STATEWIDE PRESERVATION ORGANIZATION

Preservation League of New York State
307 Hamilton Street
Albany, NY 12210
(518) 462-5658
Linda Gillies, Chair
Clark Strickland, President

NATIONAL TRUST REGIONAL OFFICE

Northeast Regional Office
7 Faneuil Hall Marketplace
5th Floor
Boston, MA 02109
(617) 523-0885
(617) 523-1199 (Fax)
Vicki Sandstead, Director

NATIONAL TRUST ADVISORS

Henry McCartney
28 Crosman Terrace
Rochester, NY 14620
(716) 546-7029

Anthony C. Wood
J.M. Kaplan Fund
30 Rockefeller Plaza
Suite 4250
New York, NY 10012
(212) 767-0633
(212) 767-0639 (Fax)

PRESERVATION ACTION COORDINATORS

Fred Cawley
Preservation League of New York State
307 Hamilton Street
Albany, NY 12210
(518) 462-5658

Susan Henshaw-Jones
New York Landmarks Conservancy
141 5th Avenue
New York, NY 10010
(212) 995-5260

AIA STATE PRESERVATION COORDINATOR

Giorgio Cavaglieri, FAIA
250 West 57th Street
New York, NY 10107
(212) 245-4207
William C. Shopsin, AIA, Deputy

NATIONAL PARK SERVICE REGIONAL OFFICE

North Atlantic Regional Office
15 State Street
Boston, MA 02109-3572
(617) 223-5001

ADVISORY COUNCIL ON HISTORIC PRESERVATION DIVISION

Eastern Office of Project Review
1100 Pennsylvania Avenue, N.W.
Suite 809
Washington, DC 20004
(202) 786-0503
Don L. Klima, Director

Accord

Historic Rochester Preservation
Commission
155 Upper Whitfield Road
Accord, NY 12404
Margaret E. Miller

Albany

Albany Planning Office
City Hall
4th Floor
Albany, NY 12207
(518)434-5190
Richard Nicholson

Historic Albany Foundation
44 Central Avenue
Albany, NY 12206
(518) 463-0622
Lynn Dunning-Vaughn

Mesick, Cohen, Waite
Architects
388 Broadway
Albany, NY 12207
(518) 463-2276
J. G. Waite

New York State Conference of Mayors
119 Washington Avenue
Albany, NY 12210
(518) 463-1185
Elizabeth R. White

Parks and Recreation Division for
Historic Preservation
Agency Building 1
Empire State Plaza
Albany, NY 12238
Rose Pascarell

Preservation League of New York State
307 Hamilton Street
Albany, NY 12210
(518) 462-5658
Fred Cawley

Program and Council Staff
442 Capitol Building
Albany, NY 12248

St. Joseph's Housing Corporation
317 Clinton Avenue
Albany, NY 12210
(518) 434-4934
Brian Danforth

Shaker Heritage Society
Shaker Meeting House
Albany Shaker Road
Albany, NY 12111
(518) 456-7890
Diane Conroy-LaCivita

State Commission on the Restoration
of the Capitol
P.O. Box 7016
A. E. Smith Building
Albany, NY 12225
(518) 473-0341
Andrea Lazarski

Washington Park Conservancy
165 Lancaster Street
Albany, NY 12210
(518) 449-3373
Flora Ninelles

Albertson

Shelter Rock Public Library
165 Searingtown Road
Albertson, NY 11507

Amagansett

Amagansett Historical Association
P.O. Drawer AS
Amagansett, NY 11930
(516) 267-8989
Peter Garnham

Arkville

Catskill Center for Conservation and
Development
Arkville, NY 12406
(914) 586-2611
Thomas H. Hatley

M-ARK Project
Box 1030
Arkville, NY 12406-1030
(914) 586-3500
Gail McEachern

Armonk

Martin H. Cohen
40 Green Valley Road
Armonk, NY 10504
(914) 273-8577

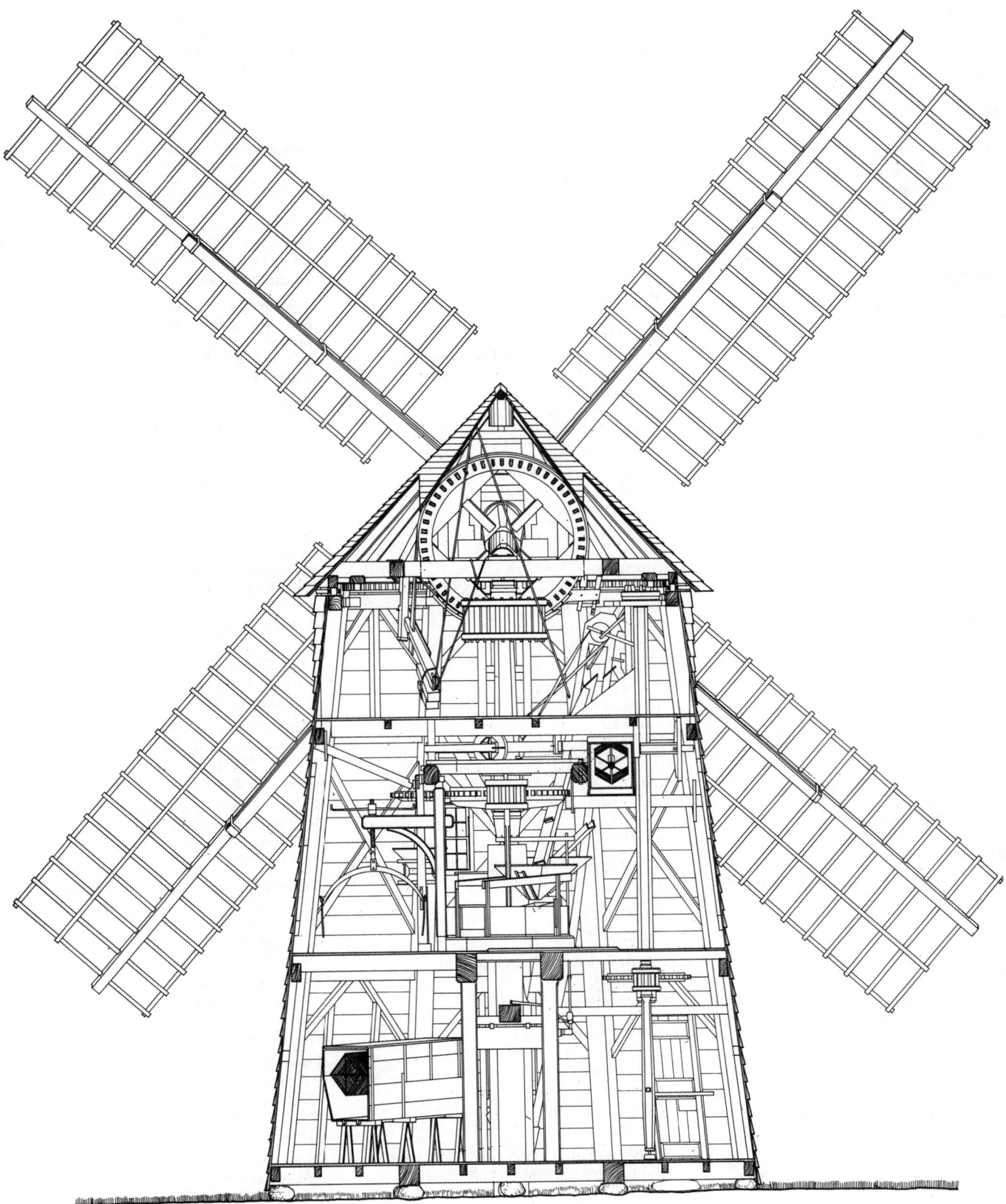

Hook Windmill, East Hampton, Long Island, New York. (K. Hoeft, HAER)

Auburn

Community Preservation Committee
P.O. Box 1021
Auburn, NY 13021
(315) 252-0339
Kathy Scollan

Avon

Village of Avon
74 Genesee Street
Avon, NY 14414
(716) 226-8118
Robert C. Hoffman

Batavia

Landmark Society of Genesee County
Box 342
Batavia, NY 14021
(716) 343-3833
Catherine Roth

Bellport

Village of Bellport
29 Bellport Lane
P.O. Box 3
Bellport, NY 11713
(516) 286-0327
Elaine Hendrie

Binghamton

City of Binghamton
Department of Planning, Housing and
Community Development
City Hall
4th Floor
Binghamton, NY 13901
(607) 772-7063
Denise M. Balkas

Preservation Association of the
Southern Tier
69 Main Street
Binghamton, NY 13903
(607) 723-4620
Richard Barons

Blauvelt

South Orangetown School District
Van Wyck Road
Blauvelt, NY 10913
Esther Korin

Bronx

City Island Historical Society
City Island Museum
190 Fordham Street
Bronx, NY 10464
(212) 885-1600
V. Gallagher

Fordham University Library
Periodicals Section
Bronx, NY 10458

Longwood Historic District
Commission Association
965 Longwood Avenue
Room 214
Bronx, NY 10459
(212) 328-4500
John W. Redic

Bronxville

Eleanor Edelman
63 Oakledge Road
Bronxville, NY 10708
(914) 337-4893

Mark S. Ingalls
9 Tanglewyde Avenue
Bronxville, NY 10708

Brooklyn

Apple Restoration and Waterproofing
132 Bedford Avenue
Brooklyn, NY 11211
(718) 326-3755
John E. Weiss

Peter Bachman
21 Montgomery Place
Brooklyn, NY 11215

Russell A. Higgins
135 Ocean Parkway
Room 17U
Brooklyn, NY 11218
(718) 633-1888

Edward Kaufman
314 State Street
Brooklyn, NY 11201
(718) 522-0126

Katherine E. Khan
18 Sidney Place
Suite 1
Brooklyn, NY 11201
(718) 875-3664

Jeffrie Lane
185 Powers Street
Suite 2
Brooklyn, NY 11211

Norman Mintz Design Association
417 6th Street
Brooklyn, NY 11215
(718) 768-8149
Norman Mintz

North Brooklyn Development
Corporation
894 Manhattan Avenue
Suite 31
Brooklyn, NY 11222
(718) 389-9044
Pearl Anish

Marjorie Pearson
850 East 23rd Street
Brooklyn, NY 11210

Pratt Institute Center for Community
and Environmental Development
379 Dekalb Street
Brooklyn, NY 11205

Prospect Park Alliance
95 Prospect Park West
Brooklyn, NY 11215
(718) 768-0227
Pam Fishman

Theodore Prudon
140 8th Avenue
Brooklyn, NY 11215
(212) 977-9696

St. Ann Center for Restoration and the
Arts
157 Montague Street
Brooklyn, NY 11201
(718) 834-8794
Susan Feldman

S. Axman and Son
1421 Utica Avenue
Brooklyn, NY 11203
(718) 629-0775
Marc Axman

Society for the Preservation of
Weeksville and Bedford Stuyvesant
History
1698 Bergen Street
Brooklyn, NY 11213
Joan Maynard

Stephen Tsou
155 Henry Street
Suite 66
Brooklyn, NY 11201
(718) 625-0684

Buffalo

Joan Bozer
27 St. Catherines Court
Buffalo, NY 14222
(716) 881-1639

Clinton E. Brown, AIA
25 East Huron Street
Buffalo, NY 14203
(716) 852-2020

Buffalo Erie Public Library
History Department
Lafayette Square
Buffalo, NY 14203

Buffalo Friends of Olmsted Parks
P.O. Box 590
Buffalo, NY 14205
(716) 649-5476
Susan West

C. McLean Faust
773 Potomac Avenue
Buffalo, NY 14209
(617) 723-6453

Preservation Coalition of Erie County
P.O. Box 768
Buffalo, NY 14213
(716) 837-8858
Scott Field

Theatre District Association/BPI
671 Main Street
Buffalo, NY 14203
(716) 856-3150
Anne E. Conable

Canandaigua

Canandaigua Lake Watershed Task
Force
482 North Main Street
Canandaigua, NY 14424
(716) 394-1341
Stephen Lewandowski

City of Canandaigua
2 North Main Street
Canandaigua, NY 14424
(716) 394-7850
Kay W. James

Granger Homestead Society
395 North Main Street
Canandaigua, NY 14424
(716) 394-1472
Patricia Boland

Sonnenberg Gardens
P.O. Box 663
Canandaigua, NY 14424
(716) 394-4922
Mary A. Bell

Cazenovia

Cazenovia Preservation Foundation
Box 627
Cazenovia, NY 13035-0627
(315) 655-3150
Donald Schwinn

Centerport

Vanderbilt Museum
180 Little Neck Road
Centerport, NY 11721
(516) 261-5656
Thomas A. Rhodes

Cheektowaga

Friends of the Canadiana
24 Honduras Lane
Cheektowaga, NY 14225
(716) 684-8012
Allen Besser

Chittenango

Village of Chittenango
222 Genesee Street
Chittenango, NY 13037
(315) 687-6369
Richard F. Sullivan

Cold Spring Harbor

Cold Spring Harbor Library
75 Goose Hill Road
Cold Spring Harbor, NY 11724
Elizabeth L. Watson
Box 100
Cold Spring Harbor, NY 11724
(516) 692-8878

Cooperstown

M B Group
P.O. Box 31
Cooperstown, NY 13326
(607) 547-4051
James B. Bernegger

Corning

Market Street Restoration Corporation
5 East Market Street
Corning, NY 14830
Elise Johnson-Schmidt

Cortland

1890 House Museum and Center for the
Victorian Arts
37 Tompkins Street
Cortland, NY 13045

Cutchogue

Old House Society
P.O. Box 361
Cutchogue, NY 11935
(516) 734-5989
William Peters

Dobbs Ferry

Dobbs Ferry Historical Society
153 Main Street
Dobbs Ferry, NY 10522
(914) 693-7766
Tema Harnik

East Aurora

Roycroft Revitalization Corporation
571 Main Street
East Aurora, NY 14052
(716) 652-6000
Jerry C. Hiller

East Hampton

Robert Hefner
53 Spring Close Highway
East Hampton, NY 11937
(516) 324-0393

Eden

Eden Historic Review Board
Box 156
2795 East Church Street
Eden, NY 14057
(716) 992-9792
Norma Webb

Elmira

Chemung County Historic Society
415 East Water Street
Elmira, NY 14901
(607) 734-4167
Constance Barone

Near Westside Neighborhood
Association
375 West Church Street
Elmira, NY 14901
(607) 733-4924
James E. Fletcher

Endwell

Monday Afternoon Club Endowment
Fund
3718 Kirk Road
Endwell, NY 13760
(607) 757-9289
Pat Weber

Essex

Essex Community Heritage
Organization
P.O. Box 260
Essex, NY 12936
(518) 963-7088
Robert J. Hammerslag

Farmingdale

Village of Farmingdale
150th Anniversary Committee
17 Paine Avenue
Farmingdale, NY 11735
(516) 845-1225
Daniel Looney

Farmingville

R. C. Chereskin Enterprises
Chereskin Restorations (tm)
80 Morris Avenue
Farmingville, NY 11738
(516) 736-6376
R. Chereskin

Flushing

Bowne House Historical Society
37-01 Bowne Street
Flushing, NY 11354
(718) 359-0528
Donna H. Russo

Forest Hills

Forest Hills Garden Corporation
2 Tennis Place
Forest Hills, NY 11375
(718) 268-2420
Susan Purcell

Fredonia

Fredonia Preservation Society
Box 422
Fredonia, NY 14063-0422
(716) 673-1011
James Boltz

Garden City

Charles Minicus
125 Clinch Avenue
Garden City, NY 11530
(516) 326-2359

Garrison-on-Hudson

Boscobel Restoration
Route 9D
Garrison-on-Hudson, NY 10521
F. W. Stanyer

Geneva

Geneva Business Improvement District
64 Seneca Street
Geneva, NY 14456
(315) 789-3838
Charles Bauder

David A. Herd
124 North Main Street
Geneva, NY 14456

Smith Opera House
82 Seneca Street
P.O. Box 58
Geneva, NY 14456
(315) 789-2221
Richard Erwin

Germantown

Friends of the Palatine Parsonage
Rural Route 2
Box 414
Germantown, NY 12526
(518) 537-4727
Alyce Cresap

Glens Falls

Champman Historical Museum
348 Glen Street
Glens Falls, NY 12801
(518) 793-2826
Kathy Allen

Gloversville

Gloversville Community Development
City Hall
Frontage Road
Gloversville, NY 12078
Dolores M. Tile

Great Neck

Village of Great Neck
Plaza Centre
P.O. Box 440
Great Neck, NY 11022
(516) 482-4500
Allan Gussack

Hamlin

Barn Preservation Society
Sandy Creek Farm
2675 Church Road
Hamlin, NY 14464
(716) 964-3754
James J. Hamme

Hempstead

Nassau County Housing
250 Fulton Avenue
Hempstead, NY 11550
Raymond H. Malone

Hudson Falls

James R. Cronkhite
12 Hudson Place
Hudson Falls, NY 12839
(518) 747-5634

Huntington

Lloyd Harbor Historical Society
41 Lloyd Harbor Road
Huntington, NY 11743

Ilion

Wrenovations
81 South 3rd Avenue
Ilion, NY 13357
(315) 894-3453
Kevin Wren

Irvington

Eric F. Hermann
Mountian Road
Irvington, NY 10533
(914) 332-1817

Ithaca

City of Ithaca
108 East Green Street
Ithaca, NY 14850
(607) 272-1713
Leslie Chatterton

Cornell University
Campus Planning Office
102 Humphreys Service Building
Ithaca, NY 14853
(607) 255-1245
Nancy H. Goody

Historic Ithaca
120 North Cayuga Street
Ithaca, NY 14850
(607) 273-6633

National Council For Preservation
Education
210 West Sibley
Cornell University
Ithaca, NY 14853
(607) 255-7261
M. A. Tomlan

Whole Duck Catalogue
141 The Commons
Ithaca, NY 14850
(607) 272-6341
Albert Fortner

Jamaica

Greater Jamaica Development
Corporation
90-04 161st Street
Jamaica, NY 11432
F. C. Towery

Jamestown

Hope's Landmark Products
95-99 Blackstone Avenue
Jamestown, NY 14701
(716) 665-6223
Mike Emules

Jamesville

Nottingham
40 Cottage Drive
Jamesville, NY 13078
Ross E. Dickinson

Katonah

Friends of John Jay Homestead
P.O. Box AH
Katonah, NY 10536
(914) 232-5651
Linda Connelly

Kingston

City of Kingston
1 Garraghan Drive
Kingston, NY 12401

Larchmont

Larchmont Public Library
121 Larchmont Avenue
Larchmont, NY 10538

Sabet Associates Architects
65 Kane Avenue
Larchmont, NY 10538
(914) 834-5411
Susan Sabet

Leroy

Leroy House
P.O. Box 176
23 East Main Street
Leroy, NY 14482
(716) 768-7433
Lynne Belluscio

Lima

Lima Historical Society
1850 Rochester Street
Box 532
Lima, NY 14485
(716) 582-2608
Richard Kennett

Long Island City

Arrow Restoration
37-15 Hunters Point Avenue
Long Island City, NY 11101
(718) 729-0411
Alan Schinderman

Mahopac Falls

Putnam County
Putnam County Office Facility
Myrtle Avenue
Mahopac Falls, NY 10542-0368
(914) 621-2302
Sallie Sypher

Maine

Bowers Foundation
Nanticoke Road
Box 23
Maine, NY 13802
(607) 862-3243
Lawrence Bothwell

Manhasset

North Hempstead Planning
Department
Town Hall
Manhasset, NY 11030
(516) 627-0590
Rhoda Becker

Melville

Susan T. Currie
P.O. Box 962
Melville, NY 11747

Half Hollow Hills Community Library
510 Sweet Hollow Road
Melville, NY 11747

Middletown

City of Middletown
City Hall
16 James Street
Middletown, NY 10940
(914) 343-3626
James Kelly

Millbrook

Dutchess County Cooperative
Extension
P.O. Box 259
Millbrook, NY 12545
(914) 677-3488
Judy L. Schneyer

Bernard H. LaLone, Jr.
P.O. Box 94
Millbrook, NY 12545
(914) 677-8440

Trout Restoration
Rural Route 1
Box 167D
Millbrook, NY 12545
(917) 677-8383
David H. Bova

New City

Historic Society of Rockland County
20 Zukor Road
New City, NY 10956
(914) 634-9629
Debra Clyde

West Branch Conservation Association
443 Buena Vista Road
New City, NY 10956
(914) 634-2327

New Paltz

Huguenot Historical Society of New
Paltz
P.O. Box 339
New Paltz, NY 12561
(914) 255-1660
Kenneth Hasbrouck

New Rochelle

Lithos International
45 Oxford Road
New Rochelle, NY 10804
(212) 490-2700

New York

Abigail Adams Smith Museum
421 East 61st Street
New York, NY 10021
(212) 838-6878
Ralph Sessions

Joseph Gabriele
516 East 78th Street
Apartment 6I
New York, NY 10021
(212) 570-9666

American Council for the Arts
1285 Avenue of the Americas
3rd Floor
New York, NY 10019

American Society of Mechanical
Engineers
345 East 47th Street
New York, NY 10017

Kurt Andersen
1271 Avenue of Americas
Room 23-30B
New York, NY 10020-1351

Antique Furniture Workroom
225 East 24th Street
5th Floor
New York, NY 10010
(212) 683-0551
Bill Olsen

Architecture and Furniture
157 Chambers Street
New York, NY 10007
(212) 619-5944
John Petrarca

Bantam Books
666 5th Avenue
New York, NY 10103
Coleen O'Shea

Bernheimer
1240 Park Avenue
Suite 1D
New York, NY 10128

Beyer Blinder Belle
41 East 11th Street
New York, NY 10003
(212) 777-7800
John Belle

Buckhurst Fish Hutton Katz
72 5th Avenue
6th Floor
New York, NY 10011
(212) 620-0050
Rachel Belsky

Nancy N. Campbell
1088 Park Avenue
New York, NY 10128
(203) 346-1160

Celebrations International
172 West 79th Street
New York, NY 10024
(212) 669-9400
Brooks Lappin

Central Park Administrative Office
830 5th Avenue
New York, NY 10021
(212) 860-1800
Timothy Marshall

Church of St. Mary the Virgin
145 West 46th Street
New York, NY 10036
(212) 869-5830
William R. Anderson

City of New York
Division of Real Estate
2 Lafayette Street
Room 2000
New York, NY 10007
T. Bolenbaugh

Cooper Union
School of Architecture
Cooper Square
New York, NY 10003
John Hejduk

David Derbyshire
375 South End Avenue
Apartment 24H
New York, NY 10280
(212) 807-7727

Dumond Chemical
1501 Broadway
New York, NY 10036
(212) 869-6350
H. Dubin

Evergreen Painting Studios
635 West 23rd Street
4th Floor
New York, NY 10011
(212) 727-9500
Jeff Greene

Fraunces Tavern Museum
54 Pearl Street
New York, NY 10004
(212) 425-1778
William Ayres

Jerrold F. Fuchs
220 East 78th Street
New York, NY 10021
(212) 628-6175

Garden Club of America
598 Madison Avenue
New York, NY 10022

Ginter-Gothan Urban History
Ruxton Towers
50 West 72nd Street
Suite 312
New York, NY 10023
(212) 496-6859
Valerian A. Ginter

Girl Scouts of the USA
National Historic Preservation Center
830 3rd Avenue
7th floor
New York, NY 10022
(212) 940-7662
Mary Levey

Henry Street Settlement
265 Henry Street
New York, NY 10002
(212) 766-9200
Catherine Cullen

Josephine E. Jones
137 West 122nd Street
New York, NY 10027
(212) 666-7816

Joseph R. Loring and Associates
1 Pennsylvania Plaza
New York, NY 10119
Ira Guterman

Mitchell Kurtz
Architect
611 Broadway
Room 542
New York, NY 10012
(212) 598-4367

KWH Associates
280 Rector Place
Suite 6F
New York, NY 10280
(212) 308-7505
Kathryn W. Howe

Mark A. Levine
Sive, Paget and Riesel
460 Park Avenue
New York, NY 10022
(212) 421-2150

Li/Saltzman Architects
375 West Broadway
3rd Floor
New York, NY 10012
(212) 941-1838
(212) 941-1834
Judith Saltzman

Lower East Side Tenement Museum
97 Orchard Street
New York, NY 10002
(212) 431-0233
Ruth J. Abram

Lionel Martinez
39 Worth Street
New York, NY 10013
(212) 966-0093

Felicia Mayro
223 West 78th Street
Suite 1R
New York, NY 10024
(212) 721-7833

Robert E. Meadows
Architects
40 Dover Street
New York, NY 10038
(212) 962-2645

Metropolitan Historic Structures
Association
1 World Trade Center, Room 2611
New York, NY 10048
(212) 432-5450
Elise M. Quasebarth

A. G. Neumann
Architect
149 5th Avenue
8th Floor
New York, NY 10010
(212) 255-6739

New York City Landmarks Preservation
Commission
225 Broadway
New York, NY 10007
(212) 553-1100
Donald A. Plotts

New York Department of Parks and
Recreation
830 5th Avenue
Room 203
New York, NY 10021
(212) 360-8123
Mary E. Hern

New York Landmarks Conservancy
141 5th Avenue
New York, NY 10010
(212) 995-5260
Susan H. Jones

New York Public Library
Division G
P.O. Box 2237
Grand Central Station
New York, NY 10017

New York Studio School
8 West 8th Street
New York, NY 10011
(212) 777-0742
D. Morgan

New York Times Magazine
Home Section
229 West 43rd Street
New York, NY 10036-3913
Stephen Drucker

Old Merchants House
29 East 4th Street
New York, NY 10003
(212) 473-5516
Margaret Gardiner

Parsons School of Design
Gimbel Library-Periodicals
2 West 13th Street
New York, NY 10011

Robert B. Pauls
Real Estate Consultants
29 West 17th Street
New York, NY 10011
(212) 255-0264

Rand Engineering
159 West 25th Street
New York, NY 10001
(212) 765-8844
Joshua Rubin

RESTORE
41 East 11th Street
New York, NY 10003
(212) 766-0120
Jan C. Anderson

St. Thomas Church
1 West 53rd Street
New York, NY 10019
(212) 757-7013
James E. Marlow

Maurice Schickler
Facade Consultants
318 East 70th Street
Suite 2RW
New York, NY 10021
(212) 288-5700

Sculpture and Arts Research
172 Norfolk Street
New York, NY 10002
(212) 255-7749
Al Orensenz

Sierra Club
New York City Group
310 West 18th Street
Suite 1B
New York, NY 10011
(212) 243-2319
R. Eadie

SITE Projects
65 Bleecher Street
New York, NY 10012
Michelle Stone

Sharyn Ann Sohlberg
5 Tudor City Place
Suite 921
New York, NY 10017
(212) 664-5277

Sons of the Revolution in the State of
New York
54 Pearl Street
Fraunces Tavern Museum
New York, NY 10004
(212) 425-4776
Patricia Kesling

South Street Seaport Corporation
207 Front Street
New York, NY 10038
(212) 669-9400
John M. Pollack

John M. Szabol
24 West 68th Street
New York, NY 10023
(212) 496-0030

Theatres of New York
4 West 109th Street
Suite 6B
New York, NY 10025
(212) 865-6949
Jane Preddy

Merin Elizabeth Urban
45 Christopher Street
Suite 11E
New York, NY 10014

Gina L. Walker
122 East 42nd Street
Suite 1700
New York, NY 10017
(212) 984-0705

E.L. Weiss Associates
211 East 18th Street
New York, NY 10003
Elaine L. Weiss

Newburgh

Historical Society of Newburgh Bay and
the Highlands
Crawford House
189 Montgomery Street
Newburgh, NY 12550
(914) 561-2586
Elizabeth Bull

Newburgh Free Library
124 Grand Street
Newburgh, NY 12550
(914) 561-1836

Niagara Falls

City Hall of Niagara Falls
745 Main Street
Niagara Falls, NY 14302
(716) 627-8820
William K. Clark

Northport

Architectural and Historic Board of
Review
P.O. Box 358
Northport, NY 11768
(516) 261-7502
Leonard Poveromo

North Salem

North Salem Historical Society
P.O. Box 1731
North Salem, NY 10560
(914) 669-5021
Thomas L. Purdy

North Tonawanda

Carousel Society of the Niagara
Frontier
P.O. Box 672
180 Thompson Street
North Tonawanda, NY 14120
(716) 693-1885
Don C. Traub

Nyack

Tappan Zee Playhouse Preservation
Association
P.O. Box 229
Nyack, NY 10960

Orchard Park

Western New York Railway Historical
Society
P.O. Box 172
Orchard Park, NY 14057
(716) 992-3425
Joe Rafter

Orient

Oysterponds Historical Society
Village Lane
P.O. Box 844
Orient, NY 11957
(516) 323-2480
Barbara Fertig

Ossining

Clemco Construction
15 Grace Lane
Ossining, NY 10562
(914) 941-0686
Thomas Clemmens

Ossining Historical Society Museum
196 Croton Avenue
Ossining, NY 10562
(914) 941-0001
Roberta Y. Arminio

Tall and Kamen Architects
4 Liberty Street
Ossining, NY 10562
(914) 762-8911
Joanne Tall

Westchester Preservation League
36 South Highland Avenue
Ossining, NY 10562
(914) 941-2750
Robin Imhoff

Oswego

The Heritage Foundation of Oswego
53 West Bridge Street
P.O. Box 405
Oswego, NY 13126
(315) 342-3354
Sally A. Sappey

Oyster Bay

Lorraine Gilligan
P.O. Box 58
Oyster Bay, NY 11771-0058
(516) 922-0479

Raynham Hall Museum
20 West Main Street
Oyster Bay, NY 11771
(516) 922-6808
Stuart A. Chase

Patterson

Putnam Preservation League
Box 534
Patterson, NY 12563
(914) 878-6169

Pelham

Jacqueline V. Prior
4 Broadside Avenue
Pelham, NY 10803
(914) 738-6973

Pittsford

Historic Pittsford
18 Monroe Avenue
Pittsford, NY 14534
(716) 381-3799

Port Jefferson

Kitt Barrett
P.O. Box 424
Port Jefferson, NY 11777
(516) 751-9636

Susan Klaffky
7 Chips Court
Port Jefferson, NY 11777
(516) 331-1649

Port Jervis

Port Jervis Development Corporation
14-18 Hammond Street
P.O. Box 3105
Port Jervis, NY 12771
(914) 856-6961
Sally T. Martinez

Port Washington

Nassau County Division of Museum
Services
Sands Point Preserve Library
95 Middleneck Road
Port Washington, NY 11050
(516) 364-1050

Poughkeepsie

City of Poughkeepsie Partnership
12 Garden Street
Poughkeepsie, NY 12601
(914) 471-9424
Kathryn LaVanche

Dutchess County Landmarks
Association
P.O. Box 944
Poughkeepsie, NY 12602
(914) 229-9894
Stephanie W. Mauri

Springside Landscape Restoration
P.O. Box 4915
Poughkeepsie, NY 12602
(914) 473-5583
Richard C. Mason

Young Morse Historic Site
P.O. Box 1649
Poughkeepsie, NY 12601
(914) 454-4500
Timothy J. Countryman

Rensselaer

Dutch Barn Preservation Society
P.O. Box 176
Rensselaer, NY 12144
(914) 432-5077
Mike Bathrick

Rhinebeck

Hudson River Heritage
Box 287
Rhinebeck, NY 12572

Wilderstein Preservation
P.O. Box 383
Rhinebeck, NY 12572
(914) 889-8265
Raymond Armater

Richmond Hill

Richmond Hill Development
Corporation
111-27 Jamaica Avenue
Richmond Hill, NY 11418
(718) 805-0533
Kate Carpenter

Ridgewood

Greater Ridgewood Historical Society
1820 Flushing Avenue
Ridgewood, NY 11385
(718) 456-1776
George Miller

Riverhead

Hallockville
163 Sound Avenue
Riverhead, NY 11901
(516) 298-5292
Courtney Burns

Rochester

Hugh Anderson
LBJ-2765
1 Lomb Memorial Drive
Rochester, NY 14623-0887
(716) 475-6428

Bureau of Neighborhood Development
30 Church Street
Rochester, NY 14614
(716) 428-6886
Donald Fuller

Charlotte Genesee Lighthouse
Historical Society
70 Lighthouse Street
Rochester, NY 14612
(716) 621-6179

City of Rochester
30 Church Street
Rochester, NY 14614
(716) 428-7053
John Spoelhof

Hafner Associates
9 South Goodman Street
Rochester, NY 14607
Beverly Hafner

Landmark Society of Western New York
133 South Fitzhugh Street
Rochester, NY 14608-2204
(716) 546-7029
Henry McCartney

Lawson and Pulver Architects
251 Park Avenue
Rochester, NY 14607
(716) 442-0396
William E. Pulver

Valerie O'Hara Murray
180 St. Paul Street
Rochester, NY 14604
(716) 546-7570

Gwendolyn K. Ream
223 B City Hall
30 Church Street
Rochester, NY 14614
(716) 428-6502

St. Stanislaus Kostka Parish
34 St. Stanislaus Street
Rochester, NY 14621
(716) 262-8373
Kathleen Urbanic

Thousand Island Park Landmarks
Society
745 Harvard Street
Rochester, NY 14610
Truda Fitelson

Rockville Centre

Deborah Conrad
655 Lakeview Avenue
Rockville Centre, NY 11570

Rome

Neighborhood Improvement Program
City Hall
Rome, NY 13440
Thomas H. Larrabee

Roslyn

Robert A. Hansen
110 Main Street
Roslyn, NY 11576
(212) 747-7788

Roslyn Landmark Society
1 Papermill Road
Roslyn, NY 11576
(516) 621-2961
Jean Chapman

Trinity Church
1759 Northern Boulevard
Roslyn, NY 11576
(516) 759-9443
Jean Henning

Rye

Paul Benowitz
Architect
29 Ellsworth Street
Rye, NY 10580
(914) 967-0557

Jay Heritage Center
P.O. Box 661
Rye, NY 10580
(9140698-8118
Rhonda Kornreich

Rye Historical Society
1 Purchase Street
Rye, NY 10580
(914) 967-7588
Susan A. Morison

Rye Brook

Ronald J. Pennella
77 Tamarack Road
Rye Brook, NY 10573
(914) 939-0823

Sackets Harbor

Sackets Harbor Historical Society
P.O. Box 78
Sackets Harbor, NY 13685
(516) 646-2459
John Burdick

Sag Harbor

David H. Cory
19 Cove Drive
Bay Point
Sag Harbor, NY 11963
(516) 725-4118

Eastville Community Historical
Society
P.O. Box 2036
Sag Harbor, NY 11963
Kathleen Tucker

Village of Sag Harbor
Main Street
P.O. Box 660
Sag Harbor, NY 11963
(516) 725-0222
Joan Feehan

Sands Point

Village of Sands Point
P.O. Box 188
Sands Point, NY 11050
(516) 883-3044
Michael Puntillo

Saranac Lake

Historic Saranac Lake Preservation
Enterprises
P.O. Box 1030
Saranac Lake, NY 12983
(518) 891-0971

Saratoga Springs

New York State Parks, Recreation and
Historic Preservation
P.O. Box W
Saratoga Springs, NY 12866
Cheryl Gold

Saratoga Associates
443 Broadway
Saratoga Springs, NY 12866
(518) 587-2550
Don Minnery

Saratoga Springs Preservation
Foundation
P.O. Box 442
Saratoga Springs, NY 12866
(518) 587-5030
Carol W. Shepard

Julia S. Stokes
429 Maple Avenue
Saratoga Springs, NY 12866
(518) 474-0468

Saugerties

Saugerties Lighthouse Conservation
P.O. Box 654
Saugerties, NY 12477
(914) 246-9170
Cliff Stren

Schaghticoke

Knickerbocker Historical Society
P.O. Box 29
Schaghticoke, NY 12154
(518) 753-7846
Hollie McNeil

Schenectady

Lamont House
Union College
Schenectady, NY 12308
(518) 370-6169
Patricia Rush

Seneca Falls

Village of Seneca Falls
Planning Department
60 State Street
P.O. Box 10
Seneca Falls, NY 13148
(315) 568-6894
Francis Caraccilo

Setauket

Society for the Preservation of Long
Island Antiquities
93 North County Road
Setauket, NY 11733
(516) 941-9444
Robert B. MacKay

Smithtown

B.O.C.E.S. III
Marion Carll Farm
P.O. Box 604
Smithtown, NY 11787
(516) 360-3652
Edward J. Zero

Smithtown Historical Society
P.O. Box 69
Smithtown, NY 11787
(516) 265-6768
Louise P. Hall

Southampton

Southampton Colonial Society Thomas
Halsey House
Box 303
Southampton, NY 11968
Robert Keene

Staten Island

Preservation League of Staten Island
P.O. Box 71
Staten Island, NY 10301
(718) 447-2036
Mitchell A. Grubler

Staten Island Historical Society
441 Clarke Avenue
Staten Island, NY 10306
Barnett Shepherd

Stony Brook

Stony Brook Community Fund
P.O. Box 572
Stony Brook, NY 11790
(516) 751-2244
G. Rocchio

Suffern

Christopher E. Ruehl
112 Haverstraw Road
Suffern, NY 10901

Syosset

Friends for Long Island's Heritage
1864 Muttontown Road
Syosset, NY 11791
(516) 364-1050
Gerald S. Kessler

Syracuse

Heritage Coalition
P.O. Box 6233
Teall Avenue Station
Syracuse, NY 13217
(315) 471-2162
Charles Bartlett

Quinlivah Pierik and Krause
Architects/Engineers
101 East Water Street
Syracuse, NY 13201
(315) 472-7806
Alfred F. Krause

Regional Conference of Historical
Agencies
1400 North State Street
Syracuse, NY 13208
(315) 475-1525
Susan Edwards Harvitn

Carl D. Stearns
212 Scottholm Boulevard
Syracuse, NY 13224
(315) 471-2162

Syracuse Landmark Preservation Board
Hills Building, 4th Floor
Syracuse, NY 13202
(315) 473-2873
Joanne B. Arany

Tappantown

Tappantown Historical Society
Box 71
Tappan, NY 10983

Tarrytown

Historic Hudson Valley Library
150 White Plains Road
Tarrytown, NY 10591
(914) 631-8200
Michael G. Carew

William F. Plunkett
45 Benedict Avenue
Tarrytown, NY 10591
(212) 575-7772

Thendara

Allyn Gardner, A.S.I.D.
12 Birch Street
Thendara, NY 13472
(315) 369-3872

Ticonderoga

Pride of Ticonderoga
A Rural Preservation Company
Cummins House
146 Montcalm Street
Ticonderoga, NY 12883
(518) 585-6366
Susan D. Rathbun

Troy

Holmes and Watson
450 Broadway
Troy, NY 12180
(518) 273-8526
Richard G. Knight

Rensselaer County Historical Society
59 2nd Street
Troy, NY 12180
(518) 272-7232
Breffny A. Walsh

Utica

Landmarks of Greater Utica
212 Rutger Street
Utica, NY 13501
(315) 732-7376
Barton Rasmus

Vails Gate

National Temple Hill Associates
Box 315
Vails Gate, NY 12584
(914) 562-6397

Wappingers Falls

Zion Episcopal Church
Satterlee Place
Wappingers Falls, NY 12590
(914) 297-3142
Diane Levitt

Wassaic

Harlem Valley Contractor
Rural Route 1
Box 123A
Wassaic, NY 12592-9801
(914) 855-9626
Stephen Sterra

Water Mill

Water Mill Museum
Old Mill Road
P.O. Box 63
Water Mill, NY 11976

Watertown

City Center Development
230 Franklin Street
Watertown, NY 13601
Bonnie Wilkinson

North Country Affordable
Development Company
120 Washington Street
Watertown, NY 13601
(315) 785-8684
Barbara H. Willis

White Plains

International Percy Grainger Society
7 Cromwell Place
White Plains, NY 10601
(914) 948-1394
Rolf K. Stang

Planned Parenthood of
Westchester/Rockland
175 Tarrytown Road
White Plains, NY 10607-1607
(914) 428-7876
Francine S. Stein

Westchester County Housing and
Community Development
418 Michaelian Office Building
148 Martine Avenue
Room 416
White Plains, NY 10601
Cynthia L. Russell

Woodhaven

Ernest J. Naples
87-83 97 Street
Woodhaven, NY 11421

Yonkers

City of Yonkers
Department of Parks, Recreation and
Conservation
Yonkers, NY 10701
(914) 964-3504
S. A. Desantis

Joseph Kozlowski
54 Hillbright Trail
Yonkers, NY 10703

Yonkers Planning Bureau
87 Nepperhan Avenue
Room 311
Yonkers, NY 10701
(914) 964-3386

NORTH CAROLINA

STATE HISTORIC PRESERVATION OFFICE

Division of Archives and History
Department of Cultural Resources
109 East Jones Street
Raleigh, NC 27601-2807
(919) 733-7305
(919) 733-5679 (Fax)
William S. Price, Jr., Director, SHPO
David Brook, Administrator
Archaeology and Historic Preservation
Section
Deputy SHPO
(919) 733-4763

STATEWIDE PRESERVATION ORGANIZATION

Preservation North Carolina
P.O. Box 27644
Raleigh, NC 27611
(919) 832-3652
(919) 832-1651
R. Beverly R. Webb, President
Myrick Howard, Executive Director

NATIONAL TRUST REGIONAL OFFICE

Southern Regional Office
456 King Street
Charleston, SC 29403
(803) 722-8552
(803) 722-8652 (Fax)
Susan A. Kidd, Director

NATIONAL TRUST ADVISORS

Robert C. Allen
Duke Power Company
P.O. Box 1006
Charlotte, NC 28201-1006
(704) 373-3224

Linda Edmisten
212 Lake Wheeler Road
Raleigh, NC 27603
(919) 821-9175

PRESERVATION ACTION COORDINATOR

J. Myrick Howard
Historic Preservation Foundation of
North Carolina
P.O. Box 27644
Raleigh, NC 27611-7644
(919) 832-3652

Juli Aulik
5F River Birch Road
Durham, NC 27705
(919) 383-5227

AIA STATE PRESERVATION COORDINATOR

Vacant

NATIONAL PARK SERVICE REGIONAL OFFICE

Southeast Regional Office
Richard B. Russell Federal Building
75 Spring Street, S.W.
Atlanta, GA 30303
(404) 331-2632

ADVISORY COUNCIL ON HISTORIC PRESERVATION DIVISION

Eastern Office of Project Review
1100 Pennsylvania Avenue, N.W.
Suite 809
Washington, DC 20004
(202) 786-0503
Don L. Klima, Director

Albemarle

Albemarle Downtown Development
Corporation
P.O. Box 190
Albemarle, NC 28002
(704) 982-0131
Vicki Coggins

Albemarle-Stanly County Historic
Preservation Commission
112 North 3rd Street
Albemarle, NC 28001
(704) 983-7316
Penny King

Asheville

Albemarle Park-Manor Grounds
Association
P.O. Box 2231
Asheville, NC 28802
(704) 258-1283
Richard Mathews

Asheville Downtown Development
29 Haywood Street
Asheville, NC 28801
(704) 251-9973
Leslie Anderson

North Carolina Division of Archives
and History
Western Office
13 Veterans Drive
Asheville, NC 28805
(704) 298-5024
Martha Fullington

Pack Memorial Public Library
67 Haywood Street
Asheville, NC 28801

Preservation Society of Asheville and
Buncombe
P.O. Box 2806
Asheville, NC 28802
(704) 254-2343
Jeanne Warner

Jim E. Samsel, AIA
60 Biltmore Avenue
Asheville, NC 28801
(704) 253-1124
YMI Cultural Center
P.O. Box 7301
Asheville, NC 28807
(704) 252-4614
Wanda H. Coleman

Beaufort

Friends of the Museum
315 Front Street
Beaufort, NC 28516

Bethania

Bethania Historical Association
P.O. Box 131
Bethania, NC 27010

Brevard

Heart of Brevard
P.O. Box 27
24 Jordan Street Center
Brevard, NC 28712
(704) 884-3278
James E. Rowe

Transylvania County Historical Society
P.O. Box 2061
Brevard, NC 28712
(704) 884-5137
Robert Gash

Transylvania County Historic
Properties Commission
28 East Main Street
Brevard, NC 28712
(704) 883-2138
Rowell Bosse

Burlington

Burlington Historic Resources
Commission
P.O. Box 1358

Canton

Canton Downtown Association
P.O. Box 136
Canton, NC 28716
(704) 648-2573
Jacquelyn B. Patrick

Carrboro

Town of Carrboro
301 West Main
P.O. Box 337
Carrboro, NC 27510
(919) 968-7706
Robert W. Morgan

Catawba

Town of Catawba Historical
Association
P.O. Box 147
Catawba, NC 28609

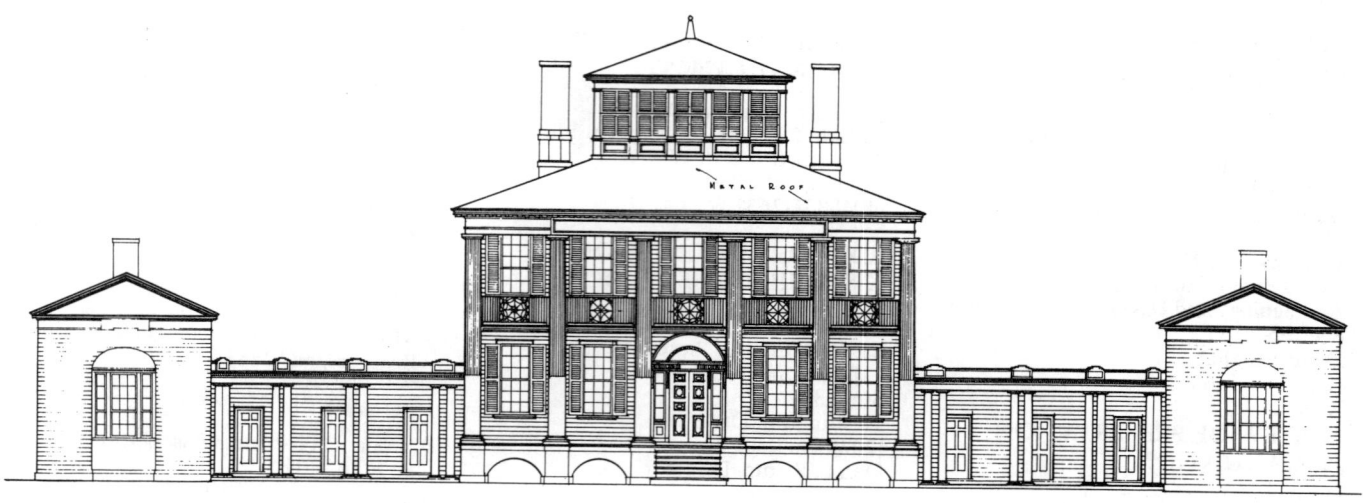

Hayes Manor, Edenton, North Carolina. (F. Nichols, T. Herman, HABS)

Chapel Hill

Chapel Hill Preservation Society
610 East Rosemary Street
Chapel Hill, NC 27514
(919) 942-7818

Diane E. Lea
2154 Lakeshore Court
Chapel Hill, NC 27514
(919) 544-6880

Robert E. Stipe
100 Pine Lane
Chapel Hill, NC 27514
(919) 967-2448

Charlotte

Jack O. Boyte
Box 6093
Charlotte, NC 28207
(709) 332-7819

Charlotte Historic District
Commission
600 East 4th Street
Charlotte, NC 28202-2853
(704) 336-2302
John Rogers

Charlotte-Mecklenburg Historic
Landmarks Commission
1225 South Caldwell Street
Box D
Charlotte, NC 28203
(704) 376-9115
Dan Morrill

Chemical Specialties
1 Woodlawn Green
Suite 250
Charlotte, NC 28217
(704) 522-0825
Frank Kicklighter

Hugh Torance House and Store
335 Eastover Road
Charlotte, NC 28207
(704) 334-4836
Raymond Jones

Richard Mattson
309 East Park Avenue
Apartment 4
Charlotte, NC 28203

Mecklenburg Historical Association
P.O. Box 35032
Charlotte, NC 28235
(704) 554-7476
Clarence O. Kuester

Myers Park Foundation
2044 Sherwood Avenue
Charlotte, NC 28207

Jack B. Weaver
P.O. Box 911
Cherryville, NC 28021
(704) 435-3056

Yelverton Architects
5200 Park Road
Suite 229
Charlotte, NC 28209
(704) 523-7834
Donald W. Yelverton

Clinton

City Of Clinton
P.O. Box 199
Clinton, NC 28328
(919) 592-1961
Lucille Yancey

Concord

City of Concord
Planning and Community
Development
66 Union Street, South
P.O. Box 308
Concord, NC 28025
(704) 786-6161
J. J. Young

Concord Downtown Redevelopment
57 Union Street
Suite 305
Concord, NC 28025
(704) 784-4208

Historic Cabarrus
P.O. Box 966
Concord, NC 28026

Dallas

Gaston County Museum
P.O. Box 429
Dallas, NC 28034
(704) 922-7681
Alan D. Waufle

Durham

Durham City/County Planning
101 City Hall Plaza
Durham, NC 27701
(919) 683-4137
Keith Luck

L. Lane Sarver
1910 Sedwick Road
Suite 200C
Durham, NC 27713-9426
(919) 942-3464
L. L. Sarver

Elizabeth City

City of Elizabeth City
P.O. Box 347
Elizabeth City, NC 27909
(919) 338-3981
John Kinsey

Fayetteville

Cumberland County Joint Planning
Board
P.O. Drawer 1829
Old Court House Building
Fayetteville, NC 28302
(919) 483-8131
George Vaughan

Flat Rock

Historic Flat Rock
P.O. Box 295
Flat Rock, NC 28731
Georgia Paxton

Franklin

Franklin Downtown Development
Association
P.O. Box 700
Franklin, NC 28734
(704) 524-4471
MaryAnn Sloan

Goldsboro

Downtown Goldsboro Development
Corporation
P.O. Box 202
Goldsboro, NC 27530
(919) 735-4959
Taffy Tamblyn

Graham

Alamance County Historic Properties
Commission
124 West Elm Street
Graham, NC 27253
(919) 228-1312
M. M. Way

Greensboro

Fisher Park Neighborhood Association
P.O. Box 2043
Greensboro, NC 27402
(919) 273-8800
Julia M. Clement

Greensboro Historic District
Commission
Department of Planning and
Community Development
Drawer W-2
Greensboro, NC 27402
Mike Cowhig

Greensboro Preservation Society
P.O. Box 13136
Greensboro, NC 27415
(919) 272-5003
John S. Acker

Guilford County Joint Historic
Properties Commission
P.O. Box 3427
Greensboro, NC 27402
(919) 373-3747
Melinda L. Faley

Old Greensborough Preservation
Society
P.O. Box 1047
Greensboro, NC 27402
(919) 272-6617
Mary C. Eubanks

Greenville

East Carolina University
Willis Building
1st and Reade Street
Greenville, NC 27858
(919) 757-6650
Mack Simpson

Eastern Office
Archives and History
117 West 5th Street
Greenville, NC 27858
(919) 752-7778

Henderson

Henderson-Vance Downtown
Development Commission
P.O. Box 1434
Henderson, NC 27536
(919) 492-2041
Carolyn Powell

Hendersonville

Downtown Hendersonville
443 North Main Street
Hendersonville, NC 28792
(704) 697-2022

Hickory

Craig M. Copper
Architect
219 1st Avenue Southwest
P.O. Box 1391
Hickory, NC 28603
(704) 322-5715

Hickory Chair Company
Trade Show Manager
P.O. Box 2147
Hickory, NC 28603
(704) 328-1801
William M. Merrill

Hickory Landmarks Society
P.O. Box 2341
Hickory, NC 28603
(704) 322-4731
(704) 324-826
Carey N. Walker

High Point

City of High Point
P.O. Box 230
High Point, NC 27261
(919) 883-3352
C. A. Cooper-Ruska

Hillsborough

Orange County Planning Department
106 East Margaret Lane
Hillsborough, NC 27278
(919) 732-8781
Marvin E. Collins

Jamestown

Historic Jamestown Society
P.O. Box 512
Jamestown, NC 27282

Lexington

Davidson County Historical Museum
Old Courthouse
2 South Main Street
Lexington, NC 27292
(704) 249-7011
Catherine M. Hoffman

Lumberton

City of Lumberton
P.O. Box 1388
Lumberton, NC 28359
(919) 671-3816
Kirk Mattson

Mill Prong Preservation
P.O. Box 1087
Lumberton, NC 28359
(919) 738-5257
William S. McLean

Marion

City of Marion
Planning Department
P.O. Drawer 700
Marion, NC 28752
(704) 652-3551
Freddie Killough

Marion Downtown Business
Association
P.O. Drawer 700
Marion, NC 28752
(704) 652-3551
Freddie Killough

Milton

Thomas Day House/Union Tavern
Restoration
P.O. Box 1996
Milton, NC 27305
(919) 234-7215
Marian Thomas

Monroe

City of Monroe
P.O. Box 69
Monroe, NC 28110
C. F. Boyd

Morganton

City of Morganton
Main Street Office
P.O. Drawer 430
Morganton, NC 28655
(704) 437-8863
Sharon Jablonski

Murfreesboro

Murfreesboro Historical Association
P.O. Box 3
Murfreesboro, NC 27855
(919) 398-5922
Kay Mitchell

New Bern

New Bern Historical Society
Foundation
P.O. Box 119
New Bern, NC 28560
(919) 638-8558
Kathy D. Beckwith

New Bern Preservation Foundation
P.O. Box 207
New Bern, NC 28560
(919) 633-6448
Barbara W. Howlett

Swiss Bear
P.O. Box 597
New Bern, NC 28560
(919) 638-5781
Susan Moffat

Tryon Palace Restoration Complex
P.O. Box 1007
New Bern, NC 28560
(919) 638-1560
Kay P. Williams

Newton

City of Newton
Main Street Center
P.O. Box 550
Newton, NC 28658
(704) 465-7425
Connie A. Kincaid

Raleigh

Arts Together
114 St. Mary's Street
Raleigh, NC 27605
(919) 828-1713
Elizabeth Boney

Capital Area Preservation
Mordecai Historic Park
1 Mimosa Street
Raleigh, NC 27604
(919) 834-4844
Sally A. Poland

Division of Community Assistance
1307 Glenwood Avenue
Suite 250
Raleigh, NC 27605
(919) 733-2850
Jamie Englund

Frank R. Gailor
P.O. Box 20045
Raleigh, NC 27619

Historic Preservation Foundation of
North Carolina
1804 Hillsborough Street
P.O. Box 27644
Raleigh, NC 27611
(919) 832-3652
J. Howard

Eric Jackson
2705½ Clark Avenue
Raleigh, NC 27607
(919) 834-5883

National Council of Preservation
Executives
Division of Archives and History
109 East Jones Street
Raleigh, NC 27601-2807
David Brook

North Carolina AIA
115 West Morgan Street
Raleigh, NC 27601
(919) 833-6656
Timothy D. Kent

North Carolina Department of Cultural
Resources
109 East Jones Street
Raleigh, NC 27611
William S. Price

North Carolina League of
Municipalities
P.O. Box 3069
Raleigh, NC 27602
David E. Reynolds

North Carolina Main Street Center
Division of Community Assistance
1307 Glenwood Avenue
Suite 250
Raleigh, NC 27605
(919) 733-2850
Rodney Swink

Raleigh Historic Properties
Commission
Century Station
P.O. Box 829
Raleigh, NC 27602
(919) 832-7238
Dan Becker

James William Smith
3015 Wake Forest Road
Raleigh, NC 27609
(919) 821-5547

Society for the Preservation of Historic
Oakwood
P.O. Box 11137
Raleigh, NC 27604

Stained Glass Associates
P.O. Box 1531
Raleigh, NC 27602
(919) 266-2493
Robert Wysocki

Rebecca Tiffany
5618 Meadowbrook Road
Raleigh, NC 27603

Research Triangle Park

Triangle J Council of Governments
100 Park Drive
P.O. Box 12276
Research Triangle Park, NC 27709
(919) 549-0551
Brian Benson

Richlands

Onslow County Museum
P.O. Box 384
Richlands, NC 28574
(704) 324-5008
Albert Potts

Rodanthe

Chicamacomico Historical Association
P.O. Box 5
Rodanthe, NC 27963
(919) 987-2203
James F. Henry

Rutherfordton

Downtown Rutherfordton
106 North Main Street
Rutherfordton, NC 28139
(704) 287-8864
Sherry B. Adams

Salisbury

Downtown Salisbury
100 East Innes Street
P.O. Box 4166
Salisbury, NC 28144-0102
(704) 637-7814
Heidi D. Singer

Historic Salisbury Foundation
P.O. Box 4221
Salisbury, NC 28144-0103
Frederick W. Lyman

Rowan Museum
2029 Robin Road
Salisbury, NC 28144
J. B. Fisher

Sanford

City of Sanford
P.O. Box 1523
Sanford, NC 27331
(919) 774-6153
Mary E. Bowen

Shelby

City of Shelby
Community Development Department
P.O. Box 207
Shelby, NC 28150
(704) 484-6829
Hal Mason

Historic Shelby Foundation
504 South Dekalb Street
Shelby, NC 28150

Uptown Shelby Association
P.O. Box 2045
Shelby, NC 28151
(704) 484-3100
Ted Alexander

Smithfield

Downtown Smithfield Development
Corporation
P.O. Box 761
Smithfield, NC 27577
(919) 934-0887
Raymond Gibbs

Southern Pines

Naomi B. Faison
22 Middleton Place
Southern Pines, NC 28387
(919) 692-4661

Statesville

Downtown Statesville Development
Corporation
P.O. Box 205
Statesville, NC 28677
(704) 878-3436
Robin Hubbell

Wake Forest

Wake Forest Historic District
Commission
401 Elm Street
Wake Forest, NC 27587
(919) 556-2024
Chip Russell

Waynesville

Downtown Waynesville Association
P.O. Box 1409
Waynesville, NC 28786
(704) 452-0113
Ronald J. Huelster

Wilmington

Greater Wilmington Chamber
Foundation
505 Nutt Street
Building B
P.O. Box 330
Wilmington, NC 28401
(919) 762-2611
Paul M. Laird

Historic Wilmington Foundation
209 Dock Street
Wilmington, NC 28401
(919) 762-2511
E. D. Scott

Lower Cape Fear Historical Society
P.O. Box 813
Wilmington, NC 28402
(919) 762-0492
Jean S. Scott

Deborah Sheetenhelm
Preservation Planner
P.O. Drawer 1810
Wilmington, NC 28402
(919) 341-7837

Wilson

City of Wilson
P.O. Box 10
Wilson, NC 27893
(919) 291-8111
LuAnn Monson

Mercy Hospital Associates
P.O. Box 121
Wilson, NC 27893
(912) 237-3664
Hattie Ellis

Wilson Downtown Redevelopment
Corporation
P.O. Box 1567
Wilson, NC 27893
(919) 237-4151
J. R. Steigerwald

Windsor

Historic Hope Plantation
P.O. Box 601
Windsor, NC 27983
(919) 794-3140
Jay Jordan

Winston-Salem

David Winslow and Associates
P.O. Box 10973
Winston Salem, NC 27108
(919) 722-7982
David Winslow

Phillips and Oppermann
1134-A Burke Street
Winston-Salem, NC 27101
(919) 723-0865
Joseph K. Oppermann

William R. Wallace, Jr.
2250 Silas Creek Parkway
Suite 202
Winston-Salem, NC 27103
(919) 722-1415

Winston-Salem/Forsyth
County/Kernersville
Historic Properties Commission
P.O. Box 2511
Winston-Salem, NC 27102
(919) 727-2087
C. L. Pegram

NORTH DAKOTA

STATE HISTORIC PRESERVATION OFFICE

State Historical Society of North Dakota
Heritage Center
612 East Boulevard Avenue
Bismarck, ND 58505
(701) 224-2667
James E. Sperry, SHPO
Louis N. Hafermehl, Deputy SHPO

STATEWIDE PRESERVATION ORGANIZATION

Preservation North Dakota
P.O. Box 2635
Bismarck, ND 581-3
(218) 233-4433
Royce Yeater, President

NATIONAL TRUST REGIONAL OFFICE

Mountains/Plains Regional Office
511 16th Street
Suite 700
Denver, CO 80202
(303) 623-1504
(303) 825-8073 (Fax)
Barbara J. Pahl, Director

NATIONAL TRUST ADVISORS

Joan Galleger
Garsten Management Corporation
223 4th Avenue
Devils Lake, ND 58301
(701) 662-4776
(701) 662-6251 (Fax)

Lowell F. Johnson
1640 North 5th Street
Wahpeton, ND 58075
(701) 671-2951
(701) 6671-2145 (Fax)

PRESERVATION ACTION COORDINATOR

Joan Galleger
Garsten Management Corporation
223 4th Avenue
Devils Lake, ND 58301
(701) 662-4776

AIA STATE PRESERVATION COORDINATOR

Robert A. Mitchell, AIA
State Historical Society
612 East Boulevard
Bismarck, ND 58505-0830
(701) 224-2672

NATIONAL PARK SERVICE REGIONAL OFFICE

Rocky Mountain Regional Office
12795 West Alameda Parkway
P.O. Box 25287
Denver, CO 80225-0287
(303) 969-2875

ADVISORY COUNCIL ON HISTORIC PRESERVATION DIVISION

Western Office of Project Review
730 Simms Street
Room 401
Golden, CO 80401
(303) 231-5320
Claudia Nissley, Director

Bismarck

Preservation North Dakota
P.O. Box 2635
Bismarck, ND 58502
(218) 233-4422
Royce Yeater

Cavalier

City of Cavalier
P.O. Box B
Cavalier, ND 58220
7012658800
Diane Vistad

Fargo

Fargo Heritage Society
P.O. Box 3161
Fargo, ND 58108-3161
(701) 293-0276
Dawn Morgan

Lightowler Johnson Association
P.O. Box 2464
Fargo, ND 58108
(701) 293-1350
Martin K. Ness

Grand Forks

Grand Forks Preservation Commission
1114 Chestnut Street
Grand Forks, ND 58201
(701) 780-1155
Marsha Gunderson

Jamestown

Jane Edwards
524 5th Street, N.E.
Jamestown, ND 58401
(701) 252-3096

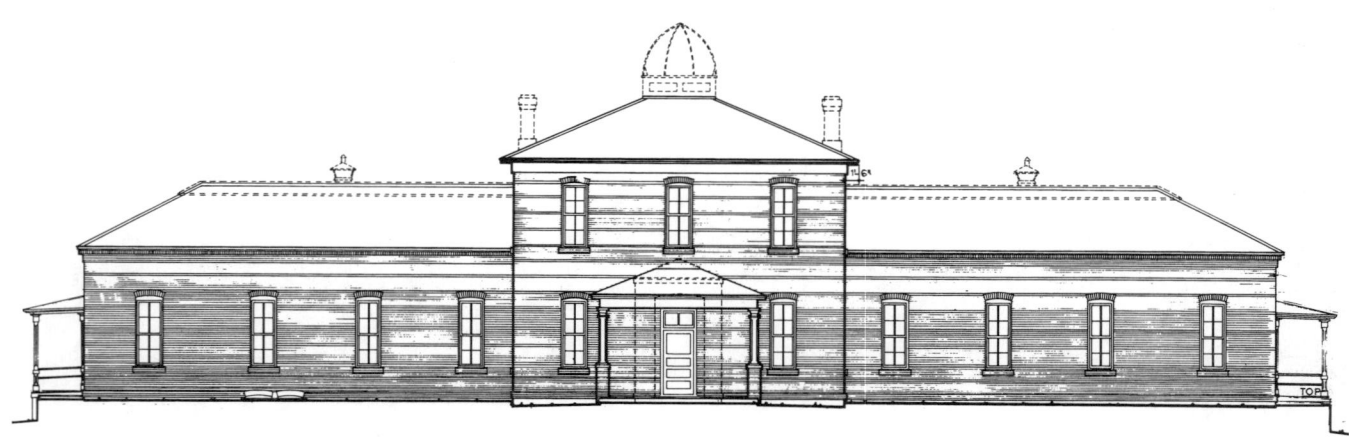

Hospital, Fort Totten, North Dakota. (K.L. Anderson, HABS)

OHIO

STATE HISTORIC PRESERVATION OFFICE

Ohio Historical Society
Historic Preservaton Division
1985 Velma Avenue
Columbus, OH 43211
(614) 297-2470
(614) 297-297-2411
W. Ray Luce, SHPO
Mr. Franco Ruffini, Deputy SHPO

STATEWIDE PRESERVATION ORGANIZATION

Ohio Preservation Alliance
65 Jefferson Avenue
Columbus, OH 43215
(614) 228-3133
C. William Hager, President

NATIONAL TRUST REGIONAL OFFICE

Midwest Regional Office
52 West Jackson Boulevard
Suite 1135
Chicago, IL 60604
(312) 939-5447
(312) 939-5651 (Fax)
Tim Turner, Director

NATIONAL TRUST ADVISORS

Mary Ann Brown
Cincinnati Preservation Association
Hamilton County Memorial Building
1225 Elm Street
Cincinnati, OH 45210
(513) 721-4506
(513) 721-6832 (Fax)

Ted J. Ligibel
2036 North Kennison
Toledo, OH 43609-1918
(313) 487-0218

PRESERVATION ACTION COORDINATOR

John Cimperman
401 Detroit Avenue
Cleveland, OH 44113
(216) 631-6274

AIA STATE PRESERVATION COORDINATOR

Bruce E. Goetzman, AIA
2606 Vine Street
Cincinnati, OH 45219-2017
(513) 281-7244

NATIONAL PARK SERVICE REGIONAL OFFICE

Midwest Regional Office
1709 Jackson Street
Omaha, NE 68102
(402) 221-3431

ADVISORY COUNCIL ON HISTORIC PRESERVATION DIVISION

Eastern Office of Project Review
1100 Pennsylvania Avenue, N.W.
Suite 809
Washington, DC 20004
(202) 786-0503
Don L. Klima, Director

Akron

Progress through Preservation
641 West Market Street
Akron, OH 44303
(216) 762-1411
Ramona Smith

Stan Hywet Hall Foundation
714 North Portage Path
Akron, OH 44303
(216) 836-5533
John F. Miller

Summit County Historical Society
550 Copley Road
Akron, OH 44320
(216) 535-1120
Stephen H. Paschen

Aurora

Mark N. Gilles
Architect
P.O. Box 431
173 South Chillicothe Road
Aurora, OH 44202
(216) 562-4111

Barberton

City of Barberton
Planning Department
Municipal Building
576 West Park
Barberton, OH 44203
(216) 848-6729
Deborah Sanborn

Barnesville

Barnesville Area Development Council
211 North Chestnut
P.O. Box 462
Barnesville, OH 43713
(614) 425-2301
Bruce A. Yarnall

Bowling Green

Bowling Green University
Center for Government Research and
Public Service
12 Williams Hall
Bowling Green, OH 43403
(419) 372-8710
Frank McKenna

City of Bowling Green
304 North Church Street
Bowling Green, OH 43402
(419) 354-6220
Carolyn M. Lineback

Wood County Historical Center
13660 County Home Road
Bowling Green, OH 43402

Brecksville

Brecksville Town Centre
7670 Chippewa Road
Suite 401
Brecksville, OH 44141
(216) 526-7350

Cambridge

Cambridge Alliance for Progress
927 Wheeling Avenue
Cambridge, OH 43725
(614) 432-6820
Kaye Gaddy

Canal Fulton

Village of Canal Fulton
155 East Market Street
Canal Fulton, OH 44614

Canton

Canton Preservation Society
P.O. Box 20174
Canton, OH 44701
Roberta Higley

Canton Tomorrow
229 Wells Avenue, N.W.
Canton, OH 44703
(216) 486-7253
Kevin P. Flaherty

Paul A. Knoch
1026 Knollwood, N.W.
Canton, OH 44708
(216) 477-4638

Stark Preservation Alliance
511 County Office Building
Canton, OH 44702
(216) 438-0400
Gerald Cody

Chagrin Falls

Maxine G. Levin
3400 Fairmount Boulevard
Chagrin Falls, OH 44022
(216) 781-5230
(216) 461-4166

Chillicothe

First Capital District
18 West Water Street
Chillicothe, OH 45601
(614) 773-6292
Susan Sherman

Cincinnati

Cincinnati Planning Department
Historic Conservation Office
801 Plum Street
Room 228
Cincinnati, OH 45202

Greyhound Terminal, Dayton, Ohio. (S. Bauer, HABS, adapted from Railroad & Bus Terminal and Station Layout, 1945)

Keith Enterprises
P.O. Box 11038
Cincinnati, OH 45211
H. Keith

Miami Purchase Association for
Historic Preservation
1225 Elm Street
Cincinnati, OH 45210
(503) 721-4506
Mary A. Brown

Miami Purchase Preservation Fund
727 Ezzard Charles Drive
Cincinnati, OH 45203
(513) 241-0504
Linda B. Fabe

W. Kevin Pape
1318 Main Street
Cincinnati, OH 45210
(513) 287-7700

John Spencer
20 Village Sqaure
Cincinnati, OH 45246
(513) 771-7131

Verdin Company
444 Reading Road
Cincinnati, OH 45202
(513) 241-4010
Suzanne Sizer

James D. Wichman
810 Plum Street
Cincinnati, OH 45202-1973
(513) 241-9933

Cleveland

City of Cleveland
116 City Hall
Cleveland, OH 44114

Cleveland Landmarks Commission
City Hall
Room 519
Cleveland, OH 44114
Robert Keiser

Cleveland Restoration Society
Statler Office Tower
1127 Euclid Avenue
Suite 463
Cleveland, OH 44115-1601
(216) 621-1498
Kathleen H. Crowther

David W. Dennis
1435 West 38th Street
Cleveland, OH 44113
(216) 651-0827

Detroit Shoreway Community
Development Organization
6516 Detroit Avenue
Room 242
Cleveland, OH 44102

Extravagant Homes of Greater
Cleveland
3544 Trent Avenue
Cleveland, OH 44109
(216) 348-3104
Maria L. Cruz

Franklin Properties Company
13212 Shaker Square
Room 100
Cleveland, OH 44120
(216) 921-5663
Joseph M. Shafran

Friends of Shaker Square
13125 Shaker Square
Room 102
Cleveland, OH 44120
(216) 751-9204
Joseph Mazzola

Historic Warehouse District
614 Superior Avenue, N.W.
Suite 714
Cleveland, OH 44113
(216) 344-3937
Steven B. Seaton

Norman Marek
27214 Bagley
Cleveland, OH 44138

Mid State Restoration
2609 Monroe Avenue
Cleveland, OH 44113
(216) 771-2112
James L. Hayes

Midtown Corridor
4614 Prospect Avenue
Cleveland, OH 44103
(216) 391-5080
Molly Donovan

North Cuyahoga Valley Corridor
2089 West Boulevard
Cleveland, OH 44102
(216) 651-9366
Tom Yablonsky

Preservation of Miles Park
P.O. Box 605005
Cleveland, OH 44105-0005
(216) 281-6649
Ken Carpenter

Stride for Pride of Ohio City
4115 Bridge Avenue
Suite 110
Cleveland, OH 44113
(216) 281-8171
Joseph G. Lucas

Western Reserve Historical Society
10825 East Boulevard
Cleveland, OH 44106
(216) 721-5722
Theodore A. Sande

Cleveland Heights

City of Cleveland Heights
Planning and Development
Department
40 Severance Circle
Cleveland Heights, OH 44118
(216) 291-4817
Cheryl L. Stephens

Columbus

City of Columbus
Planning Division
140 Marconi Boulevard
Columbus, OH 43215
(614) 222-8172
Ty Potterfield

Columbus Association for the
Performing Arts
55 East State Street
Columbus, OH 43215
(614) 469-1045
Douglas F. Kridler

Columbus Design Center
1273 West Broad Street
Columbus, OH 43222
(614) 274-4141
Bill Shaffer

Columbus Landmarks Foundation
65 Jefferson Avenue
Columbus, OH 43215-3840
(614) 221-0227

General Preservation Corporation
7760 Olentangy River Road
Suite 108
Columbus, OH 43235
(614) 888-0350
James L. Rish

GLK and Associates
2107 Concord Road
Columbus, OH 43212
(614) 486-5291
Gretchen Klimoski

Susan King
P.O. Box 2251
Columbus, OH 43216
(614) 236-2221

North Market Development Authority
30 West Spruce Street
Columbus, OH 43215
(614) 224-9325
Nancy D. Porter

Office of Local Government Services
77 South High Street
24th Floor
P.O. Box 1001
Columbus, OH 43266-0101
(614) 466-2285
Gregory M. Carr

Ohio Arts Council
727 East Main Street
Columbus, OH 43205
(614) 466-2613
Wayne P. Lawson

Ohio Historical Society
Archives-Library
1985 Velma Avenue
Columbus, OH 43211
Tom Starbuck

Ohio Historic Preservation Office
Ohio Historical Society
1985 Velma Avenue
Columbus, OH 43211
(614) 297-2300
W. R. Luce

Ohio Preservation Alliance
P.O. Box 18274
Columbus, OH 43218
(513) 296-1284
Bill Hager

Ohio State University
190 North Oval Mall
Room 108
Columbus, OH 43210
(614) 292-7970
John R. Kleberg

Ohio State University
Department of Architecture
189 Brown Hall
190 West 17th Avenue
Columbus, OH 43210
(614) 292-9062
Paul Young

Volunteers of America
379 West Broad Street
Columbus, OH 43215
(614) 224-8650
Linda Belhorn

Dayton

City of Dayton
Department of Planning
101 West 3rd Street
P.O. Box 22
Dayton, OH 45401
(513) 443-3685
Paul R. Woodie

Classics Corporation
22 Park Avenue
Dayton, OH 45419
(513) 296-1140
C. W. Hager

Dayton and Montgomery County
Public Library
215 East 3rd Street
Dayton, OH 45402
(513) 224-1651

Doug Gregory
102 Floral Avenue
Dayton, OH 45405
(513) 226-1191

Gerald A. Johnson
3839 Hillmont Avenue
Dayton, OH 45414
(513) 278-4969
(513) 228-8951

Montgomery County Community
Development
451 West 3rd Street
Dayton, OH 45422
(513) 225-6318
Kathy Fedler

Montgomery County Historical Society
7 North Main Street
Dayton, OH 45402
(513) 228-6271
Kirby Turner

Preservation Dayton
604 Hickory Street
Dayton, OH 45410
(513) 228-7234
Linda Caron

Delaware

Strategic Planning Committee
78 Elmwood Drive
Delaware, OH 43015
(614) 369-2011
Constance C. Whitaker

Elyria

Lorain County Historical Society
509 Washington Avenue
Elyria, OH 44035
(216) 322-3341
Ronald L. Burdick

Franklin

Architectural Reclamation
312 South River Street
Franklin, OH 45005
(513) 746-8964
Susan Stewart

Fremont

City of Fremont
Zoning Inspector
323 South Front Street
Fremont, OH 43420
(419) 334-8963
Fran Gierhart

Galion

Galion Area Chamber of Commerce
122 Harding Way East
P.O. Box 705
Galion, OH 44833
(419) 468-7737
Dan Ritchey

Granville

William G. Heim
Rita A. Fuerst
133 East College Street
Granville, OH 43023
(614) 587-2137

Hamilton

Citizens for Historic and Preservation
Services
365 South B Street
Hamilton, OH 45013
(513) 863-1716
Ann Antenen

City of Hamilton
Planning Department
20 High Street
Room 107
Hamilton, OH 45011
(513) 868-5878
John Lehner

Historic Hamilton
319 North 3rd Street
Hamilton, OH 45011
(513) 863-8873
Nancy Tryloff

Main Street Area Associates
445 Main Street
Hamilton, OH 45013
(513) 844-6246
John K. Roberts

Harrison

Harrison Main Street
P.O. Box 293
Harrison, OH 45030-0293
Connie Loftus

Hilliard

Kimball W. Shields
3368 Scioto Run Boulevard
Hilliard, OH 43026
(614) 459-1622

Hudson

Hudson Heritage Association
P.O. Box 2218
Hudson, OH 44236

3B Design
120-A North Main Street
Hudson, OH 44236
(216) 650-1255
Baroness Louisa Bille Brahe

Jackson

Community Development
327 East Main Street
Jackson, OH 45640
(614) 286-6433
Deanna Figlestahler

Kettering

Kettering-Moraine Museum and
Historical Society
35 Moraine Circle, South
Kettering, OH 45439
Melba Hunt

Lakeside

Lakeside Association
236 Walnut Avenue
Lakeside, OH 43440
(419) 798-4461
Philip L. Zimmerman

Lakewood

Thomas P. Wright and Sons
14414 Detroit Avenue, N.E.
Lakewood, OH 44107
(216) 521-1300

Lebanon

Kathleen McGurn
320 East Mulberry Street
Lebanon, OH 45036
(513) 932-6234

Lima

American House
P.O. Box 5283
Lima, OH 45802
(419) 224-6873
Martha S. MacDonell

Lorain

City of Lorain
Community Development Department
200 West Erie Avenue
5th Floor
Lorain, OH 44052
(216) 245-1010
Sanford A. Prudoff

Mansfield

City of Mansfield
Community Development Department
30 North Diamond Street
Mansfield, OH 44902
(419) 755-9795
Iwana Wagner

Engwiller Properties
P.O. Box 37
Mansfield, OH 44901
(419) 529-6100
John C. Fernyak

Main Street Mansfield
55 North Mulberry Street
Mansfield, OH 44902
(419) 522-0099
Carol Butera-Dutton

Marietta

James Hoy
124 Seneca Drive
Marietta, OH 45750
(614) 373-8039

Marysville

Marysville Area Chamber of Commerce
126 West 5th Street
Marysville, OH 43040
(513) 642-3922
Suzanne Boylan

Massillon

Massillon Main Street
Box 1099
Massillon, OH 44648
(216) 833-3146
Steven B. Seaton

Medina

Medina Community Design
Committee
141 South Prospect Street
Medina, OH 44256
(216) 725-7516
Janet Senkar

Middlefield

Middlefield Historical Society
P.O. Box 1100
Middlefield, OH 44062

Nelsonville

Hocking Valley Museum of Theatrical
History
P.O. Box 217
Nelsonville, OH 45764
(614) 753-1924
Shirley Seckinger

Newark

Nola Miles Rogers
309 Darlene Drive, N.E.
Newark, OH 43055
(614) 349-7480

Oakwood

C. William Hager
22 Park Avenue
Oakwood, OH 45419

Oberlin

Citizens For Appropriate Development
MPO 95
Oberlin, OH 44074
(216) 774-1138
Laura Paxton

Mary Durling
221 Elm Street
Oberlin, OH 44074
(216) 774-3411

Orrville

City Of Orrville
City Hall
Orrville, OH 44667
(216) 683-8715
Howard E. Wade

Oxford

Miami University Art Museum
Patterson Avenue
Oxford, OH 45056
(513) 529-2232
Bonnie G. Kelm

Pataskala

Jay L. Dunham
75 North Street
P.O. Box 1355
Pataskala, OH 43062
(614) 927-6886

Pepper Pike

Ursuline College
2550 Lander
Pepper Pike, OH 44124
(216) 449-4200
Leslie Pina

Perrysburg

Historic Perrysburg
P.O. Box 703
Perrysburg, OH 43551
(419) 874-2815
W. B. Ball

Peggy B. Orser
420 East Front Street
Perrysburg, OH 43551
(419) 874-8815

Portsmouth

Main Streets of Portsmouth
741 5th Street
Portsmouth, OH 45662
(614) 353-1116
David A. Horr

Ravenna

Ravenna Heritage Associations
P.O. Box 524
Ravenna, OH 44266
(216) 296-3753
E. A. Kyle

St. Clairsville

Belmont Technical College
120 Fox Shannon Place
St. Clairsville, OH 43950
(614) 695-9500
Dave Mertz

Sandusky

City of Sandusky
222 Meigs Street
Sandusky, OH 44870
Frank A. Link

Steubenville

Main Street Steubenville
630 Market Street
Steubenville, OH 43952
(614) 282-6226
Kurt A. Zende

Tallmadge

City of Tallmadge
46 North Avenue
Tallmadge, OH 44278
(216) 633-0145
Pat Sauner

Tallmadge Historical Society
718 Northeast Avenue
Tallmadge, OH 44278
C. E. Stalnaker

Tiffin

Tiffin Historic Trust
P.O. Box 333
Tiffin, OH 44883

Toledo

Collingwood Springs
Old West End Partnership
2272 Collingwood
Toledo, OH 43620
(419) 241-9682
Mary C. Rietz

Housing Commissioner
1 Government Center
18th Floor
Toledo, OH 43604-2275
Paul Z. Tecpanecatl

Lagrange Development Corporation
3108 Lagrange Street
Toledo, OH 43608
(419) 255-8406
Terry Glazer

Landmarks Committee
2305 Scottwood Avenue
Toledo, OH 43620
(419) 537-2940
Ted J. Ligibel

Toledo Warehouse District Association
110 Ottawa
Suite 500
Toledo, OH 43602
(419) 255-7100
Julie Pearson

Women of the Old West End
Owe Station
P.O. Box 4745
Toledo, OH 43620
(419) 242-2908
Marietta Hubbard

Troy

Troy Main Street
P.O. Box 486
Troy, OH 45373
(513) 339-5455
Jerry Caruso

Troy-Miami County Public Library
419 West Main Street
Troy, OH 45373

Twinsburg

Twinsburg Historical Society
P.O. Box 7
Twinsburg, OH 44087
Marjorie Percy

Warren

City of Warren
Community Development Department
418 South Main Street
Warren, OH 44481
(216) 841-2595
John Foley

Phillips Management
130 Pine Avenue, S.E.
Warren, OH 44481
(216) 395-5405
Brad Phillips

Upton Association
1522 North Road, N.E.
Warren, OH 44483-4529
(216) 856-5700
Shari Harrell

Westerville

Westerville Public Library
126 South State Street
Westerville, OH 43081

Wickliffe

Robert E. Aufuldish
28730 Ridge Road
Wickliffe, OH 44092
(216) 943-7103

Wilmington

Clinton County Historical Society
P.O. Box 529
Wilmington, OH 45177
(513) 382-4684
Rhonda L. Curtis

Woodsfield

Monroe County Park District
Courthouse
Room 34
Woodsfield, OH 43793
(614) 472-1328
Richard A. Schumacher

Wooster

Main Street Wooster
147 1/2 East Liberty Street
Wooster, OH 44691-4345
(216) 262-6222
Sandra C. Hull

Worthington

NPP Architects
7870 Olentangy River Road
Worthington, OH 43235
(614) 436-2122
Fritz Nevenschwander

Worthington Historical Society
50 West New England Avenue
Worthington, OH 43085

Yellow Springs

Antioch College
795 Livermore Street
Yellow Springs, OH 45387
(513) 767-7708
William Hooper

Youngstown

Christopher B. Owen
2260 5th Avenue
Youngstown, OH 44504
(216) 743-5278

Xenia

City of Xenia
101 North Detroit Street
Xenia, OH 45385-2926
Nimfa Simpson

Zanesville

City of Zanesville
Community Development
401 Market Street
Zanesville, OH 43701-3571
(614) 455-0668
Robert F. Guentter

OKLAHOMA

STATE HISTORIC PRESERVATION OFFICE

Oklahoma Historical Society
2100 North Lincoln Boulevard
Oklahoma City, OK 73105
(405) 521-2491
(405) 525-3272 (Fax)
J. Blake Wade, Acting Director, SHPO
Melvena Thurman Heisch, Deputy SHPO
State Historic Preservation Office
621 North Robinson
Suite 375
Oklahoma City, OK 37102
(405) 521-6249

NATIONAL TRUST REGIONAL OFFICE

Mountains/Plains Regional Office
511 16th Street
Suite 700
Denver, CO 80202
(303) 623-1504
(303) 825-8073 (Fax)
Barbara J. Pahl, Director

NATIONAL TRUST ADVISORS

Sally Ferrell
The Ferrell Company
101 West Indian Trail
Chandler, OK 74834
(405) 258-1006

Susan Guthrie
Capital Development Corporation
6440 Avondale Drive
Suite 201
Oklahoma City, OK 73116
(405) 848-5006
(405) 848-6983 (Fax)

PRESERVATION ACTION COORDINATORS

Susan Guthrie
Capital Development Corporation
6440 Avondale Drive
Suite 201
Oklahoma City, OK 73116
(405) 848-5006

AIA STATE PRESERVATION COORDINATOR

Bill E. Peavler, AIA
Office of Public Affairs
50 Northeast 23rd Street
Oklahoma City, OK 73105
(405) 521-2111
Arn Henderson, AIA, Deputy

NATIONAL PARK SERVICE REGIONAL OFFICE

Southwest Regional Office
P. O. Box 728
Santa Fe, NM 87504
(505) 988-6100

ADVISORY COUNCIL ON HISTORIC PRESERVATION DIVISION

Western Office of Project Review
730 Simms Street
Room 401
Golden, CO 80401
(303) 231-5320
Claudia Nissley, Director

Ada

Ada Main Street
P.O. Box 2066
Ada, OK 74820
(405) 436-3032
Jill Frye

Altus

Preservation Works Company
P.O. Box 8376
Altus, OK 73521
(405) 482-6647
Michelle Maahs

Ardmore

Ardmore Main Street
9 A Street, S.W.
Ardmore, OK 73401
(405) 226-6246
Melissa Legg

Bethany

Bethany Main Street
P.O. Box 1616
Bethany, OK 73008
(405) 495-1313
Lisa Mercer

Cushing

Cushing Chamber of Commerce and Industry
P.O. Box 1007
Cushing, OK 74023
(918) 225-2400
Karen Kastl

Duncan

Association of South Central Oklahoma Goverments
802 Main
P.O. Box 1647
Duncan, OK 73534
(405) 252-0595
Paul Fisher

Historic Duncan
916½ Main Street
P.O. Box 112
Duncan, OK 73533
(405) 252-8696
Vickie J. Morris

El Reno

El Reno Main Street
P.O. Box 606
El Reno, OK 73036
(405) 262-8888
Carolyn Howard

Eufaula

Eufaula Main Street Project
P.O. Box 684
Eufaula, OK 74432
(918) 689-3227
Les Hall

McAlester

McAlester in Motion
First National Center
Suite 112
McAlester, OK 74501
(918) 426-1212
Lynette Roberts

Miami

City of Miami
P.O. Box 309
Miami, OK 74355
(918) 542-6685
Sue Valliere

Miami Downtown Redevelopment
111 North Main
P.O. Box 760
Miami, OK 74354
(918) 542-4481

Muskogee

City of Muskogee
Planning Department
P.O. Box 1927
Muskogee, OK 74402
(918) 684-6232
Steve Tuck

Newkirk

Newkirk Community Historic Society
7th and Maple
Newkirk, OK 74647
(405) 362-2525
Karen Dye

Norman

Norman-Cleveland County Historical Society and Museum
P.O. Box 260
Norman, OK 73070
(405) 321-0156
Kathleen Wallis

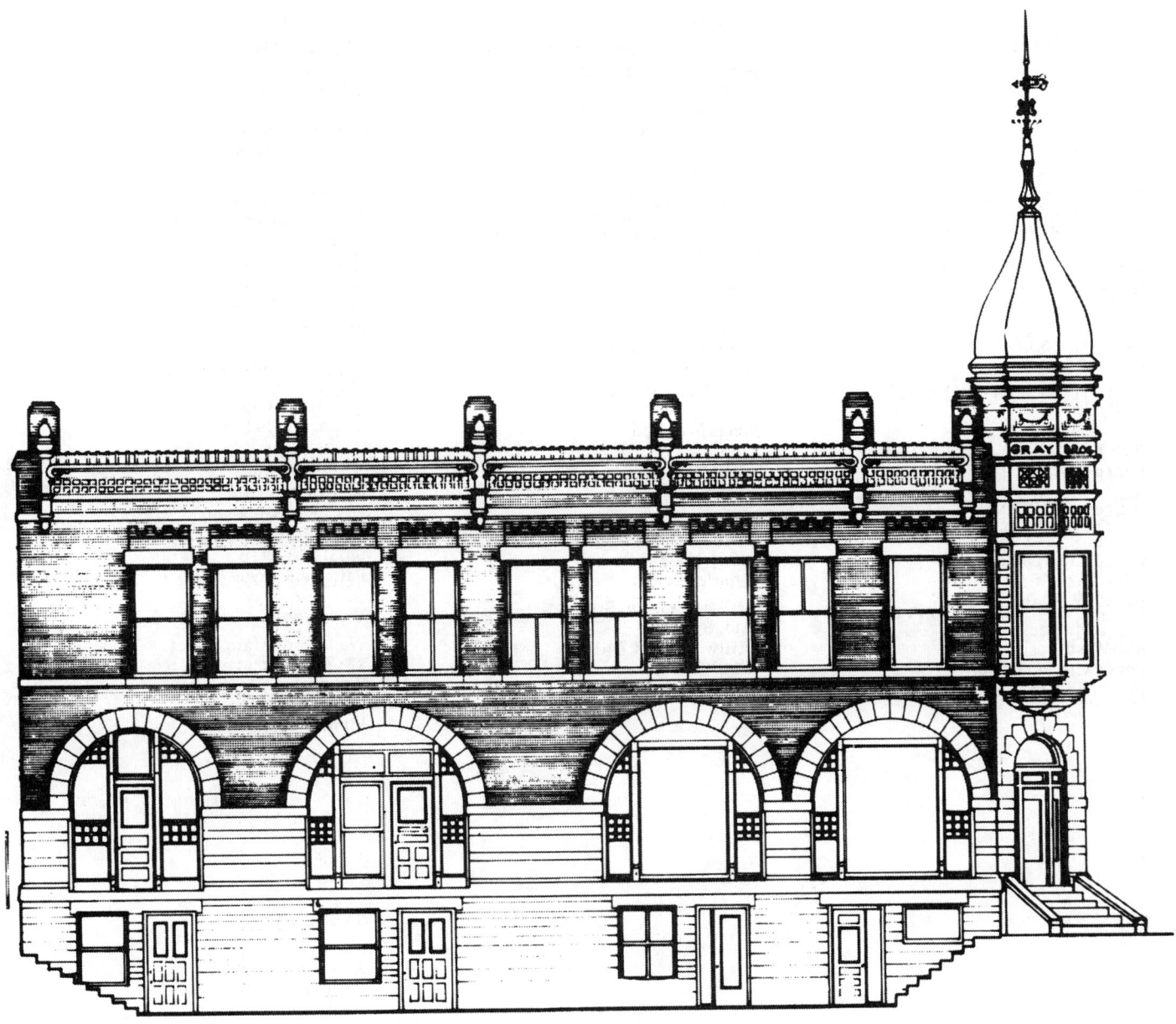

Gray Brothers Block, Guthrie, Oklahoma. (J. Robbins, HABS)

Oklahoma City

Historical Preservation
515 N.W. 15th Street
Oklahoma City, OK 73103
(405) 524-01420
Bill Gumerson

Neighborhood Development and
Conservation Center
1236 N.W. 36th Street
Oklahoma City, OK 73118
(405) 528-6322
Debbie Blackburn

Oklahoma City Urban Renewal
Authority
204 North Robinson
Suite 410
Oklahoma City, OK 73102
Tiana Douglas

Oklahoma Department of Commerce
Oklahoma Main Street Program
P.O. Box 26980
Oklahoma City, OK 73126-0980
(405) 843-9770
Susan Clinard

Oklahoma Historical Society
State Historic Preservation Office
Wiley Post Historical Buildng
Oklahoma City, OK 73105
(405) 521-2491
Melvena Heisch

Oklahoma Tourism and Recreation
Department
Planning and Development Division
500 Will Rogers Building
Oklahoma City, OK 73105

Stockyards City Main Street
P.O. Box 82446
Oklahoma City, OK 73148-0446
(405) 239-5876
Carroll Monden

Okmulgee

Robert Inglish
P.O. Box 130
Okmulgee, OK 74447
(918) 756-5274

Pawnee

Pawnee Tribe Of Oklahoma
P.O. Box 470
Pawnee, OK 74058
(918) 762-3624

Ponca City

City of Ponca City
P.O. Box 1450
Ponca City, OK 74602

Ponca City Main Street Authority
117 North 3rd
P.O. Box 2532
Ponca City, OK 74602
(405) 762-1002
David W. Keathly

Laura M. Streich
317 West Fresno
Ponca City, OK 74601
(405) 765-6337
(405) 765-6108

Purcell

McClain County Commissioners
McClain County Courthouse
Purcell, OK 73080
(405) 252-0595
Phyllis Bennett

Sapulpa

Sapulpa Main Street
101 East Dewey
Sapulpa, OK 74066
(918) 224-5709
Julie Warden

Shawnee

Downtown Shawnee
P.O. Box 534
Shawnee, OK 74802
(405) 273-1080
Leisa Mitchell

Stillwater

Oklahoma State University
Landscape Architecture Program
Stillwater, OK 74078
(405) 744-5414
Charles Leider

Stillwater Main Street
P.O. Box 1881
Stillwater, OK 74076
(405) 624-2921
Cindy Marshall

Tahlequah

Cherokee National Historic Society
Box 515
Tahlequah, OK 74465
(918) 456-6195
Myrna Moss

Tulsa

Rick Hendricks
4184 South Troost Place
Tulsa, OK 74105

Vernon S. Mills
827 South Peoria
Tulsa, OK 74120
(918) 585-5494

Neighborhood Housing Services of
Tulsa
806 North Osage Drive
Tulsa, OK 74106

Waurika

City of Waurika
107 West Anderson
Waurika, OK 73573
(405) 228-2713
Patricia Runyan

Woodward

Woodward Main Street
1219 8th Street
Woodward, OK 73801
(405) 254-8521
Linda Stinnett

OREGON

STATE HISTORIC PRESERVATION OFFICE

State Parks and Recreation Department
525 Trade Street, S.E.
Salem, OR 97310
(503) 378-5019
(503) 378-6447 (Fax)
David G. Talbot, Director, SHPO
James Hamrick, Deputy SHPO
(503) 378-6508

STATEWIDE PRESERVATION ORGANIZATION

Historic Preservation League of Oregon
P.O. Box 40053
Portland, OR 97240
(503) 243-1923
Mike Byrnes, President
Richard Matthews, Assistant Director

NATIONAL TRUST REGIONAL OFFICE

Western Regional Office
1 Sutter Street
Suite 707
San Francisco, CA 94104
(415) 956-0610
(415) 956-0837 (Fax)
Kathryn A. Burns, Director

NATIONAL TRUST ADVISORS

David Brauner
Department of Anthropology
Oregon State University
Waldo Hall 238
Corvallis, OR 97331-6403
(503) 737-3855
(503) 737-2434 (Fax)

Judith Rees
1965 S.E. Hemlock
Portland, OR 97214
(503) 823-3358
(503) 823-3368 (Fax)

PRESERVATION ACTION COORDINATOR

Sharr Prohaska
3640 S.W. Dosch Road
Portland, OR 97201
(503) 227-3307

AIA STATE PRESERVATION COORDINATOR

David Bissett, AIA
3000 Southwest Christy Avenue
Beavertown, OR 97005
(503) 245-7100
(503) 646-6603

NATIONAL PARK SERVICE REGIONAL OFFICE

Pacific Northwest Regional Office
83 South King Street
Suite 212
Seattle, WA 98104
(206) 553-5565

ADVISORY COUNCIL ON HISTORIC PRESERVATION DIVISION

Western Office of Project Review
730 Simms Street
Room 401
Golden, CO 80401
(303) 231-5320
Claudia Nissley, Director

Albany

City of Albany
250 Broadalbin
P.O. Box 490
Albany, OR 97321
(503) 967-4300
Linda Sarnoff

Astoria

Clatsop County
Historical Society
1618 Exchange Street
Astoria, OR 97103
(503) 325-2203
John P. Cooper

Columbia River Maritime Museum
1792 Marine Drive
Astoria, OR 97103
(503) 325-2323
Jerry L. Ostermiller

Aurora

Aurora Colony Historical Society
Box 202
Aurora, OR 97002
(503) 678-5754
Patrick Harris

City of Aurora
Historic Review Board
P.O. Box 100
Aurora, OR 97002
Gayelle Gregory

Baker City

Baker County Chamber of Commerce
490 Campbell Street
Baker City, OR 97814
(503) 523-5855
Ginger Savage

Historic Baker City
P.O. Box 811
Baker City, OR 97814
Gary Middleton

Gary D. Middleton
Kathleen L. Middleton
P.O. Box 811
Baker City, OR 97814

Beaverton

Kristin Kimmerling
11173 South West Davies
Suite 301
Beaverton, OR 97007

Canby

Canby Heritage League
P.O. Box 336
Canby, OR 97013
(503) 266-7664
Peggy Sigler

Corvallis

City of Corvallis
P.O. Box 1083
Corvallis, OR 97339-1083
(503) 757-6936
Peter Idema

Downtown Corvallis Association
P.O. Box 1536
Corvallis, OR 97339
(503) 754-6624
Joanne L. Van Ness

Dayton

Judy Gerrard
Leslie Miller
P.O. Box 159
Dayton, OR 97114

Eugene

Lisa Teresi-Burcham
2090 West 17th Avenue
Eugene, OR 97402
(503) 344-4964

Gresham

Parks and Recreation
1333 Northwest Eastman Parkway
Gresham, OR 97030
(503) 669-2408
Julee M. Conway

Hillsboro

Friends of Imbrie
Farmstead
P.O. Box 1744
Hillsboro, OR 97123
(503) 244-0012
Mike Byrnes

Hood River

Darin R. Gardner
1040 East Marina Way
Hood River, OR 97031
(503) 386-2276

Jefferson

City of Jefferson
163 North Main Street
P.O. Box 83
Jefferson, OR 97352
(503) 327-2768
LaQuita Stec

John Day

Historic Preservation Society
P.O. Box 190
John Day, OR 97845
(503) 575-1227
Lorene Allen

McKenzie Bridge

Wilamette National Forest
McKenzie Ranger District
Highway 126
McKenzie Bridge, OR 97413
(503) 822-3381
Pete Cecil

Medford

Southern Oregon Historical Society
106 North Central Avenue
Medford, OR 97501
(503) 773-6536
Samuel J. Wegner

Milwaukie

City of Milwaukie
10722 South East Main Street
Milwaukie, OR 97222
(503) 659-5171
Maggie Collins

Neskowin

Andres Von Foerster
P.O. Box 849
Neskowin, OR 97149
(503) 392-4600

Oakland

Louise J. Stearns
506 Green Valley Road
Oakland, OR 97462

Oregon City

Clackamas County Historic
Landmarks Commission
Department of Transportation and
Development
902 Abernathy Road
Oregon City, OR 97045
Pam Hayden

Brady Sheets
13862 South Carus Road
Oregon City, OR 97045
(503) 657-8903

Portland

Association for Portland Progress
520 South West Yamhill Street
Room 1000
Portland, OR 97204-1333
Steffeni M. Gray

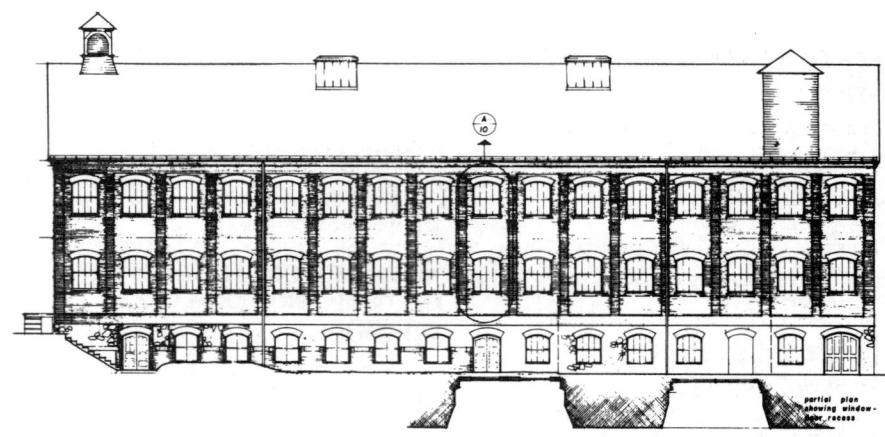

Thomas Kay Woolen Mill, Salem, Oregon. (Hanns, HABS)

Historic Preservation League of Oregon
P.O. Box 40053
Portland, OR 97240
(503) 243-1923
Eric L. Eiseman

Marque Lampert
2510 North East 11th
Apartment 1
Portland, OR 97212
(503) 281-6925

Olde House Restoration
5430 South West Ames Way
Portland, OR 97225
(503) 244-0012
Mike Byrnes

Oregon Downtown Development
Association
921 South West Morrison Street
Suite 508
Portland, OR 97205-2734
(503) 222-2182
Brian Scott

Portland Development Commission
1120 South West 5th Avenue
Suite 1102
Portland, OR 97204
(503) 796-5328
Judith Rees

W.W. Clark Memorial Library
University of Portland
P.O. Box 83017
Portland, OR 97283

Roseburg

Douglas County
Planning Department
Courthouse Annex 2
205 South East Jackson Street
Roseburg, OR 97470
(503) 440-4289
Betty L. Mack

Salem

Mission Mill Museum Association
1313 Mill Street, S.E.
Salem, OR 97301
(503) 858-7012
Bruce Wulf

Oregon State Historic Office
525 Trade Street, S.E.
Salem, OR 97310
(503) 378-5001
James Hamrick

Save the Elsinore Theatre Committee
1070 Saginaw Street, South
Salem, OR 97302
(503) 371-6171
Marian Milligan

State Historic Preservation Office
Division of State Parks Recreation
525 Trade Street, S.E.
Salem, OR 97310
James Hamrick

Springfield

Springfield Historical Commission
225 North 5th Street
Springfield, OR 97477
(503) 726-3775
Lydia Neill

Tualatin

Pilgrim John Howland Society
17835 Southwest Shasta Trail
Tualatin, OR 97062
(503) 691-2090
Merwin F. Almy

Vale

Malheur Historical Project
P.O. Box 413
Vale, OR 97918
(503) 473-3080
Marion Caputi

PENNSYLVANIA

STATE HISTORIC PRESERVATION OFFICE

Pennsylvania Historical and Museum Commission
P.O. Box 1026
Harrisburg, PA 17108
(717) 787-2891
Brent D. Glass, Executive Director, SHPO
Brenda Barrett, Director
Bureau for Historic Preservation
Deputy SHPO
(717) 787-2891
(717) 783-1073 (Fax)

STATEWIDE PRESERVATION ORGANIZATION

Preservation Pennsylvania
2470 Kissell Hill Road
Lancaster, PA 17601
(717) 569-2243
Henry Jordan, President
Grace Gary, Executive Director

NATIONAL TRUST REGIONAL OFFICE

Mid-Atlantic Regional Office
Cliveden
6401 Germantown Avenue
Philadelphia, PA 19144
(215) 438-2886
(215) 438-2892 (Fax)
Patricia D. Wilson, Director

NATIONAL TRUST ADVISOR

Mary Werner DeNadai, AIA
John Milner Architects
Route 1 and Route 100
P.O. Box 937
Chadds Ford, PA 19317
(215) 388-0111
(215) 388-0119

PRESERVATION ACTION COORDINATORS

William S. Blades
Philadelphia Historic Preservation Corporation
1616 Walnut Street
Suite 2210
Philadelphia, PA 19103
(215) 546-1146

Grace Gary
Preservation Fund of Pennsylvania
2470 Kissel Road
Lancaster, PA 17601
(717) 569-2243

AIA STATE PRESERVATION COORDINATOR

Elliot J. Rothschild, AIA
Rothschild Company Architects
616 South American Street
Philadelphia, PA 19147
(215) 928-9410

NATIONAL PARK SERVICE REGIONAL OFFICE

Mid-Atlantic Regional Office
143 South 3rd Street
Philadelphia, PA 19106
(215) 597-7018

ADVISORY COUNCIL ON HISTORIC PRESERVATION DIVISION

Eastern Office of Project Review
1100 Pennsylvania Avenue, N.W.
Suite 809
Washington, DC 20004
(202) 786-0503
Don L. Klima, Director

Allentown

City of Allentown
Bureau of Planning
City Hall
Room 320
435 Hamilton Street
Allentown, PA 18101
(215) 437-7613

John Harry
1820 Roth Avenue
Allentown, PA 18104
(215) 433-3425

Lehigh County Historical Society
Old Courthouse
Hamilton at 5th
Allentown, PA 18101
(215) 435-4664
Carol B. Wickkiser

Altoona

Altoona Main Street Program
1107 12th Street
Bankers and Barristers Building
Altoona, PA 16601
(814) 942-9083
Jane Sheffield

Ambler

Wissahickon Valley Historical Society
P.O. Box 96
Ambler, PA 19002
(215) 542-9381
Mary L. McFarland

Wissahickon Valley Watershed Association
12 Morris Road
Ambler, PA 19002
David Froehlich

Ardmore

Robert L. DeSilets
Architect
154 Sutton Road
Ardmore, PA 19003
(215) 642-5039
Bob DeSilets

Friends of the Lower Merion Academy
Lower Merion School District
301 Montgomery Avenue
Ardmore, PA 19003
Bonnie Atwood

Beaver Falls

John Cambra
Jan Cambra
3623 Ann Street
Beaver Falls, PA 15010
(412) 843-5571

Mazzant Painting
2524 9th Avenue
Beaver Falls, PA 15010
(412) 846-7733
Robert Mazzant

Bedford

Bedford County Planning Commission
203 South Julian Street
Bedford, PA 15522
Jeffry W. Kloss

Borough of Bedford
244 West Penn Street
Bedford, PA 15522
(814) 623-8192
John Montgomery

Bellefonte

Centre County Office of Housing and Community Development
105 South Allegheny Street
Bellefonte, PA 16823
(814) 355-6830
Patrick Casher

Berwick

Masonry Preservation Service
P.O. Box 324
Berwick, PA 18603
(717) 752-3607
Douglas L. Nieman

Ebenezer Maxwell House, Germantown, Philadelphia, Pennsylvania.

(A.C. Morrison, HABS)

Bethlehem

K.C.L. Associates
P.O. Box 1488
Bethlehem, PA 18016-1488
(215) 822-8340
Kenneth C. Loush

Judith D. Maish
R.R. 9
Box 254
Bingen Road
Bethlehem, PA 18015
(215) 866-4030

Transamerica Insurance Group
1560 Valley Center Parkway
Bethlehem, PA 18017
(215) 974-7760
Jim Lillegard

Bloomsburg

Pennsylvania Downtown Center
301 East 2nd Street
Bloomsburg, PA 17815
(717) 784-0456
Beth C. Spokas

Braddock

The Braddock's Field Historical Society
419 Library Street
Braddock, PA 15104

Brookville

Borough of Brookville
2 Jefferson Court
Brookville, PA 15825
(814) 849-5321
David L. Taylor

Brush Valley

Thomas R. Harley
Box 249
Brush Valley, PA 15720
(412) 479-2828

Butler

Butler County Historical Society
P.O. Box 414
Butler, PA 16003
(412) 283-8116
Diana M. Ames

Carlisle

Association of Professional Landscape
Designers
556 West Penn Street
Carlisle, PA 17013
(717) 243-8046
Dennis J. Rydberg

Historic Carlisle
3 Irvine Row
Carlisle, PA 17013

Chadds Ford

Brandywine Conservancy
P.O. Box 141
Chadds Ford, PA 19317
(215) 459-1900
Patricia A. Maley

Chadds Ford Historical Society
P.O. Box 27
Chadds Ford, PA 19317
(215) 388-7376

John Milner Architects
Route 1 and Route 100
P.O. Box 937
Chadds Ford, PA 19317
(215) 388-0111
Mary W. DeNadai

Chester

Chester Historic Preservation
Commission
2320 Chestnut Street
Chester, PA 19013
(215) 436-2066
Dave Guleke

Delaware County Historical Library
Widener University
Chester, PA 19013
(215) 874-6444

Trinity United Methodist Church
8th and Butler Streets
Chester, PA 19013
(215) 499-3701
D.E. Italiano

Chester Springs

Historic Yellow Springs
Art School Road
P.O. Box 627
Chester Springs, PA 19425
(215) 827-7414
Sandra S. Momyer

Coatsville

Coatsville Main Street
53 South 1st Avenue
City Hall
Coatesville, PA 19320
(215) 383-0584
Diane F. Bernardo

Conshohocken

Eastern National Park and Monument
Association
446 North Lane
Conshohocken, PA 19428
(215) 238-6672
George Minnucci

Cornwall

Cornwall Manor
P.O. Box 125
Cornwall, PA 17016
(717) 273-2647
Anne C. Peiffer

Devon

Jennifer E. Balson
227 Lancaster Avenue
Devon, PA 19333
(215) 688-0316

Doylestown

Bucks County Conservancy
85 Old Dublin Pike
Doylestown, PA 18901
(215) 345-8966
(215) 345-7020
Jeffrey L. Marshall

Drexel Hill

Alumni Association
Delta Omicron Chapter
Theta Chi Fraternity
719 Anderson Avenue
Drexel Hill, PA 19026
(215) 565-2896
Jeffrey H. Glisson

East Berlin

East Berlin Historical Preservation
Society
P.O. Box 73
East Berlin, PA 17316
Vivian E. Lighty

Easton

Historic Easton
Box 994
Easton, PA 18042

Jacob Bachman Easton House Tavern
P.O. Box 1414
Easton, PA 18042
Maureen McAteer

Edinboro

Edinboro University
Reeder Hall
Edinboro, PA 16444
(814) 732-2585
David O'Dessa

Erie

Erie County Historical Society
417 State Street
Erie, PA 16501
(814) 454-1813
Donald Muller

Exton

West Whiteland Historical Commission
222 North Pottstown Pike
Exton, PA 19341
(215) 363-8091
Diane S. Snyder

Fallsington

Historic Fallsington
4 Yardley Avenue
Fallsington, PA 19054
(215) 295-6567
Jacqueline V. Prior

Fort Washington

Highlands Historical Society
7001 Sheaff Lane
Fort Washington, PA 19034
(215) 641-2687
Catherine G. Hoffman-Lynch

Gettysburg

Borough of Gettysburg/HARB
34 East Middle Street
Gettysburg, PA 17325
(717) 334-1160
Walter Powell

Historic Gettysburg-Adams County
12 Lincoln Square
Gettysburg, PA 17325
(717) 334-8188
Chester S. Byers

Main Street Gettysburg
16 East Middle Street
Gettysburg, PA 17325

Gladwyne

Friends of Historic Rittenhousetown
1541 Waverly Road
Gladwyne, PA 19035
(215) 649-7335
Hugh B. Hanson

Glen Mills

Nicholas Newlin Foundation
South Cheyney Road
P.O. Box 219
Glen Mills, PA 19342

Glenside

Dennis Cline
435 Elm Avenue
Glenside, PA 19038
(215) 572-7739

Jeanne McMinoes
638 Hillcrest Avenue
Glenside, PA 19038
(215) 885-1875

Harrisburg

Bureau for Historic Preservation
Pennsylvania Historic and Museum
Commission
Box 1026
Harrisburg, PA 17120
(717) 783-9927
Toni Gilson

Bureau of Housing and Development
Forum Building
Room 506
Harrisburg, PA 17120
(717) 783-3068
Diana Kerr

Crabtree Rohrbaugh and Associates
20 North Market Square
Harrisburg, PA 17101
(717) 233-0127
Richard C. Leblanc

Friends of Pennsylvania Historical and
Museum Commission
Box 11466
Harrisburg, PA 17108
Marcia Gobrecht

Harrisburg Community and Economic
Affairs
32 Evergreen Street
Harrisburg, PA 17401
(717) 238-3185

Historic Harrisburg Association
P.O. Box 951
Harrisburg, PA 17108
(717) 233-4646
David Morrison

Pennsylvania Historic and Museum
Commission
P.O. Box 1026
Harrisburg, PA 17108
(717) 787-2891
Brent D. Glass

Hatboro

Millbrook Society
42 Harding Avenue
P.O. Box 506
Hatboro, PA 19040
(215) 675-0119
David T. Shannon

Hershey

Antique Auto Club of America
501 West Governor Road
P.O. Box 417
Hershey, PA 17033
(717) 534-1910
William H. Smith

Homestead

Mon Valley Initiative
303/305 East 8th Avenue
Homestead, PA 15120
(412) 464-4000
Blythe Merrill

Horsham

Friends of Graeme Park
859 County Line Road
Horsham, PA 19044
(215) 243-0965
Mary Bluder

Indiana

Downtown Indiana
655 Philadelphia Street
Room 209
Indiana, PA 15701
(412) 463-6110
Peg Robinson

Irwin

Downtown Commission
412 Main Street
Irwin, PA 15642
(412) 864-3100
William A. Kunkle

Lancaster

Heritage Center of Lancaster County
Center Square
Lancaster, PA 17604

Historic Preservation Trust of Lancaster
County
123 North Prince Street
Lancaster, PA 17603
(717) 291-5861
David B. Schneider

Lancaster County Historical
Society
230 North President Avenue
Lancaster, PA 17603
(717) 392-4633
Debbie Smith

Lancaster County Planning
Commission
50 North Duke Street
P.O. Box 3480
Lancaster, PA 17603
(717) 299-8333
Ronald T. Bailey

Preservation Fund of Pennsylvania
2470 Kissell Hill Road
Lancaster, PA 17601
(717) 569-2243
Grace Gary

Stevens State School of
Technology-Library
750 East King Street
Lancaster, PA 17602

Lansdale

John Ernst
400 Perkiomen Avenue
Lansdale, PA 19446
(215) 855-2937
John J. Ernst

Lansdale Borough Hall
1 Vine Street
Lansdale, PA 19446
(215) 368-1691
F. Mangan

Lewisburg

SEDA Council of Governments
Timberhaven
R.D. 1
Lewisburg, PA 17837
(717) 524-4491
Tom Grbenick

Lewistown

Johnson Design and Construction
12 Montgomery Avenue
Lewistown, PA 17044
(717) 248-9232
Cordell Johnson

Lima

Tyler Arboretum
P.O. Box 216
Lima, PA 19037
(215) 566-5431
Rick Colbert

Lock Haven

Clinton County Historical Society
362 East Water Street
Lock Haven, PA 17745
(717) 748-7254
Virginia Edmonston

Historic Directions
P.O. Box 222
317 East Main Street
Lock Haven, PA 17745
(717) 748-6220

Malvern

Marian L. Pyle
36 Yellow Springs Road
R.D. 1
Malvern, PA 19355
(215) 827-7814

Township of Willistown
Sugartown Road
Rural Delivery 4
Box 67
Malvern, PA 19355

Penelope P. Wilson
1371 North Valley Road
Malvern, PA 19355

Meadville

Redevelopment Authority
R.D. 2
Meadville, PA 16335
(814) 724-2975
Roberta G. Rourke

Mechanicsburg

Tita Eberly
143 West Main Stret
Mechanicsburg, PA 17055
(717) 766-5628

Media

Delaware County Planning Department
2nd and Orange Streets
Media, PA 19063
(215) 891-5205
Jill Kelly

Merion Station

Pamela W. Fox
511 Howe Road
Merion Station, PA 19066
(215) 667-5654

Middletown

Whittle International Group
403 Market
Middletown, PA 17057-1150
(717) 944-2508
Brett M. Whittle

Milford

USDA Forest Services
Grey Towers
P.O. Box 188
Milford, PA 18337
(717) 296-6401
Frank Swancara

Millsboro

Penn Craft Community Association
R.D. 1
Penn Craft East
Millsboro, PA 15433
Marlene Anderson

Milton

Milton Area Chamber of Commerce
P.O. Box 118
Milton, PA 17847-0118
(717) 742-7341
David L. Smith

Mont Clare

Schuylkill Canal Association
P.O. Box 3
Mont Clare, PA 19453
(215) 933-2127
Neal G. Thorpe

Morrisville

Pennsbury Society
Pennsbury Manor
400 Pennsbury Memorial Road
Morrisville, PA 19067
(215) 946-0400
Alice Hemenway

New Holland

Christian Eaby
405 Peters Road
New Holland, PA 17557

Newtown

Bucks County Community College
Social and Behavioral Science
Department
Newtown, PA 18940
(215) 968-8270
Lyle L. Rosenberger

Friends of the Farmstead
P.O. Box 844
Newtown, PA 18940-0844
(215) 968-3085
Roger Taylor

Newtown Historic Association
P.O. Box 303
Newtown, PA 18940
(215) 968-4004
Francella Smith

Norristown

Montgomery County
Planning Commission
Court House
Norristown, PA 19404
(215) 278-3757
Arthur F. Loeben

Norristown Preservation Society
P.O. Box 1851
Norristown, PA 19404
Martin Singleton

North East

North East Chamber of Commerce
2 East Main Street
P.O. Box 466
North East, PA 16428
(814) 725-4262
Scott Eckert

Philadelphia

Building Conservation International
1924 Arch Street
Philadelphia, PA 19103
(215) 567-0547
Gersil N. Kay

Ebenezer Maxwell Mansion
200 West Tulpehocken Street
Philadelphia, PA 19144
(215) 438-1861
Kathy S. Moses

Elfreth's Alley Association
126 Elfreth's Alley
Philadelphia, PA 19106

Fairmont Park Commission
Memorial Hall
West Park
Philadelphia, PA 19131
(215) 686-0046
Johanna Cairo

Free Library of Philadelphia
Serials Section
Logan Square
Philadelphia, PA 19103

Glen Foerd Conservation Corporation
5001 Grant Avenue
Philadelphia, PA 19114
(215) 632-5330

Grand Army of the Republic Memorial
Museum
4278 Griscom Street
Philadelphia, PA 19124
(215) 673-1688
E. Atkinson

Greater Germantown Housing
Development Corporation
48 East Penn Street
Geneva Larkford
Philadelphia, PA 19144

Henry George Birthplace
413 South 10th Street
Philadelphia, PA 19147
(215) 922-4278
Michael K. Curtis

David Hollenberg
255 South 44th Street
Philadelphia, PA 19104

Institute For Scientific Information
3501 Market Street
Publication Processing Department
Philadelphia, PA 19104

Jan M. Istad
1901 John F. Kennedy Boulevard
Suite 1306
Philadelphia, PA 19103
(215) 592-0900

G. Eric Johansen
213 Monroe Street
Philadelphia, PA 19147
(215) 928-8237

Laurel Hill Mansion
Edgely Road
Fairmont Park East
Philadelphia, PA 19121
(215) 235-1776
Dolores Fenlon

Robert E. Linck
725 Walnut Street
Suite 2C
Philadelphia, PA 19106-3240
(215) 574-9649

John E. McGaw
129 Beck Street
Philadelphia, PA 19147
(215) 755-4587

H. J. Magaziner
2036 Rittenhouse Square
Philadelphia, PA 19103
(215) 732-8285

Naomi Wood Collection at Woodford
Mansion
33rd and Dauphin Street
Philadelphia, PA 19132
Lawrence Berger

National Society of the Colonial Dames
of America
Stenton Mansion
1630 Latimer Street
Philadelphia, PA 19103
(215) 735-6737
Mary Costello

Old City Civic Association
156 North 3rd Street
Philadelphia, PA 19106

Lyssa Papazian
1520 South 10th Street
Philadelphia, PA 19147
(215) 271-5635

Partners for Sacred Places
1616 Walnut Street
Suite 2210
Philadelphia, PA 19103
(215) 546-1288

Philadephia City Planning Commission
Library
1515 Market Street
17th Floor
Philadelphia, PA 19102
(215) 686-4625

Philadelphia Historical Commission
1301 UGI Building
1401 Arch Street
Philadelphia, PA 19102

Philadelphia Historic Preservation
Corporation
1616 Walnut Street
Suite 2210
Philadelphia, PA 19107
(215) 546-1146
Donna A. Harris

Philadelphia Maritime Museum
321 Chestnut Street
Philadelphia, PA 19106
(215) 925-5439
John S. Carter

Philadelphia Society for Preservation of
Landmarks
321 South 4th Street
Philadelphia, PA 19106
(215) 925-2251
Margo Burnette

Preservation Coalition of Greater
Philadelphia
250 South 16th Street
Philadelphia, PA 19102
(215) 546-0531
Howard J. Kittell

Reading Terminal Market Preservation
Fund
1616 Walnut Street
Philadelphia, PA 19103
(215) 732-0775
William A. Kingsley

Ronald Shur
214 East Mount Airy Avenue
Apartment 1
Philadelphia, PA 19119
(215) 248-7963

Saint Michael Lutheran Church
6671 Germantown Avenue
Philadelphia, PA 19119
(215) 848-0199
Janet Peteman

Robert P. Thomas
3624 Hamilton Street
Philadelphia, PA 19104
(215) 387-9242

Tony Atkin and Associates
125 South 9th Street
Suite 900
Philadelphia, PA 19107
(215) 925-7812
Charles Evers

University City Historical Society
Woodlands Mansion
40th and Woodland Avenue
Philadelphia, PA 19104
(215) 387-3019
Rebecca Trumbull

University of Pennsylvania
Facilities Development Office
3451 Walnut Street
Philadelphia, PA 19104
(215) 898-5831
Titus D. Hewryk

University of Pennsylvania
Graduate School of Fine Arts
Historic Preservation Program
214 Meyerson Hall
Philadelphia, PA 19104-6311
(215) 898-3169

Upsala Foundation
6430 Germantown Avenue
Philadelphia, PA 19119
(215) 842-1798
Steven Berman

Phoenixville

Dwain Hayter
352 Washington Avenue
Phoenixville, PA 19460
(215) 933-2433

Pittsburgh

John R. Axtell
279 Main Street
Pittsburgh, PA 15201
(412) 682-7202

William E. Brocious
412 South Highland Avenue
Pittsburgh, PA 15206
(412) 363-4362

Carnegie Library of Pittsburgh
Acquisitions Department/Periodicals
4400 Forbes Avenue
Pittsburgh, PA 15213

Chatham College
Woodland Road
Pittsburgh, PA 15232
(412) 365-1135
Susan Piros

Committee for the Preservation of Sag
Harbor
612 South Dallas Avenue
Pittsburgh, PA 15217

Fallingwater
316 4th Avenue
Pittsburgh, PA 15222
(412) 288-2766
Thomas M. Schmidt

David Henderson
157 Dickson Avenue
Pittsburgh, PA 15202
(412) 394-7012

Historical Society of Western
Pennsylvania
4338 Bigelow Boulevard
Pittsburgh, PA 15213
(412) 681-5533

Henry P. Hoffstot
5057 5th Avenue
Pittsburgh, PA 15232
(412) 288-3154

Landmarks Design Associates,
Architects
400 Landmarks Building
1 Station Square
Pittsburgh, PA 15219
(412) 391-7640
Ellis L. Scmidlapp

Main Street on East Carson
1417 East Carson Street
Pittsburgh, PA 15203
(412) 481-0660

Beverly Mitchum
7118 Card Lane
Pittsburgh, PA 15208
(412) 244-8754

Gerald Lee Morosco
50 South 15th Street
Pittsburgh, PA 15203
(412) 431-4347

Pennsylvania Railway Museum
Association
Arden Trolley Museum
P.O. Box 832
Pittsburgh, PA 15230

Pittsburgh History & Landmarks
Foundation
450 Landmarks Building
1 Station Square
Pittsburgh, PA 15219
(412) 471-5808
Louise Sturgess

Society for the Preservation of the
Duquesne Heights Incline
1220 Grandview Avenue
Pittsburgh, PA 15211
(412) 381-1665
David H. Miller

South Hills Orthopedic Surgery
Associates
363 Vanadium Road
Pittsburgh, PA 15243-1498

Urban Redevelopment Authority of
Pittsburgh
200 Ross Street
Pittsburgh, PA 15219
(412) 255-6540
Evan Stoddard

Plymouth Meeting

Plymouth Meeting Historical Society
Library
Box 167
Plymouth Meeting, PA 19462
(215) 828-8111
Ed Addison

Pottstown

French and Pickering Creeks
Conservation Trust
Rural Delivery 7
Box 360
Pottstown, PA 19464
(215) 469-0150
Samuel W. Morris

Pottsville

Bureau of Economic and Community
Development
P.O. Box 50
401 North Centre Street
Pottsville, PA 17901
(717) 628-4417
Richard J. Schuettler

Punxsutawney

Punxsutawney Main Street
123 South Gilpin Street
Punxsutawney, PA 15767
(814) 938-7701
Nanci A. Puchy

Reading

Bureau of Planning
City Hall
8th and Washington Streets
Reading, PA 19601
Michelle Nicholl

Centre Park Historic District
P.O. Box 13325
Reading, PA 19612
(215) 372-4019
Michele Lauter

Friends of the Astor
P.O. Drawer 1618
Reading, PA 19603
Louis R. Perugini

Muhlenberg/Greene Architects, Ltd.
The Madison
400 Washington Street
Reading, PA 19601
(215) 376-4927
L. A. Greene

Mrs. Joseph Pendleton
1802 Hill Road
Reading, PA 19602-1505

Ridley Park

Stephen B. Laxton
427 Bartlett Avenue
Ridley Park, PA 19078
(215) 586-4749

Scottdale

Greater Scottdale Chamber of
Commerce
Box 276
Scottdale, PA 15683
(4120 887-6350
Richard R. Campbell

Scranton

Architectural Heritage Association
Box 1301
Scranton, PA 18510
(717) 347-6076
Nancy Bisignani

Lackawanna Historical Society
232 Monroe Avenue
Scranton, PA 18510
(717) 344-3841
Mary Ellen Calemmo

Sewickley

F. J. Stevenson III
616 Broad Street
Sewickley, PA 15143
(712) 741-0316

Sharon Hill

Sharon Hill Historical Commission
c/o 726 Poplar Street
Sharon Hill, PA 19079
(215) 583-2757
Kathleen Dowd

Sunbury

Philip Michael Clark
225 Market Street
Sunbury, PA 17801
(717) 286-7768

Community Development Office
City of Sunbury
225 Market Street
Sunbury, PA 17801
(717) 286-7768
Philip M. Clark

Tamaqua

Vesper Boat Club
244 East Broad Street
Tamaqua, PA 18252
(717) 668-1208
Larry A. Wittig

Trappe

Historical Society of Trappe
P.O. Box 828
Trappe, PA 19426
(215) 489-2624
John C. Shetler

Uniontown

Downtown Business District Authority
20 North Gallatin Avenue
Uniontown, PA 15401
(412) 430-2909
Laury Podolinski

Valley Forge

Valley Forge National Historical Park
Valley Forge, PA 19481
(215) 783-7700
Joan Marshall Dutcher

Vandergrift

Victorian Vandergrift Museum and
Historical Society
Municipal Building
East Wing
1895 P.O. Box 183
Vandergrift, PA 15690
Eugene Iagnemma

Villanova

Villanova Preservation Foundation
112 Radnor Avenue
Villanova, PA 19085
(215) 527-7988
Walter A. Connell

Warren

Charles Warren Stone Museum
P.O. Box 427
Warren, PA 16365
(814) 723-1795

Washington Cross

Washington Crossing Foundation
P.O. Box 1976
Washington Cross, PA 18977
(215) 493-6577
Ann H. Hutton

West Chester

Chester County Historical Society
225 North High Street
West Chester, PA 19380
(215) 692-4800
Roland H. Woodward

Chester County Historic Preservation
Office
117 West Gay Street
West Chester, PA 19380
(215) 344-6917
Jane L. Davidson

Culbertson Restoration
590 Snyder Avenue
West Chester, PA 19382
(215) 436-4455
Melissa L. Alleman

Ray H. Ott, Jr.
427 West Union Street
West Chester, PA 19382
(215) 431-7899

Thomas Comitta Associates
18 West Chestnut Street
West Chester, PA 19382
(215) 696-3896

West Chester Historical and
Architectural Review Board
Department of Building and Housing
401 East Gay Street
West Chester, PA 19380
(215) 696-1773

Whitehall

Whitehall Historical Preservation
Society
P.O. Box 39
Whitehall, PA 18052
(215) 264-2225
John C. Wieand

Wilkes-Barre

Wyoming Historical and Geological
Society
49 South Franklin Street
Wilkes-Barre, PA 18701
(717) 823-6244
Mary R. Kelly

Williamsport

Jefferson G. Porter
Debralee C. Porter
329 Maynard Street
Williamsport, PA 17701
(717) 322-5411

Wyomissing

Berks County Conservancy
960 Old Mill Road
Wyomissing, PA 19610
(215) 372-4992
Phoebe Hopkins

York

Historic York
P.O. Box 2312
York, PA 17405
(717) 843-0320
Melinda G. Higgins

Main Street York
1 East Market Street
York, PA 17405
(717) 848-4000
Anne R. Druck

RHODE ISLAND

STATE HISTORIC PRESERVATION OFFICE

Rhode Island Historical Preservation Commission
Old State House
150 Benefit Street
Providence, RI 02903
(401) 277-2678
(401) 277-2968 (Fax)
Frederick C. Williamson, SHPO
Edward F. Sanderson, Executive Director
Rhode Island Historical Preservation Commission
Deputy SHPO

STATEWIDE PRESERVATION ORGANIZATION

Heritage Trust of Rhode Island
199 Hope Street
Providence, RI 02906
(401) 253-2707
Mark Zelonis, Executive Director

NATIONAL TRUST REGIONAL OFFICE

Northeast Regional Office
7 Faneuil Hall Marketplace
5th Floor
Boston, MA 02109
(617) 523-0885
(617) 523-1199 (Fax)
Vicki Sandstead, Director

NATIONAL TRUST ADVISORS

Sean O. Coffey
22 James Street
Providence, RI 02903
(401) 421-8030

Alfred B. Van Liew
1 Regency Plaza
Suite 1
Providence, RI 02903
(401) 272-2531
(401) 272-6590 (Fax)

PRESERVATION ACTION COORDINATORS

Bonnie Warren
Preservation Action/Rhode Island
18 Homestead Avenue
Barrington, RI 02806
(401) 277-2678

AIA STATE PRESERVATION COORDINATOR

Vacant

NATIONAL PARK SERVICE REGIONAL OFFICE

North Atlantic Regional Office
15 State Street
Boston, MA 02109-3572
(617) 223-5001

ADVISORY COUNCIL ON HISTORIC PRESERVATION DIVISION

Eastern Office of Project Review
1100 Pennsylvania Avenue, N.W.
Suite 809
Washington, DC 20004
(202) 786-0503
Don L. Klima, Director

Block Island

Block Island
Historical Society
P.O. Box 79
Block Island, RI 02807
(401) 466-2481
Lisa D. Nolan

Bristol

Bristol Historic District Commission
10 Court Street
Bristol, RI 02809
(401) 253-7000
Dory Skemp

Central Falls

Historic Central Falls
507 Broad Street
Central Falls, RI 02863
(401) 726-0800
M. C. Byrnes

Cranston

Steven Baskin
34 Cliffside Drive
Cranston, RI 02920

Cranston Historical District Commission
39 Forsythia Lane
Cranston, RI 02920
Robert M. Drew

Friends of Nobska
128 Ocean Avenue
Cranston, RI 02905
Robert Cleasby

Cumberland

Blackstone Valley
Tourism Council
P.O. Box 7663
Cumberland, RI 02864
(401) 722-1839
Robert D. Billington

Cumberland Historic District
Commission
Town Hall
Cumberland, RI 02864
R. D. Billington

East Greenwich

Town of East Greenwich
Department of Planning
100 Pierce Street
P.O. Box 111
East Greenwich, RI 02818
(401) 886-8645

East Providence

Carousel Park Commission
East Providence City Hall
145 Taunton Avenue
East Providence, RI 02914
(401) 434-3311
Donna McMahon

Manville

Healy Brothers
18 New River Road
P.O. Box 4
Manville, RI 02838

Middletown

Middletown Historical Society
P.O. Box 4196
Middletown, RI 02840
Stanley Grossman

Narragansett

Town of Narragansett
25 5th Avenue
Narragansett, RI 02882
(401) 789-1044
C. A. Collins

Newport

International Tennis Hall of Fame
194 Bellevue Avenue
Newport, RI 02840
(401) 849-3990
(401) 849-6378
Jan Armstrong

Preservation Society of Newport County
118 Mill Street
Newport, RI 02840
(401) 847-1000
Frank R. Winnert

North Kingstown

Town of North Kingstown
80 Boston Neck Road
North Kingstown, RI 02852
(401) 294-3331

Chateau-sur-Mer, Newport, Rhode Island. (P. Veeder, HABS)

Pawtucket

City of Pawtucket
Department of Planning and
Development
200 Main Street
Suite 210
Pawtucket, RI 02860
(401) 724-5200
Roger F. Giraud

Preservation Society of Pawtucket
97 Walcott Street
Pawtucket, RI 02860
(401) 725-9581

Providence

City of Providence
Providence City Archives
25 Dorrance Street
Providence, RI 02903
(401) 421-7740
Carole B. Pace

Elmwood Foundation
1 Trinity Square
Providence, RI 02907
(401) 273-2330
April H. Wolf

Heritage Trust of Rhode Island
199 Hope Street
Providence, RI 02906
(401) 253-2707
Mark Zelonis

Irving B. Haynes and Associates
Architects
1 Park Row
Providence, RI 02903
(401) 274-1555
Cornelis J. de Boer

Karen Jessup
Preservation Consultant
223 Morris Avenue
Providence, RI 02906
(401) 521-0486

Providence Historical District
Commission
Department of Planning and
Development
400 Westminster Street
Providence, RI 02903
Mary P. Turkel

Providence Preservation Society
21 Meeting Street
Providence, RI 02903
(401) 831-7440
Wendy Nicholas

Providence Preservation Society
Revolving Fund
24 Meeting Street
Providence, RI 02903
(401) 272-2760
Clark Schoettle

Rhode Island Historic Preservation
Commission
150 Benefit Street
Providence, RI 02903
(401) 277-2678
Edward F. Sanderson

Rhode Island Historical Society Library
110 Benevolent Street
Providence, RI 02906
(401) 331-8575
Albert Klyberg

Rhode Island State Law Library
Frank Licht Judicial Complex
250 Benefit Street
Providence, RI 02903
(401) 277-3275
Kendall F. Svengalis

Urban Design Group
68 South Main Street
Providence, RI 02903
(401) 521-0096
Ron Wood

Warwick

Department of City Planning
Warwick City Hall
3275 Post Road
Warwick, RI 02886
(401) 738-2000
Stuart Boggs

West Kingston

Friends of Kingston Railroad Station
P.O. Box 191
West Kingston, RI 02892
(401) 295-0783
Arthur D. Strauss

Woonsocket

Main Street 2000 Development
Corporation
162 Main Street
P.O. Box 1259
Woonsocket, RI 02895
(401) 769-2322
Jeffrey O'Connell

SOUTH CAROLINA

STATE HISTORIC PRESERVATION OFFICE

Department of Archives and History
P.O. Box 11669
Columbia, SC 29211
(803) 734-8592
(803) 734-8820 (Fax)

George Vogt, Director, SHPO
Mary Watson Edmonds
Director of Historic Preservation
Deputy SHPO
(803) 734-8593

STATEWIDE PRESERVATION ORGANIZATION

Palmetto Trust for Historic
Preservation
P.O. Box 12547
Columbia, SC 29211
Bob Bainbridge, President

NATIONAL TRUST REGIONAL OFFICE

Southern Regional Office
456 King Street
Charleston, SC 29403
(803) 722-8552
(803) 722-8652 (Fax)
Susan A. Kidd, Director

NATIONAL TRUST ADVISORS

Howard E. Duvall, Jr.
South Carolina Municipal Association
Box 12109
Columbia, SC 29211
(803) 799-9574

Evelyn M. McGee
131 Church Street
Charleston, SC 29401
(803) 722-3049

PRESERVATION ACTION COORDINATOR

Lawrence A. Walker
Historic Charleston Foundation
51 Meeting Street
Charleston, SC 29401
(803) 723-1623

AIA STATE PRESERVATION COORDINATOR

Vacant

NATIONAL PARK SERVICE REGIONAL OFFICE

Southeast Regional Office
Richard B. Russell Federal Building
75 Spring Street, S.W.
Atlanta, GA 30303
(404) 331-2632

ADVISORY COUNCIL ON HISTORIC PRESERVATION DIVISION

Eastern Office of Project Review
1100 Pennsylvania Avenue, N.W.
Suite 809
Washington, DC 20004
(202) 786-0503
Don L. Klima, Director

Abbeville

Abbeville County
Historical Society
P.O. Box 12
Abbeville, SC 29620
(803) 459-2466

Trinity Episcopal Church
P.O. Box 911
Abbeville, SC 29620
(803) 459-5186
Oscar H. Reid

Aiken

Aiken County Historical Museum
433 Newberry Street, S.W.
Aiken, SC 29801
(803) 642-2015
Nana Farris

Historic Aiken Foundation
P.O. Box 959
Aiken, SC 29801

Thomas F. Maurice
P.O. Box 257
Aiken, SC 29802

Beaufort

Board of Architectural Review
P.O. Drawer 1167
Beaufort, SC 29901
(803) 524-4304
Danny E. Taylor

Historic Beaufort Foundation
P.O. Box 11
Beaufort, SC 29901
(803) 524-6334
Cynthia Cole

Bishopville

Lee County Arts Council
P.O. Box 714
Bishopville, SC 29010
(803) 484-5874
Rodney McDaniel

Charleston

Carolina Law County Girl Scouts
7257 Cross County Road
Charleston, SC 29418
(803) 552-9910
Julia B. Chaplin

Charleston County Planning
Department
2 Court House Square
Room 317
Charleston, SC 29401
(803) 723-6739
Wiiliam Miller

Charles Town Landing Foundation
1500 Old Town Road
Charleston, SC 29407
(803) 556-4450
Janson L. Cox

City of Charleston
Department of Planning and
Urban Development
116 Meeting Street
Charleston, SC 29401
(803) 724-3765
Yvonne Fortenberry

City of Charleston
Office of Cultural Affairs
133 Church Street
Charleston, SC 29401
(803) 724-7305
Diane N. Abbey

Richard Coen
49 Broad
Suite 200
Charleston, SC 29401
(803) 723-3300

Myrlle Gloscoe
125 Bull Street
Charleston, SC 29424
(803) 792-5742

Historic Charleston Foundation
51 Meeting Street
Charleston, SC 29401

Historic Charleston Foundation
Preservation Center Library
108 Meeting Street
Charleston, SC 29401
(803) 724-8485

Meadors' Construction Company
P.O. Box 21758
Charleston, SC 29413-1758
(803) 723-8585
James Meadors

National Railroad Historical Society
Charleston Chapter
P.O. Box 10081
Charleston, SC 29411-0081
J. M. Lecato

G. Legare Porcher, Jr.
P.O. Box 20986
Charleston, SC 29413
(803) 762-0803

Preservation Society of Charleston
P.O. Box 521
Charleston, SC 29402
(803) 722-4630
John W. Meffert

Revitalization Office
496 King Street
Charleston, SC 29403-5527
(803) 724-3796
John V. Deehan

Cheraw

Cheraw Appearance Commission
P.O. Box 111
Cheraw, SC 29520
(803) 537-2032
Mike Smith

Clemson

Regional Resources Development
Institute
265B Lehotsky Hall
Clemon University
Clemson, SC 29634
(803) 656-2182
R. H. Becker

Columbia

Columbia Landmarks Commission
P.O. Box 147
Columbia, SC 29217
(803) 733-8324
Nathaniel B. Land

Historic Columbia Foundation
1601 Richland Street
Columbia, SC 29201
(803) 252-7742
Daniel R. Sigmon

Municipal Association of South
Carolina
P.O. Box 12109
Columbia, SC 29211
J. M. Wray

Richland County Historic Preservation
Commission
1616 Blanding Street
Columbia, SC 29201
Nancy Vodry

South Carolina Archives and History
Department
1430 Senate Street
Columbia, SC 29211
George L. Vogt

South Carolina Downtown
Development Association
1529 Washington Street
Columbia, SC 29201
(803) 256-3560
Ben C. Boozer

Michael Trinkley
Chicora Foundation
P.O. Box 8664
Columbia, SC 29202
(803) 787-6910

University of South Carolina
Department of History
Columbia, SC 29208

Florence

Florence Heritage Foundation
P.O. Box 5802
Florence, SC 29501
Agnes Wilcox

Georgetown

Rice Museum
P.O. Box 902
Georgetown, SC 29442
(803) 546-7423

Greenwood

Steve Bailey
215 East Cambridge Avenue
Greenwood, SC 29646-223

The Museum-Greenwood
P.O. Box 3131
Greenwood, SC 29648
(803) 229-7093
Despina Panagakos

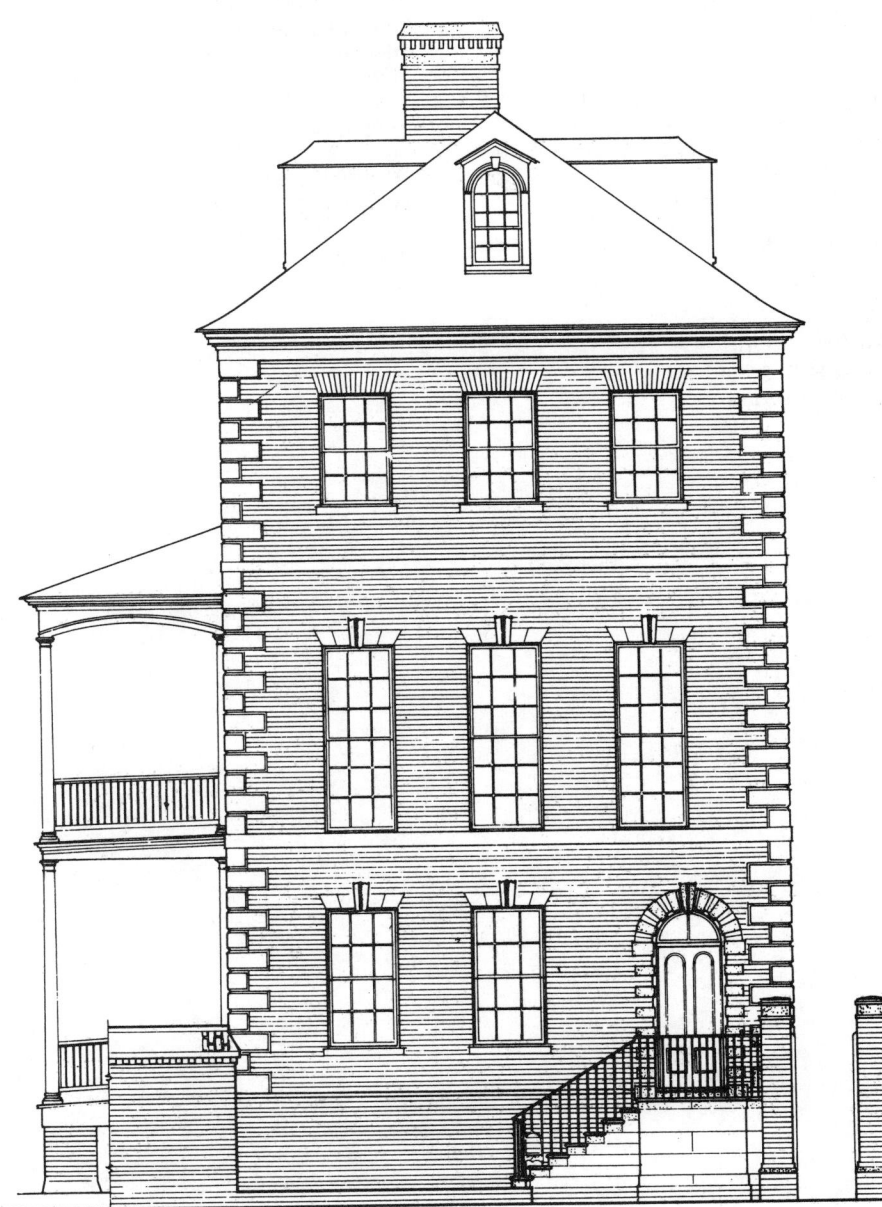

Gadsden House, Charleston, South Carolina. (J. Burnett, HABS)

Hilton Head

Vicki Warren
3 Cardinal Court
Hilton Head, SC 29928
(803) 671-1255

Laurens

Laurens County Historical Society
P.O. Box 292
Laurens, SC 29360
(803) 984-0596
Bill Cooper

McClellanville

USDA Forest Service
Francis Marion and Sumter National
Forest
P.O. Box 788
McClellanville, SC 29458
(803) 887-3257
Robert T. Morgan

McCormick

Savannah Valley Authority
P.O. Drawer K
McCormick, SC 29835
(803) 443-2168
John C. Blythe

Moncks Corner

Berkeley County Historical Society
P.O. Box 65
Moncks Corner, SC 29461
(803) 899-3220
Oliver W. Buckles

Mount Pleasant

Glenn Keyes Architects
270 West Coleman Boulevard
Suite 2B
Mount Pleasant, SC 29464-3426
(803) 722-4100
Glenn Keyes

Mary Moore Jacoby
660 Fox Pond Drive
Mount Pleasant, SC 29464
(803) 722-4630

Murrells Inlet

Duc Watts
Box 224
Murrells Inlet, SC 29576
(803) 651-6385

Pendleton

Pendleton District Historical and
Recreational Commission
125 East Queen Street
P.O. Box 565
Pendleton, SC 29670
(803) 646-3782
Hurley Badders

Rock Hill

City of Rock Hill
P.O. Box 11706
Rock Hill, SC 29730
Stephen Turner

Seneca

Community Development Program
P.O. Box 4773
Seneca, SC 29679
Frank E. Wise

Spartanburg

Elizabethine G. McClure
1040 Partridge Road
Spartanburg, SC 29302
(803) 573-9554

Summerville

Summerville Preservation Society
P.O. Box 511
Summerville, SC 29483

Town of Summerville
104 Civic Center
Summerville, SC 29483
(803) 871-6000
Joseph Christie

Winnsboro

Mary Ann Hayes
120 North Zion Street
Winnsboro, SC 29180

SOUTH DAKOTA

STATE HISTORIC PRESERVATION OFFICE

South Dakota State Historical Society
Cultural Heritage Center
900 Governors Drive
Pierre, SD 57501
(605) 773-3458
J.R. Fishburne, Director, SHPO

Paul M. Putz, Director
State Historical Preservation Center,
Deputy SHPO
P.O. Box 417
Vermillion, SD 57069
(605) 677-5314
(605) 577-5364 (Fax)

STATEWIDE PRESERVATION ORGANIZATION

Historic South Dakota Foundation
P.O. Box 2998
Rapid City, SD 57709
(605) 341-5820
William Wyatt, President
James Quinn, Executive Director

NATIONAL TRUST REGIONAL OFFICE

Mountains/Plains Regional Office
511 16th Street
Suite 700
Denver, CO 80202
(303) 623-1504
(303) 825-8073 (Fax)
Barbara J. Pahl, Director

NATIONAL TRUST ADVISORS

Spencer Ruff
207 East 23rd Street
Sioux Falls, SD 57105
(605) 331-5413
(605) 331-2101 (Fax)

William R. Wyatt
4275 Starlite Drive
Rapid City, SD 57702
(605) 342-7505

PRESERVATION ACTION COORDINATOR

Peg Lamont
P.O. Box 1415
Aberdeen, SD 57402
(605) 225-8942

AIA STATE PRESERVATION COORDINATOR

Roger G. Hartman, AIA
Rysavy Hartman Associates
2204 South Minnesota Avenue
Sioux Falls, SD 57105
(605) 336-3568

NATIONAL PARK SERVICE REGIONAL OFFICE

Rocky Mountain Regional Office
12795 West Alameda Parkway
P. O. Box 25287
Denver, CO 80225-0287
(303) 969-2875

ADVISORY COUNCIL ON HISTORIC PRESERVATION DIVISION

Western Office of Project Review
730 Simms Street
Room 401
Golden, CO 80401
(303) 231-5320
Claudia Nissley, Director

Old Fort Randall Church, Gregory County, South Dakota. (H.E. Anderson, HABS)

Belle Fourche

Belle Fourche Main Street Program
P.O. Box 341
Belle Fourche, SD 57717
Kevin Kuchenbecker

Brookings

Brookings Historic Preservation
Commission
P.O. Box 270
Brookings, SD 57006
Shari Dunn

Custer

John Gausman
Route 2
Box 201 G
Custer, SD 57730
(605) 673-4764

Deadwood

Deadwood Historic Preservation
District Commission
City Hall
3 Siever Street
Deadwood, SD 57732
(605) 578-2082
Mark Wolfe

Mitchell

Friends Of The Middle Border Pioneer
Museum
P.O. Box 1071
Mitchell, SD 57301
(605) 996-2122
Hazel Jordan

Mitchell Central Business District
P.O. Box 206
Mitchell, SD 57301
(605) 996-5567
Pat Helgeland

Oscar Howe Art Center
119 West 3rd
Mitchell, SD 57301
(605) 996-4111
Margaret Quintal

Pierre

Governor's Office of Economic
Development
South Dakota Main Street Program
711 Wells Avenue
P.O. Box 6000
Pierre, SD 57501
(605) 773-5032
Kathy Lucas

Rapid City

Black Hills Council of Local
Governments
P.O. Box 1586
Rapid City, SD 57709
(605) 394-2681
Van A. Lindquist

City of Rapid City Historic
Preservation
300 6th Street
Rapid City, SD 57701
(605) 394-4120
David Hough

Historic South Dakota Foundation
P.O. Box 2998
Rapid City, SD 57709
(605) 341-5820
Debra K. Brown

Sioux Falls

City of Sioux City
Planning Department
224 West 9th Street
2nd Floor
Sioux Falls, SD 57102
(605) 339-7130
Gregg Morris

Main Street Sioux Falls
415 South 1st Avenue
Sioux Falls, SD 57102
(605) 338-4009
Sharon Chase

Vermillion

State Historical Preservation Center
P.O. Box 417
Vermillion, SD 57069
(605) 677-5314
Paul Putz

Watertown

Watertown Main Street Program
P.O. Box 1564
103 West Kemp
Watertown, SD 57201
(605) 886-4661
Elizabeth Green

Yankton

Downtown Yankton
P.O. Box 274
Yankton, SD 57078
(605) 665-4501
Kathy Wright

Historic Yankton
200 West 3rd Street
Yankton, SD 57078
Jim Means

TENNESSEE

STATE HISTORIC PRESERVATION OFFICE

Department of Conservation
701 Broadway
Nashville, TN 37243-0442
(615) 742-6758
J.W. Luna, Commissioner, SHPO
Herbert L. Harper, Executive Director
Tennessee Historical Commission
Deputy SHPO
(615) 742-6719
(615) 742-6594 (Fax)

STATEWIDE PRESERVATION ORGANIZATION

Tennessee Heritage Alliance
c/o Rossie, Bethea, Carr, Phillips and
Luckett
530 Oak Court Drive
Suite 145
Memphis, TN 38117-3700
(901) 763-1800
Mimi Phillips, President

NATIONAL TRUST REGIONAL OFFICE

Southern Regional Office
456 King Street
Charleston, SC 29403
(803) 722-8552
(803) 722-8652 (Fax)
Susan A. Kidd, Director

NATIONAL TRUST ADVISORS

Mimi Phillips
Rossie, Bethea, Carr, Phillips and
Luckett
530 Oak Court Drive
Suite 145
Memphis, TN 38117-3700
(901) 763-1800
(901) 767-6514 (Fax)

Debby Dale Mason
Nashville Chamber of Commerce
161 4th Avenue North
Nashville, TN 37219
(615) 259-4708
(615) 256-3074 (Fax)

PRESERVATION ACTION COORDINATOR

Ann Reynolds
Metro Historical Commission
176 3rd Avenue North
Nashville, TN 37201
(615) 862-7970

AIA STATE PRESERVATION COORDINATOR

Charles Waterfield, Jr., FAIA
Charles Waterfield Architects
2416 Hillsboro Road
Nashville, TN 37212
(615) 385-3130
William H. Gaskill, AIA, Deputy

NATIONAL PARK SERVICE REGIONAL OFFICE

Southeast Regional Office
Richard B. Russell Federal Building
75 Spring Street, S.W.
Atlanta, GA 30303
(404) 331-2632

ADVISORY COUNCIL ON HISTORIC PRESERVATION DIVISION

Eastern Office of Project Review
1100 Pennsylvania Avenue, N.W.
Suite 809
Washington, DC 20004
(202) 786-0503
Don L. Klima, Director

Byrdstown

Friends of Cordell Hull
Vision Five Group
Route 1
Box 144
Byrdstown, TN 38549
(615) 864-3511
Kay Elder

Chattanooga

Garnet Chapin
James Building
735 Broad Street
Suite 1002
Chattanooga, TN 37402
(615) 756-4444

Chattanooga Area Regional Council of
Governments
South East Tennessee Development
District
216 West 8th Street
Suite 300
Chattanooga, TN 37402

Chattanooga-Hamilton County
Regional Planning Commission
200 City Hall Annex
Chattanooga, TN 37402

Clarksville

Clarksville-Montgomery County
Historical Museum
P.O. Box 383
Clarksville, TN 37040
(615) 648-5780
Robert Patterson

Cleveland

Main Street Cleveland
P.O. Box 304
Cleveland, TN 37364-0304
(615) 479-1000
Paul Kirksey

Collierville

Historic Collierville
101 Walnut Street
Collierville, TN 38017
(901) 853-3200
Nancy Kirk

Columbia

Columbia Main Street
20 Public Square
P.O. Box 1940
Columbia, TN 38402
(615) 388-3647
Paula S. Mitchell

Maury County Courthouse
Public Square
Columbia, TN 38401
(615) 381-3690
Sam Kennedy

Maury County Historical Society
P.O. Box 147
Columbia, TN 38402
(615) 388-5834
Jane Oakes

South Central Tennessee Development
District
P.O. Box 1346
Columbia, TN 38402
(615) 381-2040
Richard Quin

Cookeville

Depot Museum
City of Cookeville
P.O. Box 998
Cookeville, TN 38503

Elizabethton

Main Street Elizabethton
115 South Sycamore Street
Elizabethton, TN 37643
(615) 542-3131

Fayetteville

Fayetteville Main Street
P.O. Box 162
Fayetteville, TN 37334
(615) 433-9543
Sarah Phillips

Franklin

Carnton Association
1345 Carnton Lane
Franklin, TN 37064
(615) 794-0903
Susan Larson

Carter House
P.O. Box 555
1140 Columbia Avenue
Franklin, TN 37064
Dolores Kestner

Downtown Franklin Association
P.O. Box 807
Franklin, TN 37064
(615) 790-7094
Rudy Jordan

Heritage Foundation of Franklin and
Williamson County
P.O. Box 723
Franklin, TN 37064
(615) 790-0378
Mary Evins

Georgetown

Mrs. R. D. Shelley
12123 Ooltewah-Georgetown Road
Route 1
Box 5
Georgetown, TN 37336
(615) 961-2796

Greeneville

Main Street Greeneville
200 North College Street
Greeneville, TN 37743
(615) 639-7105
Nancy McNeese

Harriman

Hooray for Harriman
431 Devonia Street
Harriman, TN 37748
(615) 882-8570
Barbara Margiotta

Hermitage

Ladies Hermitage Association
4580 Rachel's Lane
Hermitage, TN 37076
(615) 889-2941
George M. Anderjack

Jackson

Jackson Main Street Program
206 East Main
Suite 101
Jackson, TN 38301
(901) 427-7573
Kay Ferree

Jonesboro

William E. Kennedy
400 West Main Street
Jonesboro, TN 37659
(615) 753-2224

Northeast Tennessee Tourism Council
P.O. Box 375
Jonesboro, TN 37659
(615) 753-4188
Claudia S. Moody

Kingsport

Allandale Estate
City of Kingsport
225 West Center Street
Box A-2
Kingsport, TN 37660
(615) 229-9422
Teresa Foster

Downtown Kingsport Association
235 East Center Street
1st Tennessee Bank Building
Kingsport, TN 37660
(615) 246-6550
Peggy Turner

Roane County Heritage Commission
P.O. Box 738
Kingston, TN 37763
(615) 376-9211
Jean Spille

Knoxville

Crescent Bend
Armstrong-Lockett House and W.P.
Toms Memorial Gardens
2728 Kingston Pike
Knoxville, TN 37919
(615) 637-3163
William P. Beall

Knoxville Heritage
P.O. Box 1242
Knoxville, TN 37901
(615) 523-8008
Malcolm Rogers

Metropolitan Planning Commission
Knoxville-Knox County
400 Main Street
Suite 403
Knoxville, TN 37902
(615) 521-2500
Ann K. Bennett

Old North Knoxville
Association
P.O. Box 3896
Knoxville, TN 37927

Ramsey House
Thorngrove Pike
Knoxville, TN 37914
(615) 525-6676
Joyce Fowler

Maryville

Blount County Historical Trust
P.O. Box 4093
Maryville, TN 37802-4093

Memphis

Cornerstone Foundation
545 South Highland Street
Memphis, TN 38111
(901) 452-6009
David Sample

Division of Housing and Community
Development
Conwood Building
701 North Main Street
Suite 100
Memphis, TN 38107
(901) 576-7370
Cathy Mancinko

Memphis and Shelby County
Office of Planning and Development
125 North Mid-America Mall
Room 468
Memphis, TN 38103-2084
(901) 576-7192
Clair Berry

Memphis Multi Bank Community
Development Corporation
619 North 7th Street
Memphis, TN 38107
(901) 521-9968
Tim Bolding

Memphis Museum System
Malibu Neely House
3050 Central Avenue
Memphis, TN 38111
Kate Dixon

Jack R. Tucker
81 South Front
Memphis, TN 38103
(901) 523-0900

U.S. Postal Service
Southern Regional Office
Memphis, TN 38166-0330

Murfreesboro

Friends of the Stones River National
Battlefield
2115 Shannon Drive
Murfreesboro, TN 37129
(615) 890-1400
Pennie Jekot

Main Street Murfreesboro
P.O. Box 5075
Murfreesboro, TN 37133-5075
(615) 895-1887
Dawn Eaton

Murfreesboro Historic Zoning
Commission
320 South Church Street
Murfreesboro, TN 37130
(615) 893-3750
Carla McClellan

Nashville

Association for Preservation of
Tennessee Antiquities
5025 Harding Road
Nashville, TN 37205
(615) 352-8247
Sara A. Bond

Belle Meade Mansion
5025 Harding Road
Nashville, TN 37205
(615) 356-0501
Donna Russell

St. Mary's Church, Nashville, Tennessee. (D. Woodrum, HABS)

Ben West Public Library
8th Avenue North and Union
Nashville, TN 37203
(615) 259-6125
Mary Hearne

Historic Nashville
P.O. Box 2785
Nashville, TN 37219
(615) 244-7835
Evy Ritzen

Metropolitan Development and
Housing Agency
172 2nd Avenue, North
Suite 212
Nashville, TN 37201
(615) 259-5442
Judy B. Steele

Metropolitan Historical Commission
Customs House
176 3rd Avenue, North
Nashville, TN 37201-1803
(615) 259-5027
Ann V. Reynolds

Tennessee Department of Tourist
Development
P.O. Box 23170
Nashville, TN 37202
(615) 741-7994
Carolyn Brackett

Tennessee Heritage Alliance
P.O. Box 121225
Nashville, TN 37212
(615) 383-0227

Tennessee Historical Commission
701 Broadway
Nashville, TN 37219
(615) 742-6716
Herbert L. Harper

Tennessee Main Street Program
Department of Economic and
Community Development
Rachel Jackson Building
6th Floor
Nashville, TN 37243-0405
(615) 741-2373
Cate Hamilton

Tennessee Municipal League
226 Capitol Boulevard
Room 317
Nashville, TN 37219
Joseph A. Sweat

Oak Ridge

Jane L. Shelton
1029 West Outer Drive
Oak Ridge, TN 37830
(615) 482-3911

Pulaski

Main Street: Pulaski
131 North 1st Street
Pulaski, TN 38478
(615) 363-1764
Beth Holley

Reliance

Friends of Reliance
P.O. Box 74
Childers Creek Road
Reliance, TN 37369
(615) 338-2373
Sarah M. Williams

Rogersville

Rogersville Heritage
415 South Depot Street
Rogersville, TN 37857
(615) 272-7622
John L. Campbell

Rugby

Historic Rugby
Box 8
Rugby, TN 37733
(615) 628-2441
Barbara Stagg

Sparta

Sparta Main Street Program
P.O. Box 607
Sparta, TN 38583
(615) 836-6450
John R. Lewis

Wartrace

Hulan Johnson
P.O. Box 245
Wartrace, TN 37183
(615) 389-6131
Lynn W. Hulan

Waverly

Patricia C. Sherwood
113 North Church Street
Waverly, TN 37185
(615) 396-5038

Winchester

Franklin County Activity Center
Hundred Oaks Castle
Winchester, TN 37398
(615) 967-0100
Deborah Rains

Winchester Main Street Development
Association
7 South High Street
P.O. Box 361
Winchester, TN 37398
(615) 967-4857
Gina Buckley

TEXAS

STATE HISTORIC PRESERVATION OFFICE

Texas Historical Commission
P.O. Box 12276
Capitol Station
Austin, TX 78711
(512) 463-6100
(512) 463-6095 (Fax)
Curtis Tunnell, Executive Director, SHPO

James Wright Steely, Director of Surveys,
Nominations and CLG Division
Deputy SHPO
(512) 463-6094

Stanley O. Graves
Director of Architecture Division
Deputy SHPO
(512) 463-6094

James E. Bruseth
Director of Review and Compliance Department
Deputy SHPO
(512) 463-6096

STATEWIDE PRESERVATION ORGANIZATION

Preservation Texas Alliance
712 Main Street
Suite 110
Houston, TX 77002
(713) 236-5000
Margie Elliott, President

Texas Historical Foundation
P.O. Box 530056
Austin, TX 78753
(512) 251-7423
Elizabeth Susser, President
Joan Rabins, Executive Director

NATIONAL TRUST REGIONAL OFFICE

Texas/New Mexico Field Office
500 Main Street
Suite 606
Fort Worth, TX 76102
(817) 332-4398
(817) 332-4390 (Fax)
Elizabeth R. Willis, Field Office Coordinator

NATIONAL TRUST ADVISORS

Paula Peters
Dallas Parks Foundation
400 South Record Street, 6th Floor
Dallas, TX 75202-4806
(214) 977-6653
(214) 977-7781 (Fax)

Karl Komatsu, AIA
Komatsu/Rangel
550 Bailey Avenue
Suite 102
Fort Worth, TX 76107
(817) 332-1914
(817) 877-4754 (Fax)

PRESERVATION ACTION COORDINATOR

Jill Harrison Souter
350 Wildrose Avenue
San Antonio, TX 78209
(512) 224-6163

AIA STATE PRESERVATION COORDINATOR

James G. Rome, AIA
218 Chandler Lane
Corpus Christi, TX 78404
(512) 881-9182

NATIONAL PARK SERVICE REGIONAL OFFICE

Southwest Regional Office
P. O. Box 728
Santa Fe, NM 87504
(505) 988-6100

ADVISORY COUNCIL ON HISTORIC PRESERVATION DIVISION

Western Office of Project Review
730 Simms Street
Room 401
Golden, CO 80401
(303) 231-5320
Claudia Nissley, Director

Abilene

Abilene Preservation League
P.O. Box 3451
Abilene, TX 79604
(915) 676-3775
Debra Brown

City of Abilene
Planning and Development
P.O. Box 60
Abilene, TX 79604
(915) 676-6230

Weatherl and Welch Architects
1482 North 1st Street
Abilene, TX 79601
(915) 673-6725
Rick Weatherl

Amarillo

City of Amarillo
Planning Department
P.O. Box 1971
Amarillo, TX 79186-0001
(806) 378-4222
J. D. Smith

Lueise Tyson
2104 South Polk
Amarillo, TX 79109
(806) 371-0728

Angleton

Angleton Main Street
121 South Velasco Street
Angleton, TX 77515
(409) 849-4364
Jack D. Handmacher

Arlington

City of Arlington
P.O. Box 231
Arlington, TX 76010
(817) 459-6650
Rose Jacobson

Austin

Donna Carter
817 West 11th Street
Austin, TX 78701
(512) 476-1812

Joe C. Freeman, AIA
Architect
204 Skyline Drive
Austin, TX 78746

Heritage Society of Austin
P.O. Box 2113
Austin, TX 78768
(512) 474-5198
Elaine Mayo

Ron Magnuson
11704 Whitewing
Austin, TX 78753

Darlene Marwitz
P.O. Box 9802-125
Austin, TX 78766
(512) 482-9406

Sue W. Moss
1407 Westover Road
Austin, TX 78703
(512) 474-2029

Preservation Texas Alliance
P.O. Box 12832
Austin, TX 78711
Jill Souter

Prewitt and Associates
7701 North Lamar
Suite 104
Austin, TX 78752
Elton R. Prewitt

St. Edward's University
3001 South Congress Avenue
Austin, TX 78704
(512) 448-8417
R. H. Kinsey

Alan Schumann
2000 Apricot Glen
Austin, TX 78746

Texas Downtown Association
Texas Historical Commission
1511 Colorado Street
P.O. Box 12276
Austin, TX 78711
(512) 463-6092
Anice Read

Texas Neighborhood Conservation
Fund
4604 Avenue C
Austin, TX 78751
(512) 452-1103
Terri Myers

University of Texas
School of Architecture
Austin, TX 78713
(512) 471-1766
Wayne Bell

Dan K. Utley
3802 Skipton Drive
Austin, TX 78727
(512) 836-8803

Volz and Associates
1406 Preston Avenue
Austin, TX 78703
John Volz

Kim Alan Williams
700 East 44th Street
Austin, TX 78751
(512) 454-1938

Beaumont

Beaumont Heritage Society
Community Account
2985 French Road
Beaumont, TX 77706
(409) 898-0348
Becki Stedman

City of Beaumont
Planning Division
P.O. Box 3827
Beaumont, TX 77704
(409) 880-3764
Kirt Anderson

Jennifer Marino
Debbie Bauman
P.O. Box 3786
Beaumont, TX 77704
(409) 838-2202

Blanco

Old Blanco Courthouse Preservation
Society
P.O. Box 302
Blanco, TX 78606
(512) 833-5695
Jack Kent

Brownsville

Cameron County Historical
Commission
608 East Adams Street
Brownsville, TX 78520
Sam Griffin

City of Brownsville
Planning Department
P.O. Box 911
Brownsville, TX 78520
(512) 541-6302
Mark Lund

Bryan

City of Bryan
Community Development
P.O. Box 1000
Bryan, TX 77805
(409) 779-5622
Jo Ann Powell

Downtown NOW
106 West 26th
Suite 46
Bryan, TX 77803
(409) 822-1733
Elizabeth Miller

Jon Mondrik
2827 Oakside Drive
Bryan, TX 77802
(409) 845-2801

Burton

Operation Restoration
P.O. Box 98
Burton, TX 77835
Doug Hutchinson

Canyon

Randall County Historical Commission
503 A Harrell Lane
Canyon, TX 79015
(806) 656-2261
Claire R. Kuehn

Center

Shelby Foundation
321 Shelbyville Street
Center, TX 75935
(409) 598-3682
Mary Haley

College Station

Texas Historical Foundation
Texas A & M University
College of Architecture
Center for Historical Resources
College Station, TX 77843-3137
(409) 845-6125

David G. Woodcock
Architect
1511 Wolf Run
College Station, TX 77840-3134
(409) 696-4312

Corpus Christi

Corpus Christi Area Heritage Society
P.O. Box 2532
Corpus Christi, TX 78403
Paul Altheide

James G. Rome, AIA
218 Chandler Lane
Corpus Christi, TX 78404

Corsicana

City of Corsicana
Main Street Project
P.O. Box 626
Corsicana, TX 75110
(214) 872-4811
Malinda P. Sharpley

Dallas

City of Dallas
Department of Planning and
Development
1500 Marilla
Room 50N
Dallas, TX 75201

Common Ground Community
Economic Development Corporation
5405 East Grand
Dallas, TX 75223
(214) 827-2632
John Fullinwider

Dallas County Historical Commission
634 Records Building
Dallas, TX 75202-3504
(214) 653-6238
Helen Jankowski

Dallas Landmark Commission
Dallas City Hall
5D North
Dallas, TX 75201
(214) 670-4121
Jim Anderson

Donald Allen Dorward
742 McLean Avenue
Dallas, TX 75211
(214) 331-5658

Herbert C. Hale, Jr.
4630 Cherokee Trail
Dallas, TX 75209
(214) 351-3431

Jefferson Area Association
220 West Jefferson Boulevard
Dallas, TX 75208
(214) 942-5029
Laura J. Mulry

Douglas Newby
4912 Tremont
Dallas, TX 75214
(214) 828-1472

Paula Peters
7106 La Vista Drive
Dallas, TX 75214
(214) 953-1270

Southern Methodist University
Archaelolgy Research Program
Institute for the Study of Earth and Man
Dallas, TX 75275
Randall W. Moir

Sunshine Elizabeth Chapel Restoration
Committee
3419 Michigan Avenue
Dallas, TX 75216
(214) 424-8500
David Perry

Texas Theater Historical Society
351 East Brooklyn
Dallas, TX 75203
(214) 948-8888
Dennis Hamilton

Don Weeks
2033 Elmwood Boulevard
Dallas, TX 75224
(214) 332-3424

Woodbine Development Corporation
1445 Ross Avenue
Suite 5000
Dallas, TX 75202-2733

Denton

City of Denton
100 West Oak
Suite 204
Denton, TX 76201
(817) 566-8329
Jane Finley

Denton County Historical Commission
P.O. Box 2184
Denton, TX 76202
(817) 383-8073
Leon D. Callihan

Dripping Springs

Friends of the Pound House Foundation
P.O. Box 589
Dripping Springs, TX 78620
(512) 858-4659
Dennis Cannon

Eagle Pass

Fort Duncan Restoration Association
926 Avenue A-5
Eagle Pass, TX 78852
Ward Weste

Elgin

City of Elgin
P.O. Box 591
Elgin, TX 78621
(512) 285-5721
Molly Alexander

Elgin Historical Society
P.O. Box 1234
Elgin, TX 78621
(512) 285-3311
Jeff Carter

El Paso

El Paso Mission Trail Association
P.O. Box 3789
El Paso, TX 79923
(915) 592-5252
Sheldon Hall

Office of Urban Design and Historic
Preservation
Department of Planning
2 Civic Center Plaza
8th Floor
El Paso, TX 79999
(915) 541-4192
Nat Campos

Encina

Mila L. Presno
P.O. Box 378
Encina, TX 78019

Ennis

City of Ennis
P.O. Box 220
Ennis, TX 75120
(214) 875-1234
Steve Howerton

Ennis Heritage Society
P.O. Box 189
Ennis, TX 75119
Sugar Glaspy

Fort Stockton

Fort Stockton Historical Society
Annie Riggs Memorial Museum
Fort Stockton, TX 79735
(915) 336-2167
Mary K. Shannon

Fort Worth

City of Fort Worth
Planning and Growth Management
1000 Throckmorton Street
Fort Worth, TX 76102
(817) 870-8042
Emil R. Moncivais

E. Dwain Dent
1300 Summit Avenue
Suite 700
Fort Worth, TX 76102
(817) 332-2889

Fairmount Association
2230 Lipscomb Street
Fort Worth, TX 76110
(817) 926-0874
Phil Bordeleau

Fort Worth Economic Development
Corporation
410 West 4th Street
Fort Worth, TX 76102-3705
(817) 336-6420
Franklin D. Moss

Fort Worth Public Library
Acquistions Unit
300 Taylor Street
Fort Worth, TX 76102

Historic Preservation Council for
Tarrant County
1303 Foch Street
Suite 103
Fort Worth, TX 76107-3003
(817) 338-0267
Marty Craddock

Polytechnic Main Street Project
3216 East Rosedale Street
Fort Worth, TX 76105
(817) 531-4476
Terry Meza

Texas Heritage
1509 Pennsylvania Avenue
Fort Worth, TX 76104
(817) 336-1212
Deborah Phelan

Texas Wesleyan University
1201 Wesleyan
Fort Worth, TX 76105
(817) 531-4444
W. L. Hailey

Fredericksburg

Fort Martin Scott
1606 East Main Street
Fredericksburg, TX 78624
Ted Hollingsworth

Gillespie County Historical Society
P.O. Box 765
Fredericksburg, TX 78624
(512) 997-2835
Nina N. Mender

Wagner and Klein
208 South Lelano Street
Fredericksburg, TX 78624
(512) 997-9525
Barry A. Wagner

Gainesville

City of Gainesville
200 South Rusk
Gainesville, TX 76240
(827) 665-4323
Lyle H. Dresher

Community Preservation Foundation
1301 East Pecan Street
Gainesville, TX 76240
(817) 668-8762
Margaret Hays

Cooke County Heritage Society
Morton Museum
P.O. Box 150
Gainesville, TX 76240
(817) 668-8900
Shana Powell

Galveston

The Center for Transportion and
Commerce
123 Rosenberg Avenue
Galveston, TX 77550
(409) 765-5700
Timothy M. Kingsburg

Lennie Brown
526 Eleventh
Galveston, TX 77550
(409) 762-4042

City of Galveston
Department of Urban Planning
P.O. Box 779
Galveston, TX 77553-0779
(409) 766-2106
Harold L. Holmes

Galveston Historical Foundation
2016 Strand
Galveston, TX 77550
(409) 765-7834
Betty A. Massey

Rosenberg Library
2310 Sealy
Galveston, TX 77550

Transitional Learning Center
1528 Post Office Street
Galveston, TX 77550
(409) 762-6661
Michael Inman

Georgetown

City of Georgetown
P.O. Box 409
Georgetown, TX 78627
(512) 863-5533
John J. Cregoire

Georgetown Heritage Society
101 West 7th Street
Georgetown, TX 78626
(512) 863-5598
J. C. Johnson

Gonzales

Gonzales Crystal Theater
511 St. Lawrence
P.O. Box 1956
Gonzales, TX 78629
(512) 672-6208
Noell Ince

Gordon

Thurber Historical Association
Box 192
Gordon, TX 76453

Grapevine

Main Street Project
520 South Main Street
Grapevine, TX 76051
(817) 481-0395
Ron Emrich

Houston

Ellen Beasley
7326 Staffordshire Street
Suite 3
Houston, TX 77030-5149

David J. Bebout
943½ Cortlandt
Houston, TX 77008
(713) 869-9608

Robert W. Collins
2016 Main Street
Suite 2006
Houston, TX 77002
(713) 759-1272

Greater Houston Preservation Alliance
712 Main
Suite 110
Houston, TX 77002
(713) 658-8938
Margie Elliott

Houston Archaeological and Historical
Commission
Department of Planning
P.O. Box 1562
Houston, TX 77251
(713) 247-1238
Wendy T. Jamieson

Houston Heights Main Street
545 West 19th Street
Houston, TX 77008
(713) 861-6735
Angela Kerr Smith

Mrs. Robert W. Kneebone
3435 Westheimer
Suite 311
Houston, TX 77027
(713) 627-3778

Vicki List
P.O. Box 130868
Houston, TX 77219-0868
(713) 526-4564

Graham B. Luhn, AIA
2211 Norfolk Street,
Suite 626
Houston, TX 77098

Metropolitan Transit Authority
P.O. Box 61429
Houston, TX 77208-1429
(713) 739-4628
Steve Brooks

University of Houston
College of Architecture
4800 Calhoun Road
Houston, TX 77004
(713) 749-1181
V. N. Dorian-Becnel

S. Elwood York, Jr.
8649 Village of Fondren Drive
Houston, TX 77071
(713) 229-0001

Huntsville

City of Huntsville
Planning Department
1212 Avenue M
Huntsville, TX 77340
(409) 291-5410
R. Brown

Kilgore

City of Kilgore
Main Street Project
P.O. Box 990
Kilgore, TX 75663
(903) 984-5081
Amanda Pratt

Laredo

Azteca Economic Development and
Preservation Corporation
20 Iturbide Street
Laredo, TX 78040
(512) 726-4462
Rafael I. Torres

Los Caminos del Rio
1120 Matamoros Street
Laredo, TX 78040-8007
(512) 791-4300
Gloria Canseco

Webb County Heritage Foundation
P.O. Drawer 29
Laredo, TX 78042-0029
(512) 727-0977
Gloria Z. Canseco

Longview

Gerald Bratz, AIA
P.O. Box 2723
1100 Judson Road
Suite 620
Longview, TX 75606
(214) 236-3771

Lubbock

Lubbock Centercorp
1220 Broadway
Suite 1307
Lubbock, TX 79401
(806) 765-8910
Jim R. Shearer

Lubbock Heritage Society
Box 5443
Lubbock, TX 79417
(806) 792-4799

Marshall

Audrey D. Kariel
503 Lansdowne
Marshall, TX 75670
(903) 935-5387

Menard

Menard Historical Society
Box 663
Menard, TX 76859
(915) 396-4318
Mayon Neel

Mesquite

Mesquite Historical Commission
P.O. Box 850137
Mesquite, TX 75149
(214) 216-6293
Carol Zolnerowich

Mount Vernon

Franklin County Historical Association
P.O. Box 289
Mount Vernon, TX 75457
(903) 537-2264
B. F. Hicks

Odessa

Main Street Odessa
606 North Grant
Odessa, TX 79761
(915) 332-0291
Craig Hunter

University of Texas
Permian Basin
4901 East University
Odessa, TX 79762
(915) 367-2011
J. Tillapaugh

Palestine

Palestine Junior Service League
P.O. Box 1423
Palestine, TX 75802
(214) 729-8171
Helen Hanks

Richard E. Smith
Route 6
Box 6172
Palestine, TX 75801
(214) 729-4821

Pasadena

Pasadena Historical Society
722 Fairmont Parkway
Suite 204
Pasadena, TX 77504
(713) 487-3970
Ann Thomas

Plano

Heritage Farmstead
1900 West 15th Street
Plano, TX 75075
(214) 424-7874
Peggy Riddle

Walter Gresham House, Galveston, Texas. (L. Johnston, HABS)

Port Arthur

City of Port Arthur
Planning Department
P.O. Box 1089
Port Arthur, TX 77640
(409) 983-8138
Dale L. Watson

Richmond

George Foundation
P.O. Drawer C
Richmond, TX 77469
(713) 342-6109
Ronald Adamson

Round Top

Winedale Historical Center
P.O. Box 11
Round Top, TX 78954
(409) 278-3530
Gloria Jaster

San Angelo

Historic San Angelo
1221 Paseo de Vaca
San Angelo, TX 76901
Donna Crisp

San Antonio

James Foster, AIA
101 Lindell
San Antonio, TX 78212

Killis Almond and Associates
342 Wilkens
San Antonio, TX 78210
Killis P. Almond

King William Association
222 King William Street
Suite 2
San Antonio, TX 78204
(512) 227-8786
Jean Alexander-Williams

Leor Construction Services
240 Bushnell
Suite 701
San Antonio, TX 78212
(512) 861-8516
Yonni Leor

Paul McCombs
Norma McCombs
9303 Fallworth
San Antonio, TX 78205
(512) 696-4811

Matthews and Branscomb
106 South Saint Mary's
Suite 800
San Antonio, TX 78205
(512) 226-4211
Frank Z. Ruttenberg

San Antonio Conservation Society
107 King William Street
San Antonio, TX 78204
(512) 224-6163

Jill Harrison Souter
350 Wildrose Avenue
San Antonio, TX 78209
(512) 828-8280

A. Maria Watson
213 Washington Street
San Antonio, TX 78204

Williams, Schneider, Calvetti
2955 Nacogdoches Road
San Antonio, TX 78217
(512) 828-6419
Per K. Schneider

San Marcos

The Heritage Asssociation of San
Marcos
P.O. Box 1806
San Marcos, TX 78666
R. B. Ayers

Sherman

Sherman Preservation League
915 South Crockett
P.O. Box 159
Sherman, TX 75091
(214) 892-9091
Dorothy McKee

Stanton

Martin County Convent
P.O. Box 1435
Stanton, TX 79782
(915) 756-2801
Miles Tollison

Temple

Bell County Historical Commission
202 Twelve Oak Drive
Temple, TX 76504
(819) 773-7252
Carol Kehl

Railroad and Pioneer Museum
P.O. Box 5126
Temple, TX 76501
(817) 778-6873
Mary Irving

Tyler

Historic Tyler
P.O. Box 6774
Tyler, TX 75711
(214) 595-1960
Michael Butler

Ann Etta Wells Payne
1819 Circle Drive
Tyler, TX 75703
(903) 561-2882

Tyler Main Street Project
P.O. Box 158
Tyler, TX 75710
(903) 593-6905
Claire Squibb

Vernon

Wilbarger County
Historical Commission
Route 2
Box 332
Vernon, TX 76381
(817) 552-6404
Jack White

Victoria

J. M. Joyce
P.O. Box 1758
Victoria, TX 77902
(512) 572-2795

Victoria Preservation
P.O. Box 1486
Victoria, TX 77902
(512) 578-7171
Gary Breech

Waco

Getterman Homes
92 Sugar Creek Place
Waco, TX 76712
Holt Getterman

Historic Waco Foundation
810 South 4th
Waco, TX 76706
(817) 753-5166
Pamela B. Crow

Sanger Heights Neighborhood
Association
2326 Colcord Avenue
Waco, TX 76707
(817) 753-7494
Kent Keeth

Weatherford

Parker County Heritage Society
P.O. Box 97
Weatherford, TX 76086
(817) 599-5900
Annette Carrell

Weatherford Main Street Project
P.O. Box 255
Weatherford, TX 76086
(817) 534-5441
David Clay

Wichita Falls

Midwestern University
3400 Taft Boulevard
Wichita Falls, TX 76308
(817) 692-6611
Kenneth E. Hendrickson

Wichita County Heritage Society
900 Bluff
Wichita Falls, TX 76301
(817) 723-9623
Gayla Morris

Woodville

Tyler County Heritage Society
DBA Heritage Village Museum
P.O. Box 888
Woodville, TX 75979

UTAH

STATE HISTORIC PRESERVATION OFFICE

Utah State Historical Society
300 Rio Grande
Salt Lake City, UT 84101
(801) 533-5755
Max Evans, Director, SHPO

Wilson Martin
Office of Preservation
Deputy SHPO
(801) 533-7039

STATEWIDE PRESERVATION ORGANIZATION

Utah Heritage Foundation
355 Quince Street
Salt Lake City, Utah 84103
(801) 533-0858
Wallace Cooper, President
Michael Leventhal, Executive Director

NATIONAL TRUST REGIONAL OFFICE

Western Regional Office
1 Sutter Street
Suite 707
San Francisco, CA 94104
(415) 956-0610
(415) 956-0837 (Fax)
Kathryn A. Burns, Director

NATIONAL TRUST ADVISORS

Tina Stahlke Lewis
P.O. Box 808
481 Woodside Avenue
Park City, UT 84060
(801) 649-8746
(801) 649-3757 (Fax)

Rob White
352 North Quince
Salt Lake City, UT 84103
(801) 364-6110
(801) 626-1330

PRESERVATION ACTION COORDINATORS

Mike Leventhal
Utah Heritage Foundation
355 Quince Street
Salt Lake City, UT 84103
(801) 533-0858

AIA STATE PRESERVATION COORDINATOR

Wallace Cooper, AIA
Cooper Roberts Architects
202 West 300 North
Salt Lake City, UT 84103
(801) 355-5915

NATIONAL PARK SERVICE REGIONAL OFFICE

Rocky Mountain Regional Office
12795 West Alameda Parkway
P. O. Box 25287
Denver, CO 80225-0287
(303) 969-2875

ADVISORY COUNCIL ON HISTORIC PRESERVATION DIVISION

Western Office of Project Review
730 Simms Street
Room 401
Golden, CO 80401
(303) 231-5320
Claudia Nissley, Director

Ephraim

Sanpete Trade Association
96 North Main Street
Suite 11-16
Ephraim, UT 84627

Midvale

Corporation Midvale City
80 East Center Street
Midvale, UT 84047
(807) 561-1418
Michael Siler

Ogden

City of Ogden
Department of Community
Development
2484 Washington Boulevard
2nd Floor
Ogden, UT 84401
Patricia Comarell

Egyptian Theatre Foundation
1665 Binford Street
Ogden, UT 84401
(801) 394-2730

Weber County Heritage Foundation
Eccles Community Arts Center
2580 Jefferson Avenue
Ogden, UT 84401
(801) 392-3231
Catherine Feeny

Park City

Park City Municipal Corporation
P.O. Box 1480
Park City, UT 84060
(801) 649-9321
Nora L. Seltenrich

Washington School Inn
544 Park Avenue
P.O. Box 536
Park City, UT 84060
(801) 649-3800
Nancy Beaufait

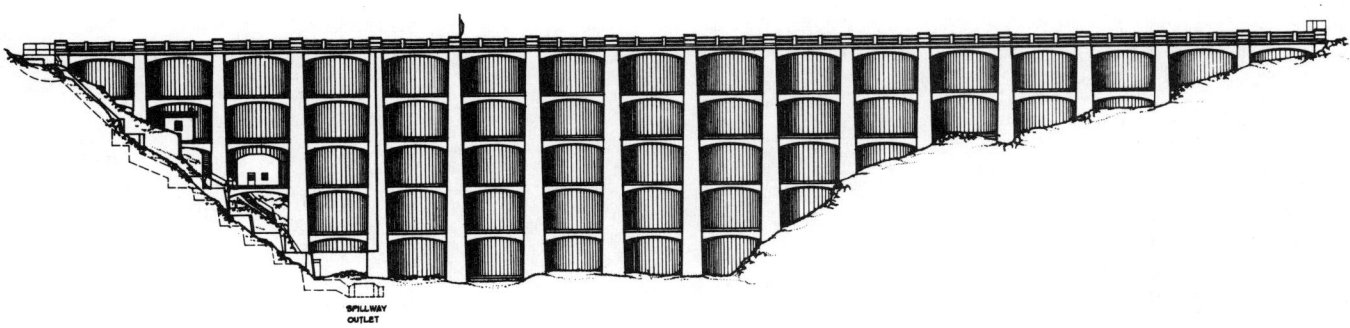

Mountain Dell Dam, Parleys Canyon, Salt Lake County, Utah. (T. Ristau, Madgen, HAER)

Provo

City of Provo
Redevelopment Agency
152 West Center
P.O. Box 1849
Provo, UT 84601
(801) 375-1822
Ron Madsen

Salt Lake City

Cooper/Roberts Architects, AIA
202 West 300 North
Salt Lake City, UT 84103
(801) 355-5915
Allen D. Roberts

Historical Arts and Casting
1939 South 4130 West
Unit F
Salt Lake City, UT 84104
(801) 974-0242
Robert A. Baird

Office of Preservation
300 Rio Grande
Salt Lake City, UT 84101-1182
(801) 533-7039
Wilson G. Martin

Salt Lake Association of Community
Councils
451 South State
Room 335
Salt Lake City, UT 84111
(801) 535-7029
Stan Penfold

Salt Lake City Corporation
451 South State
Room 306
Salt Lake City, UT 84111
(801) 535-7704
Stephen C. Oliver

University of Utah
Graduate School of Architecture
Salt Lake City, UT 84112
(801) 581-8254
Thomas Carter

Utah Heritage Foundation
355 Quince Street
Salt Lake City, UT 84103
(801) 533-0858
Michael S. Leventhal

Saint George

City of Saint George
Community Development Office
175 East 200 North
Saint George, UT 84770
(801) 634-5800
Robert Nicholson

VERMONT

STATE HISTORIC PRESERVATION OFFICE

Agency of Development and
Community Affairs
109 State Street
Montpelier, VT 05609-0501
(802) 828-3211
Barbara Ripley, Agency Counsel, SHPO

Eric Gilbertson, Director
Vermont Division for Historic
Preservation
Deputy SHPO
58 East State Street
Montpelier, VT 05602
(802) 828-3226
(802) 828-3233 (Fax)

STATEWIDE PRESERVATION ORGANIZATION

Preservation Trust of Vermont
104 Church Street
Burlington, VT 05401
(802) 658-6647
Judy Hayward, President
Paul Bruhn, Executive Director

NATIONAL TRUST REGIONAL OFFICE

Northeast Regional Office
7 Faneuil Hall Marketplace
5th Floor
Boston, MA 02109
(617) 523-0885
(617) 523-1199 (Fax)
Vicki Sandstead, Director

NATIONAL TRUST ADVISOR

Thomas P. Salmon
P.O. Box 535
Bellows Falls, VT 05101
(802) 463-4507

PRESERVATION ACTION COORDINATOR

Judy Hayward
Historic Windsor
Main Street, Box 1777
Windsor, VT 05089-0021
(802) 674-6752

AIA STATE PRESERVATION COORDINATOR

Jay White, AIA
The Burley Partnership
P.O. Box 150
Waitsfield, VT 05673
(802) 496-3900

NATIONAL PARK SERVICE REGIONAL OFFICE

North Atlantic Regional Office
15 State Street
Boston, MA 02109-3572
(617) 223-5001

ADVISORY COUNCIL ON HISTORIC PRESERVATION DIVISION

Eastern Office of Project Review
1100 Pennsylvania Avenue, N.W.
Suite 809
Washington, DC 20004
(202) 786-0503
Don L. Klima, Director

Barre

Trow and Holden Company
45-57 South Main Street
P.O. Box 475
Barre, VT 05641
(802) 476-7221
Norman Akley

Bennington

Bennington Region Preservation Trust
P.O. Box 1478
Bennington, VT 05201
(802) 442-3285
Charles R. Putney

Town of Bennington
205 South Street
Bennington, VT 05201
(802) 442-1037
Paul W. Bohne

Brandon

Brandon Free Public Library
Brandon, VT 05733

Brattleboro

Barbara George
12 Cherry Street
Brattleboro, VT 05301

Burlington

Flynn Theatre for the Performing Arts
153 Main Street
Burlington, VT 05401
(802) 863-8778
Andrea Rogers

Preservation Trust of Vermont
104 Church Street
Burlington, VT 05401
(802) 658-6647
Paul Bruhn

Retrovest
286 College Street
Burlington, VT 05401
(802) 863-8323
David Scheuer

University of Vermont
Historic Preservation Program
Architectural Conservation and
Education Service
Burlington, VT 05405
(802) 656-3180
Thomas Visser

University of Vermont
Historic Preservation Program
Wheeler House
Burlington, VT 05401
(802) 656-3180

Vermont Housing Finance Agency
1 Burlington Sqaure
Burlington, VT 05401
(802) 864-5743
Allan Hunt

Charlotte

New England Monthly
Ferry Road
Charlotte, VT 05445

Hardwick

Michael Gohl
Architect
P.O. Box 1070
10 Main Street
Hardwick, VT 05843

Ludlow

Black River Academy Museum
High Street
P.O. Box 73
Ludlow, VT 05149
(802) 228-5050
Gordon M. Crandall

Manchester

Friends of Hildene
P.O. Box 377
Manchester, VT 05254
Gerrit W. Kouwenhoven

Middlebury

Fire Safety Institute
P.O. Box 674
Middlebury, VT 05753
(802) 462-2663
John M. Watts

Preservation Investments
Box 567
Middlebury, VT 05753
Townsend H. Anderson

Montpelier

Nancy E. Boone
466 Elm Street
Montpelier, VT 05602

David Schutz
Curator of State Buildings
1 Baldwin Street
Montpelier, VT 05602

Vermont Division for Historic
Preservation
Agency of Development and
Community Affairs
Pavilion Office Building
Montpelier, VT 05602
(802) 828-3226
Eric Gilbertson

North Bennington

Park McCullough House Associates
P.O. Box 366
North Bennington, VT 05257
(802) 442-5441
Joseph C. King

Newport

Northeast Kingdom Community
Action
10 Main Street
Newport, VT 05855
Margaret Nicely

Norwich

Douglas Gest Restorations
Box 832
Norwich, VT 05055
(802) 649-2928
Douglas Gest

Orleans

Orleans County Historical Society
Old Stone House Museum
Orleans, VT 05860
(802) 754-2022
Reed Cherington

Richmond

Locomotive and Railway Preservation
P.O. Box 95
Richmond, VT 05477
(802) 434-2351
Mark Smith

State of Vermont Capital Complex, Montpelier, Vermont. (P. Borchers, M. Melragon, M. Fazlo, HABS)

Rutland

Rutland Area Art Association
Chaffee Art Center
16 South Main Street
Rutland, VT 05701
(802) 773-7385
Susan Farrow

Rutland Partnership
56½ Merchants Row
Rutland, VT 05701
(802) 773-9380
Richard Courcelle

South Royalton

Vermont Law School
Law Library
South Royalton, VT 05068

Sudbury

Edward Dihos
Route 30
Box 1269
Sudbury, VT 05733
(802) 623-7940

Windsor

Historic Windsor
Windsor House
P.O. Box 1777
Windsor, VT 05089-0021
(802) 674-6752
Judy L. Hayward

VIRGINIA

STATE HISTORIC PRESERVATION OFFICE

Department of Historic Resources
Commonwealth of Virginia
221 Governor Street
Richmond, VA 23219
(804) 786-3143
Hugh C. Miller, Director, SHPO

H. Bryan Mitchell, Deputy SHPO
(804) 786-3143
(804) 225-4261 (Fax)

STATEWIDE PRESERVATION ORGANIZATION

Association for the Preservation of
Virginia Antiquities
2300 East Grace Street
Richmond, VA 23223
(804) 648-1889
Mrs. John H. VanLandingham III,
President
Peter Dun Grover, Executive Director

Preservation Alliance of Virginia
P.O. Box 1407
Staunton, VA 24401
(703) 886-4362
Robert B. Lambeth, Jr., President
David J. Brown, Executive Director

NATIONAL TRUST REGIONAL OFFICE

Mid-Atlantic Regional Office
Cliveden
6401 Germantown Avenue
Philadelphia, PA 19144
(215) 438-2886
(215) 438-2892 (Fax)
Patricia D. Wilson, Director

NATIONAL TRUST ADVISORS

John Zehmer
Historic Richmond Foundation
2407 East Grace Street
Richmond, VA 23223
(804) 643-7407
(804) 788-4244 (Fax)

Patricia L. Zontine
614 Tennyson Avenue
Winchester, VA 22601

PRESERVATION ACTION COORDINATORS

David J. Brown
Preservation Alliance of Virginia
P.O. Box 295
Charlottesville, VA 22902-0295
(804) 979-3899

AIA STATE PRESERVATION COORDINATOR

Scott M. Spence, AIA
Colonial Williamsburg Foundation
P.O. Box 1776
Williamsburg, VA 23187
(804) 220-7406
Charles Richard Bierce, AIA, Deputy

NATIONAL PARK SERVICE REGIONAL OFFICE

Mid-Atlantic Regional Office
143 South 3rd Street
Philadelphia, PA 19106
(215) 597-7018

ADVISORY COUNCIL ON HISTORIC PRESERVATION DIVISION

Eastern Office of Project Review
1100 Pennsylvania Avenue, N.W.
Suite 809
Washington, DC 20004
(202) 786-0503
Don L. Klima, Director

Aldie

Joseph Prendergast
Oak Hill Farm
Route 1
Box 271
Aldie, VA 22001

Alexandria

Alexandria Association
P.O. Box 21322
Alexandria, VA 22320-2322
(703) 548-1922
Joseph A. Ziemba

Alexandria Library
Lloyd House
717 Queen Street
Alexandria, VA 22314
(703) 838-4555

Alexandria Board of Architectural
Review
Department of Planning and
Community Development
301 King Street
Room 2100
Alexandria, VA 22314

C. Richard Bierce, AIA
7932 Bolling Drive
Alexandria, VA 22308
(703) 836-9085

Leslie Blythe
4768 Southland Avenue
Alexandria, VA 22312
(703) 285-2653

Anne Carman
104 Duke Street
Alexandria, VA 22314

Emily L. Cooper
206 East Oak Street
Alexandria, VA 22301
(703) 549-8064

P. L. Fitzgerald
315 South Pitt Street
Alexandria, VA 22314
(202) 778-5522

Historic Alexandria Foundation
P.O. Box 19252
Alexandria, VA 22320-0252
(703) 549-5811
Penny C. Morrill

Office of Historic Alexandria
City Hall
Box 178
Alexandria, VA 22313
(703) 838-4554
Jean T. Federico

Charles Parran
706 Duke Street
Alexandria, VA 22314
(703) 519-8650

David Wormser
Janet Hawkins
2420 Farm Road
Alexandria, VA 22302
(703) 836-8867

Amherst

Amherst County Historical Museum
P.O. Box 741
Amherst, VA 24521
(804) 946-9348
Tom Mackie

Annandale

Carolyn C. Abbott
9107 Colt Lane
Annandale, VA 22003
(703) 239-9352

Arlington

Susan Anthony
2362 North Oakland Street
Arlington, VA 22207
(703) 841-0651

Arlington County Department of
Community Planning, Housing and
Development
Community Improvement Division
2100 Clarendon Boulevard
Suite 701
Arlington, VA 22201
(703) 558-2358

John Campbell
4252 North Vacation Lane
Arlington, VA 22207

Glencarlyn Branch Library
300 South Kensington Street
Arlington, VA 22204

Norene Halvonik
4375-D Lee Highway
Arlington, VA 22207
(703) 527-1696

Antoinette J. Lee
4141 North Henderson Road
Apartment 210
Arlington, VA 22203-0282
(703) 525-0943

New England Antique Furniture Repair
Shop
1118 North Jackson Street
Arlington, VA 22201

Tayon Papovich
2707 Arlington Boulevard
Suite 202
Arlington, VA 22201
(703) 528-6172

Jeffrey B. Werner
3511 North Potomac Street
Arlington, VA 22213
(703) 538-6145

Bedford

Bedford Main Street-Centertown
P.O. Box 405
Bedford, VA 24523
(703) 586-2148
Linda Kirkendorfen

Berryville

Clark County Historic Preseration
Commission
102 North Church Street
P.O. Box 169
Berryville, VA 22611
(703) 955-3269
Chuck Johnston

Town of Berryville
23 East Main Street
Berryville, VA 22611
(703) 955-1099
R. J. Hogan

Bluemont

Mark M. Newland and Company
Route 1
Box 129
Bluemont, VA 22012
(703) 554-8531

Boydton

Boyd Tavern Foundation
P.O. Box 183
Boydton, VA 23917
(804) 738-6226
Roberta B. Shelton

Brandy Station

Brandy Station Foundation
Box 165
Brandy Station, VA 22714
(703) 825-9433
B. B. Mitchell

Bristol

Main Street: Bristol
22 6th Street
P.O. Box 1782
Bristol, VA 24203
(615) 764-1486
Stephanie Large

Brookneal

Patrick Henry Memorial Foundation
Red Hill
Route 2
Box 27
Brookneal, VA 24528
(804) 376-2044
James M. Elson

Charlottesville

Charlottesville Downtown Foundation
111 East Main Street
P.O. Box 2472
Charlottesville, VA 22902
(804) 296-8548
James Batten

IIRW Consulting Engineers
800 East Jefferson Street
Charlottesville, VA 22901
(804) 296-2116
Robert R. Somers

Preservation Alliance of Virginia
P.O. Box 295
Charlottesville, VA 22902-0295
(804) 979-3899
David J. Brown

Thomas Jefferson Memorial Foundation
Monticello
P.O. Box 316
Charlottesville, VA 22902
(804) 293-2158
Daniel P. Jordan

University of Virginia
Law Library
Charlottesville, VA 22901
(804) 924-3384

University of Virginia
School of Architecture
Campbell Hall
Charlottesville, VA 22903
(804) 924-6448
Mary C. Crockett

Walter Wadlington
1620 Keith Valley Road
Charlottesville, VA 22901
(804) 293-5261

Chesapeake

Chesapeake Public Library
300 Cedar Road
Chesapeake, VA 23320

City of Chesapeake
P.O. Box 15225
Chesapeake, VA 23320
(804) 547-6191
Robert L. Copeland

Christiansburg

Barbara Capozzi
201 East Main Street
Christiansburg, VA 24073
(703) 382-7372

Clarksville

Roanoke River Museum
Prestwould Foundation
P.O. Box 872
Clarksville, VA 23927
(804) 374-8672
J. D. Hudson

Culpeper

Culpeper Renaissance
162 East Davis Street
2nd Floor
Culpeper, VA 22701
(703) 825-4416
Patricia Seiter

Town of Culpeper
Historic and Cultural Conservation
Board
118 West Davis Street
Culpeper, VA 22701
(703) 825-4703
Jerry Davenport

Danville

Danville Public Library
511 Patton Street
Danville, VA 24541
(804) 799-5195
Janet Haymore

Downtown Danville Associates
P.O. Box 205
Danville, VA 24543
(804) 791-4944
Mary Perry

Dayton

Renaissance Stone Masonry
Route 2
Box 262
Dayton, VA 22821
(703) 879-2678
James L. Flory

Town of Dayton
Dayton Planning Commission
P.O. Box 187
Dayton, VA 22821

Dumfries

Historic Dumfries Virginia
P.O. Box 26
Dumfries, VA 22026
(703) 221-3346
Ann Hoagland

Dunnsville

Frederick Hudson Ecker, II
Aspen Grove Farm
P.O. Box 1066
Dunnsville, VA 22454
(804) 443-6657
Page Ecker

Emporia

Emporia Downtown Revitalization
336 South Main Street
2nd Floor
Emporia, VA 23847
(804) 634-9470

Fairfax

City of Fairfax
Department of Community
Development and Planning
10455 Armstrong Street
Fairfax, VA 22030
(703) 385-7930
Peggy T. Wagner

R. A. Clement, Jr.
10197 Black Rock Court
Fairfax, VA 22032
(703) 271-2719

Erle Nielsen
3800 Ridgelea Drive
Fairfax, VA 22031
(703) 591-4219

Emma Jane Saxe
12975 Hampton Forest Court
Fairfax, VA 22030
(703) 830-3318

Falls Church

Larry H. Claussen
6534 Oakwood Drive
Falls Church, VA 22041
(703) 684-6166

DeTeel Patterson Tiller
7313 Hughes Court
Falls Church, VA 22046
(703) 560-0723

Fairfax County History Commission
Heritage Resources
2855 Annandale Road
Falls Church, VA 22042

Christopher Gribbs
2760 Summerfield Road
Falls Church, VA 22042

Falmouth

Elsa W. L. Schemmer
309 Colonial Avenue
Falmouth, VA 22405
(703) 373-1672

Forest

Corporation for Jefferson's Poplar
Forest
P.O. Box 419
Forest, VA 24551
(804) 525-1806
Lynn A. Beebe

Franklin

Franklin Department of Downtown
Development
207 West 2nd Street
P.O. Box 179
Franklin, VA 23851
(804) 562-8511
Carolyn Bell

Fredericksburg

Association for Preservation
Technology International
P.O. Box 8178
Fredericksburg, VA 22404
(703) 373-1621
(703) 373-6050 (Fax)
Susan F. Johnson

City of Fredericksburg
Planning Department
P.O. Box 7447
Fredericksburg, VA 22404
Eric Nelson

Historic Fredericksburg Foundation
1002 Princess Anne Street
Fredericksburg, VA 22401
(703) 371-4504
Catharine M. Gilliam

Kenmore
1201 Washington Avenue
Fredericksburg, VA 22401
(703) 373-3381
W. V. Edenfield

Mary Washington College
Center for Historic Preservation
Fredericksburg, VA 22401-5358
(703) 899-4037
Carter L. Hudgins

Our Town Fredericksburg
803 Caroline Street
Fredericksburg, VA 22401
(703) 373-5399
Mike Tringale

Galax

Galax Downtown Association
P.O. Box 544
Galax, VA 24333
(703) 236-0668
Wendy Turney

Gloucester

Rosewell Foundation
Division of Gloucester Historical
Society of Virginia
Box 1456
Gloucester, VA 23061
(804) 693-6458
George H. Whiting

Gordonsville

Historic Green Springs
Route 2
Box 375
Gordonsville, VA 22942

Hampton

Hampton Heritage Foundation
P.O. Box 536
Hampton, VA 23669
(804) 722-0969
K. A. McDonald

Phoebus Improvement League
P.O. Box 3549
15 South Mallory Street
Hampton, VA 23663
(804) 727-0808
Arleen Crittenden

Hanover

Hanover County Planning Department
P.O. Box 470
Hanover, VA 23069
(804) 537-6171
Carol Corker

Heathsville

Northumberland Preservation
P.O. Box 88
Heathsville, VA 22473
(703) 580-2828
Henry L. Hull

Herndon

Downtown Herndon
P.O. Box 1879
Herndon, VA 22070
(703) 481-8898
John P. Stanford

Karen T. France
500 Kensington Place
Herndon, VA 22070

Isle of Wight

Isle of Wight County
Economic Development
P.O. Box 80
Isle of Wight, VA 23397
(804) 357-3191
Judy A. Begland

Leesburg

County of Loudoun
750 Miller Drive S.E.
Suite 300
Leesburg, VA 22075-8919
Teckla Cox

Loudoun Restoration and Preservation
Society
P.O. Box 351
Leesburg, VA 22075

Thomas L. McConnell
Route 1
Box 273
Leesburg, VA 22075
(703) 327-4393

Town of Leesburg
Planning Department
15 West Market Street
P.O. Box 88
Leesburg, VA 22075
(703) 478-1821
(703) 777-2420
(703) 771-2727 (Fax)
Kristie Lalire

Lexington

Historic Lexington Foundation
Stonewall Jackson House
8 East Washington Street
Lexington, VA 24450

Lexington Downtown Development
Association
101 South Main Street
P.O. Box 1078
Lexington, VA 24450
(703) 463-7191
Dianne W. Herrick

Rockbridge Alum Springs Association
P.O. Box 894
Lexington, VA 24450
(503) 783-2507
Jim Brown

Washington and Lee University
Lee Chapel
Lexington, VA 24450
(703) 463-8768
R. C. Peniston

Lynchburg

Jones Memorial Library
2311 Memorial Avenue
Lynchburg, VA 24501

Lynchburg Community Planning and
Development
P.O. Box 60
Lynchburg, VA 24505
Frederick K. Ickes

Lynchburg Historical Foundation
P.O. Box 248
Lynchburg, VA 24505
(804) 528-5353

McLean

Colonial Dames of America
Charter III
1701 Crescent Lane
McLean, VA 22101

Oldham Historic Properties
904 Turkey Run Road
McLean, VA 22101
(703) 893-3219
Sally Oldham

Maidens

Tanglewood Ordinary
2210 River Road, West
Maidens, VA 23102
James G. Hardwick

Manassas

Culbertson Company of Virginia
12923 Balls Ford Road
Manassas, VA 22110
(703) 631-0502
Daniel Thorp

Historic Manassas
9108 Church Street
Manassas, VA 22110
(703) 361-6599

Kidde Consultants
8832 Rixlew Lane
Manassas, VA 22110
Rhett Whitlock

Middleburg

Lewis Whitesell
Route 1
Box 320
Middleburg, VA 22117
(703) 687-8811

Middletown

Cedar Creek Battlefield Foundation
P.O. Box 229
Middletown, VA 22645

New Market

New Market Battlefield Historical Park
P.O. Box 1864
New Market, VA 22844
(703) 740-3101
Ed Merrell

Newport News

City of Newport News
2400 Washington Avenue
Newport News, VA 23607
(804) 247-8428
Paul F. Miller

Norfolk

Louise E. Andrews, AIA
801 Colonial Avenue
Norfolk, VA 23507
(804) 623-3774

Cavalier Land
P.O. Box 3175
Norfolk, VA 23514
(804) 625-3502
Marc Poutasse

Chrysler Museum Library
Olney Road and Mowbray Arch
Norfolk, VA 23510
(804) 622-1211
T. Brennan

Hampton Roads Naval Museum
Norfolk Naval Base
Pennsylvania Building, G-29
Norfolk, VA 23511-6002
(804) 444-3827
E. Poulliot

Oakton

Susan Robinette
10205 Willow Mist Court
Suite D
Oakton, VA 22124
(703) 255-2158

Petersburg

B. David Canada
Planning and Community
Development
City Hall
Room 208
Petersburg, VA 23803
(804) 733-2308

City of Petersburg
Tabb and Union Streets
Petersburg, VA 23803
(804) 733-2300

Historic Blandford Cemetery
Foundation
250 South Sycamore Street
Petersburg, VA 23803
(804) 732-3717
Charlotte A. Irving

Historic Petersburg Foundation
P.O. Box 691
Petersburg, VA 23804
(804) 732-2096
Karen Graham

Petersburg Museums
Department of Tourism
15 West Bank Street
Petersburg, VA 23803
(804) 733-2402

Poquoson

Col. and Mrs. R. Holt Evans
675 Poquoson Avenue
Poquoson, VA 23662
(804) 868-0510

Port Royal

Alexander Long
P.O. Box 217
Port Royal, VA 22535
(804) 742-5612

Portsmouth

Commission of Architectural Review
Planning Department
801 Crawford Street
Portsmouth, VA 23704
(804) 393-8836
June Brooks

Prince William

Prince William County Historical
Commission
Planning Office
1 County Complex Court
Prince William, VA 22192
(703) 335-6830
Jan Townsend

Pulaski

Pulaski Main Street
P.O. Box 2121
Pulaski, VA 24301
(703) 980-8343
Doug Phelps

Radford

Main Street Radford
1126 Norwood Street
Radford, VA 24141
(703) 731-3656
Charlie Whitescarver

Reston

American Planning Association
National Capital Area Chapter
11302 Fairway Drive
Reston, VA 22090
(703) 358-3694
Patricia Nicoson

Richmond

Agecroft Association
4305 Sulgrave Road
Richmond, VA 23221
Dennis H. Halloran

Association for the Preservation of
Virginia Antiquities
2300 East Grace Street
Richmond, VA 23223
(804) 648-1889
Peter D. Grover

Department of Community
Development
900 East Broad Street
Richmond, VA 23261
(804) 780-6308
John Albers

Department of Historic Resources
221 Governor Street
Richmond, VA 23219
(804) 786-3143
Hugh C. Miller

Henrico Division of Recreation and
Parks
Historic Preservation Program
P.O. Box 27032
Richmond, VA 23273
(804) 672-5100
Susan A. Hanson

Historic Richmond Foundation
2407 East Grace Street
Richmond, VA 23223
John Zehmer

Lower James River Association
P.O. Box 110
Richmond, VA 23201
(804) 730-2898
Patricia A. Jackson

Marcellus Wright Cox and Smith
100 Shockoe Slip
Richmond, VA 23219
(804) 780-9067
(804) 285-7182
Frederic H. Cox

Museum of the Confederacy
1201 East Clay Street
Richmond, VA 23219

Mrs. James H. Parsons
612 West Franklin Street
Richmond, VA 23220

Rawlings and Wilson
Architects
1100 North Thompson Street
Richmond, VA 23230
(804) 358-9141
James S. Rawlings

Task Force on Historic Preservation and
the Minority Community
P.O. Box 25604
Richmond, VA 23260
(804) 788-1709

Valentine Museum
1015 East Clay Street
Richmond, VA 23219

Virginia Main Street Program
Department of Housing and
Community Development
205 North 4th Street
5th Floor
Richmond, VA 23219
(804) 786-4966
Teresa Lynch

Virginia State Library and Archives
Serial Section
11th Street at Capitol Square
Richmond, VA 23219
(212) 254-4454

Roanoke

Sueann Brown
519 King George Avenue, S.W.
Roanoke, VA 24016

City of Roanoke
Office of Community Planning
Municipal Building
215 Church Avenue, S.W.
Room 355
Roanoke, VA 24011
(703) 981-2344
Evelyn Gunter

Hollingsworth House, Winchester, Virginia. (R. Rogers, HABS)

Downtown Roanoke Incorporated
310 1st Street, S.W.
Roanoke, VA 24011
(703) 342-2028
Franklin D. Kimbrough, III

Gary Miller
3907 Rachel Drive
Roanoke, VA 24019

Roanoke Regional Preservation Office
1030 Penmar Avenue, S.E.
Roanoke, VA 24013
John R. Kern

Roanoke Valley Preservation
Foundation
P.O. Box 1558
Roanoke, VA 24007
(703) 342-9289
Martha Boxley

Salem

Salem Historical Society
P.O. Box 201
Salem, VA 24153
(703) 389-0407
David L. Foster

Smithfield

Historical Saint Luke's Restoration
331 South Church Street
Smithfield, VA 23430
(804) 357-7107
Nancy D. Fortier

Springfield

Capitol Office Furniture Company
6629 Iron Place
Springfield, VA 22151-4307
(703) 914-1666
Ronald Gerhardt

Stafford

Aquia Episcopal Church
P.O. Box 275
Stafford, VA 22554
(703) 659-4007
Joseph R. Kerr

Staunton

City of Staunton
P.O. Box 58
Staunton, VA 24401
(703) 885-2839
George W. Edwards

Frazier Associates
121 South Augusta Street
Staunton, VA 24401
(703) 886-6230
William T. Frazier

Historic Staunton Foundation
120 South Augusta Street
Staunton, VA 24401
(703) 885-7676
Deneen Dameron

Woodrow Wilson Birthplace
P.O. Box 24
20 North Coalter Street
Staunton, VA 24401
(703) 885-0897
Katharine L. Brown

Strasburg

Wayside Museum of American History
P.O. Box 31
Strasburg, VA 22657
(703) 465-5884
Linda K. Faris

Stratford

Robert E. Lee Memorial Association
Stratford Hall Plantation
Stratford, VA 22558
(804) 493-8038
John F. Wall

Suffolk

City of Suffolk
P.O. Box 1858
Suffolk, VA 23434
(804) 934-3111
Anna D'Antonio

Downtown Suffolk Association
P.O. Box 404
Suffolk, VA 23434
(804) 539-3592
Kate Van Pelt

Sweet Briar

Mary Helen Cochran Library
Sweet Briar College
Sweet Briar, VA 24595
(804) 381-6138
John Jaffe

The Plains

Save Railroad Station
P.O. Box P
The Plains, VA 22171
(703) 754-8564
T. Lawrence

Virginia Beach

Central Library
Periodicals
4100 Virginia Beach Boulevard
Virginia Beach, VA 23452

David L. Vevoda
1440 Talisman Circle
Virginia Beach, VA 23464-6118
(804) 474-2963

Warrenton

Partnership for Warrenton
P.O. Box 3528
Warrenton, VA 22186
(703) 349-8606
Christopher Wilson

Piedmont Environmental Council
28 Main Street
P.O. Box 460
Warrenton, VA 22186

Washington

Waters Craftsmen
Middle Street
P.O. Box 329
Washington, VA 22747
(800) 232-1395
Dale Waters

Waterford

Waterford Foundation
P.O. Box 142
Waterford, VA 22190
(703) 882-3018
Catherine Ladd

Williamsburg

Murrel Dee Hobt, AIA
173 Merrimac Trail
Apartment 10
Williamsburg, VA 23185
(804) 220-0767

James City and County
P.O. Box JC
Williamsburg, VA 23187
(804) 566-0363
Phyllis Cody

Winchester

Downtown Development Board
15 North Cameron Street
Suite 415
Winchester, VA 22601
(703) 665-0079
Sonya L. Tolley

Preservation of Historic Winchester
530 Amherst Street
Winchester, VA 22601
(703) 667-3577
Anna Thomson

Yorktown

Yorktown Victory Center
P.O. Box 1976
Yorktown, VA 23690
(804) 887-1776
Ed Ayres

WASHINGTON

STATE HISTORIC PRESERVATION OFFICE

Office of Archeology and Historic Preservation
111 West 21st Street
KL-11
Olympia, WA 98504
(206) 753-4011
(206) 586-0250 (Fax)
Mary Thompson, Director, SHPO
David M. Hansen, Chief
Office of Archeology and Historic Preservation
Deputy SHPO
(206) 753-4117

STATEWIDE PRESERVATION ORGANIZATION

Washington Trust for Historic Preservation
204 1st Avenue, South
Seattle, WA 98104
(206) 624-7880
Valerie Sivinski, President

NATIONAL TRUST REGIONAL OFFICE

Western Regional Office
1 Sutter Street
Suite 707
San Francisco, CA 94104
(415) 956-0610
(415) 956-0837 (Fax)
Kathryn A. Burns, Director

NATIONAL TRUST ADVISORS

Anthony H. Anderson
Eastern Washington University
West 421 Riverside
Suite 421
Spokane, WA 99201
(509) 458-6219
(509) 623-4230 (Fax)

Karen Gordon
1531 38th Avenue
Seattle, WA 98122
(206) 684-0381
(206) 233-5142 (Fax)

PRESERVATION ACTION COORDINATOR

Valerie A. Sivinski
747 Market Street
Suite 900
Tacoma, WA 98402
(206) 591-5220

AIA STATE PRESERVATION COORDINATOR

Larry Johnson, AIA
The Johnson Partnership
1212 Northeast 65th Street
Seattle, WA 98115
(206) 523-1618
Mark Mordan, AIA, Deputy

NATIONAL PARK SERVICE REGIONAL OFFICE

Pacific Northwest Regional Office
83 South King Street
Suite 212
Seattle, WA 98104
(206) 553-5565

ADVISORY COUNCIL ON HISTORIC PRESERVATION DIVISION

Western Office of Project Review
730 Simms Street
Room 401
Golden, CO 80401
(303) 231-5320
Claudia Nissley, Director

Auburn

City of Auburn
25 West Main Street
Auburn, WA 98001
(206) 931-3033

Bainbridge Island

Bainbridge Island Historical Society
7823 North East High School Road
Bainbridge Island, WA 98110
(206) 842-2773
G. W. Elfendahl

Arlene Hetherington
P.O. Box 11349
Bainbridge Island, WA 98110
(206) 842-9575

Beaver

Susan Goff
P.O. Box 387
Beaver, WA 98305
(206) 327-3827

Belfair

A. H. Parker
East 17271 Highway 106
Belfair, WA 98528

Bellevue

City of Bellevue
Planning Department
Suite 38875
P.O. Box 90012
Bellevue, WA 98009
(206) 455-6880
Bruce Freeland

Bothell

City of Bothell
Department of Community Development
18305 101st Avenue, N.E.
Bothell, WA 98011
(206) 486-8152
Gordon Y. Ericksen

Coupeville

Island County Historical Society
Box 305
Coupeville, WA 98239
(206) 678-6854
Del Benneti

Everett

Everett Community Planning Department
City Hall
3002 Wetmore Avenue
Everett, WA 98201
(206) 259-8731
David Koenig

Newland Development Company
2911 1/2 Hewitt Avenue
Everett, WA 98201
(206) 774-7264
Doris Newland

Kennewick

Benton Franklin Riverfront Trailway and Bridge Committee
P.O. Box 6975
Kennewick, WA 99336
(509) 943-1571
Homer Moulthrop

City Of Kennewick
P.O. Box 6108
Kennewick, WA 99336
(509) 586-4181
William Kennedy

Longview

Longview Historic Preservation Commission
P.O. Box 128
Longview, WA 98632
(206) 577-3330
Julie Hourcle

Metaline Falls

North County Theatre
Box 133
Metaline Falls, WA 99153
(509) 442-3256
Eva G. Six

Mount Vernon

Lincoln Theatre Center Foundation
P.O. Box 2312
Mount Vernon, WA 98273
(206) 293-9610
Peter Heffelfinger

Olympia

Deskoba
112 East 4th Avenue
Suite 204
Olympia, WA 98501
(206) 352-4861
F. S. Desner

Downtown Revitalization Program
111 West 21st Avenue
KL-11
Olympia, WA 98504
(206) 753-5669
Dick Larman

Office of Historic Preservation
111 West 21st Avenue
KL-11
Olympia, WA 98504
(206) 753-4011
Jacob E. Thomas

Thurston Regional Planning Council
2000 Lakeridge Drive
Olympia, WA 98502
(206) 786-5480

Pasco

Pasco Downtown Development
Association
P.O. Box 842
Pasco, WA 99301
(509) 545-0738
Jeanene Landby

Port Angeles

Port Angeles Downtown Association
P.O. Box 582
Port Angeles, WA 98362
(206) 452-3871
Loretta Martin

Port Townsend

City of Port Townsend
City Hall
540 Water Street
Port Townsend, WA 98368
(206) 385-3000
Katherine Johnson

Jefferson County Historical Society
210 Madison
Port Townsend, WA 98368
(206) 385-1003
Dixie Romadka

Pink House Committee
City of Port Townsend
210 Water Street
Port Townsend, WA 98368
Joseph Vleck

Port Townsend Civic Trust
211 Taylor
Suite 4
Port Townsend, WA 98368
(206) 385-7911
Carolyn Niemeyer

Puyallup

Ezra Meeker Museum
P.O. Box 103
Puyallup, WA 98371
(206) 848-1770
Sharone Ketterman

Puyallup Main Street Association
P.O. Box 476
301 Meridian Street
Puyallup, WA 98371
(206) 840-2631
Paula J. Braaten

Seattle

Adak Community Musuem
Box 5244
Naval Station
FPO Seattle, WA 98791
(907) 592-3221
Steve Hutchens

Boyle-Wagoner Architects
911 Western Avenue
Suite 300
Seattle, WA 98104
(206) 382-9651
Susan Boyle

The Bumgardner Architects
101 Stewart Street
Suite 200
Seattle, WA 98101
(206) 223-1361
Donald T. Brubeck

Cardwell Thomas
1221 2nd Avenue
Suite 300
Seattle, WA 98101
Rich Cardwell

Foundation for Historic Preservation
and Adaptive Reuse
14705 27th Avenue, N.E.
Seattle, WA 98155
(206) 362-6247
Sue C. Jennings

Historic Seattle Preservation and
Development Authority
207 1/2 1st Avenue, South
Seattle, WA 98104
(206) 622-6952
Catherine Galbraith

King County Historic Preservation
Program
506 2nd Avenue
1115 Smith Tower
Seattle, WA 98104
Kristine Lund

Northwest Seaport
1002 Valley Street
Seattle, WA 98109
(206) 447-9800
Al Elliot

Jeffrey Karl Ochsner
415 Wheeler
Seattle, WA 98109
(206) 285-6634

Office of Urban Conservation
Seattle Division
700 3rd Avenue
8th Floor
Seattle, WA 98104
(206) 684-0228
Karen Gordon

Tonkin/Cook Architects
204 1st Avenue, South
Seattle, WA 98104
Les Tonkin

Trinity Parish Church Restoration
609 8th Avenue
Seattle, WA 98104
(206) 323-1492
John Cosby

Washington Trust for Historic
Preservation
204 1st Avenue, South
Seattle, WA 98104
Becky Day

Spokane

ALSC Architects
700 Old National Building
Spokane, WA 99201
(509) 838-8568
Sue W. Bonstrom

Eastern Washington University
West 421 Riverside
Suite 421
Spokane, WA 99201
Anthony H. Anderson

Friends of the Davenport
P.O. Box 785
Spokane, WA 99210
(509) 838-8568
Dave Shockley

Historic Preservation Office
Spokane Regional Council
West 808 Spokane Falls Boulevard
Room 627
Spokane, WA 99201
(509) 456-4378

Northwest Regional Foundation
East 525 Mission
Spokane, WA 99202

Wells and Company
911 East 20th Avenue
Spokane, WA 99203
R. R. Wells

Steilacoom

Town of Steilacoom
1715 Lafayette
Steilacoom, WA 98388
(206) 581-1900
Steven Fischer

Tacoma

Pantages Centre
901 Broadway
Tacoma, WA 98402
(206) 591-5890
Eli D. Ashley

Pierce County Planning
Suite 175
2401 South 35th Street
Tacoma, WA 98409
Caroline Gallacci

Garry Schalliol
219 Tacoma Avenue North
Suite 106
Tacoma, WA 98403-2626
(206) 586-0583

Tacoma Historical Preservation Office
747 Market Street
Suite 1036
Tacoma, WA 98402
(206) 591-5220
Valerie Sivinski

Toppenish

Yakima Nation Museum
P.O. Box 151
Toppenish, WA 98948
Brycene Neaman

Vancouver

Downtown Vancouver Association
303 East Evergreen Boulevard
Vancouver, WA 98660
(206) 693-2978
Karen Ciocia

Friends of Fort Vancouver
612 East Reserve Street
Vancouver, WA 98661

Heritage Trust of Clark County
1351 Officers Row
Vancouver, WA 98661
(206) 699-2361
Glenda J. Choate

Walla Walla

Walla Walla Main Street
14 East Main
Suite 3
Walla Walla, WA 99362
(509) 529-8755
Bruce A. Buchanan

Yakima

Yakima Valley Museum and Historic
Association
2105 Tieton Drive
Yakima, WA 98902
Versa C. K'ang

Yelm

Town of Yelm
P.O. Box 479
Yelm, WA 98597
(206) 458-3244
Gene Borges

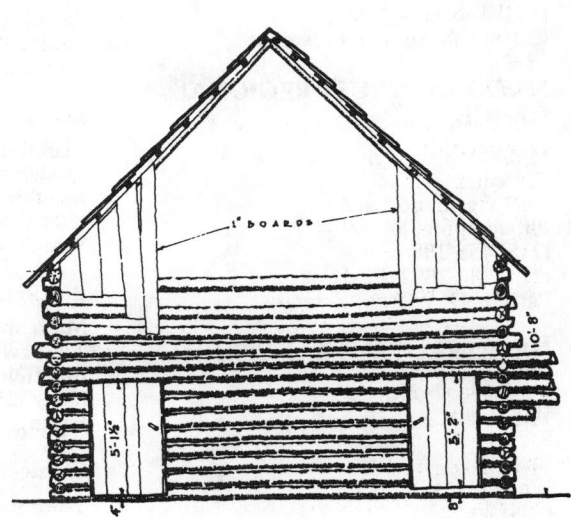

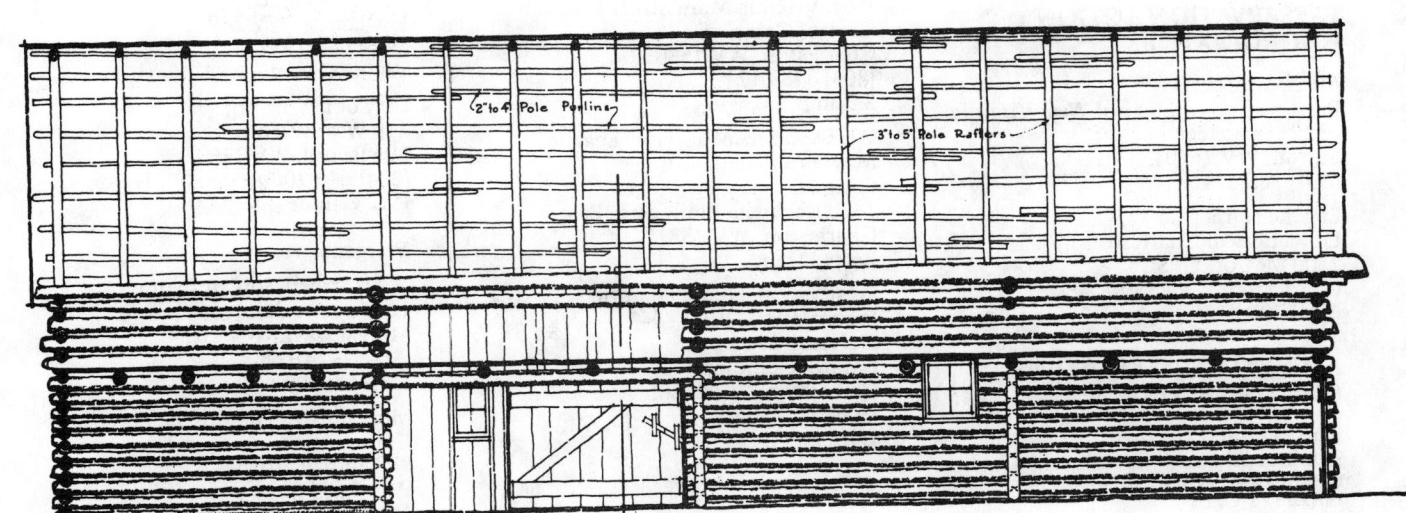

Clugston Barn, Colville, Washington. (F. Dings, A. Welch HABS)

WEST VIRGINIA

STATE HISTORIC PRESERVATION OFFICE

West Virginia Division of Culture and History
Historic Preservation Office
Cultural Center
1900 Kanawha Boulevard East
Charleston, WV 25305-0300
(304) 558-0220
William M. Drennan, Jr.,
Commissioner, SHPO
William G. Farrar, IV, Deputy
Commissioner, Deputy SHPO

STATEWIDE PRESERVATION ORGANIZATION

Preservation Alliance of West Virginia
c/o Matewan Development Center
P.O. Box 368
Matewan, WV 25678
(304) 426-4239
C. Paul McAllister, President

NATIONAL TRUST REGIONAL OFFICE

Mid-Atlantic Regional Office
Cliveden
6401 Germantown Avenue
Philadelphia, PA 19144
(215) 438-2886
(215) 438-2892 (Fax)
Patricia D. Wilson, Director

NATIONAL TRUST ADVISOR

Betty Woods Nutting
Windfield
R.D. 4
Box 101
Wheeling, WV 26003
(304) 242-2700

PRESERVATION ACTION COORDINATORS

Michael Gioulis
Preservation Alliance of West Virginia
612 Main Street
Sutton, WV 26601
(304) 765-5716

Eugene Harper
University of Charleston
2300 McCorkel, SE
Charleston, WV 25304
(304) 357-4770

AIA STATE PRESERVATION COORDINATOR

Paul D. Marshall, AIA
915 Breezemont Drive
Charleston, WV 25302-3323
(304) 343-5310
David M. Marshall, AIA, Deputy

NATIONAL PARK SERVICE REGIONAL OFFICE

Mid-Atlantic Regional Office
143 South 3rd Street
Philadelphia, PA 19106
(215) 597-7018

ADVISORY COUNCIL ON HISTORIC PRESERVATION DIVISION

Eastern Office of Project Review
1100 Pennsylvania Avenue, N.W.
Suite 809
Washington, DC 20004
(202) 786-0503
Don L. Klima, Director

Arthurdale

Arthurdale Heritage
P.O. Box 850
Arthurdale, WV 26520
(304) 864-5286
Glenna Williams

Beckley

Beckley Main Street Program
Drawer A J
Beckley, WV 25802
(304) 256-1776
Sara Pleasant

Bluefield

Main Street Bluefield
224 Law and Commerce Building
Bluefield, WV 24701
(304) 325-5442
Karen Simmons

Charleston

Community and Industrial Development
West Virginia Main Street Program
Capitol Complex
Charleston, WV 25305
(304) 348-0121
Susan F. Salisbury

West Virginia Municipal League
Mayors Village
Suite 1B
1620 Kanawha Boulevard, East
Charleston, WV 25311
Betty Dean

West Virginia State Historic Preservation Office
Cultural Center
Capitol Complex
Charleston, WV 25305
(304) 348-0240
William G. Farrar

Elkins

Davis and Elkins College
100 Sycamore Street
Elkins, WV 26241
(304) 636-1900
William Johnston

Fairmont

Friends of High Gate
P.O. Box 957
Fairmont, WV 26554

Pricketts Fort Memorial Foundation
Route 3
Fairmont, WV 26554
(304) 363-3030
David Elkinton

Hinton

Hinton Main Street Program
P.O. Box 477
Hinton, WV 25951
(304) 466-4971
Karla T. Gunnoe

Huntington

Huntington Main Street
945 4th Avenue
Suite 216
Huntington, WV 25701
(304) 529-0053
Renee Maass

Keyser

Keyser Main Street Program
115 Armstrong Street
Keyser, WV 26726
(304) 788-2823
Charles H. Peterson

Lewisburg

Carnegie Hall
105 Church Street
Lewisburg, WV 24901
(304) 645-7917
Vivian Conly

City of Lewisburg
119 West Washington Street
Lewisburg, WV 24901
(304) 645-2080
P. L. Gainer

Morgantown

Main Street Morgantown
160 Fayette Street
Morgantown, WV 26505
(304) 292-0168
Terri Cutright

Morgantown Historic Landmarks Commission
389 Spruce Street
P.O. Box 1620
Morgantown, WV 26507
(304) 291-7433

West Virginia University
Department of History
Morgantown, WV 26506
(304) 293-2421
Barbara J. Howe

Parkersburg

Junior League of Parkersburg
Cook House Committee
1301 Murdock Avenue
Parkersburg, WV 26101

Philippi

City of Philippi
108 North Main Street
Philippi, WV 26416
(304) 457-3701
Joseph P. Mattaliano

Main Street Philippi
5 South Main Street
Philippi, WV 26416
(304) 457-5237
Rex Otey

Joseph P. Mattaliane
108 North Main Street
Philippi, WV 26416
(304) 457-3700

Point Pleasant

Point Pleasant Main Street
305 Main Street
Point Pleasant, WV 25550
(304) 675-3844
Al Alderfer

Ranson

Corporation of Ranson
312 South Mildred Street
Ranson, WV 25438
(304) 725-1010
John Defries

Salem

Fort New Salem
Salem-Teikyo University
Salem, WV 26426
(304) 782-5245
Carol A. Schweiker

Salem Area Chamber of Commerce
P.O. Box 191
Salem, WV 26426

Shepherdstown

Historic Shepherdstown Commission
P.O. Box 1786
Shepherdstown, WV 25443
(304) 876-6555

Joseph J. Snyder
2008 Ashley Drive
Shepherdstown, WV 25443
(304) 876-9431

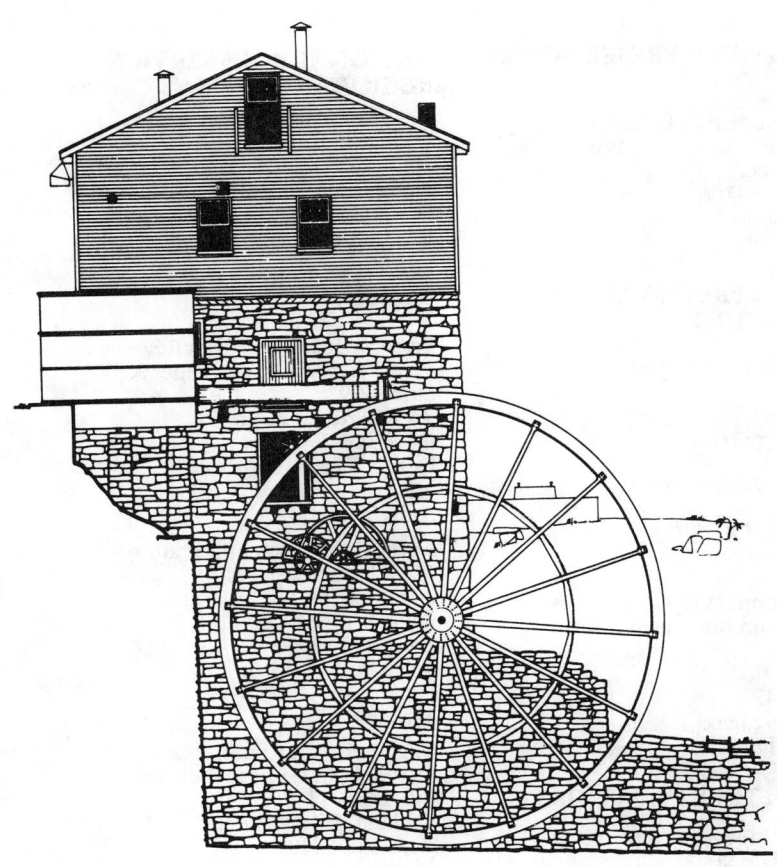

Thomas Shepherd's Grist Mill, Shepherdstown, West Virginia. (B. Freeman, HABS)

Summersville

Nicholas County Historic Landmark
Commission
Office of Nicholas County Commission
Summersville, WV 26651
(304) 872-5726
Wilma Richardson

Weirton

Weirton Main Street
3300 Main Street
Suite D
Weirton, WV 26062
(304) 797-1604
Dawn Ohalek

Wellsburg

City of Wellsburg
City Hall
70 7th Street
Wellsburg, WV 26070
(304) 737-2104
Mark S. Baldwin

Wheeling

Friends Of Wheeling
P.O. Box 889
Wheeling, WV 26003
(304) 233-3511
George McLaughlin

Mount de Chantal Visitation Academy
410 Washington Avenue
Wheeling, WV 26003-6296
(304) 233-3771
Joanne Gonter

Preservation Alliance of West Virginia
R.D. 4
Box 120
Wheeling, WV 26003
(304) 234-3701
Hydie Hopkins

Richard A. Smith
725 Main Street
Wheeling, WV 26003
(304) 223-1954

Victorian Wheeling Landmarks
Foundation
P.O. Box 666
Wheeling, WV 26003
(304) 242-2700
Snookie Nutting

White Sulphur Springs

Main Street White Sulphur
Springs
34 West Main Street
White Sulphur Springs, WV 24986
(304) 536-4007
Erica Gillespie

WISCONSIN

STATE HISTORIC PRESERVATION OFFICE

Historic Preservation Division
State Historical Society of Wisconsin
816 State Street
Madison, WI 53706
(608) 264-6500
(608) 264-6404 (Fax)

STATEWIDE PRESERVATION ORGANIZATION

Wisconsin Trust for Historic
Preservation
P.O. Box 825
Merrill, WI 54452-0825
(715) 536-6220
Alice Krueger, President

NATIONAL TRUST REGIONAL OFFICE

Midwest Regional Office
52 West Jackson Boulevard
Suite 1135
Chicago, IL 60604
(312) 939-5447
(312) 939-5651 (Fax)
Tim Turner, Director

NATIONAL TRUST ADVISORS

George L. N. Meyer
312 East Buffalo Street
Suite 55
Milwaukee, WI 53202
(414) 277-8501
(414) 261-5088

Brian Rude
206-S State Capitol
Box 7882
Madison, WI 53707-7882
(608) 452-3305

PRESERVATION ACTION COORDINATOR

Dawn F. Goshorn Schumann
Hatterhorn Enterprises
5550 North Kent
Whitefish Bay, WI 53217
(414) 961-7103

AIA STATE PRESERVATION COORDINATOR

Charles J. Quigliana, AIA
DOSFM
P.O. Box 7866
Madison, WI 53707
(608) 266-1458

NATIONAL PARK SERVICE REGIONAL OFFICE

Midwest Regional Office
1709 Jackson Street
Omaha, NE 68102
(402) 221-3431

ADVISORY COUNCIL ON HISTORIC PRESERVATION DIVISION

Eastern Office of Project Review
1100 Pennsylvania Avenue, N.W.
Suite 809
Washington, DC 20004
(202) 786-0503
Don L. Klima, Director

Antigo

Antigo Main Street Program
612 Clermont Street
P.O. Box 165
Antigo, WI 54409
(715) 623-3595
Keith Voss

Ashland

Chequamegon Center
Ashland Main Street Program
P.O. Box 782
Ashland, WI 54806
(715) 682-3020
Stacey Adams

Baraboo

Al Ringling Theatre Friends
P.O. Box 381
136 4th Avenue
Baraboo, WI 53913
(608) 356-8080
Robert R. Dippel

Beloit

Beloit Main Street Program
P.O. Box 291
Beloit, WI 53511
(608) 365-0150
Andrew Janke

City of Beloit
100 State Street
Beloit, WI 53511
(608) 364-6700
Steven Lere

Cashton

Cashton Area Development
Corporation
P.O. Box 1
Cashton, WI 54619
(608) 654-5121
Scott Wall

Chippewa Falls

Chippewa Falls Main Street
P.O. Box 554
Chippewa Falls, WI 54729
(715) 723-6661
Kathy LaPlante

Cleveland

Centreville Settlement
P.O. Box 247
Cleveland, WI 53015
(414) 693-8558
Dorothy Anderson

Cottage Grove

Indeco Design
2035 Uphoff Road
Cottage Grove, WI 53527
(608) 839-5467
Alice R. Miller

P. Ohlson
286 North Main Street
Cottage Grove, WI 53527

DePere

Main Street DePere
DePere Development Corporation
P.O. Box 3121
DePere, WI 54115
(414) 433-7767
Sandra Duckett

Eagle

Old World Wisconsin
S103 W37890 Highway 67
Eagle, WI 53119
(414) 594-2116
Hugh Gurney

Eau Claire

Chippewa Valley Museum
P.O. Box 1204
Eau Claire, WI 54702
(715) 834-7871
Susan McLeod

City of Eau Claire
Planning and Development
203 South Farwell Street
Call Box 5148
Eau Claire, WI 54702
(715) 839-4947
Michael Huggins

Eau Claire Main Street Association
306 Main Street
Eau Claire, WI 54701
(715) 839-0011
Vickie L. McCluskey

Fond Du Lac

City of Fond Du Lac
Redevelopment Division
160 South Macy Street
P.O. Box 150
Fond Du Lac, WI 54936-0150
(414) 929-3311
John Angeli

Germantown

Germantown Historical Society
N110W16867 Ashbury Circle
Suite 7
Germantown, WI 53022
(414) 251-6378
Irene M. Blau

Greenburgh

Alan C. Pape Consulting
Box 31
Greenburgh, WI 53026
(414) 526-3433

Hartford

Hartford Historic Preservation and
Enhancement Committee
109 North Main Street
Hartford, WI 53027
(414) 673-8265
John Cohen

Janesville

Judy Berg
429 Propect Avenue
Janesville, WI 53545

Janesville Historic Commission
18 North Jackson
Janesville, WI 53545
(608) 755-3180
Judith Adler

Kenosha

City of Kenosha
625 52nd Street
Kenosha, WI 53140
(414) 656-8030
Ray Forgianni

Kemper Center
6501 3rd Avenue
Kenosha, WI 53140
(414) 657-6005
Peggy Gregorski

King

Russell D. Bressler
Wisconsin Veterans Home
306 Burns Clemens Hall
King, WI 54946

La Crosse

Architectural Researchers
2540 Sherwood Drive
La Crosse, WI 54601
(608) 788-5932
Joan Rausch

City of La Crosse
Planning Department
City Hall
400 La Crosse Street
La Crosse, WI 54601
(608) 798-7512
John W. Florine

Downtown Main Street
712 Main Street
La Crosse, WI 54601
(608) 782-2467
Margaret Weinzierl

Schute-Larson Architects
125 North 4th Street
P.O. Box 2496
La Crosse, WI 54602
(608) 785-2217
Valentine J. Schute

Madison

City of Madison
Department of Planning and
Development
215 Martin Luther King Boulevard
P.O. Box 2985
Madison, WI 53701
(608) 266-6552
Katherine H. Rankin

Dane County Cultural Affairs
Commission
City County Building
Room 421
Madison, WI 53709
(608) 266-5915
Lynn Eich

Historic Preservation Division
State Historical Society of Wisconsin
816 State Street
Madison, WI 53706
(608) 262-1339
Larry Reed

Diane Schwartz
1316 Mounno
Madison, WI 53715
(608) 255-6904

Wisconsin Heritage Tourism
123 West Washington
6th Floor
Madison, WI 53707
(608) 266-7299
Sharon Folcey

James Fraser House, Honey Creek Falls, Wisconsin. (E. Kusseraw, HABS)

Wisconsin Main Street Program
Department of Development
123 West Washington Avenue
P.O. Box 7970
Madison, WI 53703
(608) 267-3855
Alicia L. Goehring

Manitowoc

Association of the Great Lakes
Maritime History
75 Maritime Drive
Manitowoc, WI 54220-6845
(414) 684-0218
Bert Logan

Marinette

Marinette Renaissance
1812 Hall Avenue
P.O. Box 434
Marinette, WI 54143
(715) 732-0909
Vivian M. Haight

Marshfield

Main Street Marshfield
P.O. Box 551
Marshfield, WI 54449
(715) 387-3299
Jay Schlinsog

Mazomanie

Mazomanie Historical Society
Mazomanie, WI 53560
(608) 795-4733
Frank E. Wolf

Menomonee Falls

Village of Menomonee Falls
W156 N8480 Pilgrim Road
P.O. Box 100
Menomonee Falls, WI 53051
(414) 255-8323
William E. Freisleben

Menomonie

City of Menomonie
800 Wilson Avenue
Menomonie, WI 54751-2795
(715) 232-2369
Chuck Stokke

Mequon

City of Mequon Landmarks
Commission
11333 North Cedarburg Road
60 West
Mequon, WI 53092
(414) 228-9100
Joanne Simone

Mequon Landmarks
11333 North Cedarburg Road
Mequon, WI 53092
(414) 242-6370
Joanne Simone

Merrill

Wisconsin Trust For Historic
Preservation
P.O. Box 825
Merrill, WI 54452-0825

Milton

City of Milton Historic Commission
P.O. Box 188
Milton, WI 53563

Milwaukee

George Meyer Family Foundation
312 East Buffalo Street
Suite 55
Milwaukee, WI 53202
(414) 277-8501

Historic Milwaukee
P.O. Box 2132
Milwaukee, WI 53201
(414) 277-7795
Sandy Ackerman

Lee C. Jensen
Building Inspection Department
841 North Broadway
Room 1001
Milwaukee, WI 53202

Millen Roofing Company
2247 North 31st Street
Milwaukee, WI 53208
(414) 442-1424
Matt Millen

Milwaukee Preservation Commission
Department of City Development
809 North Broadway
Milwaukee, WI 53202
(414) 223-5705
Les Vollmert

Milwaukee Redevelopment Authority
P.O. Box 324
Milwaukee, WI 53201
(414) 223-5820
John L. Czarnezki

Plankinton Trust
161 West Wisconsin
Milwaukee, WI 53203
(414) 271-0071
(407) 655-2528
E. P. MacKintosh

Renner Architects
626 North Water Street
Milwaukee, WI 53202
(414) 273-6637
Allyson Dahlen

Robert Schweininger
5421 North 106th Street
Milwaukee, WI 53225
(414) 464-3877

Walkers Point Development
Corporation
734 South 5th Street
Milwaukee, WI 53204

Westside Conservation Corporation
3209 West Highland Boulevard
Milwaukee, WI 53208
(414) 933-2300
Perry Harenda

William Manley Associates
302 North Jackson Street
Milwaukee, WI 53202
(414) 291-5200

Wisconsin Heritage
2000 West Wisconsin Avenue
Milwaukee, WI 53233
(414) 931-0808
Shawn Graff

Wisconsin Preservation Federation
205 East Wisconsin Avenue
Milwaukee, WI 53202
Bruce T. Block

Mineral Point

City of Mineral Point
137 High Street
P.O. Box 269
Mineral Point, WI 53565
(608) 987-2361
Myron Remington

New Berlin

Conrad Schmitt Studios
2405 South 162nd Street
New Berlin, WI 53151
(414) 786-3030
Heidi Gruenke

North Freedom

David L. Henke
S5765 Seeley Lane
North Freedom, WI 53951
(608) 785-1925

Oshkosh

City of Oshkosh
City Planning Office
P.O. Box 1130
Oshkosh, WI 54902
(414) 236-5059

Portage

Portage Canal Society
528 West Cook Street
Portage, WI 53901
(608) 742-2889
Frederica Kleist

Racine

City of Racine
730 Washington Avenue
Racine, WI 53403

Preservation Racine
P.O. Box 383
Racine, WI 53401
(414) 634-5748
Don Rintz

Racine Landmarks Preservation
Commission
730 Washington Avenue
Racine, WI 53403
(414) 636-9151
Larry G. Vaile

Richland Center

City of Richland Center
P.O. Box 248
Richland Center, WI 53581
(608) 647-3466
Luella Edwards

Ripon

City of Ripon
100 Jackson Street
Ripon, WI 54971
(414) 748-7771

Ripon Main Street Program
301 Watson Street
Ripon, WI 54971
(414) 748-7466
Robert Pauls

River Falls

River Falls Main Street Program
220 South Main Street
River Falls, WI 54022
(715) 425-8901
Dick Rieg

Shawano

Main Street Shawano
213 East Green Bay Street
Shawano, WI 54166
(715) 524-2105
Mike Ascher

Sheboygan

Sheboygan Area School District
830 Virginia Avenue
Sheboygan, WI 53081

Sheboygan Falls

Sheboygan Falls Main Street
110 Pine Street
Sheboygan Falls, WI 53085
(414) 467-6206
Rhonda Luniak

Siren

Burnett County Historical Society
100 East Johnson Street
P.O. Box 31
Siren, WI 54872
(715) 349-2219
E. S. Oerichbauer

Sparta

Sparta Main Street
P.O. Box 232
Sparta, WI 54656
(608) 269-4080
Brian Doudna

Stevens Point

Historic Preservation Design Review
1515 Strongs Avenue
Stevens Point, WI 54481
(715) 346-1567
John Gardner

Stoughton

Stoughton Main Street Program
P.O. Box 318
Stoughton, WI 53589
(608) 873-7743
Kevin Pomeroy

Superior

Fairlawn Mansion and Museum
Douglas County Historical Society
Harbor View Parkway, East
Superior, WI 54880
(715) 394-5712
Rachael E. Martin

Viroqua

Viroqua Revitalization Association
124 1/2 South Main Street
Viroqua, WI 54665
(608) 637-2666
Theresa Washburn

Waukesha

Waukesha Business Improvement
District
217 Wisconsin Avenue,
Suite B16
Waukesha, WI 53186
(414) 549-6154
Penny A. Van Fleet

West Bend

Charles Mayhew III
6871 Hickory Road
West Bend, WI 53095
(414) 675-2341

Whitefish Bay

Hatterhorn Enterprises
5550 North Kent
Whitefish Bay, WI 53217
(414) 961-7103
Dawn F. Goshorn

WYOMING

STATE HISTORIC PRESERVATION OFFICE

Wyoming State Historic Preservation Office
Barrett Building
2301 Central Avenue
4th Floor
Cheyenne, WY 82002
(307) 777-7013
(307) 777-6005
John Keck, SHPO

NATIONAL TRUST REGIONAL OFFICE

Mountains/Plains Regional Office
511 16th Street
Suite 700
Denver, CO 80202
(303) 623-1504
(303) 825-8073 (Fax)
Barbara J. Pahl, Director

NATIONAL TRUST ADVISORS

David C. Clark
P.O. Box 953
Rawlins, WY 82301

Richard D. Wilder
P.O. Box 1866
Cody, WY 82414
(307) 587-5297

PRESERVATION ACTION COORDINATOR

Rheba Massey
AMH-SHPO
Barrett Building
State of Wyoming
Cheyenne, WV 82002
(307) 777-6694

AIA STATE PRESERVATION COORDINATOR

Vacant

NATIONAL PARK SERVICE REGIONAL OFFICE

Rocky Mountain Regional Office
12795 West Alameda Parkway
P. O. Box 25287
Denver, CO 80225-0287
(303) 969-2875

ADVISORY COUNCIL ON HISTORIC PRESERVATION DIVISION

Western Office of Project Review
730 Simms Street
Room 401
Golden, CO 80401
(303) 231-5320
Claudia Nissley, Director

Big Horn

Bradford Brinton Memorial Museum
Box 460
Big Horn, WY 82833
(307) 672-3773
Kenneth L. Schuster

Cheyenne

Dennis Auker
320 East 1st Avenue
Cheyenne, WY 82001

Economic Development and
Stabilization Department
Wyoming Main Street Project
2nd West
Cheyenne, WY 82002
(307) 777-7285
Rick Hunnicutt

Eleanor D. MacMillan
717 Mitchell Court
Cheynenne, WY 82007-2801
(307) 632-1217

State Historic Preservation Office
1825 Carey Avenue
Cheyenne, WY 82002
Rheba Massey

Evanston

Evanston Urban Renewal Agency
236 9th Street
Evanston, WY 82930
(307) 789-1633
James Davis

Laramie

Laramie Downtown Developement
Authority
316 South 3rd Street
Laramie, WY 82070
(307) 721-8881
Tim Rubald

Laramie Plains Museum
603 Ivinson Avenue
Laramie, WY 82070
(307) 742-4448
Daniel A. Nelson

Rawlins

Old Pen Joint Powers Board
5th and Walnut
P.O. Box 1331
Rawlins, WY 82301
(307) 324-4111
Margaret Brown

Sheridan

Sheridan Heritage Center
P.O. Box 6352
Sheridan, WY 82801
(307) 674-4674
Della Hubst

Uptown Sheridan Association
P.O. Box 13
Sheridan, WY 82801
(307) 672-8881
Lori Wingerter

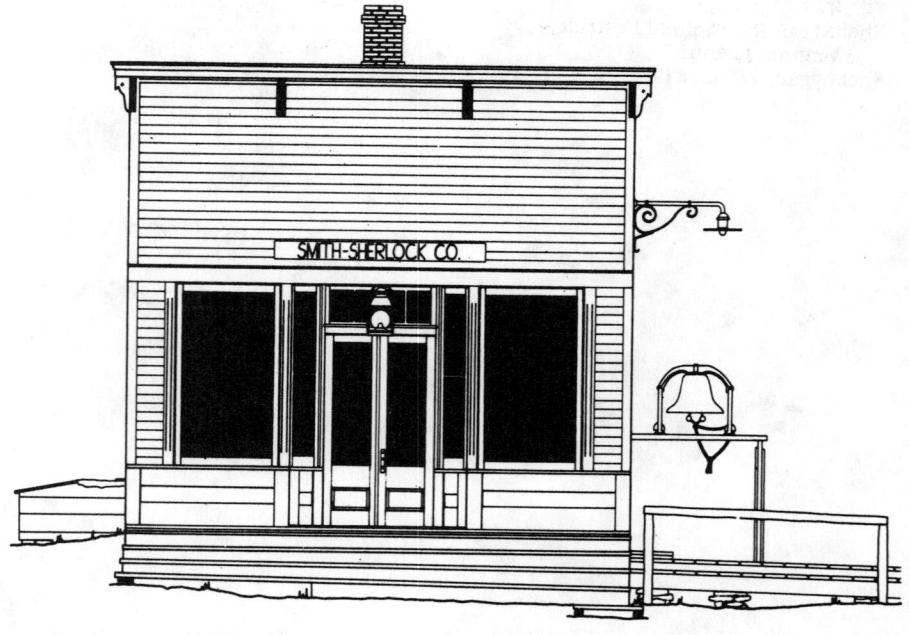

Smith-Sherlock Store, South Pass City, Wyoming. (C. Fraser, HABS)

AMERICAN SAMOA

HISTORIC PRESERVATION OFFICE

Stan Sorensen, HPO
Department of Parks and Recreation
Government of American Samoa
Pago Pago, American Samoa 96799
011-684-699-9614
011-684-699-4427 (Fax)
David J. Herdrich, Deputy
011-684-699-9513

NATIONAL TRUST REGIONAL OFFICE

Western Regional Office
1 Sutter Street
Suite 707
San Francisco, CA 94104
(415) 956-0610
(415) 956-0837 (Fax)

NATIONAL PARK SERVICE REGIONAL OFFICE

Western Regional Office
600 Harrison Street
Suite 600
San Francisco, CA 94107
(415) 556-7741

ADVISORY COUNCIL ON HISTORIC PRESERVATION DIVISION

Western Office of Project Review
730 Simms Street
Room 401
Golden, CO 80401
(303) 231-5320
Claudia Nissley, Director

FEDERATED STATES OF MICRONESIA

HISTORIC PRESERVATION OFFICE

Federated States of Micronesia National
Government
Division of Archives and Historic
Preservation
Office of Administrative Services
P.O. Box PS 35
Palikir, Pohnpei, FSM 96941
011-691-320-2343
011-619-320-2597, 320-5634 (Fax)
Teddy John, HPO

Yap

Office of the Governor
Colonia, Yap, FSM
West Caroline Islands 96943
011-691-350-2194
011-691-350-2381 (Fax)
Andrew Kugfas, SHPO

Chuuk

Department of Resources and
Development
Moen, Chuuk, FSM
East Caroline Islands 96942
011-691-330-3309
011-591-330-2232 (Fax)
Elvis Killion O'Sonia, SHPO

Pohnpei

Department of Land
Pohnpei State Government
P. O. Box 158
Kolonia, Pohnpei, FSM
East Caroline Islands 96941
011-691-320-2715
011-691-320-2505 (Fax)
Emensio Eperiam, SHPO

KOSRAE

DIVISION OF HISTORY AND CULTURAL PRESERVATION

Department of Conservation and Development
Kosrae, FSM
East Caroline Islands 96944
011-691-370-3078
011-691-370-3003 (Fax)
Berlin Sigrah, Administrator, SHPO

NATIONAL TRUST REGIONAL OFFICE

Western Regional Office
1 Sutter Street
Suite 707
San Francisco, CA 94104
(415) 956-0610
(415) 956-0837 (Fax)

NATIONAL PARK SERVICE REGIONAL OFFICE

Western Regional Office
600 Harrison Street
Suite 600
San Francisco, CA 94107
(415) 556-7741

ADVISORY COUNCIL ON HISTORIC PRESERVATION DIVISION

Western Office of Project Review
730 Simms Street
Room 401
Golden, CO 80401
(303) 231-5320
Claudia Nissley, Director

GUAM

HISTORIC PRESERVATION OFFICE

Guam Historic Preservation Office
Department of Parks and Recreation
490 Naval Hospital Road
Agana Heights, Guam 96910
011-671-477-9620
011-671-477-2822 (Fax)
Richard D. Davis, SHPO

NATIONAL TRUST REGIONAL OFFICE

Western Regional Office
1 Sutter Street
Suite 707
San Francisco, CA 94104
(415) 956-0610
(415) 956-0837 (Fax)

AIA PRESERVATION COORDINATOR

Stephen F. Lander, AIA
P.O. Box 10571
Tamuning, GU 96931
011-671-646-1101

NATIONAL PARK SERVICE REGIONAL OFFICE

Western Regional Office
600 Harrison Street
Suite 600
San Francisco, CA 94107
(415) 556-7741

ADVISORY COUNCIL ON HISTORIC PRESERVATION DIVISION

Western Office of Project Review
730 Simms Street
Room 401
Golden, CO 80401
(303) 231-5320
Claudia Nissley, Director

MARSHALL ISLANDS

HISTORIC PRESERVATION OFFICE

Secretary of Interior and Outer Islands
Affairs
P.O. Box 1454
Majuro Atoll
Republic of the Marshall Islands 96960
011-692-625-3413, 625-3240, 625-3264
011-692-625-3412 (Fax)
Carmen Bigler, HPO
Dirk Spennemann, Acting Deputy

NATIONAL TRUST REGIONAL OFFICE

Western Regional Office
1 Sutter Street
Suite 707
San Francisco, CA 94104
(415) 956-0610
(415) 956-0837 (Fax)

NATIONAL PARK SERVICE REGIONAL OFFICE

Western Regional Office
600 Harrison Street
Suite 600
San Francisco, CA 94107
(415) 556-7741

NORTHERN MARIANA ISLANDS

HISTORIC PRESERVATION OFFICE

Department of Community and
Cultural Affairs
Commonwealth of the Northern
Mariana Islands
Saipan, Mariana Islands 96950
011-670-322-9722, 322-9556
011-670-322-4058, 322-5096 (Fax)
Michael Fleming, HPO
Scott Russell, Deputy

NATIONAL TRUST REGIONAL OFFICE

Western Regional Office
1 Sutter Street
Suite 707
San Francisco, CA 94104
(415) 956-0610
(415) 956-0837 (Fax)

NATIONAL PARK SERVICE REGIONAL OFFICE

Western Regional Office
600 Harrison Street
Suite 600
San Francisco, CA 94107
(415) 556-7741

PALAU

HISTORIC PRESERVATION OFFICE

Chief of Cultural Affairs
Ministry of Community and Cultural
Affairs
P.O. Box 100
Koror, Republic of Palau 96940
011-680-488-2489
011-680-488-1725, 488-1662 (Fax)
Victoria N. Kanai, HPO

NATIONAL TRUST REGIONAL OFFICE

Western Regional Office
1 Sutter Street
Suite 707
San Francisco, CA 94104
(415) 956-0610
(415) 956-0837 (Fax)

NATIONAL PARK SERVICE REGIONAL OFFICE

Western Regional Office
600 Harrison Street
Suite 600
San Francisco, CA 94107
(415) 556-7741

PUERTO RICO

HISTORIC PRESERVATION OFFICE

Office of Historic Preservation
Box 82, La Fortaleza
San Juan, Puerto Rico 00901
(809) 721-2676, 721-3727
(809) 723-0957 (Fax)
Mariano G. Coronas Castro, HPO
Luis F. Irizarry Ramirez, Deputy

NATIONAL TRUST REGIONAL OFFICE

Mid-Atlantic Regional Office
Cliveden
6401 Germantown Avenue
Philadelphia, PA 19144
(215) 438-2886
(215) 438-2892 (Fax)
Patricia D. Wilson, Director

NATIONAL TRUST ADVISORS

Pablo Ojeda-O'Neill
1529 Loiza Street
Box 707
San Juan, PR 00911-1850

Jose R. Coleman-Davis Pagan
Apartado Postal 3912
Old San Juan, PR 00902-3912
(809) 728-6660
(809) 728-4455 (Fax)

AIA PRESERVATION COORDINATOR

Beatriz del Cueto, AIA
Architects and Historic Preservation
Consultants
Valencia No 11
Torrimar
Guaynabo, PR 00966
(809) 792-2456

NATIONAL PARK SERVICE REGIONAL REGIONAL OFFICE

Southeast Regional Office
Richard B. Russell Federal Building
75 Spring Street, S.W.
Atlanta, GA 30303
(404) 331-2632

ADVISORY COUNCIL ON HISTORIC PRESERVATION DIVISION

Eastern Office of Project Review
1100 Pennsylvania Avenue, N.W.
Suite 809
Washington, DC 20004
(202) 786-0503
Don L. Klima, Director

Humacao

Iglesia Parroquial
Dulce Nombre de Jesus
P.O. Box 546
Humacao, PR 00661
Eric R. Buermann

San German

Corporacion Para La Conservacion y
Preservacion de Monumentos
1 Estrella Street
Historicos de Puerto Rico
San German, PR 00753
(809) 892-4988
Luis J. Oliver

San Juan

Conservation Trust of Puerto Rico
Box 4747
San Juan, PR 00905
F. J. Blanco

Estado Libre Asociado de Puerto Rico
Departamento de Hacienda
Area del Tesoro
Apartment S-4515
San Juan, PR 00905-4515

Oficina Estatal de Preservacion
Historica
Calle San Jose
Suite 109
San Juan, PR 00901

M. Lopez del Valle
G.F. El Caribe Building
San Juan, PR 00901
(809) 721-0350

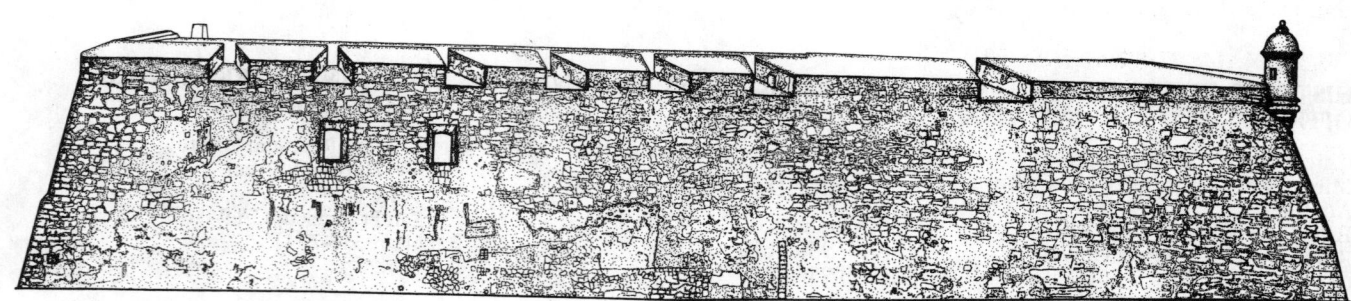

San Carlos Ravelin, San Juan National Historic Site, Old San Juan, Puerto Rico. (J. Blanco, HABS)

VIRGIN ISLANDS

HISTORIC PRESERVATION OFFICE

Department of Planning and Natural Resources
Nisky Center
45A Estate Nisky
Suite 231
St. Thomas, VI 00802
(809) 774-3320
Roy E. Adams, Commissioner, HPO
Claudette C. Lewis, Assistant Director,
Deputy HPO

TERRITORIAL PRESERVATION ORGANIZATIONS

St. Croix Landmarks Society
P.O. Box 2855
Frederiksted
St. Croix, VI 00841
(809) 772-0598
Barbara Hagan-Smith, Executive
Director

St. Thomas Historical Trust
P.O. Box 11849
Charlotte Amalie
St. Thomas, VI 00801
(809) 774-9193
Douglas White, President

NATIONAL TRUST REGIONAL OFFICE

Mid-Atlantic Regional Office
Cliveden
6401 Germantown Avenue
Philadelphia, PA 19144
(215) 438-2886
(215) 438-2892 (Fax)
Patricia D. Wilson, Director

NATIONAL TRUST ADVISORS

Barbara M. Hagan-Smith
St. Croix Landmarks Society
P.O. Box 2855
Frederiksted
St. Croix, VI 00840
(809) 772-0598
(809) 772-9446 (Fax)

George Tyson
9-33 Estate Nazareth
St. Thomas, VI 00802
(809) 775-6109

NATIONAL PARK SERVICE REGIONAL OFFICE

Southeast Regional Office
Richard B. Russell Federal Building
75 Spring Street, S.W.
Atlanta, GA 30303
(404) 331-2632

ADVISORY COUNCIL ON HISTORIC PRESERVATION DIVISION

Eastern Office of Project Review
1100 Pennsylvania Avenue, N.W.
Suite 809
Washington, DC 20004
(202) 786-0503
Don L. Klima, Director

St. Croix

Jean-Robert Alfred
3 Hill Street
Christiansted
St. Croix, VI 00820
(809) 773-6628

Our Town Frederiksted
523 Strand Street
Frederiksted
St. Croix, VI 00840
(809) 772-3550
Susan Lipsky

St. Croix Landmarks Society
P.O. Box 2855
Frederiksted
St. Croix, VI 00840
(809) 772-0598
Barbara Hagan-Smith

T.K. Properties
P.O. Box 5968
St. Croix, VI 00824
(809) 778-5626
Tracey K. Kirkman

St. Thomas

Department of Planning and Natural
Resources
Nisky Center
45A Estate Nisky
Suite 231
St. Thomas, VI 00802
Roy E. Adams

Frederick C. Gjessing
P.O. Box 1844
Charlotte Amalie
St. Thomas, VI 00803
(809) 775-5674

Christopher Haig
Serenity, Morningstar
P.O. Box 262CA
St. Thomas, VI 00804
(809) 776-2121

St. Thomas Historical Trust
P.O. Box 11849
Charlotte Amalie
St. Thomas, VI 00801
(809) 774-1400
David Bornn

Tri-Island Economic Development
Council
P.O. Box 838
St. Thomas, VI 00801
(809) 774-7215
Flavius A. Ottley

Former Lutheran Church of Our Lord of Zebaoth, Christiansted, St. Croix, Virgin Islands. (F.C. Gjessing, HABS)

FOREIGN COUNTRIES

Representatives of Preservation Forums in Foreign Countries

BARBADOS

Caribbean Conservation Association
Savannah Lodge
The Garrison
St. Michael
Eastern Caribbean
Barbados

Museums Association of the Caribbean
Secretariat at Caribbean Conservation
Association
Savannah Lodge
The Garrison
St. Michael
Eastern Caribbean
Barbados

CANADA

Alberta

City of Calgary
Planning and Building Department
8108 Information Center
P.O. Box 2100
Station M
Calgary, Alberta T2P 2M5

W. Jamieson
Environmental Design
University of Calgary
2500 University Drive, N.W.
Calgary, Alberta T2N 1N4
(403) 220-7426

British Columbia

City of Victoria Planning Department
1 Centennial Square
Victoria, British Columbia V8W 1P6

Kalico Developments Limited
1307 West Georgia Street
Vancouver, British Columbia V6E 3K5

Nova Scotia

Development and Planning
Departmnet
City of Halifax
P.O. Box 1749
Halifax, Nova Scotia B3J 3A5
(902) 421-7787

Heritage Canada
P.O. Box 2024, Station M
Halifax, Nova Scotia 82Y 3J2
(902) 421-1889
Peter Hyndman

Ontario

Tacoma Steckley and Associates
350 Speedvale Avenue, West
Suite 10
Guelph, Ontario N1H 7M7

City of Kingston Planning Department
216 Ontario Street
Kingston, Ontario
(613) 546-4291
R. Dobbin

T.K. Woods and Partners
Chartered Surveyors
112A Main Street
Markham, Ontario L3P 1Y1
(416) 474-4747
Ian K. Woods

Corporation of the Town of Markham
101 Town Centre Boulevard
Markham, Ontario L3R 9W3

Huronia Historical Parks
Parcs Historique de la Huronie
P.O. Box 160, C.P. 160
Midland, Ontario L4R 4K8

Property and Economic Development
Department
City of North York
5100 Yonge Street
North York, Ontario M2N 5V7

Baldwin and Franklin Architects
167 Richmond Street, East
Toronto, Ontario M5A 1N9
(416) 634-9521
M. Franklin

Adam Carr
141 Lyon Court
Apartment 904
Toronto, Ontario M6B 3H2
(416) 787-0598

Hannivan and Company
132 Maplewood Avenue
Toronto, Ontario M6C 1J6
(416) 597-0298
Patti L. Burner

Quebec

Centre Canadien d'Architecture
3013 A/S Bibliotheque
1920 Rue Baile
Montreal, Quebec H3H 2S6

Ministere Affairs Culturelles
Centre de Documentation
225 Grande Allee Est
Sous Sol, Quebec G1R 5G5

Simpa
Centre de Documentation
164 Est Rue Notre Dame
Montreal, Quebec H2Y 1C2
(514) 872-1704

France

Naomi Barry
61 Quai D'Orsay
Paris, 75007
France

Ireland

Frank L. Benson and Associates
38 Dawson Street
Dublin 2,
Ireland
01-679-9286

Italy

Biblioteca Centrale Facolta di
Architettura
Politecnico Via Bonardi, 3
20133 Milano
Italy

Mauritius

Hans Dwarka
Manager
IBL Decoration Center
Chaussee Street
Port Louis
Mauritius

New Zealand

Wanganui District Council
101 Guyton Street
P.O. Box 637
Wanganui
New Zealand
06458529

West Indies

Patricia Green
P.O. Box 8949
Kingston CSO
Jamaica
West Indies
(809) 925-6147
Terry Fleming

INDEX